Plant Oxidative Stress and Management

Plant Oxidative Stress and Management

Edited by Siena Watson

SYRAWOOD
PUBLISHING HOUSE
New York

Published by Syrawood Publishing House,
750 Third Avenue, 9th Floor,
New York, NY 10017, USA
www.syrawoodpublishinghouse.com

Plant Oxidative Stress and Management
Edited by Siena Watson

International Standard Book Number: 978-1-64740-420-8 (Hardback)

Cataloging-in-publication Data

Plant oxidative stress and management / edited by Siena Watson.
p. cm.
Includes bibliographical references and index.
ISBN 978-1-64740-420-8
1. Plants--Effect of stress on. 2. Oxidative stress. 3. Plant physiology.
4. Stress management. I. Watson, Siena.
QK754 .P53 2023
581.7--dc23

TABLE OF CONTENTS

PREFACE

Every book is initially just a concept; it takes months of research and hard work to give it the final shape in which the readers receive it. In its early stages, this book also went through rigorous reviewing. The notable contributions made by experts from across the globe were first molded into patterned chapters and then arranged in a sensibly sequential manner to bring out the best results.

Oxidative stress in plants is a complex chemical and physiological phenomenon that accompanies virtually all biotic and abiotic stresses in higher plants. There are various environmental factors such as extreme temperature, air pollutants, ultraviolet-B (UV-B) radiation, salinity, draught, pathogen infection and pesticides that can lead to the development of oxidative stress in plants. Exposure to all these factors affects a range of biological processes due to the generation of reactive oxygen species (ROS). It is produced during mitochondrial oxidative metabolism and due to the cellular response to xenobiotics, cytokines, and bacterial invasion. An excessive production and accumulation of ROS causes an imbalance between the production of ROS and the neutralization of free radicals by antioxidants. Plant oxidative stress can be managed either by developing transgenic plants overexpressing one or more antioxidants, or by developing mutants that are resistant towards certain stresses. This book includes some of the vital pieces of works being conducted across the world, on various topics related to oxidative stress in plants as well as its management. It will serve as a valuable source of reference for graduate and postgraduate students.

It has been my immense pleasure to be a part of this project and to contribute my years of learning in such a meaningful form. I would like to take this opportunity to thank all the people who have been associated with the completion of this book at any step.

Editor

1

Zinc-Lysine Supplementation Mitigates Oxidative Stress in Rapeseed (*Brassica napus* L.) by Preventing Phytotoxicity of Chromium, when Irrigated with Tannery Wastewater

Ihsan Elahi Zaheer [1,†], Shafaqat Ali [1,2,*], Muhammad Hamzah Saleem [3,*,†], Muhammad Arslan Ashraf [4], Qurban Ali [5], Zohaib Abbas [1], Muhammad Rizwan [1], Mohamed A. El-Sheikh [6], Mohammed Nasser Alyemeni [6] and Leonard Wijaya [6]

1 Department of Environmental Sciences and Engineering, Government College University Allama Iqbal Road, Faisalabad 38000, Pakistan; ihsankhanlashari@gmail.com (I.E.Z.); zohaib.abbas83@gmail.com (Z.A.); mrizwan@gcuf.edu.pk (M.R.)
2 Department of Biological Sciences and Technology, China Medical University, Taichung 40402, Taiwan
3 MOA Key Laboratory of Crop Ecophysiology and Farming System Core in the Middle Reaches of the Yangtze River, College of Plant Science and Technology, Huazhong Agricultural University, Wuhan 430070, China
4 Department of Botany, Government College University Faisalabad, Faisalabad 38000, Pakistan; marslanashraf@gcuf.edu.pk
5 Key laboratory of Plant Pathology, College of Plant Science & Technology, Huazhong Agricultural University, Wuhan 430070, China; rattarqurban@hotmail.com
6 Department of Botany and Microbiology, College of Science, King Saud University, 11451 Riyadh, Saudi Arabia; melsheikh@ksu.edu.sa (M.A.E.-S.); mnyemeni@ksu.edu.sa (M.N.A.); leon077@gmail.com (L.W.)
* Correspondence: shafaqataligill@gcuf.edu.pk (S.A.); saleemhamza312@webmail.hzau.edu.cn (M.H.S.)
† These authors contributed equally to this work.

Abstract: Contamination of soil and water with metals and metalloids is one of the most serious problems worldwide due to a lack of a healthy diet and food scarcity. Moreover, the cultivation of oilseed crops such as rapeseed (*Brassica napus* L.) with tannery wastewater could contain a large amount of toxic heavy metals [e.g., chromium (Cr)], which ultimately reduce its yield and directly influence oilseed quality. To overcome Cr toxicity in *B. napus*, a pot experiment was conducted to enhance plant growth and biomass by using newly introduced role of micronutrient-amino chelates [Zinc-lysine (Zn-lys)], which was irrigated with different levels [0% (control), 33%, 66%, and 100%] of tannery wastewater. According to the results of present findings, very high content of Cr in the wastewater directly affected plant growth and composition as well as gas exchange parameters, while boosting up the production of reactive oxygen species (ROS) and induced oxidative damage in the roots and leaves of *B. napus*. However, activities of antioxidants initially increased (33% of wastewater), but further addition of tannery wastewater in the soil caused a decrease in antioxidant enzymes, which also manifested by Zn content, while the conscious addition of wastewater significantly increased Cr content in the roots and shoots of *B. napus*. To reduce Cr toxicity in *B. napus* plants, exogenous supplementation of Zn-lys (10 mg/L) plays an effective role in increasing morpho-physiological attributes of *B. napus* and also reduces the oxidative stress in the roots and leaves of the oilseed crop (*B. napus*). Enhancement in different growth attributes was directly linked with increased in antioxidative enzymes while decreased uptake and accumulation of Cr content in *B. napus* when cultivated in wastewater with the application of Zn-lys. Zn-lys, therefore, plays a protective role in reducing the Cr toxicity of *B. napus* through an increase in plant growth and lowering of Cr uptake in various plant organs. However, further studies at field levels are required to explore the mechanisms of Zn–lys mediated reduction of Cr and possibly other heavy metal toxicity in plants.

Keywords: antioxidant enzymes; growth; heavy metals; oil seed crop; micronutrient-amino chelates

---

1. Introduction

The rapid industrial and urbanization caused by the population's overly growing threat posed a threat to humanity by air, water, and land contamination with different heavy metals [1]. Tanneries have played a significant role in soil and water contamination among industries. Farming practices utilize approximately 70% of the water drains, and farmers should use domestic or urban wastewater as a source of water in times and regions of water scarcity [2]. Due to the scarcity of fresh water, wastewater cultivation became a common practice in arid and semi-arid areas, which has serious impacts on plants and human health [3]. The two major cities with a large number of tanneries in Pakistan are Sialkot and Kasur. In these areas, different factories emit large amounts of untreated wastewater and contaminate the region's freshwater resources [4]. Pakistan is also now facing a severe shortage of fresh water due to a large number of factories and manufacturers, and thus farmers often use industrial wastewater to cultivate fields [2]. Agricultural soils mainly contain low heavy metals and soil contamination with these toxic pollutants through fertilization or tannery wastewater irrigation [5]. The wastewater does, however, contain a low content of heavy metals; due to metal uptake and accumulation, the plant had higher metal contents in their body organs [6]. Chromium (Cr) is the most significant among different heavy metals found in tannery wastewater due to its large concentration [7]. Cr in abundance is harmful to plants; plant growth and composition are directly affected, with significant impacts on crop quality and decreased soil mineral uptake [8]. In addition, excess Cr has also been caused by the generation of extreme amounts of reactive oxygen species (ROS) and also the alteration of the many antioxidative enzymes in plants [3,9]. Therefore, the exploration of various technologies and different management methods is extremely important to restrict the accumulation of Cr by several plants in soil polluted by Cr and consequently to reduce the human health risks associated with Cr toxicity.

Recently, various studies have indicated the imperious role of different amino acids in alleviating heavy metal stress in the various plants [10]. Zinc (Zn)-chelated fertilizer with amino acid, like lysine (lys), is documented to boost crop yield and production [3,11]. Combining Zn with lysine significantly improves morpho-physiological attributes of the plants and nutrient uptake [9]. In addition, Zn-lys has also documented increased use of essential nutrients to increase the gas exchange attributes and improve the photosynthetic apparatus of the plants [11]. It has also been noted previously that the combination of essential micronutrients with lysine decreased the salinity effect in tomato plants and increased plant growth and crop yield [12]. Earlier research also indicated that exogenous Zn supplementation with lysine decreased Cr toxicity in many important agronomic plants, including wheat [9] and Cd toxicity in rice [11].

Oil rape (*Brassica napus* L.) is the world's second most important source of vegetable oil and has the capacity to accumulate in the edible portion of toxic heavy metals from heavy-metals soil polluted by irrigation of wastewater [2,13]. *B. napus* is cultivated extensively around the worldwide and the stress relief of heavy metals has also been widely investigated in different environmental conditions [2,14]. However, it showed severe damage in different organs when cultivated in the soil, which was polluted with various heavy metals in terms of grain quality and widespread productivity [14]. The irrigation of *B. napus* with wastewater is a very common practice in the world due to the availability of fresh water [2,14]. The threat of public exposure to these toxic pollutants occurs rapidly when contaminated food crops are being consumed in agricultural fields irrigated by wastewater. Furthermore, *B. napus* is also used to cultivate in metal-polluted soil and showed resistance against heavy metals stress conditions [15,16]. In addition, various chelators or phytohormones have been used exogenously to increase the phytoextraction of heavy metals with *B. napus* to increase the metal uptake/accumulation [14,15,17,18]. Although very large amounts of heavy metals in plant organs seriously affected the quality of their oilseeds and could not be used for oilseed rape. The uniqueness of *B. napus* due to its high biomass production and tolerance towards Cr can be valuable traits; however, sufficient information is not available regarding Cr tolerance, gas exchange traits, oxidative stress and antioxidative defense system, and Cr accumulation, when grown as oilseed rape. In order to improve plant growth and composition by the foliar application of Zn-lys, when cultivated on different levels of tannery wastewater (Cr-polluted soil), the current experiment was designed. The results from this study would add to our knowledge about (i) the effect of different levels of tannery wastewater in the soil on plant growth and biomass, (ii) photosynthetic pigments and gas exchange attributes, (iii) oxidative stress and antioxidative defense system, and (iv) Zn and Cr uptake and accumulation in various parts of the plants, when supplied externally with the foliar application of Zn-lys. To the best of our knowledge, this study is among the few studies that focus on the metal tolerance among oilseed crops in order to investigate their suitability for metal-contaminated sites with the exogenous application of micronutrient chelation with amino acid. Findings from the present study will add to our understanding of the mechanism of Cr tolerance in *B. napus*, when externally supplied with the foliar application of Zn-lys.

2. Materials and Methods

2.1. Collection and Analysis of Wastewater and Soil

Clay loam sand used in this analysis was collected from the botanical garden of the University of Punjab Lahore Pakistan (31.4015° N, 74.3070° E) from an average depth of 0–20 cm. The 2 mm sieve was used to separate mud and debris from the dried soil intentionally. A comprehensive analysis of organic soil content was performed by the process Walkley and Black [19] described. For the precise size of the soil elements, the hydrometer was used [20]. The physiochemical characteristics of the sand under study are given in Table 1. Tannery wastewater used for the pot experiment was collected from the tannery industries based in Kasur, Punjab, Pakistan. The physicochemical properties of tannery wastewater used in the presented study were estimated according to the set methods of Apha [21]. Comprehensive details of the major characteristics of tannery wastewater used in this experiment are presented in Table 2. The same soil with the tannery wastewater we already used in our previous study by Zaheer et al. [3].

Table 1. Physicochemical properties of loam sand used in the pot experiment.

Texture	Clay Loam
Silt	12.9
Sand	63.4
Clay	22.3
pH (H_2O mixture)	7.1
Electrical conductivity (dS m^{-1})	3.83
Cation exchange capacity (cmol kg^{-1}) 4.9	4.89
Soluble CO_3^{-2} (mmol L^{-1})	0.87
Soluble HCO_3 (mmol L^{-1})	3.78
Soluble Cl^- (mmol L^{-1})	6.31
Soluble Ca^{2+} + Mg^{2+} (mmol L^{-1})	15.89
Organic matter (%)	0.49
Ni (mg kg^{-1})	0.21
Cu (mg kg^{-1})	0.35
Zn (mg kg^{-1})	0.84
Cr (mg kg^{-1})	0.24

Table 2. Characteristics of tannery wastewater used for irrigation of the soil used in the pot experiment.

Parameters	Values	Permissible Limits *
EC (dS m^{-1})	1.41	<1.5
Sodium absorption rate (mmol $L^{-1})^{1/2}$	4.02	<7.5
Residual sodium carbonate (mmol c L^{-1})	2.24	<2.0
Ni (mg L^{-1})	0.09	0.20
Cd (mg L^{-1})	0.04	0.01
Pb (mg L^{-1})	1.24	5.0
Co (mg L^{-1})	0.02	0.05
Cr (mg L^{-1})	4.03	0.10
Zn (mg L^{-1})	1.95	2.00

* Ayers and Westcot [22].

2.2. *Pot Experiment*

Healthy and mature seeds of rapeseed (*Brassica napus* L.) were collected from Ayub Agriculture Research Institute Faisalabad, Pakistan. Ten seeds of *B. napus* were carefully sowed in experimental pots filled with approximately 5 kg of soil. Seeds were surface-sterilized with 10% (*v/v*) commercial bleach for 15 min, followed by thorough washing in distilled water. Thirty days old baby plants were applied with foliar application of Zn-lys (10 mg L^{-1}) alongside with various levels of tannery wastewater (0%, 33%, 66%, and 100%) applied gently. Tannery wastewater was added in the pots, and all pots were given 0%, 33%, 66%, and 100% wastewater on the basis of the treatments when needed with other intercultural operations such as weeding. Moreover, with the addition of tannery wastewater (at least) once a week, we also sprayed Zn-lys regularly, on the basis of demand, on whole plant parts. The experiment was arranged in a way that 8 treatments were as follows: (Ck) Cr 0%, Zn-lys 0 mg L^{-1} (T1) Cr 0%, Zn-lys 10 mg L^{-1} (T2) Cr 33%, Zn-lys 0 mg L^{-1} (T3) Cr 33%, Zn-lys 10 mg L^{-1} (T4) Cr 66%, Zn-lys 0 mg L^{-1} (T5) Cr 66%, Zn-lys 10 mg L^{-1} (T6) Cr 100%, Zn-lys 0 mg L^{-1} (T7) Cr 100%, Zn-lys 10 mg L^{-1}. The total time duration of the treatments was 60 days, and we used all the samples of the plants from each treatment for the analysis of morpho-physiological traits of the plants.

A hand sprayer was used for the exogenous supplementation of Zn-lys to the *B. napus*. In the whole experiment, we used a 4 L volume of Zn-lys for all the pots/plants, which were applied at different times of the experiment. Every time when we applied tannery wastewater, the plant was also sprayed exogenously with the application of Zn-lys. The experiment was executed in 3 replications of each treatment, and 5 plants were grown in a single pot. The present experimental work was conducted in the wire house Department of Botany, the University of the Punjab Lahore, Pakistan, in an open environment, protected from human and animal interactions. The rainfall was controlled, or plants were protected from rainfall by covering the whole wire house with a plastic sheet. Fertilizers such as urea (250 mg/pot), diammonium phosphate (100 mg/pot), and potassium sulfate (50 mg/pot) were applied to each pot to avoid macronutrient deficiency in the plants. The pots used in this study were rotated regularly in order to avoid environmental effects on the plants.

2.3. *Plant Harvesting*

Plants of *B. napus* were carefully rooted-up after 30 days of experimental treatment (precisely 60 days of the whole experiment) and washed gently with the help of distilled water to eliminate the aerial dust and deposition of soil and silt particles. All the harvested plants were divided into 2 parts, i.e., roots and shoots, to study different biological traits. Plant height (whole plant length) was measured straightway after the harvesting using a measuring scale, and the number of leaves per plant was also counted after it. The number of leaves and leaf area was measured, and the fresh and dry biomass of the leaves were measured, and then the remaining were analyzed for Cr and Zn content from the shoots, after mixing with stems and other shoots parts. All the plant samples (after harvest) were attentively washed with the help of de-ionized water, and after that, oven-dried at 70 °C (for 3 days) and then grounded to a very fine powder to complete further investigation.

2.4. Analysis Items and Methods

For all physiological and Cr and Zn determination from different parts of the plants, we used 3 replications for each sample. The mass of the leaf or root samples was collected/taken as recommended in different protocols, which are mentioned below. All the physiological traits were performed using fresh biomass of the plants, which was stored in the refrigerator (−80 °C). Total chlorophyll content (Chl *a* + Chl *b*) and carotenoid content were measured by following Nagata and Yamashita [23]. In 80% acetone, 1 g fresh plant of leaves was pulverized using a pestle and mortar, and the resolution was centrifuged at 3000 rpm for 10 min. Calculations were made using the following formulas:

$$\text{Chlorophyll } a \text{ (mg/g FW)} = 0.999\ A_{663} - 0.0989\ A_{645}$$

$$\text{Chlorophyll } b \text{ (mg/g FW)} = 0.328\ A_{663} + 1.77\ A_{645}$$

$$\text{Carotenoid (mg/g FW)} = 1000\ A_{470} - 2.270\ C_a - 81.4\ C_b/227$$

where C_a is chlorophyll *a* content and C_b is chlorophyll *b* content

While at a bright sunny day (9:00 a.m.–11:00 a.m.), gas exchange parameters were measured using a portable IRGA (Infra-Red Gas Analyzer, Hoddesdon, UK) before harvesting the plants. For different gas exchange parameters such as transpiration rate (T*r*), stomatal conductance (Gs), net photosynthesis (P*n*), and water use efficiency (W*i*), 3 leaves were selected randomly from each treatment and from 3 different plants of a single treatment. Moreover, we selected 3 different leaves from each treatment, which were alike (means no difference in age were observed).

The content of malondialdehyde (MDA) and hydrogen peroxide (H_2O_2) was measured by following the method of Heath and Packer [24] and Jana and Choudhuri [25], respectively. The stress-induced electrolyte leakage (EL) was measured by the method presented by Dionisio-Sese and Tobita's [26] by using the following formula:

$$EL = (EC_1/EC_2) \times 100$$

The activities of Antioxidants such SOD, POD, APX and CAT measured by following the method of Chen and Pan [27], Sakharov and Ardila [28], Aebi [29] and Nakano and Asada [30] respectively.

Plant samples were vigilantly digested via the di-acid (HNO_3-$HClO_4$) technique. 0.5 g dry sample (3 replications for each treatment) of roots and shoots of the plants were taken into the flask, having 10 mL of HNO_3-$HClO_4$ (3:1, *v:v*), this collection was then retained overnight. The final digestion of these plants' samples was completed after the addition of HNO_3 (5 mL) and then placed on the hot plate for complete digestion, as described by Rehman et al. [31]. Atomic absorption spectrophotometer (AAS) was used to investigate the correct amount of Cr and Zn in shoots and roots of the plant.

2.5. Statistical Analysis

All the data were assayed with the help of SPSS software through a multivariate post hoc test, in order to determine the interaction among significant values. One-way analysis of variance (ANOVA) was used to assess the significance of the variations of Cr among the different plant parts, followed by HSD tests. Considerable variation among different observed values was calculated at $p < 0.05$ and displayed with various letters. The graphical presentation was carried out using Origin Pro (2017) (Systat Software Inc., San Jose, CA, USA). The Pearson correlation coefficients between the measured variables of *B. napus* were also calculated. The plots of principal component analysis on rapeseed parameters were carried out by using the Rstudio software. Moreover, heatmap was also presented for the different variables of *B. napus* by using the Rstudio software.

3. Results

3.1. *Effect of Foliar Application of Zn-lys on Plant Growth and Biomass under Different Levels of Tannery Wastewater*

Data regarding different morphological parameters of *B. napus* are presented in Table 3. According to the results, the toxic content of Cr in the wastewater caused a significant decrease in all morphological traits of *B. napus* studied in this study (Table 3). Compared to the control, maximum decreased in plant height, the number of leaves, leaf area, root length, fresh root weight, root dry weight, fresh leaf weight, dry leaf weight by 68.6%, 54.5%, 84.5%, 38.4%, 89.7%, 89.5%, 78.6%, and 82.3%, respectively, in T6 plants (irrigated 100% tannery wastewater without the application of Zn-lys). However, the application of Zn-lys increased plant height, the number of leaves, leaf area, root length, fresh root weight, root dry weight, fresh leaf weight, dry leaf weight at all levels of tannery wastewater (33%, 66%, and 100%) in the soil and increased by 10%, 20%, 11.8%, 25%, 258.2%, 17.1%, 16.7%, and 17.1% at 100% addition of tannery wastewater treatment in the soil with the application of Zn-lys compared to 100% addition of tannery wastewater treatment in the soil without the application of Zn-lys. Although the data regarding (T7) were statistically non-significant with (T6), which was showing that the application of Zn-lys at 100% wastewater level did not significantly affect the plant growth and biomass in *B. napus*.

Table 3. Effect of different levels of tannery wastewater on plant growth and biomass under the same concentrations of Zinc-lysine application in *B. napus*.

Treatments	Plant Height (cm)	Number of Leaves	Leaf Area (cm^2)	Root Length (cm)	Root Fresh Weight (mg)	Root Dry Weight (mg)	Leaf Fresh Weight (mg)	Leaf Dry Weight (mg)
Ck	31 ± 0.6ab	11 ± 1ab	110 ± 7ab	13 ± 0.4cd	4025 ± 55b	837 ± 36b	19,336 ± 740b	2457 ± 50b
T1	34 ± 1.0a	13 ± 1a	126 ± 5a	17 ± 1a	5038 ± 70a	999 ± 46a	23,675 ± 1200a	2658 ± 64a
T2	22 ± 1.6de	6 ± 0.6bc	72 ± 5d	12 ± 0.6de	1661 ± 57d	334 ± 13d	11,991 ± 467d	1573 ± 56d
T3	29 ± 1.2bc	8 ± 0.6b	91 ± 4c	15 ± 0.7ab	2191 ± 64c	470 ± 18c	15,599 ± 561c	1804 ± 50c
T4	17 ± 0.6ef	6 ± 1c	38 ± 3f	10 ± 0.3e	798 ± 37f	198 ± 12e	5005 ± 100f	729 ± 30f
T5	24 ± 1.7cd	8 ± 0.6b	53 ± 3e	14 ± 0.8bc	1179 ± 53e	292 ± 19d	6872 ± 170e	909 ± 30e
T6	10 ± 0.6g	5 ± 0.6c	17 ± 1g	8 ± 0.5f	414 ± 13g	88 ± 6f	3142 ± 70g	433 ± 24g
T7	11 ± 2.8fg	6 ± 0.6bc	19 ± 3g	10 ± 0.6e	505 ± 19g	103 ± 10f	3665 ± 75fg	507 ± 28g

Values are demonstrated as means of three replicates along with standard deviation (SD; $n = 3$). One-way ANOVA was performed and mean differences were tested by highest significant deviation (HSD) at $p < 0.05$. Different lowercase letters indicate a significant difference between the treatments. Different abbreviations used in the table are as follows: Ck (without irrigation with wastewater + 0 mg/L Zn-lysine), T1 (without irrigation with wastewater + 10 mg/L Zn-lysine), T2 (33% irrigation with wastewater + 0 mg/L Zn-lysine), T3 (33% irrigation with wastewater + 10 mg/L Zn-lysine), T4 (66% irrigation with wastewater + 0 mg/L Zn-lysine), T5 (66% irrigation with wastewater + 10 mg/L Zn-lysine), T6 (100% irrigation with wastewater + 0 mg/L Zn-lysine) and T7 (100% irrigation with wastewater + 10 mg/L Zn-lysine).

3.2. *Effect of Foliar Application of Zn-lys on Photosynthetic Pigments and Gas Exchange Attributes under Different Levels of Tannery Wastewater*

In the current experiment, we have noticed that total chlorophyll content (chl *a* + chl *b*), carotenoid content, and various gas exchange parameters were decreased significantly ($p < 0.05$) in *B. napus*, when irrigated with different levels of tannery wastewater with the application of Zn-lys. The increasing level of tannery wastewater (33%, 66%, and 100%) in the soil decreased photosynthetic pigments and gas exchange parameters of *B. napus* when compared to the control treatment. The data regarding these parameters are presented in Figure 1. According to the given data, we have noticed that the maximum content of the total chlorophyll (chl *a* + chl *b*), carotenoid, transpiration rate (T*r*), stomatal conductance (Gs), net photosynthesis (P*n*), and water use efficiency (W*i*) were decreased in T6 treated plants by 74.4%, 53.8%, 32%, 52.9%, 61.1%, and 53.4%, respectively, compared to the plants grown in the control treatment (without the addition of tannery wastewater and without the application of Zn-lys). Although, the exogenous supplementation of Zn-lys increased chl, carot, T*r*, Gs, P*n*, and W*i* by 37.5%, 36.4%, 6.5%, 40%, 5.7%, and 126.3%, respectively, in T7 plants when compared to the plants grown in T6 (100% wastewater without application of Zn-lys). Although the data regarding (T7) were statistically non-significant with (T6), which was showing that the application of Zn-lys at 100%

wastewater level did not significantly affect photosynthetic pigments and gas exchange attributes in *B. napus*.

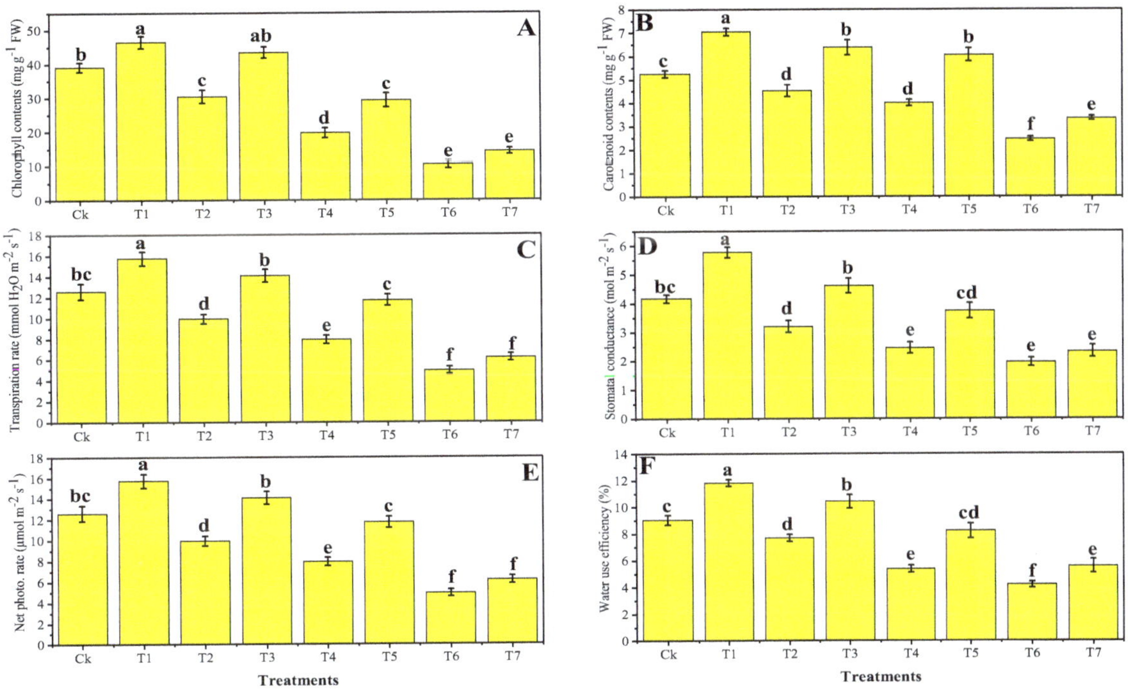

Figure 1. The effect of different levels of tannery wastewater on total chlorophyll contents (**A**), carotenoid contents (**B**), transpiration rate (**C**), stomatal conductance (**D**), net photosynthesis (**E**) and water use efficiency (**F**) under the same concentration of Zinc-lysine application in *B. napus*. Values are demonstrated as means of three replicates along with the standard deviation (SD; $n = 3$). One-way ANOVA was performed, and the mean differences were tested by HSD ($p < 0.05$). Different lowercase letters on the error bars indicate a significant difference between the treatments. Different treatments used in the figures are as follows: Ck (without irrigation with wastewater + 0 mg/L Zn-lysine), T1 (without irrigation with wastewater + 10 mg/L Zn-lysine), T2 (33% irrigation with wastewater + 0 mg/L Zn-lysine), T3 (33% irrigation with wastewater + 10 mg/L Zn-lysine), T4 (66% irrigation with wastewater + 0 mg/L Zn-lysine), T5 (66% irrigation with wastewater + 10 mg/L Zn-lysine), T6 (100% irrigation with wastewater + 0 mg/L Zn-lysine), and T7 (100% irrigation with wastewater + 10 mg/L Zn-lysine).

3.3. Effect of Foliar Application of Zn-lys on Oxidative Stress and Antioxidant Response under Different Levels of Tannery Wastewater

Increased in the content of MDA, H_2O_2 initiation, and EL (%) indicating that a high content of metal in the soil, used for the pot experiment, caused oxidative damage in *B. napus*. Although, activities of various antioxidant enzymes were initially increased due to Cr toxicity and then decreased at extreme concentration of Cr concentration in the soil. The data regarding oxidative stress and antioxidant response are presented in Figures 2 and 3, respectively. In our results, we analyzed roots and shoots differently for their statistics. Compared to the control treatment, the maximum content of MDA, H_2O_2 initiation, and EL (%) were showed in T6 plants, which were increased by 4.3, 2.5, and 4 times, respectively, in the roots and 3.1, 3.7, and 3.4 times, respectively. While the exogenous application of Zn-lys decreased the content of MDA, H_2O_2 initiation, and EL (%) in the roots as well as in the leaves of *B. napus*. Similarly, activities of antioxidants initially increased (33% of wastewater in the soil) compared to the control treatment, but further addition of wastewater in the soil caused a significant decrease in antioxidant activities. Compared to the plants grown in the control treatment, the maximum activities of SOD, POD, CAT, and APX were increased by 14.2%, 12.5%, 23%, and 30.3%, respectively, in the roots and 22.2%, 19.3%, 38.8%, and 23.2%, respectively, in the leaves in T1 plants. Our results also showed that the foliar application of Zn-lys increased the activity of antioxidant enzymes when compared with those plants, which were grown without the foliar application of Zn-lys in all treatments of wastewater, including the control treatment (Figure 3). Moreover, the plants grown in the T7 showed a non-significant decrease in the activities of antioxidants (in both roots and shoots of the plants), compared with the plants grown in T6, suggesting that application of Zn-lys showed a non-significant decrease in antioxidants when grown in 100% addition of tannery wastewater in the soil.

3.4. *Effect of Foliar Application of Zn-lys on Uptake and Accumulation of Cr and Zn under Different Levels of Tannery Wastewater*

The results regarding Cr and Zn uptake in different organs (roots and shoots) of *B. napus* were also determined in this study. In our results, we analyzed roots and shoots differently for their statistics with one-way ANOVA. The data regarding Cr and Zn uptake are presented in Figure 4 and suggested that increasing levels of wastewater in the soil caused a significant increase in Cr uptake in all organs of the plants and a decrease in the content of Zn in all organs of the plant. However, Zn was accumulated at a higher concentration up to a level (33%). At the same time, further addition of wastewater caused a significant decrease in Zn content in all organs of the *B. napus* (Figure 4). According to the results, the maximum content of Cr was observed in T6 plants, which were 141 mg kg^{-1} in the roots and 40 mg kg^{-1} in the shoots of *B. napus*. In contrast, Zn content was found to be maximum in T1 plants 47 mg kg^{-1} in the roots and 37 mg kg^{-1} in the shoots compared with the plants grown without the application of Zn-lys without the addition of tannery wastewater in the soil. However, the addition of tannery wastewater (33%, 66%, and 100%) in the soil, significantly decreased the Zn content in the roots and shoots of the plants, when compared with the plants grown without the application of Zn-lys without the addition of tannery wastewater in the soil. In contrast, foliar application of Zn-lys significantly decreased Cr content while increasing Zn content in the roots and shoots of the plants compared with those treatments in which plants were grown without the application of Zn-lys. However, non-significant results were also obtained in the Zn content of roots in T6 and T7 plants.

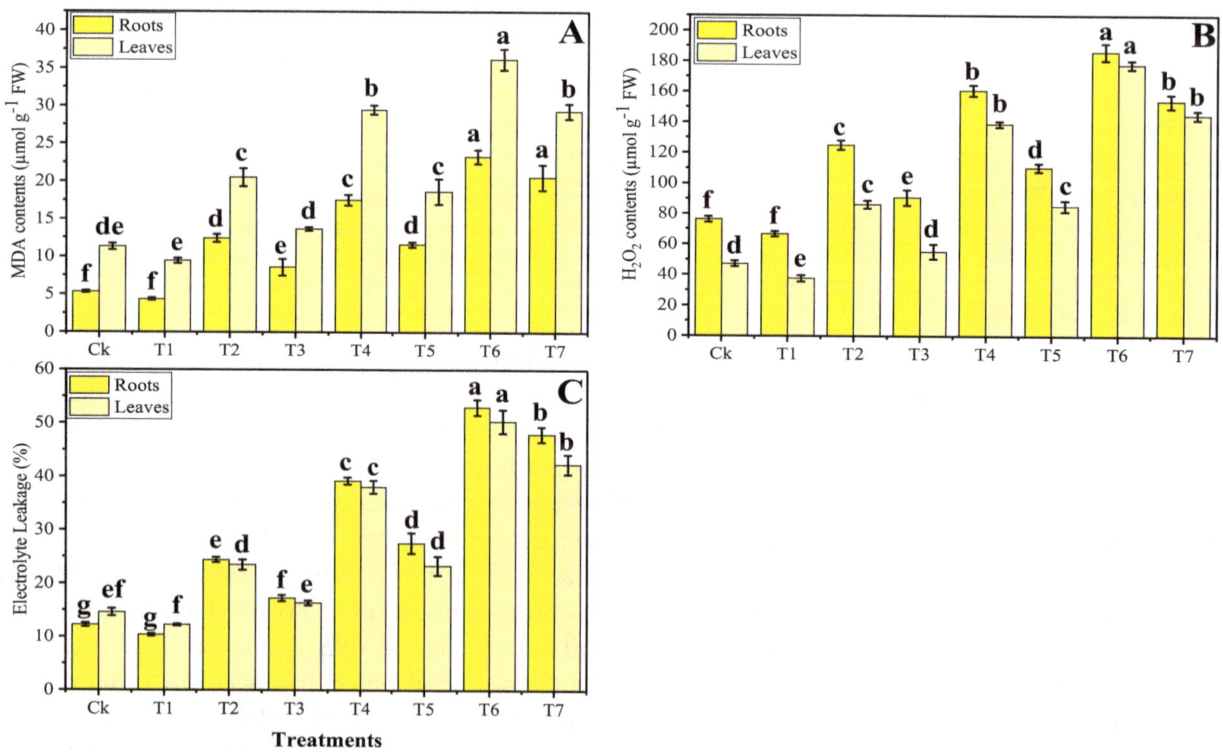

Figure 2. The effect of different levels of tannery wastewater on MDA contents (**A**) H_2O_2 contents (**B**) and EL leakage (**C**) in the roots and leaves of *B. napus* under the same concentrations of Zinc-lysine application. Values are demonstrated as means of the three replicates along with the standard deviation

(SD; $n = 3$). One-way ANOVA was performed, and mean differences were tested by HSD ($p < 0.05$). Different lowercase letters on the error bars indicate a significant difference between the treatments. Different treatments used in the figures are as follows: Ck (without irrigation with wastewater + 0 mg/L Zn-lysine), T1 (without irrigation with wastewater + 10 mg/L Zn-lysine), T2 (33% irrigation with wastewater + 0 mg/L Zn-lysine), T3 (33% irrigation with wastewater + 10 mg/L Zn-lysine), T4 (66% irrigation with wastewater + 0 mg/L Zn-lysine), T5 (66% irrigation with wastewater + 10 mg/L Zn-lysine), T6 (100% irrigation with wastewater + 0 mg/L Zn-lysine), and T7 (100% irrigation with wastewater + 10 mg/L Zn-lysine).

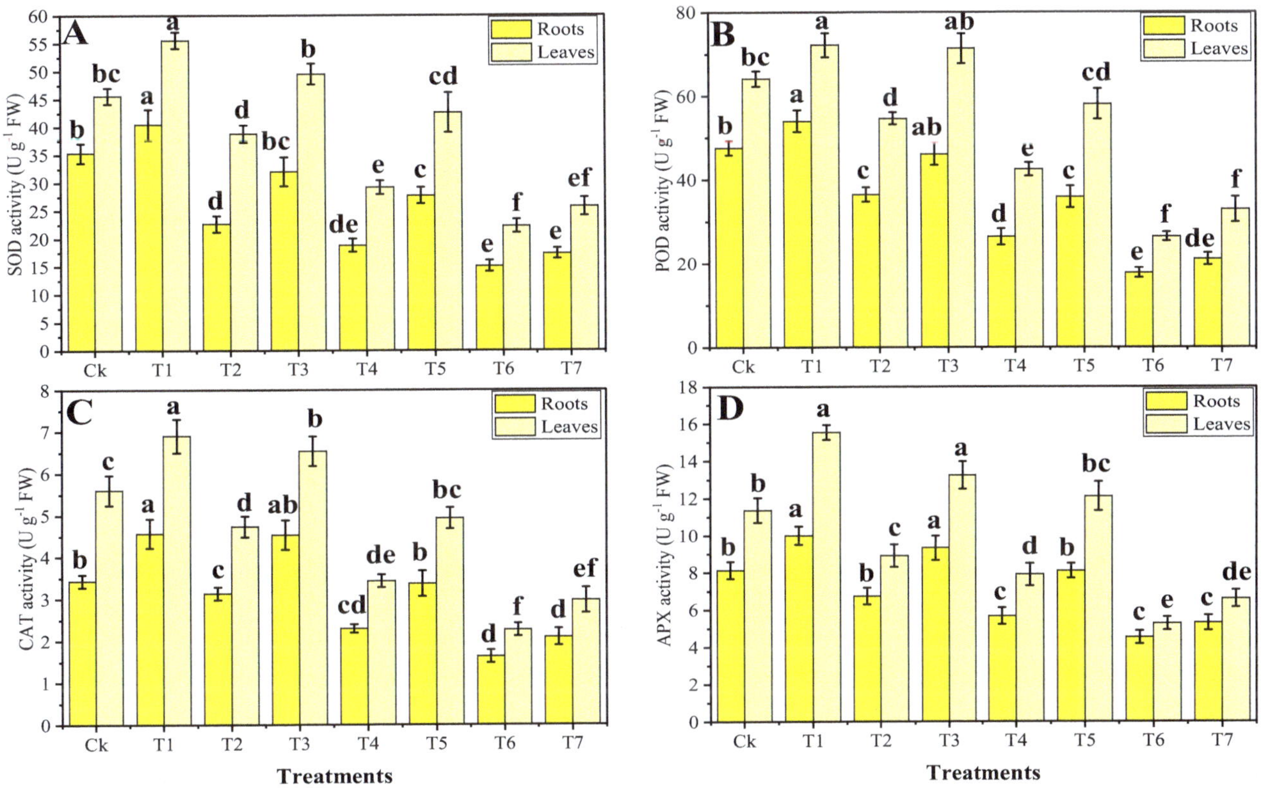

Figure 3. The effect of different levels of tannery wastewater on the activities of SOD (**A**), POD (**B**), CAT (**C**) and APX (**D**) in the roots and leaves under the same concentrations of Zinc-lysine application in *B. napus*. Values are demonstrated as the means of three replicates along with the standard deviation (SD; $n = 3$). One-way ANOVA was performed, and the mean differences were tested by HSD ($p < 0.05$). Different lowercase letters on the error bars indicate a significant difference between the treatments. Different treatments used in the figures are as follow: Ck (without irrigation with wastewater + 0 mg/L Zn-lysine), T1 (without irrigation with wastewater + 10 mg/L Zn-lysine), T2 (33% irrigation with wastewater + 0 mg/L Zn-lysine), T3 (33% irrigation with wastewater + 10 mg/L Zn-lysine), T4 (66% irrigation with wastewater + 0 mg/L Zn-lysine), T5 (66% irrigation with wastewater + 10 mg/L Zn-lysine), T6 (100% irrigation with wastewater + 0 mg/L Zn-lysine) and T7 (100% irrigation with wastewater + 10 mg/L Zn-lysine).

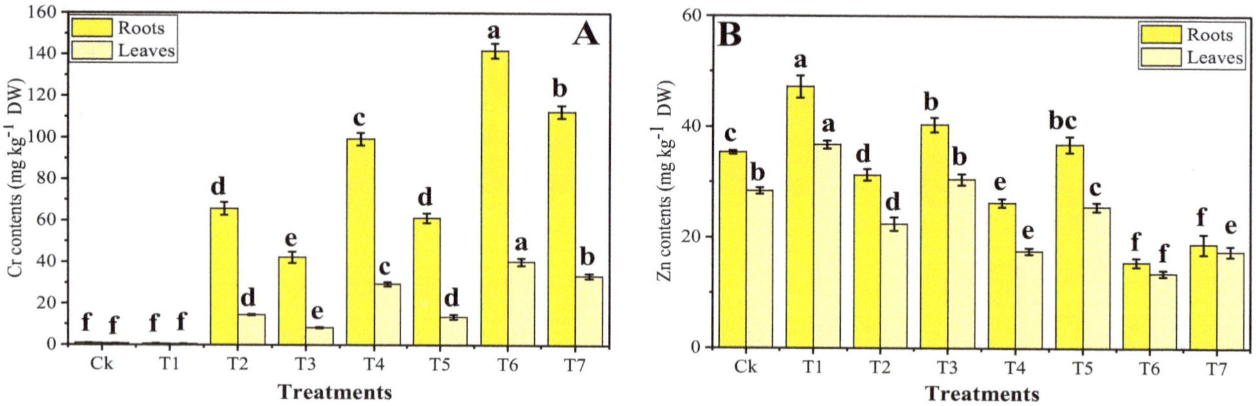

Figure 4. The effect of different levels of tannery wastewater on the uptake of Cr (**A**) and Zn (**B**) content in the roots and leaves under the same concentrations of Zinc-lysine application in *B. napus*. Values are demonstrated as the means of three replicates along with standard deviation (SD; $n = 3$). One-way ANOVA was performed, and the mean differences were tested by HSD ($p < 0.05$). Different lowercase letters on the error bars indicate a significant difference between the treatments. Different treatments used in the figures are as follow: Ck (without irrigation with wastewater + 0 mg/L Zn-lysine), T1 (without irrigation with wastewater + 10 mg/L Zn-lysine), T2 (33% irrigation with wastewater + 0 mg/L Zn-lysine), T3 (33% irrigation with wastewater + 10 mg/L Zn-lysine), T4 (66% irrigation with wastewater + 0 mg/L Zn-lysine), T5 (66% irrigation with wastewater + 10 mg/L Zn-lysine), T6 (100% irrigation with wastewater + 0 mg/L Zn-lysine) and T7 (100% irrigation with wastewater + 10 mg/L Zn-lysine).

3.5. Correlation between Different Parameters Studied in This Experiment

A correlogram was analyzed to quantify the relationship between different morpho-physiological attributes with Cr and Zn uptake by the plant organs (Figure 5). According to the results, it was suggested that Cr content in the roots was positively correlated with Cr concentration in the shoots and oxidative stress indicators while negatively correlated with plant growth and biomass, physiological traits and antioxidative enzymes, and Zn content in all organs of the plants. Similarly, Cr concentration in the shoots was positively correlated with Cr concentration in the roots, oxidative stress indicators while negatively correlated with all other morpho-physiological traits and Zn uptake. Moreover, the correlation graph also showed that Cr content in the roots was significantly increased with the Cr content in the shoots and also oxidative stress indicators and vice versa. However, Cr content in the roots was significantly decreased plant morphological and physiological attributes studied in this experiment.

A heatmap-histogram analysis was also constructed to explore the relationship between the different growth and Cr uptake attributes (Figure 6). Only significant differences were observed in the Cr uptake in T4, T6, and T7, while all other selected traits were showing non-significant differences with all treatments that were used in this study. The color yellow in Figure 6 indicates the non-significant results of our study, and the red color showed significant results in this figure. This histogram depicted a clear difference between the Cr uptake abilities and growth attributes of *B. napus*.

3.6. Principal Component Analysis

The scores and loading plots of PCA to check the effect of different levels of tannery wastewater with the exogenous application of Zn-lys on some selected attributes of *B. napus* plants are presented in Figure 7. Of all the main components, the first two components-Dim1 and Dim2-comprised more than 98% of the whole database and make up the largest portion of all components (Figure 7a). Among this, Dim1 contributed 93.7%, and Dim2 contributed 4.4% of the whole dataset. According to the results, all the respected treatments were dispersed successfully in the whole dataset (Figure 7a). The distribution of all the components in the dataset gave a clear indication that Cr toxicity in the soil significantly affected all the treatments used in this study with the exogenous application of Zn-lys. The control (1) treatment was most displaced from all other treatments of tannery wastewater in the soil, indicating that Cr toxicity in the soil significantly affected morpho-physiological traits of *B. napus* plants with the exogenous application of Zn-lys. However, Figure 7b shows that MDA and Cr content in the roots were positively correlated in the dataset from all the variables. In contrast, SOD activity in the roots, Zn content in the roots, plant height, and total chlorophyll content in the leaves were negatively correlated in PCA with all other variables (Figure 7b).

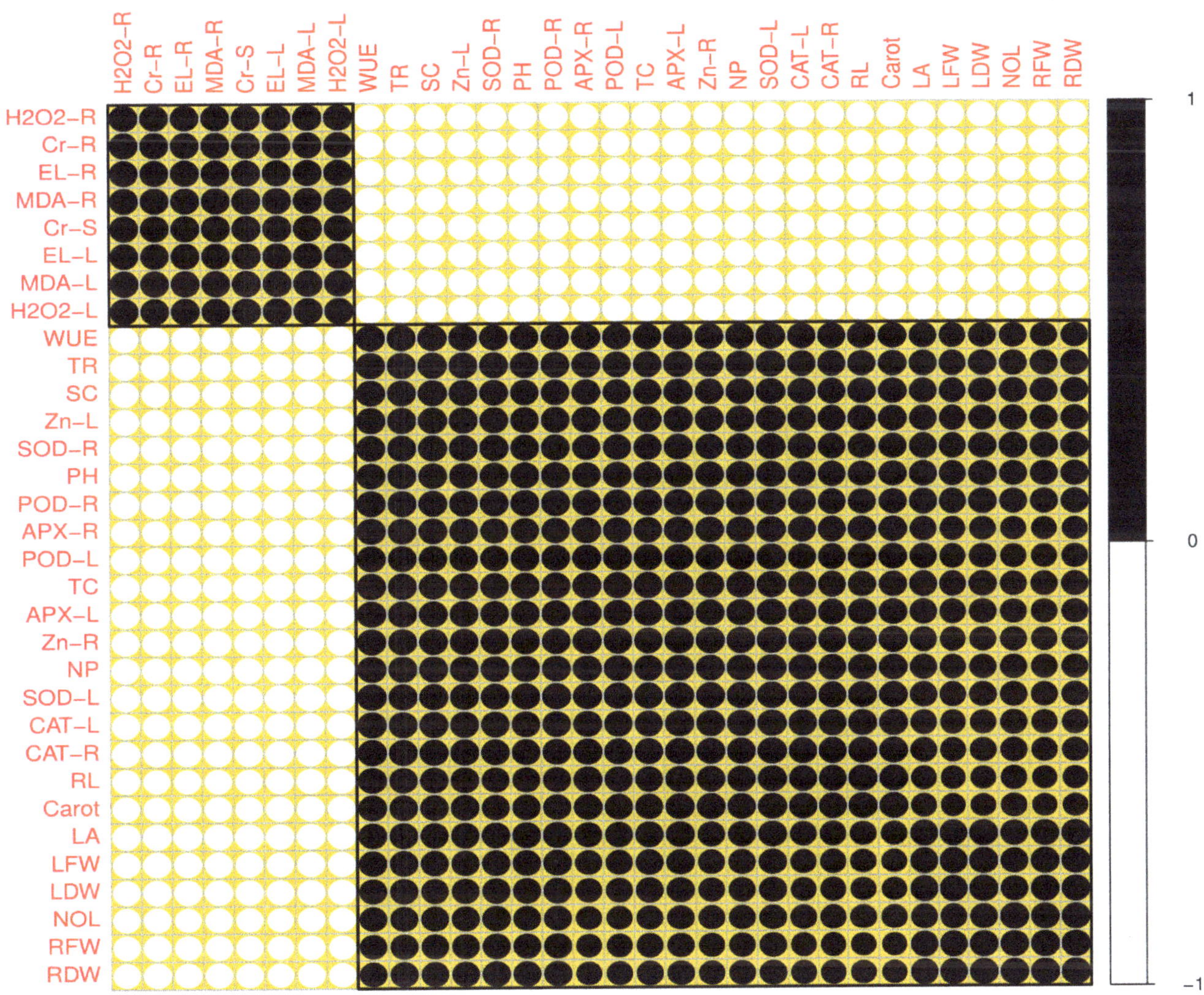

Figure 5. Correlation between different attributes studied in this study. H2O2-R (H_2O_2 initiation in roots), CR-R (Cr content in roots), EL-R (electrolyte leakage in roots), MDA-R (MDA content in roots), CR-S (Cr content in shoots), EL-L (electrolyte leakage in leaves), MDA-L (MDA content in leaves), H2O2-L (H_2O_2 initiation in leaves), Zn-R (zinc content in roots), NP (net photosynthesis), SOD-L (SOD activity in leaves), WUE (water use efficiency), TR (transpiration rate), SC (stomatal conductance), Zn-L (zinc content in leaves), SOD-R (SOD activity in roots), PH (plant height), POD-R (POD activity in roots), CAT-L (CAT activity in leaves), CAT-R (CAT activity in roots), RL (root length), Carot (carotenoid content), APX-R (APX activity in roots), POD-L (POD activity in leaves), TC (total chlorophyll content), APX-L (APX activity in leaves), LA (leaf area), LFW (leaves fresh weight), LDW (leaves dry weight), NOL (number of leaves), RFW (root fresh weight), RDW (root dry weight).

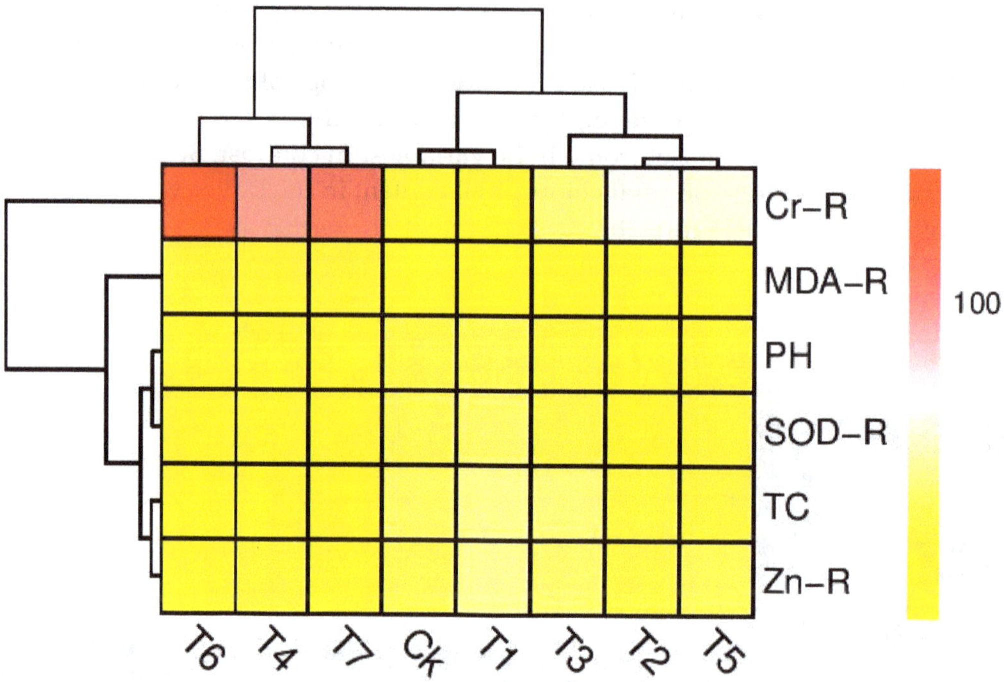

Figure 6. Heatmap histogram correlation between different parameters studied in this experiment. (Ck) Cr 0%, Zn-lys 0 mg L^{-1} (T1) Cr 0%, Zn-lys 10 mg L^{-1} (T2) Cr 33%, Zn-lys 0 mg L^{-1} (T3) Cr 33%, Zn-lys 10 mg L^{-1} (T4) Cr 66%, Zn-lys 0 mg L^{-1} (T5) Cr 66%, Zn-lys 10 mg L^{-1} (T6) Cr 100%, Zn-lys 0 mg L^{-1} (T7) Cr 100%, Zn-lys 10 mg L^{-1}. The abbreviations are as follow: CR-R (Cr content in roots), MDA-R (MDA content in roots), PH (plant height), SOD-R (SOD activity in roots), TC (total chlorophyll content), and Zn-R (zinc content in roots).

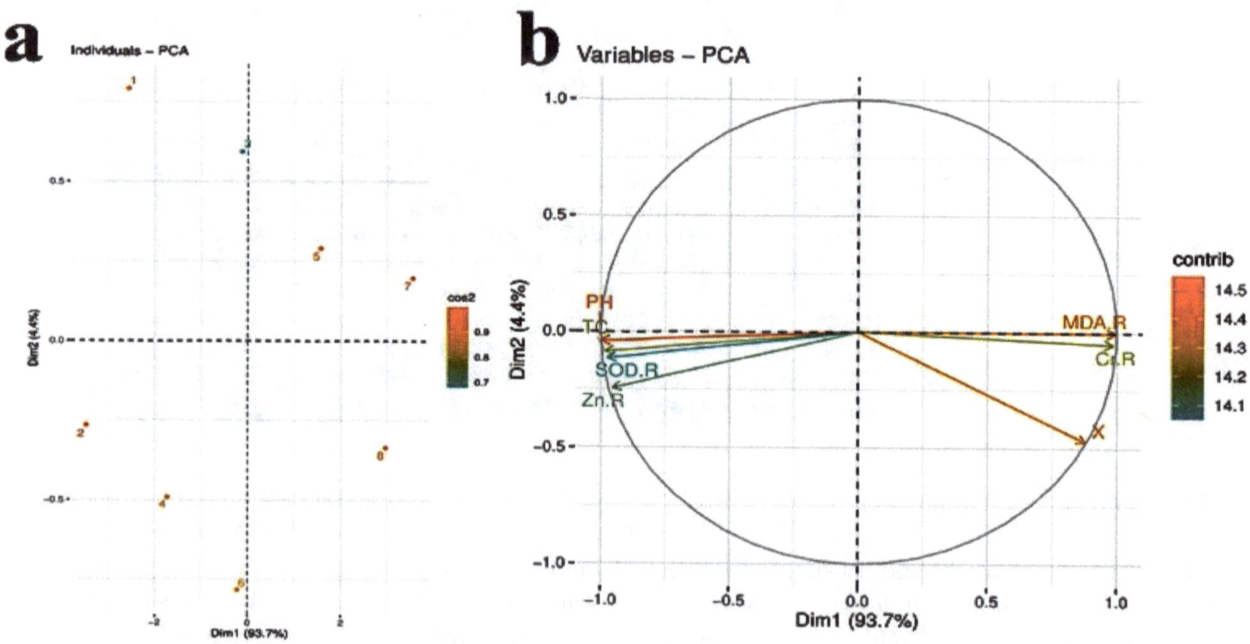

Figure 7. Score (**a**) and loading plots (**b**) of principal component analysis (PCA) on different studied attributes of *B. napus* grown in tannery wastewater soil. Score plot represents separation of treatments (1) Cr 0%, Zn-lys 0 mg L^{-1} (2) Cr 0%, Zn-lys 10 mg L^{-1} (3) Cr 33%, Zn-lys 0 mg L^{-1} (4) Cr 33%, Zn-lys 10 mg L^{-1} (5) Cr 66%, Zn-lys 0 mg L^{-1} (6) Cr 66%, Zn-lys 10 mg L^{-1} (7) Cr 100%, Zn-lys 0 mg L^{-1} (8) Cr 100%, Zn-lys 10 mg L^{-1}. The abbreviations are as follows: Zn-R (zinc content in roots), TC (total chlorophyll content), PH (plant height), SOD-R (SOD activity in roots), CR-R (Cr content in roots), and MDA-R (MDA content in roots).

4. Discussion

The harmful effects of heavy metals on food safety and marketability, plant growth because of plant-toxicity, and the ecosystem health of soil organisms cause serious metal accumulation in soils in agricultural production [32,33]. Like all living organisms, plants are very often both susceptible to the shortage and abundance of the available heavy metal ions as the key micronutrient, although they are also very toxic to normal plant growth and development at higher concentrations and even more ions such as Cr [7,10]. In our previous study, we also observed that the toxic concentration of Cu in the nutrient solution changed growth and other morphological attributes in the 40-day-old *B. napus* seedlings [18]. Although, in the current study, increasing levels of wastewater (33%, 66%, and 100%) in the soil displayed inhibitory effects on all morphological traits of the *B. napus* by declining different parameters of plant growth and biomass when compared to the control treatment (Table 1). Under the toxic level of Cr in the soil, the similar findings were showed by Li et al. [16] and Zaheer et al. [14], when they studied *B. napus* under different levels of Cr in the soil. This is because Cr can accumulate quickly in different organs of the plants, as suggested by Hussain et al. [9], which induced toxicity in the plants. The results from the rapid uptake of Cr content by the plant organs caused stunned growth and declined biomass in all parts of the plants [34]. Furthermore, it was also noticed that a very high concentration of Cr in the soil decreased some essential nutrients in the soil, which are important for normal growth and development of a plant body [10,34].

Photosynthetic pigments and gas exchange attributes were also affected with the addition of wastewater (33%, 66%, and 100%) in the soil (Figure 1), and it was observed that Cr toxicity in the soil significantly decreased these attributes in *B. napus*. Reduction in chlorophyll concentrations may lead to the removal of Mg needed for chlorophyll biosynthesis and/or ultra-structural modification to metal toxicity of chloroplast [18,35]. In addition to that, Cr toxicity influences and decreases the composition of photosynthetic pigments, which has already been shown by Yu et al. [36], PSII photochemical activity. In the leaves of *B. napus*, a remarkable decrease in photosynthetic pigments and gas exchange parameters were also noticed by Hasanuzzaman et al. [37] under Cd stress and Zhang et al. [17] under Cu stress environment. Our results coincide with the findings of Yasmeen et al. [38], which also noticed that the excess Cr concentration in the soil could cause a significant decrease in photosynthesis of the plants, including *B. napus*. Moreover, the disturbance in the transport of the water from the roots to the aboveground parts of the plants resulted in a decline in stomatal conductance and transpiration rate, which ultimately destroy the gas exchange parameters of the plants [39].

The high concentration of pollutants (heavy metals) in the soil induce oxidative stress [2], however, plants under Cr stress face severe oxidative damage owing to the overproduction of ROS and electrolyte leakage [9,16]. Phytotoxicity of Cr damaged many membrane-bounded organelles of the plant's results in oxidative stress in the plants and also a factor in lowering the biomass of the plants [34]. The extra production of ROS is extremely toxic for the plants and can cause the death of the plant, however, plants have an antioxidative defense system to scavenge these ROS [33,40]. *B. napus* is resistant to heavy metal stress but showed much higher production of O^{-2}, MDA, and H_2O_2 when grown in metal-polluted soil [16,18]. However, several studies also suggested that *B. napus* has a very strong defense system (antioxidants), which ultimately increased their activities to protect the plant against the metal stress environment [2,13,14]. In the current study, Cr toxicity caused oxidative stress in the *B. napus* by increasing malondialdehyde (MDA) content in the roots and leaves of the plants (Figure 2). This is because Cr toxicity disrupts the electron transport chain, producing much higher levels of H_2O_2 and O^{-2} radicals and increasing electrolyte leakage afterwards [10]. Extremely high contents of Cr in the plants can decrease the activities of antioxidant enzymes, which was also noticed in the current experiment with the addition of 66% and 100% tannery wastewater in the soil, which decreased the activities of various antioxidant enzymes in B. napus significantly (Figure 3). This alteration in the activities of different antioxidants might be due to the alterations in different genetic expressions and functions of some important proteins, which take part in this mechanism [8]. The decrease in the activities of antioxidants under elevating the level of Pb (300 mg kg^{-1} and 600 mg kg^{-1}) in the soil was also observed in the castor beans [41].

The absorption of heavy metals from the soil and transportation to various plant parts depends primarily on metal supplies and growth conditions. Heavy metals from the soil are absorbed by the root cell membrane and transferred and detoxified throughout the plant [42,43]. Plants with a higher Cr accumulation are not able to absorb the required range of essential nutrients, which will eventually lead to normal growth and development differences due to distorting photosynthesis and structural distortions [36]. In the current experiment, our objective is to reduce Cr content in different organs of the *B. napus* using a micronutrient-amino chelator. However, a very large amount of wastewater in the soil increased the Cr content in the various parts of the plant (Figure 4). There are several studies that suggest that increased soil Cr levels have caused a significant increase in Cr uptake and accumulation in different parts of the plants [34,44]. However, a reduction in the Zn uptake (which is an essential micronutrient) might be linked with the inability of the roots to absorb it from the soil under the high content of Cr in the soil [9]. It was also noticed that plants, when cultivated with tannery wastewater (which has a high content of toxic pollutants such as Cr), are unable to absorb/uptake a vital range of essential nutrients, which not only reduced micronutrients in the plant organs but also reduced different morpho-physiological attributes in the plants [6].

Various plant growth hormones have been widely used recently to mitigate oxidative stress and prompt plant growth and composition in different plant species when grown in metal-polluted soil [1,18,40]. Although, the usage of micronutrients with the amino acid complex is an incredibly new idea and showed beneficial results such as improved plant growth and composition and restrict the plant for uptake/accumulate toxic content in their body parts. There is also some literature on various plant species, which, when grown on metal contaminated soil, alleviates metal toxicity and improves plant growth and biomass [12,14,45]. In our previous study, we also noticed that irrigation with wastewater caused metal toxicity due to a large amount of Cr content in the soil, while an application with Zn-lys alleviates metal toxicity by increasing plant growth and biomass in spinach plants [3]. The reduced Cr concentration in different parts of *B. napus* could be due to the higher Zn concentration in plants, as Zaheer et al. [3] previously reported antagonistic effects of Cr and Zn. This is because the application of Zn-lys gives plants space to increase/absorb essential nutrients from the soil and a more secure option for plants in the stressful environment. In addition, a large amount of amino acid in a plant's various organs plays a crucial role in reducing the movement of metal pollutants in plants by forming different complexes [45]. Amino acids also take part in the ultra-structure constitution, which is also another factor in improving plant growth and photosynthesis in the plants [46]. Exogenous supplementation of Zn-lys has shown the ability to boost up the Wi of the plant by enhancing the nutrient uptake, which as a result increases plant growth by governing the processes of transpiration and photosynthesis [9,47]. Our results also showed that the application of Zn-lys reduced oxidative stress in the plants by enhancing the activities of different enzymatic antioxidants (Figures 2 and 3). The decrease in oxidative stress in *B. napus* might be due to the accumulation of a large amount of amino acids, which have the ability to scavenge increased production of ROS, as suggested by Teixeira et al. [48]. In the current experiment, exogenous supplementation of Zn-lys caused a significant increase in Zn content while a noticeable decrease in the Cr content in all organs (roots and shoots) of the plants was observed (Figure 4). This is because of external application of Zn-lys, which not only increased Zn content in roots and shoots of the plants and also decreased Cr content because amino acids have the ability to form complexes with extra heavy metals and restrict their mobility in the plant organs [12,45].

5. Conclusions

Irrigation of *B. napus* with tannery wastewater is dangerous because it may contain toxic pollutants (such as Cr), which significantly affect its growth and yield. Although, phytotoxicity of Cr also decreased their photosynthetic pigments and gas exchange parameters and induced oxidative damage due to the overproduction of ROS in the cell/tissues of the plants. Moreover, Cr toxicity also decreased essential nutrients (such as Zn) in the plant organs, which is essential for normal growth and development of the plant body. The exogenous application of Zn-lys can reduce the phytotoxicity of Cr by increasing morphological characteristics and activity of antioxidant enzymes, which helps to mitigate oxidative stress in plants. Taken together, foliar supplementation of Zn-lys amoriates Cr stress in *B. napus* by improving plant growth and yield. Therefore, Zn-lys chelate can be efficiently used for improving *B. napus* growth and Zn contents while reducing Cr concentration in plants in a Zn deficient and Cr-contaminated soil.

Author Contributions: Conceptualization, S.A.; data curation, I.E.Z., M.H.S., M.A.A., Q.A., and M.R.; formal analysis, I.E.Z., S.A., M.H.S., Q.A., Z.A., and L.W.; funding acquisition, M.R., M.A.E.-S., M.N.A., and L.W.; investigation, S.A. and Z.A.; methodology, Z.A., M.A.E.-S. and L.W.; project administration, I.E.Z., M.A.E.-S., M.N.A., and L.W.; resources, S.A., M.A.A., Q.A., M.A.E.-S., M.N.A., and L.W.; software, M.H.S.; supervision, S.A. and M.H.S.; validation, Z.A.; visualization, Q.A. and M.R.; writing—original draft, I.E.Z. and M.H.S.; writing—review and editing, I.E.Z., S.A., M.H.S., and M.R. All authors have read and agreed to the published version of the manuscript.

Acknowledgments: The authors highly acknowledge the Government College University, Faisalabad, Pakistan for its support. The authors would like to extend their sincere appreciation to the Researchers Supporting Project Number (RSP-2020/180), King Saud University, Riyadh, Saudi Arabia. The findings of this study are a part of Doctorate studies of Ihsan Elahi Zaheer.

References

1. Saleem, M.H.; Ali, S.; Rehman, M.; Hasanuzzaman, M.; Rizwan, M.; Irshad, S.; Shafiq, F.; Iqbal, M.; Alharbi, B.M.; Alnusaire, T.S. Jute: A Potential Candidate for Phytoremediation of Metals—A Review. *Plants* **2020**, *9*, 258. [CrossRef]
2. Kamran, M.; Malik, Z.; Parveen, A.; Huang, L.; Riaz, M.; Bashir, S.; Mustafa, A.; Abbasi, G.H.; Xue, B.; Ali, U. Ameliorative Effects of Biochar on Rapeseed (*Brassica napus* L.) Growth and Heavy Metal Immobilization in Soil Irrigated with Untreated Wastewater. *J. Plant Growth Regul.* **2019**, *39*, 266–281. [CrossRef]
3. Zaheer, I.E.; Ali, S.; Rizwan, M.; Abbas, Z.; Bukhari, S.A.H.; Wijaya, L.; Alyemeni, M.N.; Ahmad, P. Zinc-lysine prevents chromium-induced morphological, photosynthetic, and oxidative alterations in spinach irrigated with tannery wastewater. *Environ. Sci. Pollut. Res.* **2019**, *26*, 28951–28961. [CrossRef] [PubMed]
4. Maqbool, A.; Ali, S.; Rizwan, M.; Ishaque, W.; Rasool, N.; ur Rehman, M.Z.; Bashir, A.; Abid, M.; Wu, L. Management of tannery wastewater for improving growth attributes and reducing chromium uptake in spinach through citric acid application. *Environ. Sci. Pollut. Res.* **2018**, *25*, 10848–10856. [CrossRef] [PubMed]
5. Saleem, M.H.; Ali, S.; Hussain, S.; Kamran, M.; Chattha, M.S.; Ahmad, S.; Aqeel, M.; Rizwan, M.; Aljarba, N.H.; Alkahtani, S. Flax (*Linum usitatissimum* L.): A Potential Candidate for Phytoremediation? Biological and Economical Points of View. *Plants* **2020**, *9*, 496. [CrossRef]
6. Chowdhary, P.; Yadav, A.; Singh, R.; Chandra, R.; Singh, D.; Raj, A.; Bharagava, R.N. Stress response of *Triticum aestivum* L. and *Brassica juncea* L. against heavy metals growing at distillery and tannery wastewater contaminated site. *Chemosphere* **2018**, *206*, 122–131. [CrossRef]
7. Ding, H.; Wang, G.; Lou, L.; Lv, J. Physiological responses and tolerance of kenaf (*Hibiscus cannabinus* L.) exposed to chromium. *Ecotoxicol. Environ. Saf.* **2016**, *133*, 509–518. [CrossRef]
8. Tang, R.; Li, X.; Mo, Y.; Ma, Y.; Ding, C.; Wang, J.; Zhang, T.; Wang, X. Toxic responses of metabolites, organelles and gut microorganisms of Eisenia fetida in a soil with chromium contamination. *Environ. Pollut.* **2019**, *251*, 910–920. [CrossRef]
9. Hussain, A.; Ali, S.; Rizwan, M.; ur Rehman, M.Z.; Hameed, A.; Hafeez, F.; Alamri, S.A.; Alyemeni, M.N.; Wijaya, L. Role of zinc–lysine on growth and chromium uptake in rice plants under Cr stress. *J. Plant Growth Regul.* **2018**, *37*, 1413–1422. [CrossRef]

10. Zaheer, I.E.; Ali, S.; Saleem, M.H.; Noor, I.; El-Esawi, M.A.; Hayat, K.; Rizwan, M.; Abbas, Z.; El-Sheikh, M.A.; Alyemeni, M.N. Iron–Lysine Mediated Alleviation of Chromium Toxicity in Spinach (*Spinacia oleracea* L.) Plants in Relation to Morpho-Physiological Traits and Iron Uptake When Irrigated with Tannery Wastewater. *Sustainability* **2020**, *12*, 6690. [CrossRef]
11. Rizwan, M.; Ali, S.; Hussain, A.; Ali, Q.; Shakoor, M.B.; Zia-ur-Rehman, M.; Farid, M.; Asma, M. Effect of zinc-lysine on growth, yield and cadmium uptake in wheat (*Triticum aestivum* L.) and health risk assessment. *Chemosphere* **2017**, *187*, 35–42. [CrossRef] [PubMed]
12. Ghasemi, S.; Khoshgoftarmanesh, A.H.; Afyuni, M.; Hadadzadeh, H. Iron (II)–amino acid chelates alleviate salt-stress induced oxidative damages on tomato grown in nutrient solution culture. *Sci. Hortic.* **2014**, *165*, 91–98. [CrossRef]
13. Saleem, M.H.; Rehman, M.; Fahad, S.; Tung, S.A.; Iqbal, N.; Hassan, A.; Ayub, A.; Wahid, M.A.; Shaukat, S.; Liu, L.; et al. Leaf gas exchange, oxidative stress, and physiological attributes of rapeseed (*Brassica napus* L.) grown under different light-emitting diodes. *Photosynthetica* **2020**, *58*, 836–845. [CrossRef]
14. Zaheer, I.E.; Ali, S.; Saleem, M.H.; Imran, M.; Alnusairi, G.S.H.; Alharbi, B.M.; Riaz, M.; Abbas, Z.; Rizwan, M.; Soliman, M.H. Role of iron–lysine on morpho-physiological traits and combating chromium toxicity in rapeseed (*Brassica napus* L.) plants irrigated with different levels of tannery wastewater. *Plant Physiol. Biochem.* **2020**, *155*, 70–84. [CrossRef]
15. Mohamed, A.; Ibrahim, A.; Shalby, N.; Bai, C.; Qin, M.; Agami, R.A.; Jie, K.; Wang, B.; Zhou, G. Stomatal and Photosynthetic Traits Are Associated with Investigating Sodium Chloride Tolerance of *Brassica napus* L. Cultivars. *Plants* **2020**, *9*, 62. [CrossRef]
16. Li, L.; Zhang, K.; Gill, R.A.; Islam, F.; Farooq, M.A.; Wang, J.; Zhou, W. Ecotoxicological and Interactive Effects of Copper and Chromium on Physiochemical, Ultrastructural, and Molecular Profiling in Brassica napus L. *BioMed Res. Int.* **2018**, *2018*, 9248123. [CrossRef]
17. Zhang, M.; Ran, R.; Nao, W.S.; Feng, Y.; Jia, L.; Sun, K.; Wang, R.; Feng, H. Physiological effects of short-term copper stress on rape (*Brassica napus* L.) seedlings and the alleviation of copper stress by attapulgite clay in growth medium. *Ecotoxicol. Environ. Saf.* **2019**, *171*, 878–886. [CrossRef]
18. Zaheer, I.E.; Ali, S.; Rizwan, M.; Farid, M.; Shakoor, M.B.; Gill, R.A.; Najeeb, U.; Iqbal, N.; Ahmad, R. Citric acid assisted phytoremediation of copper by *Brassica napus* L. *Ecotoxicol. Environ. Saf.* **2015**, *120*, 310–317. [CrossRef]
19. Walkley, A.; Black, I.A. An examination of the Degtjareff method for determining soil organic matter, and a proposed modification of the chromic acid titration method. *Soil Sci.* **1934**, *37*, 29–38. [CrossRef]
20. Bouyoucos, G.J. Hydrometer method improved for making particle size analyses of soils 1. *Agron. J.* **1962**, *54*, 464–465. [CrossRef]
21. Apha, A. American Public Health Association/American Water Works Association/Water Environment Federation. In *WPCF, Standard Methods for the Examination of Water and Wastewater*; The Water Environment Federation: Washington, DC, USA, 1995.
22. Ayers, R.; Westcot, D. *Water Quality for Agriculture*; FAO Irrigation and Drainage Paper 29 Rev.1; Food and Agriculture Organization of the United Nations: Roma, Italy, 1985; Volume 15, p. 2016.
23. Nagata, M.; Yamashita, I. Simple method for simultaneous determination of chlorophyll and carotenoids in tomato fruit. *Nippon Shokuhin Kogyo Gakkaishi* **1992**, *39*, 925–928. [CrossRef]
24. Heath, R.L.; Packer, L. Photoperoxidation in isolated chloroplasts: I. Kinetics and stoichiometry of fatty acid peroxidation. *Arch. Biochem. Biophys.* **1968**, *125*, 189–198. [CrossRef]
25. Jana, S.; Choudhuri, M.A. Glycolate metabolism of three submersed aquatic angiosperms: Effect of heavy metals. *Aquat. Bot.* **1981**, *11*, 67–77. [CrossRef]
26. Dionisio-Sese, M.L.; Tobita, S. Antioxidant responses of rice seedlings to salinity stress. *Plant Sci.* **1998**, *135*, 1–9. [CrossRef]
27. Chen, C.-N.; Pan, S.-M. Assay of superoxide dismutase activity by combining electrophoresis and densitometry. *Bot. Bull. Acad. Sin.* **1996**, *37*, 107–111.
28. Sakharov, I.Y.; Ardila, G.B. Variations of peroxidase activity in cocoa (*Theobroma cacao* L.) beans during their ripening, fermentation and drying. *Food Chem.* **1999**, *65*, 51–54. [CrossRef]
29. Aebi, H. Catalase in vitro. *Methods Enzymol.* **1984**, *105*, 121–126.

30. Nakano, Y.; Asada, K. Hydrogen peroxide is scavenged by ascorbate-specific peroxidase in spinach chloroplasts. *Plant Cell Physiol.* **1981**, *22*, 867–880.
31. Rehman, M.Z.; Rizwan, M.; Ghafoor, A.; Naeem, A.; Ali, S.; Sabir, M.; Qayyum, M.F. Effect of inorganic amendments for in situ stabilization of cadmium in contaminated soils and its phyto-availability to wheat and rice under rotation. *Environ. Sci. Pollut. Res.* **2015**, *22*, 16897–16906. [CrossRef]
32. Saleem, M.; Ali, S.; Rehman, M.; Rana, M.; Rizwan, M.; Kamran, M.; Imran, M.; Riaz, M.; Hussein, M.; Elkelish, A.; et al. Influence of phosphorus on copper phytoextraction via modulating cellular organelles in two jute (*Corchorus capsularis* L.) varieties grown in a copper mining soil of Hubei Province, China. *Chemosphere* **2020**, *248*, 126032. [CrossRef]
33. Saleem, M.H.; Fahad, S.; Khan, S.U.; Ahmar, S.; Khan, M.H.U.; Rehman, M.; Maqbool, Z.; Liu, L. Morpho-physiological traits, gaseous exchange attributes, and phytoremediation potential of jute (*Corchorus capsularis* L.) grown in different concentrations of copper-contaminated soil. *Ecotoxicol. Environ. Saf.* **2020**, *189*, 109915. [CrossRef] [PubMed]
34. Singh, H.P.; Mahajan, P.; Kaur, S.; Batish, D.R.; Kohli, R.K. Chromium toxicity and tolerance in plants. *Environ. Chem. Lett.* **2013**, *11*, 229–254. [CrossRef]
35. Saleem, M.H.; Ali, S.; Rehman, M.; Rizwan, M.; Kamran, M.; Mohamed, I.A.; Bamagoos, A.A.; Alharby, H.F.; Hakeem, K.R.; Liu, L. Individual and combined application of EDTA and citric acid assisted phytoextraction of copper using jute (*Corchorus capsularis* L.) seedlings. *Environ. Technol. Innov.* **2020**, *19*, 100895. [CrossRef]
36. Yu, X.-Z.; Lu, C.-J.; Li, Y.-H. Role of cytochrome c in modulating chromium-induced oxidative stress in Oryza sativa. *Environ. Sci. Pollut. Res.* **2018**, *25*, 27639–27649. [CrossRef]
37. Hasanuzzaman, M.; Nahar, K.; Anee, T.I.; Fujita, M. Exogenous silicon attenuates cadmium-induced oxidative stress in Brassica napus L. by modulating AsA-GSH pathway and glyoxalase system. *Front. Plant Sci.* **2017**, *8*, 1061. [CrossRef]
38. Yasmeen, T.; Ali, Q.; Islam, F.; Noman, A.; Akram, M.S.; Javed, M.T. Biologically treated wastewater fertigation induced growth and yield enhancement effects in *Vigna radiata* L. *Agric. Water Manag.* **2014**, *146*, 124–130. [CrossRef]
39. Mohamed, I.A.; Shalby, N.; MA El-Badri, A.; Saleem, M.H.; Khan, M.N.; Nawaz, M.A.; Qin, M.; Agami, R.A.; Kuai, J.; Wang, B. Stomata and Xylem Vessels Traits Improved by Melatonin Application Contribute to Enhancing Salt Tolerance and Fatty Acid Composition of *Brassica napus* L. Plants. *Agronomy* **2020**, *10*, 1186. [CrossRef]
40. Saleem, M.H.; Ali, S.; Kamran, M.; Iqbal, N.; Azeem, M.; Tariq Javed, M.; Ali, Q.; Zulqurnain Haider, M.; Irshad, S.; Rizwan, M. Ethylenediaminetetraacetic Acid (EDTA) Mitigates the Toxic Effect of Excessive Copper Concentrations on Growth, Gaseous Exchange and Chloroplast Ultrastructure of *Corchorus capsularis* L. and Improves Copper Accumulation Capabilities. *Plants* **2020**, *9*, 756. [CrossRef]
41. Mallhi, Z.I.; Rizwan, M.; Mansha, A.; Ali, Q.; Asim, S.; Ali, S.; Hussain, A.; Alrokayan, S.H.; Khan, H.A.; Alam, P. Citric Acid Enhances Plant Growth, Photosynthesis, and Phytoextraction of Lead by Alleviating the Oxidative Stress in Castor Beans. *Plants* **2019**, *8*, 525. [CrossRef]
42. Ashraf, M.A.; Hussain, I.; Rasheed, R.; Iqbal, M.; Riaz, M.; Arif, M.S. Advances in microbe-assisted reclamation of heavy metal contaminated soils over the last decade: A review. *J. Environ. Manag.* **2017**, *198*, 132–143. [CrossRef]
43. Lajayer, B.A.; Moghadam, N.K.; Maghsoodi, M.R.; Ghorbanpour, M.; Kariman, K. Phytoextraction of heavy metals from contaminated soil, water and atmosphere using ornamental plants: Mechanisms and efficiency improvement strategies. *Environ. Sci. Pollut. Res.* **2019**, *26*, 8468–8484. [CrossRef] [PubMed]
44. Medda, S.; Mondal, N.K. Chromium toxicity and ultrastructural deformation of Cicer arietinum with special reference of root elongation and coleoptile growth. *Ann. Agrar. Sci.* **2017**, *15*, 396–401. [CrossRef]
45. Ghasemi, S.; Khoshgoftarmanesh, A.H.; Hadadzadeh, H.; Jafari, M. Synthesis of iron-amino acid chelates and evaluation of their efficacy as iron source and growth stimulator for tomato in nutrient solution culture. *J. Plant Growth Regul.* **2012**, *31*, 498–508. [CrossRef]
46. Kosar, F.; Akram, N.; Ashraf, M. Exogenously-applied 5-aminolevulinic acid modulates some key physiological characteristics and antioxidative defense system in spring wheat (*Triticum aestivum* L.) seedlings under water stress. *South Afr. J. Bot.* **2015**, *96*, 71–77. [CrossRef]

47. Saleem, M.H.; Kamran, M.; Zhou, Y.; Parveen, A.; Rehman, M.; Ahmar, S.; Malik, Z.; Mustafa, A.; Anjum, R.M.A.; Wang, B.; et al. Appraising growth, oxidative stress and copper phytoextraction potential of flax (*Linum usitatissimum* L.) grown in soil differentially spiked with copper. *J. Environ. Manag.* **2020**, *257*, 109994. [CrossRef]
48. Teixeira, W.F.; Fagan, E.B.; Soares, L.H.; Umburanas, R.C.; Reichardt, K.; Neto, D.D. Foliar and seed application of amino acids affects the antioxidant metabolism of the soybean crop. *Front. Plant Sci.* **2017**, *8*, 327. [CrossRef]

2

Effect of Arbuscular Mycorrhizal Colonization on Cadmium-Mediated Oxidative Stress in *Glycine max* (L.) Merr.

Alicia S. Molina [1,*], Mónica A. Lugo [2,*], María V. Pérez Chaca [1], Silvina Vargas-Gil [3], Fanny Zirulnik [1], Jorge Leporati [4], Nuria Ferrol [5] and Concepción Azcón-Aguilar [5]

1 Facultad de Química, Bioquímica y Farmacia (FQByF), Universidad Nacional de San Luis (UNSL), Área Química Biológica, Ejército de los Andes 950, Bloque I, 1er Piso, San Luis, Argentina; perezch@unsl.edu.ar (M.V.P.C.); fzirul@unsl.edu.ar (F.Z.)

2 IMIBIO-CONICET-FQByF-UNSL, Área Ecología, Ejército de los Andes 950, Bloque I, 2do Piso, Box 4 San Luis, Argentina

3 Instituto de Patología Vegetal (IPAVE, CIAP-INTA), CONICET, Camino 60 Cuadras Km. 5.5, C.P. 5119 Córdoba, Argentina; vargasgil.silvina@inta.gob.ar

4 Facultas de Ingeniería y Ciencias Agropecuarias (FICA)-UNSL, Área Matemática, Campus Universitario, Ruta 148 Ext. Norte, 2do Piso Ala Este, Villa Mercedes, San Luis, Argentina; jlleporati@unsl.edu.ar

5 Estación Experimental del Zaidín, CSIC, Profesor Albareda 1, 18008 Granada, Spain; nuria.ferrol@eez.csic.es (N.F.); conchi.azcon@eez.csic.es (C.A.-A.)

* Correspondence: amolina@unsl.edu.ar (A.S.M.); lugo@unsl.edu.ar (M.A.L.)

Abstract: Cadmium is a heavy metal (HM) that inhibits plant growth and leads to death, causing great losses in yields, especially in Cd hyperaccumulator crops such as *Glycine max* (L.) Merr. (soybean), a worldwide economically important legume. Furthermore, Cd incorporation into the food chain is a health hazard. Oxidative stress (OS) is a plant response to abiotic and biotic stresses with an intracellular burst of reactive oxygen species (ROS) that causes damage to lipids, proteins, and DNA. The arbuscular mycorrhizal fungal (AMF) association is a plant strategy to cope with HM and to alleviate OS. Our aim was to evaluate the mitigation effects of mycorrhization with AMF *Rhizophagus intraradices* on soybean growth, nutrients, Cd accumulation, lipid peroxidation, and the activity of different antioxidant agents under Cd (0.7–1.2 mg kg^{-1} bioavailable Cd) induced OS. Our results suggest that glutathione may act as a signal molecule in a defense response to Cd-induced OS, and mycorrhization may avoid Cd-induced growth inhibition and reduce Cd accumulation in roots. It is discussed that *R. intraradices* mycorrhization would act as a signal, promoting the generation of a soybean cross tolerance response to Cd pollution, therefore evidencing the potential of this AMF association for bioremediation and encouragement of crop development, particularly because it is an interaction between a worldwide cultivated Cd hyperaccumulator plant and an AMF–HM–accumulator commonly present in soils.

Keywords: glutathione; heavy metal toxicity; mycorrhization; non-protein thiols; plan antioxidant defenses; *Rhizophagus intraradices*; soybean

---

1. Introduction

The environmental concentration of cadmium (Cd) as well as other heavy metals (HMs) has increased due to anthropogenic activities. Inorganic phosphate fertilizers contain HM traces and represent an important input of Cd to agricultural soils [1]. Cd is not an essential element for plants but can be easily taken up by the roots through plasma membrane transporters of essential metals and transported to the leaves via xylem. The content of Cd in plant tissues can be transferred to the food chain, thereby harming human health [2]. It is well known that Cd produces plant metabolic alterations that lead to a decrease in growth rate, causing a reduction in crop production [3]. Moreover, Cd has been shown to overproduce reactive oxygen species (ROS), usually leading to uncontrolled oxidative stress [4,5]. This effect has been observed in both roots and leaves [2,6].

Reactive oxygen species levels are regulated by the coordinated action of enzymatic and non-enzymatic antioxidant systems. Superoxide dismutase (SOD), catalase (CAT), and peroxidase (APX) participate in the enzymatic antioxidant defense system, whereas glutathione (GSH) is an essential molecule involved in the non-enzymatic antioxidant defense [4]. GSH is synthesized from non-protein thiols (NPTs), and it is a precursor of phytochelatins (PCs) [2,4,5]; these compounds participate in the chelation and the storage of Cd in vacuoles. Different studies suggested a role for GSH in plant redox signaling processes [7,8]. Furthermore, plant adaptation to grow in the presence of Cd was linked to the capacity to maintain a high intracellular GSH concentration [9].

Arbuscular mycorrhizal fungi (AMF) are ubiquitous soil microorganisms that establish a mutualistic symbiotic association with most land plants that is known as arbuscular mycorrhiza (AM). In this association, the fungus assists the plant in uptaking soil low mobility mineral nutrients, especially phosphate, thereby increasing plant nutrition and growth [10–12]. The development of this symbiosis also increases plant tolerance to abiotic and biotic stresses, including metal stress [13,14]. AMF can improve plant tolerance to HMs through different mechanisms, e.g., enhancement of nutrient supply or decrease of water stress [15], metal sequestration through the production of binding proteins such as glomalin [16], and bioaccumulation in the inner root cells colonized by fungal structures [14] as well as in the extraradical mycelium [17,18]. Toxic elements such as Cd can travel from soil to roots through AMF hyphae; however, the fungus may also function as a biological barrier against Cd transfer from root tissues to the shoot [19]. Several studies have shown the role of AMF in reducing Cd stress via the attenuation of its negative effects on shoot biomass and the activity of photosystem II, both processes including enzymatic and non-enzymatic antioxidant and defense pathways in *Trifolium repens* and genotypes of *Cajanus cajan* [20,21]. Moreover, AM formed by *Pisum sativum* and *Rhizophagus intraradices* has been proposed as a buffering system association in Cd-induced stress, with differences in the level of the response to Cd stress between pea genotypes and involving the expression and the transcription of genes related to heavy metal chelation pathways as part of the Cd tolerance strategies acting during the symbiosis [22,23].

Soybean [*Glycine max* (L.) Merr.] is an important crop worldwide whose biomass production and nutritional quality are significantly reduced whenever being grown in metal contaminated soils. Previous studies indicate that Cd severely inhibits soybean growth and that both the amount of non-enzymatic antioxidants and the activities of enzymatic antioxidants are altered in response to Cd exposure [9]. Soybean is a well known AMF host that is cultivated in large areas around the world. Therefore, mycorrhizal soybean plants could be employed to remove Cd from metal contaminated soils.

The aim of this work was to evaluate the stress mitigation effect of the AMF *Rhizophagus intraradices* root colonization in soybean plants under Cd-induced oxidative stress. In order to do so, plant growth was examined together with Cd accumulation in plants and soil, plant oxidative stress marker (lipid peroxidation), non-enzymatic defense (GSH and non-protein thiols), and enzymatic defense (ascorbate peroxidase activity). It was hypothesized that soybean mycorrhization with *R. intraradices* can mitigate the effect of Cd-induced stress with the intervention of antioxidant defense processes, and the working hypotheses were:

(a) Cd decreases AMF colonization of soybean roots;
(b) The two components of this experimental system (soybean and the *R. intraradices* isolate) behave as efficient accumulators of Cd;
(c) The mycorrhization with *R. intraradices* improves the soybean growth under Cd stress; and
(d) The symbiotic association with the AMF avoids the increase of lipid peroxidation in soybean caused by Cd oxidative stress.

2. Results

2.1. Effect of Cd on Soybean Mycorrhizal Colonization

Cadmium addition did not modify AMF colonization in soybean roots at the end of the experiment at the final time (tf) (Table 1). Percentage of root colonization by *R. intraradices* was not modified after 40 days of Cd treatment (tf) with respect to the values of the control mycorrhized (M) plants (91.00% ± 3.59, 85.67% ± 3.09, respectively; $p = 0.1436$) without Cd. Similarly, the addition of Cd did not significantly modify the percentages of arbuscules (72.00% ± 5.14, 70.83% ± 5.02, respectively; $p = 0.4371$) and vesicles (39.33% ± 3.76, 35.67% ± 0.80, respectively; $p = 0.1918$) in the colonized roots (Table 1). Percentage of root colonization, arbuscules, and vesicles also increased significantly from the initial time (t0) to t(f) (Table 1).

Table 1. Mycorrhizal colonization of soybean plants inoculated with *Rhizophagus intraradices* supplemented or not with Cadmium (Cd).

Treatment	Harvest Time	Root Colonization (%)	Arbuscules (%)	Vesicles (%)
Without Cd	t0	66.76 ± 6.21a	36.75 ± 6.21a	1.750 ± 1.10a
	tf	85.66 ± 3.09b	70.83 ± 5.02b	35.67 ± 0.80b
With Cd	tf	91.00 ± 3.59b	72.00 ± 5.1b	39.33±3.76b [1]

[1] Data are the means of six replicates ± SE for root colonization, arbuscules, and vesicles. Different letters indicate significant differences ($p \leq 0.05$); data were analyzed by a one way ANOVA with Tukey post-test between treatments.

2.2. Soil and Tissues Cd, P, and Fe Contents in Presence and Absence of AMF after Exposure to Cd

None of the soil element (Cd, P, or Fe) concentrations were significantly different between non-mycorrhyzal (nM) and mycorrhizal (M) soybean plants (Table 2). Bioavailable fractions of Cd (1.20 ± 0.07 ppm), P (39.40 ± 0.80 ppm), and Fe (14.48 ± 0.54 ppm) were determined in the soil used for all treatments at the beginning of the trial.

Table 2. Soil Cd, P, and Fe contents (ppm of soil) of non-mycorrhizal and mycorrhizal soybean plants after 40 days of Cd addition.

Soil Elements Concentration (ppm)	Non-Mycorrhizal Soybean + Cd	Mycorrhizal Soybean+ Cd
Cd	10.97 ± 5.06	17.72 ± 8.44
P	529.68 ± 46.3	490.53 ± 49.84
Fe	10375 ± 1742	13582 ± 1441 [1]

[1] Data are the means of three replicates ± SE for soil Cd, P, and Fe contents. Average values were not significantly different between mycorrhizal and non-mycorrhizal soybean plants; data were analyzed by an unpaired t test ($p \leq 0.05$).

Soybean roots inoculated with *R. intraradices* took up about half as much Cd (53%) than roots of nM plants. The presence of AMF in soybean roots significantly reduced root Cd concentration (Table 3). The amount of Cd accumulated in both new and old leaves did not differ significantly between M and nM plants. However, considering old and new leaves separately, in the inoculated plants, Cd was more concentrated in the old leaves, while in the nM plants, it was accumulated in the new ones (Table 3).

Under pollution conditions, M plants had a significantly higher P concentration compared to nM soybean roots and old leaves (Table 3). Further, in each treatment, P was significantly more abundant in the roots than in all the leaves (*Leaves) and the old leaves. The accumulation of Fe, an essential micronutrient for the plant, was much higher in the roots than in the leaves of both M and nM plants (Table 3). Furthermore, the presence of mycorrhizal colonization significantly increased root Fe concentration in M plants with respect to nM soybean. The amount of Fe accumulated in leaves (*Leaves) of M and nM plants was not different, but P content was significantly higher in M plants (Table 3).

Table 3. Phosphorus, Fe, and Cd concentrations in tissues of non-mycorrhizal (nM) and mycorrhizal (M) soybean plants after 40 days of Cd addition.

Tissues	P (ppm)		Fe (ppm)		Cd (ppm)	
	nM	M	nM	M	nM	M
Roots	2176.66 ± 140.18ac	2583.33 ± 59.33be	3884.20 ± 46.83ac	5181.20 ± 160.57be	65.52 ± 0.76ac	30.27 ± 0.50be
Old leaves	1187.16 ± 54.87adg	1980.22 ± 204.9bfi	158.56 ± 11.88adg	130.34 ± 29.26afi	0.726 ± 0.020adg	1.056 ± 0.076bfi
New leaves	2073.35 ± 15.43ach	2257.14 ± 173.7aei	123.85 ± 10.11adg	79.16 ± 1.15afi	1.600 ± 0.005adh	0.740 ± 0.046bfj
Leaves *	1630.25 ± 199.7ad	2118.68 ± 119.95bf	141.20 ± 10.43ad	104.75 ± 17.38af	1.163 ± 0.195ad	0.898 ± 0.081af [1]

[1] Data are the means of six replicates ± SE for concentrations of P, Fe, and Cd in soybean tissues. *Leaves are the total values obtained by adding each element concentration in old and new leaves of soybean. Different letters indicate significant differences at $p \leq 0.05$ by a one way ANOVA with the Tukey post-test. The first letter indicates nM vs. M significance level for each tissue (letters a–b). The second letter represents the significance level of roots vs. all other tissues in nM (letters c–d) and M plants (letters e–f). The third letter shows the significance level of old leaves vs. new leaves in nM (letters g–h) and M plants (letters i–j).

In the same Cd stress conditions, overall Cd, P, or Fe contents of nM and M soybean tissues after 40 days of addition of Cd were estimated from the roots data plus those of total leaves. These data were averages of six replicates ± SE, and the differences between treatments were analyzed by unpaired t-test. Total Cd content of M soybean (32.06 ± 0.47 ppm) was strikingly lower (53.3%, $p \leq 0.0001$) than that of nM plants (67.84 ± 0.77 ppm), while the total content of both P and Fe increased (P 19%, $p = 0.050$ and Fe 24%, $p = 0.026$) in soybean plants colonized by AMF (P 6820.80 ± 364.56 ppm and Fe 5390.70 ± 288.71 ppm) compared to non-colonized plants (P 5435.20 ± 68.15 ppm and Fe 4166.70 ± 48.60 ppm).

2.3. *Growth Parameters of Soybean Plants after Cd Exposure*

At the end of Cd treatment (tf), total fresh weight (TFW) of nM+Cd plants was significantly lower than that of nM plants (Figure 1a). By contrast, TFW of M plants was not modified by Cd exposure (Figure 1a). Furthermore, root fresh weight (RFW) of M plants was significantly lower than that of nM (Figure 1b). However, M plants in the presence of Cd (M+Cd) yielded significantly higher RFW than control plants without Cd (M) (Figure 1b). At t0 of Cd addition, stem length (SL) of M plants was not significantly different than SL of nM soybean (Figure 1c). Non-mycorrhizal (nM) plants showed a significant decrease in SL at tf of Cd treatment; however, SL of M plants was not modified after exposure to Cd (M vs M+Cd). Moreover, SL of M soybean plants (with or without Cd) was similar to that added with Cd nM soybean and significantly lower than that of nM (Figure 1c). Comparing nM and M plants, shoot fresh weight (SFW) was equivalent to that at t0 as well as between the untreated controls at the end of the treatment. However, 40 days after treatment, Cd caused a significant decrease in SFW in both nM and inoculated plants (Figure 1d).

2.4. *Oxidative Damage Response of Soybean Plants to Cd Exposure*

The highest level of lipid peroxidation was found in colonized plants at tf. Among colonized soybean plants, M+Cd plants showed a decrease of oxidative damage measured as malon-di-aldehyde (MDA) after 40 days of treatment. Furthermore, lipid oxidative level was significantly higher in M plants than in nM plants; a similar trend was recorded at t0 but without significant differences. Cd treated soybean exhibited similar lipid oxidative level, measured as MDA content, in nM and M plants. Under Cd stress conditions, MDA was not significantly affected in nM plants, whereas it decreased in M plants (Figure 2a).

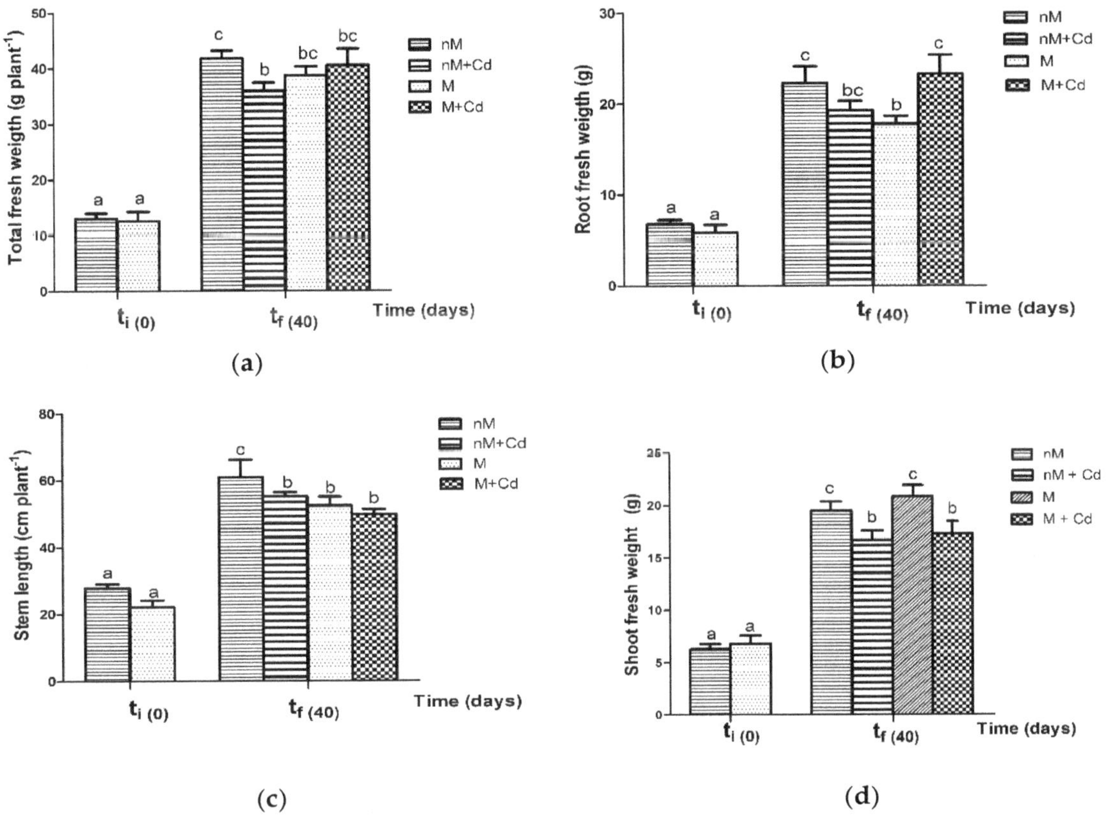

Figure 1. Effect of Cd stress on (**a**) total fresh weight; (**b**) root fresh weight; (**c**) stem length; (**d**) shoot fresh weight of non-mycorrhizal (nM) and mycorrhizal plants (M) inoculated with *Rhizophagus intraradices* at initial time (t0) and after 40 days (tf) of Cd addition. Data are the means of six replicates ± SE for each dependent variable. Different letters indicate significant differences at $p \leq 0.05$ using a two way ANOVA according to the least significant difference (LSD) post hoc test between treatments.

At t0, GSH content was significantly lower in M than in nM plants. In contrast, at tf, GSH level significantly increased in M plants (Figure 2b). At the end of the experiment, Cd significantly increased GSH level in nM plants (nM vs. nM+Cd). Moreover, in M plants, this high GSH level was similar both in the presence and in the absence of Cd. Furthermore, GSH levels in nM soybean plants were equal at t0 and tf (Figure 2b).

Activity of APX varied non-significantly with AMF colonization at t0 (Figure 2c). However, at tf, Cd exposure and AMF colonization reduced APX activity. The highest inhibition of APX activity was found in M plants exposed to Cd; furthermore, Cd addition of M soybean was more detrimental on APX activity than colonization or Cd addition individually. At tf, M plants had lower APX activity than nM soybean (Figure 2c).

In the absence of Cd, M plants had a significantly higher level of NPTs than nM plants. Cd significantly increased levels of NPTs of nM plants. Conversely, in M plants, the levels NPTs decreased in the presence of Cd. Furthermore, Cd treatment and AMF colonization individually had the same level of NPTs (Figure 2d).

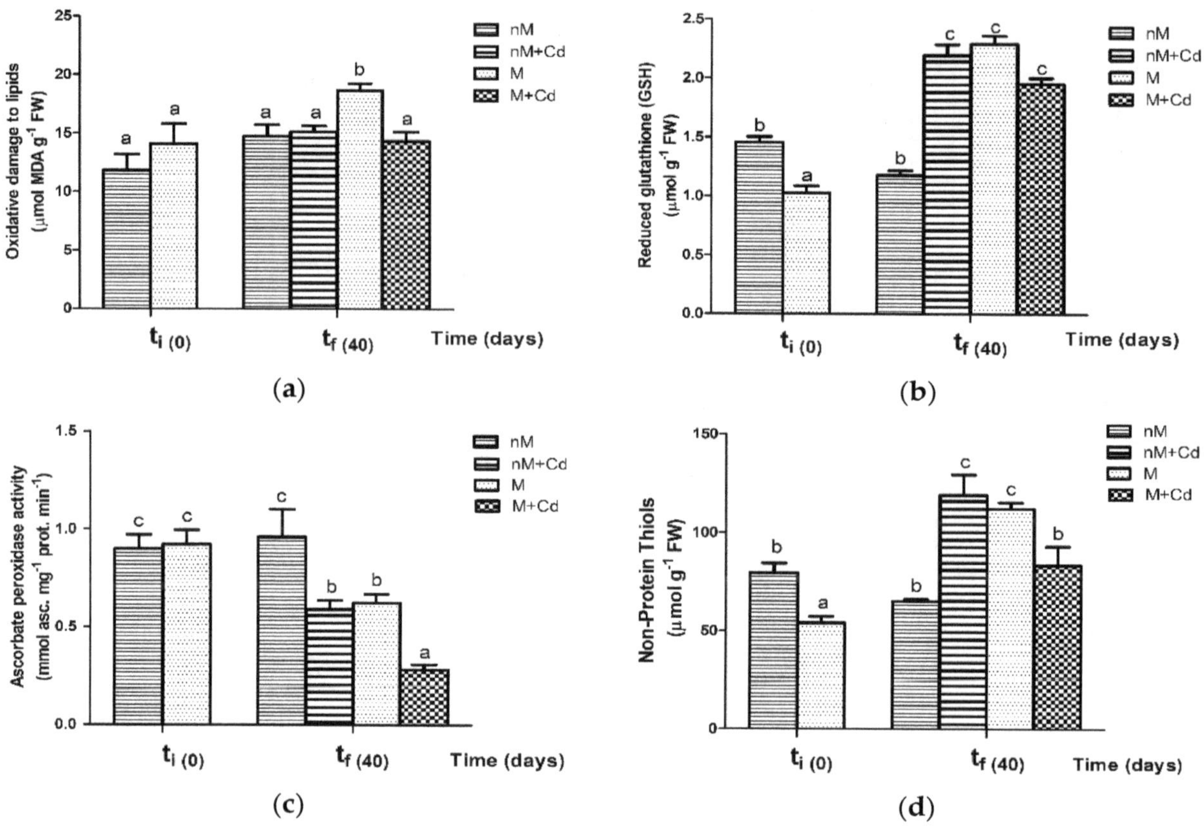

Figure 2. Effect of Cd stress on (**a**) oxidative damage to lipids [malon-di-aldehyde (MDA) content]; (**b**) glutathione (GSH) content; (**c**) peroxidase (APX) activity; (**d**) non-protein thiols (NPTs) content in leaves of non-mycorrhizal (nM) and mycorrhizal (M) soybean plants inoculated with *Rhizophagus intraradices* at initial time (t0) and after 40 days (tf) of Cd exposure. Data are the means of six replicates ± SE for each dependent variable. Different letters indicate significant differences at $p \leq 0.05$ using a two way ANOVA to the least significant difference (LSD) post hoc test between treatments.

3. Discussion

3.1. Colonization of Soybean Roots by Rhizophagus Intraradices

The results of this study show that Cd did not alter the AMF colonization of soybean roots, indicating that this metal did not inhibit symbiosis establishment. Although AM fungal colonization usually occurs in highly HM contaminated soils [24], positive (*Zea mays–Glomus* sp. isolated from high HM contaminated soils and this work), negative (*Cajanus cajan–Funneliformis mosseae*), and neutral effects (*Pisum sativum–Rhizophagus intraradices, Zea mays–Funneliformis mosseae*) of Cd on root colonization have been reported, depending on Cd concentration and plant and fungal species involved in the association [22–27].

Although the total Cd concentration supplied to the soil was within the range of a high pollution level (10–50 mg kg^{-1} soil), Cd bioavailability values (0.7–1.2 mg kg^{-1} soil) fell within the values of a slightly polluted soil (1–3 mg kg^{-1} soil) for Argentinian Regulatory Decree 831 of 1993 and National Law 24051 of Hazardous Wastes or the maximum level of an unpolluted soil (0–1 mg kg^{-1} soil) for the European Union and Great Britain Laws. This result may explain the lack of Cd effects on fungal growth and root colonization. Hence, Cd bioavailability rather than total concentration should be considered in order to determine the legal limits of soil contamination relative to AMF colonization effects [18].

3.2. Soil and Plant Cd, P, and Fe Contents

No differences in Cd, P, or Fe contents were found between nM and M soybean soils. This result is in agreement with the Cd concentration values in soil reported for *Lonicera japonica* colonized by *Glomus versiforme* and *R. intraradices* [28]. Based on our results, we can only propose some possible explanations for this lack of differences in Cd, P, and Fe soil concentrations between M and nM soybean. On the one hand, both partners of our experimental soybean–*R. intraradices* system are recognized as extremely efficient Cd accumulators [13,29,30] and therefore have the capacity to capture and transport metals. AMF Cd accumulation capacity may have considerable ecological implications such as effects on AMF diversity, forcing possible changes in the richness and the abundance of these fungi by promoting the dominance of the resilient species and the detriment of those with less or without Cd accumulation capacity. Consequently, these possible modifications in the AMF communities could modify plant communities, depending on the degree of association of plant species with the AMF (non-host, facultative mycorrhizal, and mycorrhizal dependent species), which would also depend on the metals accumulated in the soil. The similar soil content of Cd or Fe found in nM and M plants might be attributed to the co-transport/capture/translocation of these metals by soybean from the soil due to their importance as nutrients in plant growth, even in the absence of AMF. On the other hand, the efficiency of metal/HM capture in the plant–AMF interactions varies with AMF isolates (isolates of native AMF collected from contaminated soil are more efficient in the capture of these metals than native isolates from uncontaminated soil), host genetic diversity, and metal content [13,22,27]. Thus, the *R. intraradices* strain we used is native but was not isolated from metal-contaminated soils; this *R. intraradices* isolate likely does not capture metals more efficiently than the soybean variety [*Glycine max* (L.) Merr. cv. Don Mario 4800] used in our experimental system with the metal concentrations applied in this study.

Regarding the soybean capture of P, our results agree with previous reports. Mycorrhizal plants have two Pi uptake pathways: a direct pathway through the epidermal root cells and a mycorrhizal pathway. It has been shown that the contribution of the mycorrhizal pathway ranges from a small percentage to about the entire P of the plant and that, during the symbiosis, the plant changes the strategy of Pi uptake and favors the acquisition of P through the AM pathway over the direct root path [11,31]. The total amount of P obtained through the AM pathway depends on the plant, on the fungal genotypes involved in the symbiosis, and on the concentration of P in the soil solution, and it correlates with the abundance of extraradical hyphae in the soil [32]. Nowadays, it is known that fertilization with Pi increases productivity and stress tolerance and that AMF supplies Pi to the plant, thereby improving not only its nutrition but also its tolerance to multiple stresses such as drought, salinity, and pathogen attack [10,32]. Therefore, under stressful environmental conditions, as in our experimental model of HM contaminated soils, the establishment of the AM symbiosis could increase not only the efficient use of Pi by the plant but also the tolerance to other stresses. Soybean roots inoculated with native *R. intraradices* took up about half of the Cd taken up by roots of nM plants, showing that AMF colonization reduced root Cd concentration, as previously reported [22]. Extra and intraradical mycorrhizal mycelia have been proposed as a potential barrier to the entry of HM into roots through mechanisms such as binding, superficial adsorption, complexation, precipitation, and crystallization of metals on hyphal walls [33]. In particular, AMF extraradical hyphae have been shown to be highly efficient in HM binding, capturing, and translocation in soils [27]. Furthermore, the glomalin released by AMF extraradical hyphae may have been able to chelate Fe, Mn, Pb, and Cd, catching (with low affinity) these HMs to soil surrounding the hyphae [33].

Rhizophagus intraradices extraradical mycelia successfully detoxified Cd and another HM such as Cu and Zn [13]. Accordingly, *R. intraradices* tolerance to HM stress has been attributed to the strategy that includes the development of a differential AMF mycelium with extensive growth in less stressed areas [29]. Surprisingly, in our study, *R. intraradices* significantly reduced Cd concentration in soybean roots, whereas root colonization in M plants exposed to Cd pollution did not differ in soil Cd concentrations observed between M and nM soybean soils. Thus, Cd appears to be trapped in the interface between soil and roots, suggesting that the hyphal web of *R. intraradices* in soil could function as a soil–root intermediary, holding Cd inside extraradical hyphae or surrounding Cd by glomalin binding; thus, Cd could never reach the soybean roots despite their high colonization. Furthermore, the vermiculite used in the growth pot culture of soybean has been reported as a substrate that improves the sorption capacity of Al and Fe [34] and HMs such as Cu, Ni, and Cd in water solutions and irrigation systems [35,36]. However, vermiculite has low [35] or middle [34] Cd adsorption capacity compared to other HMs, depending on the pH in the media. Thus, vermiculite adsorption of metal ions increases with increasing pH [34]. Therefore, the low bioavailable Cd values could also be due to the retention of vermiculite in pot substrates.

Considering the total (old and new) leaves, the accumulated Cd did not differ between inoculated and non-inoculated plants. However, when old and new leaves were considered separately, mycorrhized plants concentrated more Cd in old leaves, but nM plants accumulated more Cd in young leaves, as previously reported [37]. Furthermore, in this work, plants did not accumulate Cd in the mycorrhizal roots. Thus, Cd content in leaves does not seem to be related to root colonization by AMF, as previously reported for other Fabaceae [22]. Moreover, it has been shown that Cd accumulation in roots of nM soybean is 10-fold higher than in shoots, with small amounts of Cd being translocated to pods and seeds at low Cd concentration in the soil [30]. In our study, Cd concentration in roots was half in M plants, which could have been due to Cd translocation towards the leaves; however, our results showed a 30-times smaller Cd concentration in the leaves than in the roots, with this ratio being double in nM plants. Therefore, these results evidence that soybean is highly efficient in Cd capture in the roots, where this HM is sequestered, preventing translocation to stem and leaves.

In our experimental model of exposure to Cd, the highest overall concentrations of P and Fe as macro- and micronutrients were found in the roots and the leaves of mycorrhized plants, showing a better nutritional condition for the host plants. This host nutritional improvement through the enhanced nutrient supply by the fungal symbiont is one of the most frequently reported mechanisms through which AMF can improve plant tolerance to HMs [15]. Thus, the soybean–*R. intraradices* plant–AMF model is a successful system to crop in soils with Cd.

3.3. Response of Soybean Growth Parameters to Cd Exposure

Considering growth parameters, soybean root colonization by *R. intraradices* negatively affected RFW of plants not treated with Cd. This negative effect could have been due to the energy cost of AMF establishment in the soybean roots also affecting stem growth. In general, AMF have been considered to be predominantly beneficial for plant growth [10]; however, it has been suggested that, when functioning as a parasitic partner, their requirement of carbohydrates from the host plant may depress host growth during the first seedling growth period in non- or facultative-mycorrhizal plants and when hosts grow in soils rich in nutrients [38]. Consequently, the costs of AMF colonization do not have deleterious effects on the host if plant growth is not affected by reduced carbon sources [10]. Therefore, soybean growth reduction by *R. intraradices* may be considered a functional rather than a damaging effect. Surprisingly, soybean TFW was preserved, and M plants were qualitatively and morphologically in better conditions than nM plants (data not shown). Moreover, soybean exposed to Cd also showed an opposite trend in RFW, which decreased in nM and increased in M plants. Thus, mycorrhizal colonization promoted root growth despite Cd exposure. Comparing plant TFW with respect to root and SFW, it is possible that, with Cd exposure, TFW of inoculated plants was the same as that of their controls due to the greater fresh weight of the root. Therefore, the association with *R. intraradices* improved soybean growth under Cd stress, showing the beneficial effects of AMF colonization in plants growing under HM pollution conditions, as previously reported [39].

3.4. Oxidative Damage Response to R. intraradices Colonization of Soybean Not Exposed to Cd

Oxidative damage was observed in M soybean plants without Cd addition, with the highest level of MDA/lipid peroxidation being recorded in M plants throughout the bioassay. These results may suggest that colonization per se with *R. intraradices* may produce oxidative damage in soybean. Furthermore, reduced glutathione (GSH) is the most important reducing agent in the non-enzymatic antioxidant defense of plants, and its level indicates the overall plant redox state. In this work, soybean mycorrhization changed GSH content, with low values being recorded with *R. intraradices* colonization at the start of the assay and a significant increase of GSH 40 days later. Thus, GSH content increased in correlation with the decrease of lipid oxidation and MDA levels, which was likely a strategy to afford the plant costs for symbiosis establishment [10,40,41]. Furthermore, there is evidence supporting that the induction/suppression mechanisms associated with plant defense play a key role in AMF colonization and the plant–fungus compatibility in the context of this mutualistic association [40], especially at early stages of AMF–plant interaction. These processes include changes in the permeability of the plasma membrane, activation of plasma membrane-bound enzymes and of kinases, phosphatases, phospholipases, and production of signal molecules, including reactive oxygen species. Regarding abiotic stresses, AMF colonization is also involved in defense mechanisms and processes, possibly with effects on the induction of stress resistance [40]. Moreover, several studies showed that root penetration and colonization by AMF involve a complex cytological and biochemical sequence of events and intracellular changes, including ROS increase and anti-oxidative damaging response [42–44].

While a decrease in oxidative damage—measured as MDA after 40 days of treatment—was observed, the activity of APX did not vary significantly with AMF colonization at the start of the assay. The NPTs showed similar patterns of concentration at the start and at the end of treatment. Although AMF colonization per se may produce oxidative burst, M plants are able to react by decreasing lipid oxidation levels in the presence of HM. Moreover, H_2O_2 is one of the well-known ROS, functioning as an important signaling compound regarding the interaction of plants and AMF root colonization [10,33,42]. Thus, AMF colonization would regulate oxidative damage [44], improving the defensive response of plants against HM, in agreement with different plant–fungus systems polluted by Cd and other HMs [22].

3.5. Oxidative Damage Response to R. intraradices Colonization of Soybean with Cd Exposure

When soybean was cultivated in Cd-polluted soil, APX activity decreased significantly in both M soybean (55%) and nM control (36%), whereas mycorrhization did not show differences between treatments. Our results suggest that, under Cd stress, mycorrhizal soybean regulates oxidative damage, maintaining GSH synthesis higher than APX activity as a possible defense strategy. In the presence of Cd pollution, the lack of differences in oxidative damage measured as MDA between M and nM soybean was also accompanied with a significant increase in the contents of GSH and NPTs in nM soybean. In contrast, soybean colonized by *R. intraradices* maintained GSH levels but reduced the content of NPTs with Cd addition. These results are in agreement with findings reported by Rivera-Becerril et al. [23], who showed that the genes coding for glutamil cisteinil synthetase and GSH synthetase, responsible for GSH synthesis, were activated by Cd in M and nM plants. Here, there was a high increase in NPTs in nM soybean with Cd supply, as expected, because it is well known that Cd promotes the synthesis of these molecules. Instead, M soybean did not react to Cd pollution and presented lower levels of these detoxification compounds than nM plants. Furthermore, these results indicate that, although the trace of HM chelation is involved in soybean responses to Cd stress, plants could not make a great contribution to the Cd tolerance strategies operating in AM symbiosis. In soybean root under Cd stress, mycorrhizal colonization per se might be able to increase NTP levels, suggesting that mycorrhization would not enhance plant defense through NTPs compared to plants growing without symbiosis. Thus, the positive effect of AMF on soybean defensive response through NTPs was reduced in the presence of Cd, suggesting that colonization by *R. intraradices* would not have a protective effect on soybean via NTPs at this Cd level. Even though the frequency of AMF colonization fluctuated under different experimental combinations of plant–AMF–HM as well as with soil types and bioavailability of nutrients, the protective effect of AMF consistently prevailed over the stress caused by HM [33]; our soybean–Cd–*R. intraradices* system is not an exception. The protective effect of AMF against Cd stress has also been observed in a Cd-sensitive genotype of *Funneliformis mosseae* through a differential display of root target protein, e.g., accumulation of ROS when *R. intraradices* colonizes roots of *Medicago truncatula*, *Nicotiana tabacum*, and *Zea mays* [42]. AMF are effective in plant protection against abiotic stress at high HM concentrations [45]. However, many factors such as metal concentration may affect AMF–plant tolerance response to HM during AMF–plant association [39]. When soybean was cultivated in Cd-polluted soil, several biochemical plant responses were in agreement with the first defensive response to mycorrhization and with the secondary buffering response against Cd in mycorrhizal soybean throughout the experimental period. Thus, it is possible that mycorrhizal colonization first predisposes the plant to a reduced defense state but then, once mycorrhization is established, it acts as a buffer to Cd contamination. Therefore, AMF colonization induces stress resistance [40], functioning as a strategy against oxidative stress caused by herbicides [46] or drought [47]. Thus, plants use common pathways and components in the stress biotic–abiotic/response relationship, which are known as cross-tolerance.

Overall, *R. intraradices* colonization in soybean plants without Cd pollution reduced some growth parameters, increased lipid oxidation levels, and produced an initial reduction in GSH levels as an anti-oxidant, suggesting that mycorrhization per se implies an oxidative stress for soybean. These phenomena allow plants to adapt to a range of different stresses after being exposed to one specific stress, with cross-tolerance having a direct and useful application in agriculture [48]. Cross-tolerance occurred by means of the redox signals in the plant cell signaling pathways and the accelerated production of ROS [48,49], which is the most important and common response of plants to various abiotic and biotic stresses (including mycorrhizal colonization). These redox-ROS mechanisms are concomitant with the two major antioxidant plant cell defenses, ascorbate and GSH, as well as the antioxidant enzymes using these antioxidants [49].

4. Materials and Methods

4.1. Biological Material and Experimental Design

The experiment was carried out considering two factors: AMF colonization and Cd exposure. During an initial culture period of 30 days, the soybean plants inoculated with *Rhizophagus intraradices* (Schenck N. C. & G. S. Sm.) Walker C. & A. Schüßler as well as the control plants showed good growth. This period of culture to determine the initial time (t0) was established according to a previous bioassay to evaluate the necessary conditions for mycorrhizal growth. At the 30th day of culture, Cd was added to the substrate for the first time, and this was considered the initial time (t0) of the Cd addition experiment; from t0 and over 40 days (final time = tf), plants were exposed to Cd. Four experimental treatments were assayed, including six replications (one for each plastic pot) per treatment: (1) plants not inoculated with AMF or non-mycorrhizal plants (nM), (2) non-mycorrhizal plants supplied with Cd (nM + Cd), (3) plants inoculated with AMF or mycorrhizal plants (M), and (4) mycorrhizal plants supplied with Cd (M+Cd). Each pot contained 2 soybean plants.

Cultures were established between *Glycine max* (L.) Merr. cv. Don Mario 4800 (hereafter soybean) and an autochthonous species of *R. intraradices* isolated from soil without Cd of the semiarid areas of Córdoba province in Central Argentina. The AMF spores were extracted from 50 g of each soil sample (n = 6), according to Daniels and Skipper [50], and identified under a light microscope using the morphotaxonomic criteria of the International Collection of Vesicular and Arbuscular Mycorrhizal Fungi (http://invam.caf.wvu.edu/).

Trap cultures using *Sorghum halepense* L. Pers. as host plant and autochthonous *R. intraradices* were established over 6 months to obtain the AMF inocula. The inocula were named root inoculum (consisting of the roots of the trap plants colonized by *R. intraradices*) and soil inoculum (composed of soil remaining after removal of colonized roots, spores, and mycelia of *R. intraradices*). Root samples were taken from trap plants to measure the percentage of AMF colonization using the gridline intersection method [51]. Trap cultures were used as inocula when root colonization by *R. intraradices* reached a minimum of 50%.

Soybean seeds were surface sterilized in a 1% NaClO solution for 30 s, rinsed in distilled water, placed on sterilized filter paper, and incubated at 28 °C over 3 days. Three-day-old seedlings were transferred to plastic pots (two plants per pot) containing 1.25 kg of substrate (1 soil:1 sand:1 vermiculite:1 perlite). The substrate was previously steamed-sterilized twice at 120 °C during 20 min after an interval of 24 h, and its determined physicochemical characteristics were the following: pH (5.83), electrical conductivity (4.3 dSm-1), total organic carbon (6.33%), total nitrogen (0.39%), available phosphorus (39.40 ppm), and organic matter (10.92%). Pots were supplied daily with distilled water to maintain soil moisture close to saturation capacity (e.g., 40% of the volume of the mixture) by keeping a volume and frequency of irrigation of 250 ml water/pot/week. Plants were grown in a greenhouse with a 16 h light/8 darkness photoperiod at 15–30 °C without fertilization.

Three-day-old seedlings were inoculated with *R. intraradices* by adding 1.4 g of root inoculum plus 310 g of soil inoculum per pot; nM plants received 1.4 g of autoclaved root inoculum per pot with the addition of 25 ml of root inoculum washing filtrate to restore the microorganisms associated with soil [22].

Cadmium concentration for irrigation and its final availability to plants were selected by a preliminary test of soil Cd adsorption [52]; Cd distribution coefficient (Kd) in substrate was also calculated [53]. Cadmium concentration remaining in the substrate was determined by inductively coupled plasma optical emission spectrometry (ICP-OES). During Cd treatments, plants received Cd by irrigation with 25 mg CdC12 L-1 solution; the calculated Cd adsorption was 98.04%, and the calculated final bioavailability was 0.7–1.2 ppm Cd in each pot. Plants (nM and M) were supplied with Cd solution over 40 days [22]. At the end of the experiment, Cd content was measured in the roots and the leaves of nM and M plants.

4.2. Evaluation of AMF Colonization

The presence of *R. intraradices* colonization in soybean roots was checked under microscope after root clearing and staining [54]. The percentage of root colonization by *R. intraradices* was calculated under microscope using the intersection method [55].

4.3. Determination of Soil and Biomass Element Content

In order to bring the metals into the solution, partial and then total digestion of the soil were conducted using a microwave digester (Microwave System Brand: Milestone, Model: START D, ETHOS lab station with easy wave or easy control software HPR1000/10S, high pressure segmented rotor).

Samples (0.5 g) of dried and sieved (in 2 mm mesh) soil of Cd treatments were treated with two acid digestions. For partial digestion, acids HNO_3 65%, HCl 37%, and HF 40% were used with an 8 min 160 °C, 5 min 210 °C, and 20 min 210 °C microwave program. For total digestion, the acids 40% HF and 5% H_3BO_3 were used with an 8 min 160 °C and 10 min 160 °C microwave program. Both digestions were performed at microwave power up to 1000 Watt using up to 500 Watt for operations with 3 or fewer vessels simultaneously. Determinations were made directly on the digested samples. A reagent blank was prepared with 5% milli Q water with Nitric Acid.

Soil Cd and P contents were quantified by mass spectrometry using inductively coupled plasma (ICP-MS) with a concentric flow nebulizer (Pneumatic). Iron determinations were made by optical emission spectrometry with inductively coupled plasma (ICP-OES) using a spectrometer (model ICP- 2070 Baird, Bedford, MA, USA). Calibration was carried out following the direct method with certified standards, according to Standard Methods for the Examination of Water and Wastewater (21st edition, 2005).

Bioavailable fractions of Fe and Cd in soil were determined by extraction with DTPA (0.005 M diethylenetriaminepentaacetic acid, 0.1 M triethanolamine, and 0.01 M $CaCl_2$) [56]. The available P was determined by the Bray and Kurtz method [57] with an acid extraction mixture (0.025 M HCl and 0.03 M NH_4F). In the respective extracts obtained, P and Cd available using ICP-MS and Fe available by ICP-OES were analyzed.

The accumulation of Cd, P, and Fe in plants was measured. From the Cd-treated plants, six complete root systems and three groups of old (3rd to 4th) and new (6th to 7th trifoliate pairs) leaves of nM and M plants were extracted and dried at 80 °C to achieve constant weight. Then, 0.1 g of biomass samples was digested with 1 mL of concentrated HNO_3 in covered polyethylene tubes and placed in an ultrasound bath for 30 seconds; then, 0.5 ml of H_2O_2 were added and placed in a thermostatic bath for 1 h. Finally, 5 temperature strokes of 30 s were given at 5 min intervals (reaching the boiling of the samples each time) to reach higher temperatures and achieve the total digestion of the samples. Cadmium, P, and Fe concentration in roots and leaves was determined by mass spectrometry with inductively coupled plasma (ICP-MS) and a concentric flow nebulizer (Pneumatic). A control without biomass was performed following the same procedure.

4.4. *Growth Parameters*

Cd toxicity was evaluated by measuring the total fresh weight (TFW), the root fresh weight (RFW), the shoot fresh weight (SFW), and the stem length (SL). Mean values were obtained from six replicates. Each replicate was a pot with 2 soybean plants.

4.5. *Measurements of Oxidative Damage and Antioxidant Defenses*

Oxidative damage, evaluated as lipid peroxidation levels, and non-enzymatic and enzymatic antioxidant defenses were measured in the leaves of plants in all treatments. Lipid peroxidation was assessed as MDA (malon-di-aldehyde) content [58], and a control value without thiobarbituric acid was determined according to Hodges et al. [59]. MDA was quantified using the molar extinction coefficient of 155 mM^{-1} cm^{-1}, and the results were expressed as nmol MDA g^{-1} LFW (leaf fresh weight).

Non-enzymatic antioxidant defense was measured as reduced glutathione content (GSH) according to Akerboom and Sies [60]. The assays were performed using acid extracts of leaves for protein precipitation. The oxidation of GSH was produced by DTNB (5.5′-di-thio-bis-nitrobenzoic acid) to obtain GSSG (oxidized glutathione) and TNB (5-thio-2-nitrobenzoic acid), which were quantified. The homogenate was prepared in liquid air with fresh leaf tissue, 0.1M HCl (10% w/v), and it was centrifuged at 3000× *g* rpm at 4 °C for 20 min. A reaction mixture containing potassium phosphate buffer 0.5 M (pH 7) and the samples were analyzed in tandem with a standard GSH (0.4 mM in 0.1 M HCl) and incubated at 60 °C for 5 min. DTNB (0.100 mL of 1.5 mg/mL in 0.5 M potassium phosphate buffer, pH 7) was added, mixed, and centrifuged at 10,000× *g* rpm for 2 min. The increase in absorbance at 412 nm (TNB) was measured, and the GSH concentration was expressed in moles GSH gr-1 LFW.

To evaluate antioxidant enzyme activity, the total activity of ascorbate peroxidase (APX) was measured as the decrease in the ascorbate absorbance at 290 nm [61]. The mixture reaction contained appropriate dilutions of the samples in 50 mM potassium phosphate buffer (pH 7.4), 0.5 mM ascorbic acid, and 0.1 mM H_2O_2. APX activity was expressed as nmol ascorbate min^{-1} mg^{-1} protein in LFW. Protein content was determined according to Bradford [62] using bovine serum albumin as the standard.

Non-protein thiols were measured using soybean homogenates prepared with 0.250 g fresh soybean leaves homogenized in a mortar with 0.1 g of PVP (polyvinyl pyrrolidone) and 2.5 mL of 0.1 N HCl (pH 2). Homogenates were centrifuged at 3000× *g* rpm for 10 min. The reaction mixture containing 0.5 M potassium phosphate buffer (pH 8), 1 mM EDTA (ethylene diamine tetraacetic acid), soybean homogenate, and 1.5 mg/mL DTNB in 0.1N HCl was bath-incubated at 30 °C for 12 to 15 min and centrifuged at 10,000× *g* rpm for 2 min. Absorbance was determined in a spectrophotometer at 412 nm and compared with the absorbance of DTNB (6 mM) as the standard. The amount of non-protein thiols in the sample was expressed as μM of DTNB g^{-1} LFW [63].

4.6. Statistical Analysis

In general, mean values were obtained from six replicates over two independent experiments (each replicate was a pool composed by 2 different plants). Prior to statistical analyses, data normality was analyzed by Kolmogorov–Smirnov or Shapiro–Wilks tests, and for the homoscedasticity of variances, Levene and Bartlett tests were used. The percentage of colonization, arbuscules, and vesicles of *R. intraradices* in soybean roots with and without Cd addition as well as Cd, P, and Fe tissue contents were analyzed by a one way ANOVA with Tukey post-test to evaluate the differences between treatments (nM vs. M plants) and tissues (roots vs. all other tissues and old leaves vs. new leaves) in nM and M soybean plants. An unpaired t test was performed only when a pair of means were compared, as it was done with soil Cd, P, and Fe contents. Data on growth parameters and oxidative damage response were analyzed by a two way ANOVA (additive model), with Cd treatment and mycorrhizal inoculation as independent factors. The post-hoc test of the least significant difference (LSD) was used to assess the differences between treatments using a bidirectional ANOVA. Results were considered significantly different at $p \leq 0.05$. Data were analyzed using the R Project [64] for Statistical Computing 3.6.1 for Windows, free language, 2019.

5. Conclusions

A lower concentration of Cd in the entire plant, a higher concentration of nutrients such as P and Fe, and a better overall reduction status (GSH and NPTs) in mycorrhizal soybean plants are advantages under Cd exposure. However, these advantages were not reflected in the growth parameters evaluated (TFW and SFW), which were similar for M and nM soybean plants. This lack of increased growth in mycorrhizal plants could be related to a detrimental oxidative effect of mycorrhization per se, which is evident in high levels of lipid peroxidation. Therefore, our results suggest that soybean root colonization by *R. intraradices* could be the signal that, as a redox trigger, initiates cross tolerance to Cd addition.

In Argentina, the Cd concentration allowed in agricultural soils is 3 mg Cd kg^{-1} soil (Regulatory Decree 831 of 1993; National Law 24051 of Hazardous Wastes), which is three times the maximum value allowed in the European Union and Great Britain. Under these conditions, the soybean–AMF system evaluated in this work would be a good bioremediation tool to remove Cd from the soil with dual socio-economic and ecological functions.

Considering that the efficiency of metal/HM capture in the plant–AMF interactions varies with AMF isolates, host genetic diversity, and soil metal content, some studies must be developed to elucidate the possible soybean–AMF mechanisms involved in transport, capture, and translocation of soil nutrients. Furthermore, future research on field investigations in agronomic systems should be carried out considering cross tolerance together with studies on the most effective crop–AMF species combinations in different levels of contamination. It will be necessary to use sustainable agronomic systems that counteract the soil HM accumulation, which is mainly generated in large quantities by the use of agrochemical inputs such as phosphate fertilizers.

Author Contributions: A.S.M. conceived and designed the project and the experimental work. M.A.L. was responsible for the root colonization quantification of AMF and main discussion of results. A.S.M., M.V.P.C., S.V.-G., and F.Z. participated in the sample processing, trap plants culture of AMF used as inoculum, and physiological analysis of plants. J.L. carried out the statistical analysis and contributed to write the results section. M.A.L. made up the slides of roots and quantified root colonization by AMF in all bioassays. A.S.M. and M.A.L. initiated and promoted the writing of the manuscript. All the authors interpreted the results and wrote the manuscript in collaboration; further, N.F. and C.A.-A. critically revised the manuscript. All authors have read and agreed to the published version of the manuscript.

Acknowledgments: The authors formally acknowledge the participation of Matilde Crespo (ID 39621950) in the revision of this publication. Her contributions as an English translator were extremely helpful and are deeply

appreciated. Silvina Vargas Gil and Mónica A. Lugo are researchers of Consejo Nacional de Investigaciones Científicas y Tecnológicas (CONICET), Argentina.

References

1. Aloui, A.; Recorbet, G.; Robert, F.; Schoefs, B.; Bertrand, M.; Henry, C.; Gianinazzi-Pearson, V.; Dumas-Gaudot, E.; Aschi-Smiti, S. Arbuscular mycorrhizal symbiosis elicits shoot proteome changes that are modified during cadmium stress alleviation in *Medicago truncatula*. *BMC Plant Biol.* **2011**, *11*, 75. [CrossRef]
2. Singh, S.; Parihar, P.; Singh, R.; Singh, V.P.; Prasad, S.M. Heavy Metal Tolerance in Plants: Role of Transcriptomics, Proteomics, Metabolomics, and Ionomics. *Front. Plant Sci.* **2016**, *6*, 1143. [CrossRef]
3. Sandalio, L.M.; Dalurzo, H.C.; Gomez, M.; Romero-Puertas, M.C.; del Río, L.A. Cadmium-induced changes in the growth and oxidative metabolism of pea plants. *J. Exp. Bot.* **2001**, *52*, 2115–2126. [CrossRef] [PubMed]
4. Bhadury, A.M.; Fulekar, M.H. Antioxidant enzyme responses of plants to heavy metal stress. *Rev. Environ. Sci. Bio/Technol.* **2012**, *11*, 55–69. [CrossRef]
5. Romero-Puertas, M.C.; Corpas, F.J.; Rodríguez-Serrano, M.; Gomez, M.; del Río, L.A.; Sandalio, L.M. Differential expression and regulation of antioxidative enzymes by cadmium in pea plants. *J. Plant Physiol.* **2007**, *164*, 1346–1357. [CrossRef] [PubMed]
6. Molina, A.S.; Nievas, C.; Pérez Chaca, M.V.; Garibotto, F.; Gonzalez, U.; Luna, C.; Giménez, M.S.; Zirulnik, F. Cadmium-induced oxidative damage and antioxidative defense mechanisms in roots and leaves of *Vigna mungo*. *Plant Growth Regul.* **2008**, *56*, 285–295. [CrossRef]
7. Michelet, L.; Zaffagnini, M.; Marchand, C.; Collin, V.; Decottignies, P.; Tsan, P.; Lancelin, J.-M.; Trost, P.; Miginiac-Maslow, M.; Noctor, G.; et al. Glutathionylation of chloroplast thioredoxin f is a redox signaling mechanism in plants. *Proc. Natl. Acad. Sci. USA* **2005**, *102*, 16478–16483. [CrossRef]
8. Foyer, C.H.; Noctor, G. Redox homeostasis and antioxidant signaling: A metabolic interface between stress perception and physiological responses. *Plant Cell* **2005**, *17*, 1866–1875. [CrossRef]
9. Balestrasse, B.K.; Gallego, S.M.; Tomaro, M.L. Oxidation of the enzymes involved in nitrogen assimilation plays an important role in the cadmium-induced toxicity in soybean plants. *Plant Soil* **2006**, *284*, 187–194. [CrossRef]
10. Smith, S.E.; Read, D.J. *Mycorrhizal Symbiosis*, 3rd ed.; Academic Press: London, UK, 2008; p. 750.
11. Smith, S.E.; Smith, F.A. Roles of arbuscular mycorrhizas in plant nutrition and growth: New paradigms from cellular to ecosystem scales. *Annu. Rev. Plant Biol.* **2011**, *62*, 227–250. [CrossRef]
12. Ferrol, N.; Tamayo, E.; Vargas, P. The heavy metal paradox in arbuscular mycorrhizas: From mechanisms to biotechnological applications. *J. Exp. Bot.* **2016**, *67*, 6253–6265. [CrossRef]
13. Hildebrandt, U.; Regvar, M.; Bothe, H. Arbuscular mycorrhiza and heavy metal tolerance. *Phytochemistry* **2007**, *68*, 139–146. [CrossRef]
14. Kaldorf, M.; Kuhn, A.J.; Schörder, W.H.; Hildebrandt, U.; Bothe, H. Selective element deposits in maize colonized by a heavy metal tolerance conferring arbuscular mycorrhizal fungus. *J. Plant Physiol.* **1999**, *154*, 718–728. [CrossRef]
15. Meharg, A.A.; Cairney, J.W.C. Co-evolution of mycorrhizal symbionts and their hosts to metal-contaminated environments. *Adv. Ecol. Res.* **2000**, *30*, 69–112. [CrossRef]
16. González-Chávez, M.C.; Carrillo-Gónzalez, R.; Wright, S.F.; Nichols, K. The role of glomalin, a protein produced by arbuscular mycorrhizal fungi, in sequestering potentially toxic elements. *Environ. Pollut.* **2004**, *130*, 317–323. [CrossRef]
17. Joner, E.J.; Briones, R.; Leyval, C. Metal-binding capacity of arbuscular mycorrhizal mycelium. *Plant Soil* **2000**, *226*, 227–234. [CrossRef]
18. Leyval, C.; Turnau, K.; Haseldwandter, K. Effect of heavy metal pollution on mycorrhizal colonization and function: Physiological, ecological and applied aspects. *Mycorrhiza* **1997**, *7*, 139–153. [CrossRef]

19. Joner, E.J.; Leyval, C. Uptake of 109Cd by roots and hyphae of a *Glomus mosseae/Trifolium subterraneum* mycorrhiza from soil amended with high and low concentrations of cadmium. *New Phytol.* **1997**, *135*, 353–360. [CrossRef]
20. Azcón, R.; Perálvarez, M.C.; Biró, B.; Roldán, A.; Ruíz-Lozano, J.M. Antioxidant activities and metal acquisition in mycorrhizal plants growing in a heavy-metal multi contaminated soil amended with treated lignocellulosica growaste. *Appl. Soil. Ecol.* **2009**, *41*, 168–177. [CrossRef]
21. Garg, N.; Kaur, H. Response of antioxidant enzymes, phytochelatins and glutathione production towards Cd and Zn stresses in *Cajanus cajan* (L.) Millisp. genotypes colonized by arbuscular mycorrhizal fungi. *J. Agron. Crop Sci.* **2013**, *199*, 118–133. [CrossRef]
22. Rivera-Becerril, F.; Calantzis, C.; Turnau, K.; Caussanel, J.-P.; Belimov, A.; Gianizzani, S.; Strasser, R.J.; Gianizzani-Pearson, V. Cadmium accumulation and buffering of cadmium-induced stress by arbuscular mycorrhiza in three *Pisum sativum* L. genotypes. *J. Exp. Bot.* **2002**, *53*, 1177–1185. [CrossRef]
23. Rivera-Becerril, F.; van Tuinen, D.; Martin-Laurent, F.; Metwally, A.; Dietz, K.-J.; Gianizzani, S.; Gianizzani-Pearson, V. Molecular changes in *Pisum sativum* L. roots during arbuscular mycorrhiza buffering of cadmium stress. *Mycorrhiza* **2005**, *16*, 51–60. [CrossRef]
24. Leung, H.M.; Wang, Z.W.; Ye, Z.H.; Yung, K.L.; Peng, X.L.; Cheung, K.C. Interactions between arbuscular mycorrhizae and plants in phytoremediation of metal-contaminated soils: A review. *Pedosphere* **2013**, *23*, 549–563. [CrossRef]
25. Chen, B.D.; Liu, Y.; Shen, H.; Li, X.L.; Christie, P. Uptake of cadmium from an experimentally contaminated calcareous soil by arbuscular mycorrhizal maize (*Zea mays* L.). *Mycorrhiza* **2004**, *14*, 347–354. [CrossRef]
26. Garg, N.; Chandel, S. Role of arbuscular mycorrhizal (AM) fungi on growth, cadmium uptake, osmolyte, and phytochelatin synthesis in *Cajanus cajan* (L.) Millisp. under NaCl and Cd stresses. *J. Plant Growth Regul.* **2012**, *31*, 292–308. [CrossRef]
27. Bano, S.A.; Ashfaq, D. Role of mycorrhiza to reduce heavy metal stress. *Nat. Sci.* **2013**, *5*, 16–20. [CrossRef]
28. Jiang, Q.Y.; Zhuo, F.; Long, S.H.; Zhao, H.D.; Yang, D.J.; Ye, Z.H.; Li, S.S.; Jinga, Y.X. Can arbuscular mycorrhizal fungi reduce Cd uptake and alleviate Cd toxicity of *Lonicera japonica* grown in Cd-added soils? *Sci. Rep.* **2016**, *6*, 21805. [CrossRef]
29. Pawlowska, T.E.; Charvat, I. Heavy-metal stress and development patterns on arbuscular mycorrhizal fungi. *Appl. Environ. Microbiol.* **2004**, *70*, 6643–6649. [CrossRef]
30. Shute, T.; Macfie, S.M. Cadmium and zinc accumulation in soybean: A threat to food safety? *Sci. Total Environ.* **2006**, *371*, 63–73. [CrossRef]
31. Smith, S.E.; Smith, F.A.; Jakobsen, I. Mycorrhizal fungi can dominate phosphate supply to plants irrespective of growth responses. *Plant Physiol.* **2003**, *133*, 16–20. [CrossRef]
32. Ferrol, N.; Azcón-Aguilar, C.; Pérez-Tienda, J. Arbuscular mycorrhizas as key players in sustainable plant phosphorus acquisition: An overview on the mechanisms involved. *Plant Sci.* **2019**, *280*, 441–447. [CrossRef]
33. González-Guerrero, M.; Benabdellah, K.; Ferrol, N.; Azcón-Aguilar, C. Mechanisms underlying heavy metals tolerance in arbuscular mycorrhizas. In *Mycorrhizas-Functional Processes and Ecological Impact*; Azcón-Aguilar, C., Barea, J., Gianinazzi, S., Gianinazzi-Pearson, V., Eds.; Springer: Berlin/Heidelberg, Germany, 2009; pp. 107–122. [CrossRef]
34. Malandrino, M.; Abollino, O.; Giacomino, A.; Aceto, M.; Mentasti, E. Adsorption of heavy metals on vermiculite: Influence of pH and organic ligands. *J. Colloids Interface Sci.* **2006**, *299*, 537–546. [CrossRef]
35. Álvarez-Ayuso, E.; García-Sánchez, A. Removal of heavy metals from waste waters by vermiculites. *Environ. Technol.* **2003**, *24*, 615–625. [CrossRef]
36. Najafi, P.; Asgari, K.; Samadi, N. Heavy metal elimination from industrial wastewater using natural substrate on pitcher irrigation. *Int. J. Recycl. Org. Waste Agric.* **2016**, *5*, 333–337. [CrossRef]
37. Dunbar, K.R.; McLaughlin, M.J.; Reid, R. The uptake and partitioning of cadmium in two cultivars of potato (*Solanum tuberosum* L.). *J. Exp. Bot.* **2003**, *381*, 349–354. [CrossRef]
38. Johnson, N.C.; Graham, J.H.; Smith, F.A. Functioning of mycorrhizal associations along the mutualism-parasitism continuum. *New Phytol.* **1997**, *135*, 575–585. [CrossRef]
39. Andrade, S.A.L.; Gratão, P.L.; Schiavinato, M.A.; Silveira, A.P.D.; Azevedo, R.A.; Mazzafera, P. Zn uptake, physiological response and stress attenuation in mycorrhizal jack bean growing in soil with increasing Zn concentrations. *Chemosphere* **2009**, *75*, 1363–1370. [CrossRef]

40. García-Garrido, J.M.; Ocampo, J.A. Regulation of the plant defence response in arbuscular mycorrhizal symbiosis. *J. Exp. Bot.* **2002**, *53*, 1377–1386. [CrossRef]
41. Pozo, M.J.; Verhage, A.; García-Andrade, J.; García, J.M.; Azcón-Aguilar, C. Priming plant defence against pathogens by arbuscular mycorrhizal fungi. In *Mycorrhizas—Functional Processes and Ecological Impact*; Azcón-Aguilar, C., Barea, J.M., Gianinazzi, S., Gianinazzi-Pearson, V., Eds.; Springer: Berlin/Heidelberg, Germany, 2009; pp. 123–135. [CrossRef]
42. Fester, T.; Hause, B. Accumulation of reactive oxygen species in arbuscular mycorrhizal roots. *Mycorrhiza* **2005**, *15*, 373–379. [CrossRef]
43. Gutjahr, C.; Parniske, M. Cell and developmental biology of arbuscular mycorrhizal symbiosis. *Ann. Rev. Cell Dev. Biol.* **2013**, *29*, 593–617. [CrossRef]
44. Wu, Q.S.; Zou, Y.-N.; Abd-Allah, E.F. Chapter 15—Mycorrhizal association and ROS in plants. In *Oxidative Damage to Plants*; Ahmad, P., Ed.; Elsevier: Amsterdam, The Netherlands, 2014; pp. 453–475. [CrossRef]
45. Hrynkiewicz, K.; Baum, C. The potential of rhizosphere microorganisms to promote the plant growth in disturbed soils. In *Environmental Protection Strategies for Sustainable Development. Strategies for Sustainability*; Malik, A., Grohmann, E., Eds.; Springer: Dordrecht, The Netherlands, 2012; pp. 35–64. [CrossRef]
46. Bressano, M.; Curetti, M.; Giachero, L.; Vargas Gil, S.; Cabello, M.; March, G.; Ducasse, D.A.; Luna, C.M. Mycorrhizal fungi symbiosis as a strategy against oxidative stress in soybean plants. *J. Plant Physiol.* **2010**, *167*, 1622–1626. [CrossRef]
47. Grumberg, B.; Urcelay, C.; Schroeder, M.A.; Vargas-Gil, S.; Luna, C.M. The role of inoculum identity in drought stress mitigation by arbuscular mycorrhizal fungi in soybean. *Biol. Fertil. Soils* **2014**, *51*, 1–10. [CrossRef]
48. Pastori, G.M.; Foyer, C. Common components, networks, and pathways of cross-tolerance to stress. The central role of "redox" and abscisic acid-mediated controls. *Plant Physiol.* **2002**, *129*, 460–468. [CrossRef]
49. Noctor, G.; Foyer, C. Ascorbate and glutathione: Keeping active oxygen under control. *Annu. Rev. Plant Physiol. Mol. Biol.* **1998**, *49*, 249–279. [CrossRef]
50. Daniels, B.A.; Skipper, H.D. Methods for the recovery and quantitative estimation of propagules from the soil. In *Methods and Principles of Mycorrhizal Research*; Schenck, N.C., Ed.; American Phytopathological Society and Press: St. Paul, MN, USA, 1982; pp. 29–35.
51. Giovannetti, M.; Mosse, B. An evaluation of techniques for measuring vesicular arbuscular mycorrhizal infection in roots. *New Phytol.* **1980**, *84*, 489–500. [CrossRef]
52. Gomes, P.C.; Fontes, M.P.F.; da Silva, A.G.; Mendonça, E.; Netto, A.R. Selectivity sequence and competitive adsorption of heavy metals by brazilian soils. *Soil Sci. Soc. Am. J.* **2001**, *65*, 1115–1121. [CrossRef]
53. Reddy, M.R.; Dunn, S.J. Distribution coefficients for nickel and zinc in soils. *Environ. Pollut.* **1986**, *11*, 303–313. [CrossRef]
54. Phillips, J.M.; Hayman, D.S. Improved procedure of clearing roots and staining parasitic and vesicular-arbusuclar mycorrhizal fungi for rapid assessment of infections. *Trans. Br. Mycol. Soc.* **1970**, *55*, 159–161. [CrossRef]
55. McGonigle, T.P.; Miller, M.H.; Evans, D.G.; Fairchild, D.L.; Swam, J.A. Anew methods which gives an objective measure of colonization of roots by vesicular-arbuscular mycorrhizal fungi. *New Phytol.* **1990**, *115*, 495–501. [CrossRef]
56. Lindsay, W.L.; Norvell, W.A. Development of a DTPA soil test for Zinc, Iron, Manganese, and Copper. *Am. Soc. Agron.* **1978**, *42*, 421–428. [CrossRef]
57. Bray, R.; Kurtz, L.T. Determination of total, organic and available forms of phosphorus in soils. *Soil Sci.* **1945**, *59*, 39–46. [CrossRef]
58. Heath, R.L.; Packer, L. Photoperoxidation in isolated chloroplasts. Kinetics and stoichiometry of fatty acid peroxidation. *Arch. Biochem. Biophys.* **1968**, *125*, 189–198. [CrossRef]
59. Hodges, D.M.; Delong, J.M.; Forney, C.F.; Prange, R.K. Improving the thiobarbituric acid-reactive-substances assay for estimating lipid peroxidation in plant tissues and other interfering compounds. *Planta* **1999**, *207*, 604–611. [CrossRef]
60. Akerboom, T.; Sies, H. Assay of glutathione, glutathione disulfide and glutathione mixed disulfides in biological samples. *Methods Enzymol.* **1981**, *77*, 373–382. [CrossRef]
61. Nakano, Y.; Asada, K. Hydrogen peroxide is scavenged by ascorbate specific peroxidase in spinach chloroplasts. *Plant Cell Physiol.* **1981**, *22*, 867–880.

62. Bradford, M. A rapid and sensitive method for the quantitation of microgram quantities of protein utilizing the principle of protein dye binding. *Anal. Biochem.* **1976**, *72*, 248. [CrossRef]
63. Gallego, S.; Benavides, M.; Tomaro, M. Effect of cadmium ions on antioxidant defense system in sunflower cotyledons. *Biol. Plant.* **1999**, *42*, 49–55. [CrossRef]
64. R Project for Statistical Computing 3.6.1 for Windows. Free Language; GNU General Public License. 2019. Available online: https://www.r-project.org/ (accessed on 10 July 2019).

3

The Ageing Process Affects the Antioxidant Defences and the Poly (ADPribosylation) Activity in *Cistus Incanus* L. Leaves

Carmen Arena [1,†], Luca Vitale [2,†], Anna Rita Bianchi [1], Carmela Mistretta [2], Ermenegilda Vitale [1], Costantino Parisi [1], Giulia Guerriero [1], Vincenzo Magliulo [2] and Anna De Maio [1,*]

1 Dipartimento di Biologia, Università degli Studi di Napoli Federico II, Via Cinthia, 80126 Napoli, Italy; c.arena@unina.it (C.A.); annarita.bianchi@unina.it (A.R.B.); ermenegilda.vitale@unina.it (E.V.); parisi@unina.it (C.P.); guerrier@unina.it (G.G.)

2 Istituto per i Sistemi Agricoli e Forestali del Mediterraneo (CNR-ISAFoM), Via Patacca 85, 80056 Ercolano (NA), Italy; luca.vitale@cnr.it (L.V.); carmela.mistretta@isafom.cnr.it (C.M.); vincenzo.magliulo@cnr.it (V.M.)

* Correspondence: andemaio@unina.it

† Co-first author.

Abstract: The ageing process in living organisms is characterised by the accumulation of several deleterious changes occurring in cells and tissues. The increase of reactive oxygen species with the advancement of age is responsible for the oxidative damage to proteins, lipids and DNA, enhancing the risk of diseases. The antioxidant response and the activation of the poly(ADP-ribosyl)ation process represent the first defences activated by organisms at all life stages to counteract damage to cell structures and genomic material. The regulation of poly(ADP ribosyl)ation with age is little known in plants, especially in combination with antioxidant defences modulation. In this study, the relationships between poly (ADP-ribose) polymerase (PARP) activity and enzymatic and non-enzymatic antioxidant pool have been studied together with the photosynthetic apparatus efficiency in the Mediterranean species *Cistus incanus* L., examining leaves at different developmental stages: young, mature and senescent. The photosynthetic performance was evaluated by chlorophyll *a* fluorescence measurement, the total soluble and fat-soluble antioxidant capacity, as well as the activities of enzymes superoxide dismutase (SOD), catalase (CAT), peroxidase (POD) and glutathione-S-transferase (GST), were determined by spectrophotometer, PARP activity was assessed by radioactive labelling. The highest photochemical activity was observed in young leaves, together with the highest GST activity. With the progress of the ageing process, the non-enzymatic antioxidant pool (namely ascorbic acid, α-tocopherol) declined, reaching the lowest value in senescent leaves, whereas PARP activity rose significantly. The overall results indicate that the decline of photosynthetic apparatus efficiency during senescence is due to the reduction of specific defences against oxidative damages, which increase the damages to DNA, as demonstrated by PARP activity rise.

Keywords: *Cistus incanus* L.; photosynthetic apparatus; antioxidants; poly (ADP-ribose) polymerase (PARP) activity

---

1. Introduction

Ageing is a natural process associated with the time-dependent general decline in the physiological function of an organism. It represents a multifactorial phenomenon, including genetic, physiological and biochemical changes related to the natural process of growth, to genetic defects and to the relationship between genotype and environmental conditions [1–3].

In plants, leaf ageing is a tightly regulated process with a crucial biological purpose: during senescence, metabolic changes and an ordered degradation of structures occur in cells. The decrease of stomatal conductance and photosynthetic rates in the leaves [4] are followed by the degradation of chlorophyll molecules responsible for changes of the leaf colour [5]. By contrast with animals, evident alterations are not observed in the mitochondria and the nucleus that remain intact until the final stages of leaf senescence. Metabolic changes include the hydrolysis of proteins, lipids, nucleic acids and pigments, that are accumulated during the growth phase [6]. At the cellular level, oxidative stress plays an essential role in the ageing process, which seems to be strongly associated with the changes in the prooxidant/antioxidant balance. The oxidative stress occurs when the reactive oxygen species (ROS) generation is transiently or chronically enhanced, and the antioxidant protection system does not counteract the disturbed physiological condition [7]. ROS are generally short-lived highly reactive molecules, derived from the partial reduction of oxygen; they are endogenously generated from healthy cellular metabolism or produced from exogenous sources, including pesticides, ultraviolet (UV) light, metal ions, smoke, ionizing radiation [8–10]. ROS represent a continuous challenge for eukaryotic cells, which may maintain under control their excess to avoid apoptosis, necrosis, autophagy and senescence [11–14]. The cell has particular defence mechanisms in protecting against ROS excess, including the enzymatic and non-enzymatic antioxidants. The main antioxidant enzymes are superoxide dismutase (SOD), catalase (CAT), peroxidase (POD), glutathione peroxidase (GPX), glutathione reductase (GR), and glutathione S transferase (GST) [15]. The non-enzymatic antioxidant defence system includes ascorbic acid (vitamin C), α-tocopherol (vitamin E), glutathione (GSH) and β-carotene [16]. The capacity of cells to counteract the overproduction of free radicals declines with age. The consequences are cumulative damages to important biological macromolecules, like DNA, proteins, and lipids [17,18]. The DNA may be damaged as single-strand breaks (SSBs), double-strand breaks (DSBs), oxidized bases and cross-linking sites [19,20]. The failure in damage repair leads to the cell surviving with altered genetic information. Alternatively, a severe mutational load may cause cell death [21]. In plants such as in animals, several DNA repair mechanisms have evolved to guarantee the integrity of genetic information. The DNA lesions produced by ROS are mainly restored by the base excision repair pathway [22–24]. During the process of base excision repair, two nuclear proteins, the poly(ADPribose) polymerase 1 (PARP-1) and the poly(ADPribose) polymerase 2 (PARP-2), regulate the accessibility of nicked DNA to other repair enzymes [25]. The poly(ADPribosyl)ation(PARylation) process represents one of the first cellular responses to oxidative and other types of DNA damages being "sensors" of DNA damaged and involved in the maintenance of genomic stability [26]. Under normal conditions, PARPS have a shallow basal enzymatic activity, which increases dramatically under conditions of cellular stress [27,28]. PARPs activation induces the synthesis of poly(ADP-ribose) (PAR) from nicotinamide adenine dinucleotide (NAD^+) and the release of nicotinamide as reaction by-product [29,30]. Defects in DNA repair lead to PARP activation and progressive oxidative DNA damage with ageing [31]. A strong correlation has been found between resistance to different stressors, including oxidative stress and longevity in mammalian cells [32,33]. The highest PARylation levels were in long-lived species, in which similar levels of PARP protein were expressed [32]. Thus, poly(ADP-ribosyl)ation capacity was correlated with the ageing process and species-specific longevity [26,34]. Besides, during senescence, in eukaryotes, the activity, stability and localization of several proteins are affected by different post-translational modifications, such as phosphorylation, glycosylation, ubiquitylation, methylation and acetylation [4]. Histone acetylation, methylation, phosphorylation, and PARylation are known to influence the remodelling of the chromatin structure, regulating DNA replication and accessibility of the transcriptional machinery to it with indirect control of gene expression. Plant PARylation was first described in the 1970s, and since then nine proteins with PARP signature were characterized [35–38]. Besides performing similar functions to those described for the animal counterparts, plant PARPs [39] are also implicated in response to abiotic and biotic stresses [40–42], in stress tolerance [38,43] and in developmental processes [44]. The role of poly(ADPribosyl)ation and its implication in the ageing process has been addressed in several studies concerning animals, but not in plants.

The aim of the present research was to study the modulation of the antioxidant defences (enzymatic and non-enzymatic antioxidant pool) and poly(ADPribosyl)ation in response to leaf development to understand the mechanisms at the basis of age-related prooxidant/antioxidant balance and decline of photosynthetic efficiency. To address this, we measured simultaneously oxidative stress markers, namely total soluble and fat-soluble antioxidant capacity, and enzymatic activity of catalase (CAT), superoxide dismutase (SOD), peroxidase (POD) and glutathione-S-transferase (GST) together with measurements of chlorophyll fluorescence emission and PARP activity in young (10-day-old, Y), mature (30 day old; M) and old (45 day old; S) summer leaves of *Cistus incanus* L. In this species the occurrence of a poly (ADP-ribose) polymerase enzyme of 80 kDa, able to synthesize a poly (ADP-ribose) of about 8–10 ADP-ribose unit was already demonstrated [41]. The Mediterranean perennial shrub used in this study is typical of the Mediterranean maquis ecosystems [45] and, from the physiological point of view, is particularly interesting because during the summertime it develops summer leaves especially suitable to cope with drought condition typical of Mediterranean areas [43,46,47]. These leaves exhibit peculiar traits compared to winter ones appearing on the branches in early spring and remaining until the beginning of the winter; during their long lifespan, these leaves experiment and withstand many environmental constraints.

2. Materials and Methods

2.1. *Plant Material and Growth Conditions*

We conducted the experiment on summer leaves of *Cistus incanus* L. subsp. *incanus*, a seasonally dimorphic species with small xeromorphic leaves (summer habitus) produced by plants from the beginning of May until beginning of June as opposed to large mesomorphic leaves (winter habitus) produced from the end of October until beginning of November [43,45,46].

In early May 2014, 10 healthy plants of *C. incanus*, approximately three years old, were screened in the field (Castelvolturno Nature Reserve, Tyrrhenian coast of southern Italy, Naples) for size and uniformity, excavated from field and immediately transplanted in 15 L pots filled with native soil, transported at Department of Biology of University of Naples Federico II and grown in glasshouse in semi-controlled conditions at a mean temperature of 23 ± 2 °C, 65–70% Relative Humidity (RH), 13 h of photoperiod. In order to avoid any additional stress, plants were watered three times a week with variations depending on the evaporative demand of the environment. At the end of June, after 20 days of acclimation at the glasshouse environmental condition, young (Y—10 days old), mature (M—30 days old) and senescent (S—45 days old) summer leaves were selected for fluorescence emission analyses and determination of enzymatic and non-enzymatic antioxidants and PARP activity.

2.2. *Chlorophyll* a *Fluorescence Measurement*

Chlorophyll *a* fluorescence measurements were performed on attached young (Y), mature (M) and senescent (S) leaves of *C. incanus* using the Mini-Pam fluorometer (Walz, Germany) equipped with a leaf-clip holder (Leaf-Clip Holder 2030-B, Walz, Germany) which allows the simultaneous recording of the incident Photosynthetic Photon Flux Density (PPFD) on the leaf surface and abaxial leaf temperature. The maximal PSII photochemical efficiency (F_v/F_m) was measured in early morning on 30 min dark-adapted leaves [48]. For the quantum yield of PSII linear electron transport (Φ_{PSII}) determination [49], the fluorescence measurements were conducted at midday under natural environmental conditions of PPFD of 800 ± (20) μmol photons m^{-2} s^{-1} and air temperature of 23 (±2) °C.

2.3. *Isolation of Nuclei and Western Blotting Analysis*

Nuclei were isolated as described in Arena et al. [41]. All operations were performed on ice or at 4 °C. Samples of (Y), (M), and (S) leaves (0.1 g) were homogenized at low speed by an Ultra Turrax T8 homogenizer (IKA-WERKE) for 30–40 s. The buffer of homogenization (buffer A) contained 10 mM Tris-HCl pH 8.0, 1 mM ethylenediaminetetraacetic acid (EDTA), 1 mM egtazic acid (EGTA), 1 mM phenylmethylsulfonyl fluoride ($PhMeSO_2F$), 10 mM $MgCl_2$, 5 mM 2-mercaptoethanol, 1% Nonidet-P40 (1:4, *w*/*v*), and protease inhibitor cocktail (5–10 μg/mL). The homogenates were filtered through three layers of cheesecloth and centrifuged at 1500 g for 30 min at 4 °C. This last procedure was repeated four times. Finally, the pellets, which represent the nuclear fractions, were resuspended in a small volume of buffer A without NP 40.

Y, M, and S leaf nuclear fractions (20 μg) were analysed on 10% polyacrylamide gels in the presence of 0.1% sodium dodecyl sulphate (SDS) tension-active according to Arena et al. [41]. Gels were stained in 0.1% Coomassie G in 10% acetic acid and 30% methanol. For immunoblotting, the electrophoresed proteins were transferred onto a polyvinylidene fluoride (PVDF) membrane (Biorad) at 200 V for 1.5 h at 4 °C in the same buffer used for the electrophoretic run. Filter was placed in blocking solution, containing 50 mM Tris-HCl buffer, pH 8.0; 150 mM NaCl; 0.5% (*v*/*v*) Tween 20 and 3% gelatine for 1.5 h. Subsequently, the filter was incubated for 2 h at room temperature in the same solution integrated with 0, 3% gelatine, in the presence of commercial polyclonal anti-PARP human antibodies (H-250, Santa Crutz), diluted 1:2000, *v*/*v*. After repeated washings in TBS-Tween 0.5%, the filter was incubated for 1 h, at room temperature, in TBS-Tween 0.5% and 0.3% gelatine, containing anti-rabbit secondary antibodies (Thermo Scientific diluted 1:2000) conjugated with peroxidase. The detection of peroxidase activity was conducted by chemiluminescence, using the kit Super Signal West Dura provided by Pierce. The acquisition and analysis of the images was performed using by Quantity One program in a Chemi-doc apparatus (Bio-Rad).

The same filter was incubated in a stripping buffer containing 62.5 m MTris-HCl (pH 6.8), 2% SDS, a final concentration of 0.1 M 2-mercaptoethanol, for 30 min at 50–60 °C according to the methods [41]. Afterward the filter was washed in TBST and analysed by immunoblotting with anti-poly(ADPR) (anti-PAR) polyclonal primary antibodies (H000085-05-B01, Alexis, 1:1000) and horseradish peroxidase-conjugated goat anti-mouse secondary antibody (Pierce, 1:2000) as previously described. Poly(ADPR) antibody is able to identify polymers of ADPR at the last of 5 units.

2.4. *Poly (ADP-Ribose) Polymerase (PARP) Assay Activity*

The assay to measure the ADP-ribosylating activity was carried outfor 10 min at 25 °C. The nuclear fractions (20 μg of proteins) were incubated in presence of 0.4 mM [^{32}P] NAD^+ (10.000 cpm/nmole) in a 500 mM Tris-HCl buffer, pH 8.0; 50 mM $MgCl_2$ and 10 mM DTT (reaction mixture) [50]. The reaction was stopped by adding of ice-cold 20% trichloroacetic acid(*w*/*v*). To isolate the product of the reaction, the mixture was filtered on Millipore filters (HAWPP0001, 0.45 μm) and subjected to various washes using 7% trichloroacetic acid. The radioactivity of insoluble acid material associated with the filter was measured in a liquid phase scintillator (Bechman LS 1701). The enzyme activity is expressed in enzymatic milliunit; 1 mU catalyses the synthesis of 1 nmol ADP-ribose/min, at the optimum of pH and temperature.

2.5. *Determination of Total Soluble and Fat-Soluble Antioxidant Capacity*

Stock solutions of ascorbic acid and α-tocopherol were prepared in water and in ethanol respectively, just before use. The concentrations were determined by spectrophotometer considering the absorption coefficients from the literature. All samples were frozen under liquid nitrogen and ground with pestle and mortar to a fine powder. After the addition of 1 mL/g of solvent (water or ethanol), the suspensions were homogenized, transferred to polypropylene tubes, and shaken for 1 h at room temperature in the dark. The suspensions were then centrifuged at 10,000 g for 15 min and the supernatant collected and kept at 4 °C (first extract). The pellets were resuspended, homogenized in another volume of solvent and centrifuged. Finally, the supernatant was joint with the first extract and kept at 4 °C until determinations. Total soluble and fat-soluble antioxidant capacity of Y, M and S leaves were measured according to Prieto et al. [51]. An aliquot of 0.1 mL of supernatant was combined with 1 mL of reagent solution (0.6 M sulfuric acid, 28 mM sodium phosphate, and 4 mM ammonium molybdate). The soluble extracts were incubated at 95 °C, while the fat-soluble extracts were incubated at 37 °C for 90 min in a water bath under a constant shaking. The absorbance was measured at 695 nm against a blank. The blank solution contained 1 mL of reagent solution and the proper volume of the same solvent used for the sample Water-soluble and fat-soluble antioxidant capacity was calculated using a standard curve of ascorbic acid (extinction coefficient $3.4 \pm 0.1 \times 103\ M^{-1}\ cm^{-1}$) and α-tocopherol (extinction coefficient $4 \pm 0.1 \times 103\ M^{-1}\ cm^{-1}$) respectively and was expressed as equivalents/g of leaves.

2.6. Antioxidant Enzyme Analysis

Catalase (CAT), peroxidase (POD) and superoxide dismutase (SOD) activities were determined in Y, M and S leaves. All operations were conducted in ice-cold. The leaves (1 g) were resuspended and homogenized in 0.1 M phosphate buffer (pH 7.5) containing 0.5 mM EDTA. After centrifugation at 4 °C and 15,000 rpm for 15 min., the supernatants were used to measure the activity of antioxidant enzymes. CAT activity was determined according to Wong and Whitaker [52] and Chance and Maehly [53]. The reaction mixture consisted of 0.1 M phosphate buffer (pH 7.0), 200 mM of H_2O_2 and enzymatic extract. The activity assay was performed by adding of 50–150 μL enzymatic extract to reaction mixture (150 μL H_2O_2, 2 mL phosphate buffer, and water up to final volume of 3 mL). The decrease of H_2O_2 was monitored at 240 nm and quantified by its molar extinction coefficient ($36\ M^{-1}\ cm^{-1}$) [54].

POD (EC 1.11.1.7) activity was determined in according to Yuan and Jiang [55] and Chance and Maehly [53]. The reaction mixture contained 0.1 M of phosphate buffer (pH 7.0), 200 mM H_2O_2, 135mM of guaiacol. SOD assay was determined by adding 500–1000 μL of enzymatic extract to the reaction mixture (1.2 mL of phosphate buffer30 μL of H_2O_2, 200 μL of guaiacol and water in a total volume of 3 mL). Activity was determined by the increase in absorbance at 420 nm caused to guaiacol oxidation ($E = 26.6\ mM^{-1}\ cm^{-1}$).

SOD (EC 1.15.1.1) activity was assayed according to Sun and Zigman [56] and Khopde et al. [57]. Enzymatic assay (final volume of 3 mL) was conducted by adding a reaction mixture (0.1 M carbonate buffer pH 10.0, 5mM epinephrine pH 2.0) to enzymatic extract. Activity was determined by absorbance increase at 320 nm for 1 min. One unit of SOD activity is expressed as the amount of enzyme required to cause 50% inhibition of epinephrine oxidation under the experimental conditions.

The absorbance increase was recorded at 320 nm for 1 min. The linear portion of the reaction progress curve (product versus time) at 320 nm was calculated (Δ Abs320 min^{-1}). The slope was used to determine the rate of epinephrine oxidation, which was calculated in the same conditions but without enzymatic extract. One unit of SOD was defined as the amount of enzyme required to reduce the epinephrine auto-oxidation rate by 50%.

The glutathione S-transferase (GST) activity has been estimated according to Habig et al. [58]. Young (Y), mature (M) and senescent (S) leaf samples (5 g) were homogenized in liquid nitrogen to obtain fine powder and suspended in cold potassium phosphate buffer 0.1 M pH 6.5, EDTA 1 mM. The solution was centrifuged at 4 °C at 7000 g for 30 min; the supernatants were then separated and used for enzyme assays. 25 μL of each sample was added to phosphate-buffered saline (PBS) buffer, containing 1-chloro-2,4-dinitrobenzene (CDNB as substrate for GST) at final concentrations of 1 mM reduced glutathione, 1 mM CDNB and 100 mM potassium phosphate buffer (pH 6.5). The formation of the conjugate complex was monitored by spectrophotometer for 5 min, 25 °C at 340 nm. The enzyme activity was expressed as μM/min/g of fresh tissue.

2.7. Statistical Analysis

The software package Sigma-Plot 12.0 (Jandel Scientific, San Rafael, CA, USA) was used for graphical and statistical data processing. Statistically significant differences were checked by one-way analysis of variance (ANOVA) based on a significance level of $p < 0.05$. The normal distribution of data was verified by Shapiro–Wilk and Kolmogorov–Smirnov tests. Percent data were transformed through the arcsine function before statistical analysis. The data are mean ± standard deviation (SD) ($n = 5$).

3. Results

3.1. Photochemical Apparatus Efficiency

The response of photochemical apparatus in *C. incanus* leaves of different age was assessed by chlorophyll fluorescence emission measurements analysing the indexes: maximal PSII photochemical efficiency (F_v/F_m) and quantum yield of PSII linear electron transport (Φ_{PSII}). As expected, compared to young and mature, senescent leaves showed a strong decrease ($p < 0.05$) of F_v/F_m (Figure 1a) and Φ_{PSII} (Figure 1b). On the contrary, no difference in photochemistry was found between young and mature leaves.

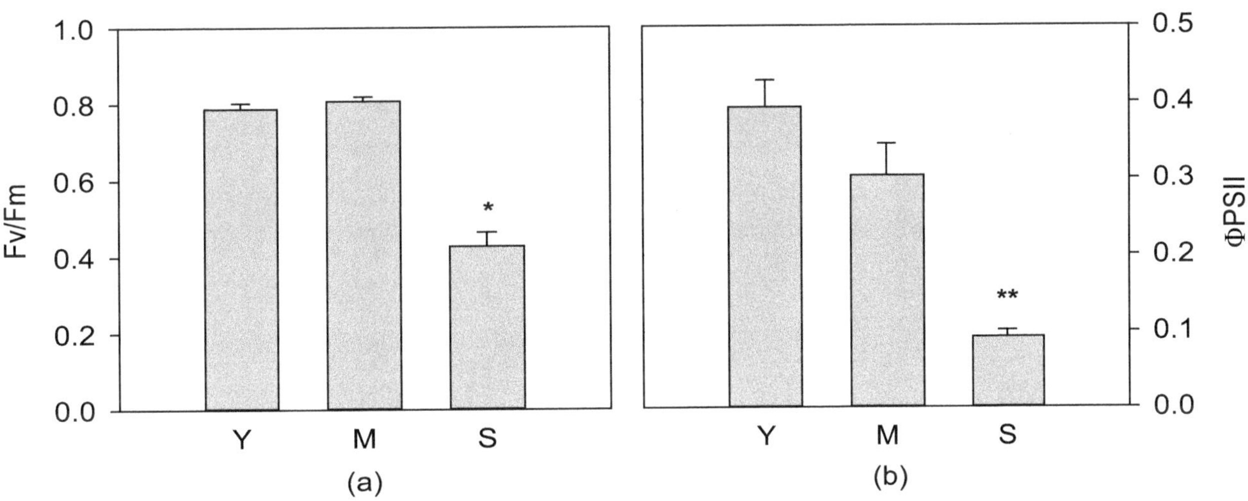

Figure 1. Maximal photochemical efficiency, F_v/F_m (**a**) and quantum yield of linear electron transport, Φ_{PSII} (**b**), of young (Y), mature (M) and senescent (S) leaves of *Cistus incanus* L. Data are mean ± SD ($n = 5$). Results were analysed by one-way analysis of variance (ANOVA) with Holm–Sidak post hoc test. Asterisks represent different levels of significance (** $0.001 < p \leq 0.01$, * $0.01 < p \leq 0.05$) compared to Y leaves.

3.2. Poly(ADP)ribosylation Characterization

No significant difference in PARP activity was found between S and Y leaves; on the contrary, PARP activity was three times lower ($p < 0.01$) in M compared to Y and S leaves (Figure 2).

Moreover, no qualitative and quantitative difference in protein patterns was evidenced (Figure 3a). A single and clear band, corresponding to a protein with molecular weight of about 80 kDa, resulted immunopositive to anti-PARP, which was able to recognize the highly conserved catalytic site of the enzyme (Figure 3b). Western blotting with anti-PAR antibodies used to identify ADPR polymers, consisting of at least five units, showed only one immunopositive signal of about 90 kDa, corresponding to a covalent ADPR protein acceptor (Figure 3c).

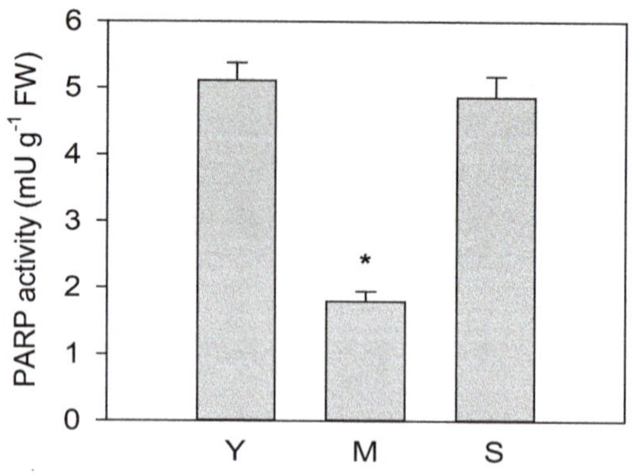

Figure 2. Poly (ADP-ribose) polymerase (PARP) activity measured in young (Y), mature (M) and senescent (S) leaves of *Cistus incanus* L. Data are mean ± SD (n = 5). Results were analysed by one-way ANOVA with Holm–Sidak post hoc test. Asterisks represent different levels of significance (* $0.01 < p \leq 0.05$) compared to Y leaves.

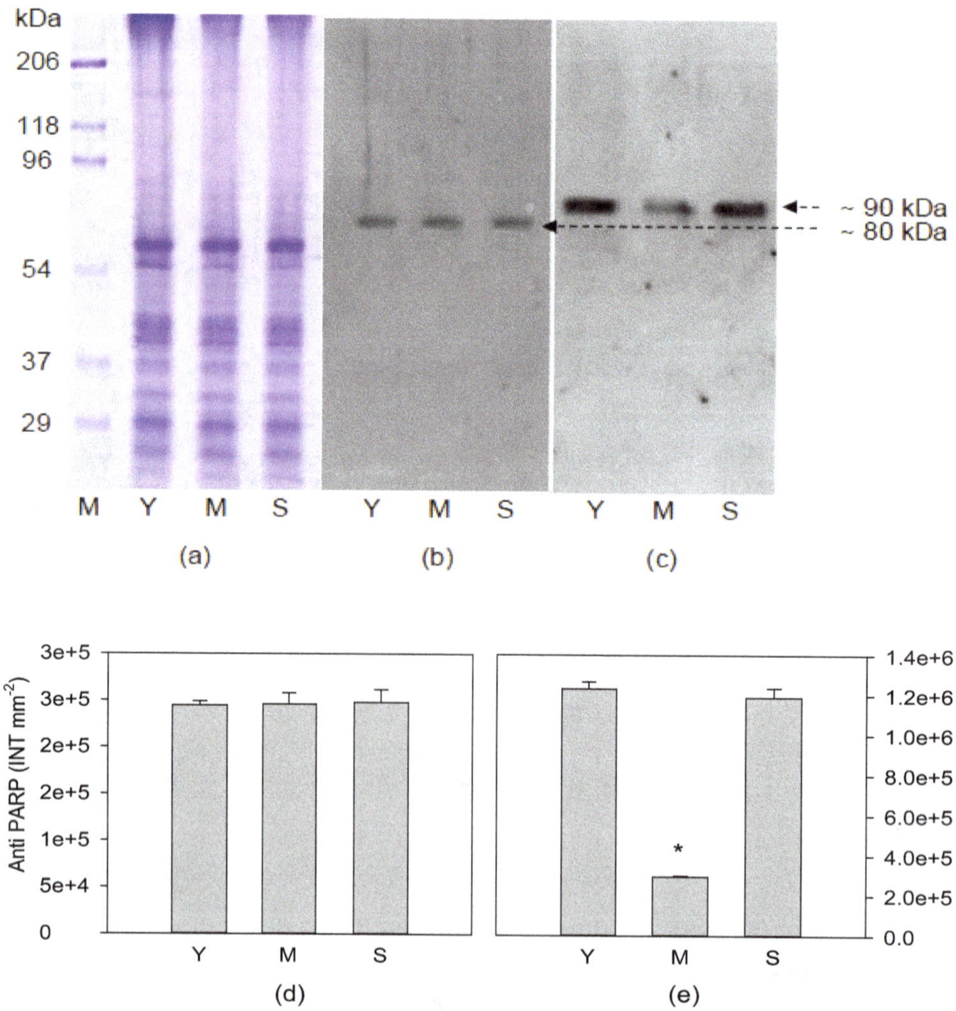

Figure 3. Sodium dodecyl sulfate polyacrylamide gel electrophoresis (SDS-PAGE) 10% (**a**), immunoblotting with anti-PARP (**b**) and immunoblotting with anti-PAR (**c**), in nuclear fractions of young (Y), mature (M) and senescent (S) leaves of *Cistus incanus* L.; densitometric analysis of immunopositive bands to anti-PARP (**d**) and to anti-PAR (**e**). For densitometric analysis data are mean ± SD (n = 3). Results were analysed by one-way ANOVA with Holm–Sidak post hoc test. Asterisks represent different levels of significance (* $0.01 < p \leq 0.05$) compared to Y leaves.

Densitometric analysis of bands immunopositive to anti-PARP (Figure 3d) and anti-PAR (Figure 3e) was conducted to correlate the intensity of 80kDa proteins, corresponding to their expression, with poly(ADPR) synthesis. Data showed that the intensity of 80kD protein did not change in examined samples, while the intensity of 90kD protein observed in Y and S leaves was significantly higher than that measured in M leaves (Figure 3e). These results suggest that, although the PARP expression was not affected by leaf age, on the contrary the levels of poly(ADPR) produced in the Y and S leaves were significantly higher than those found in M leaves.

3.3. *Total Soluble and Fat Soluble Antioxidant Capacity*

Total soluble and fat-soluble antioxidant capacity was evaluated to assess if and how the antioxidant activity may be affected by leaf age. The results showed that in Y and M leaves, soluble (Figure 4a) and fat-soluble (Figure 4b) antioxidant capacities did not show any differences, while in S leaves a significant decrease ($p < 0.01$) of both antioxidant capacities occurred.

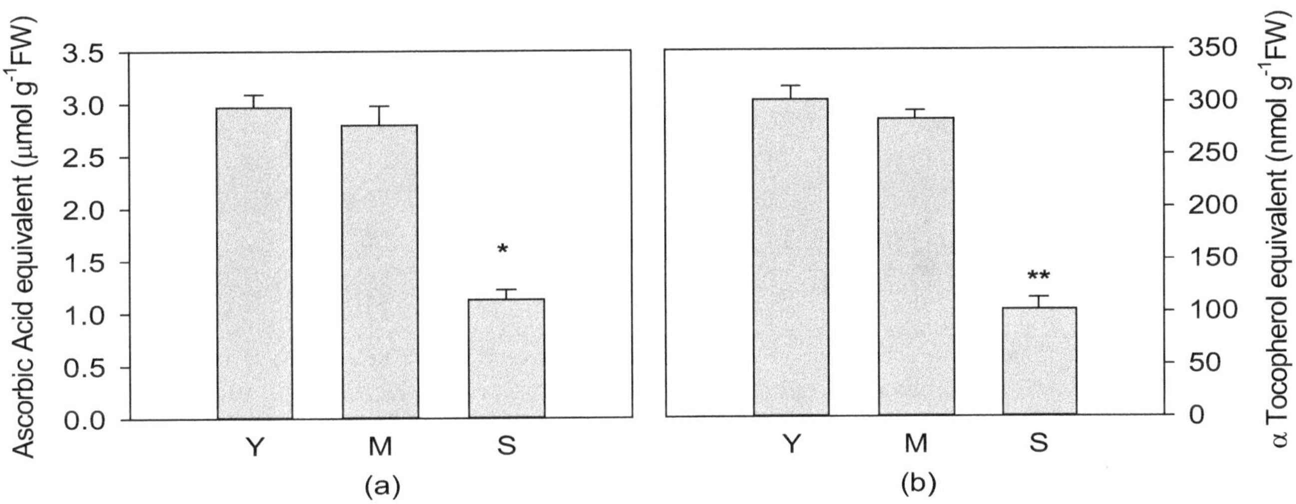

Figure 4. Total soluble antioxidant capacity expressed as ascorbic acid equivalents (**a**) and fat-soluble antioxidant capacity expressed as α tocopherol equivalents (**b**), of young (Y), mature (M) and senescent (S) leaves of *Cistus incanus* L. Data are mean ± SD ($n = 5$). Results were analysed by one-way ANOVA with Holm–Sidak post hoc test. Asterisks represent different levels of significance (** $0.001 < p \leq 0.01$, * $0.01 < p \leq 0.05$) compared to Y leaves.

3.4. *Activity of Antioxidant Enzymes*

The activities of catalase (CAT), superoxide dismutase (SOD), peroxidase (POD) and glutathione S-transferase (GST) were analysed to examine particular changes in enzymatic activity related to plant aging. The highest ($p < 0.01$) activity for CAT and SOD enzyme was found in S leaves. A significant difference in CAT and SOD activities was also measured in M compared to Y leaves, with higher values ($p < 0.05$) for Y leaves (Figure 5a,b). No significant variation of POD activity was found between Y and M leaves (Figure 5c). GST activity was statistically higher ($p < 0.01$) in young, compared to mature and senescent leaves. GST activity further decreased ($p < 0.001$) in senescent leaves, reaching the lowest value (Figure 5d).

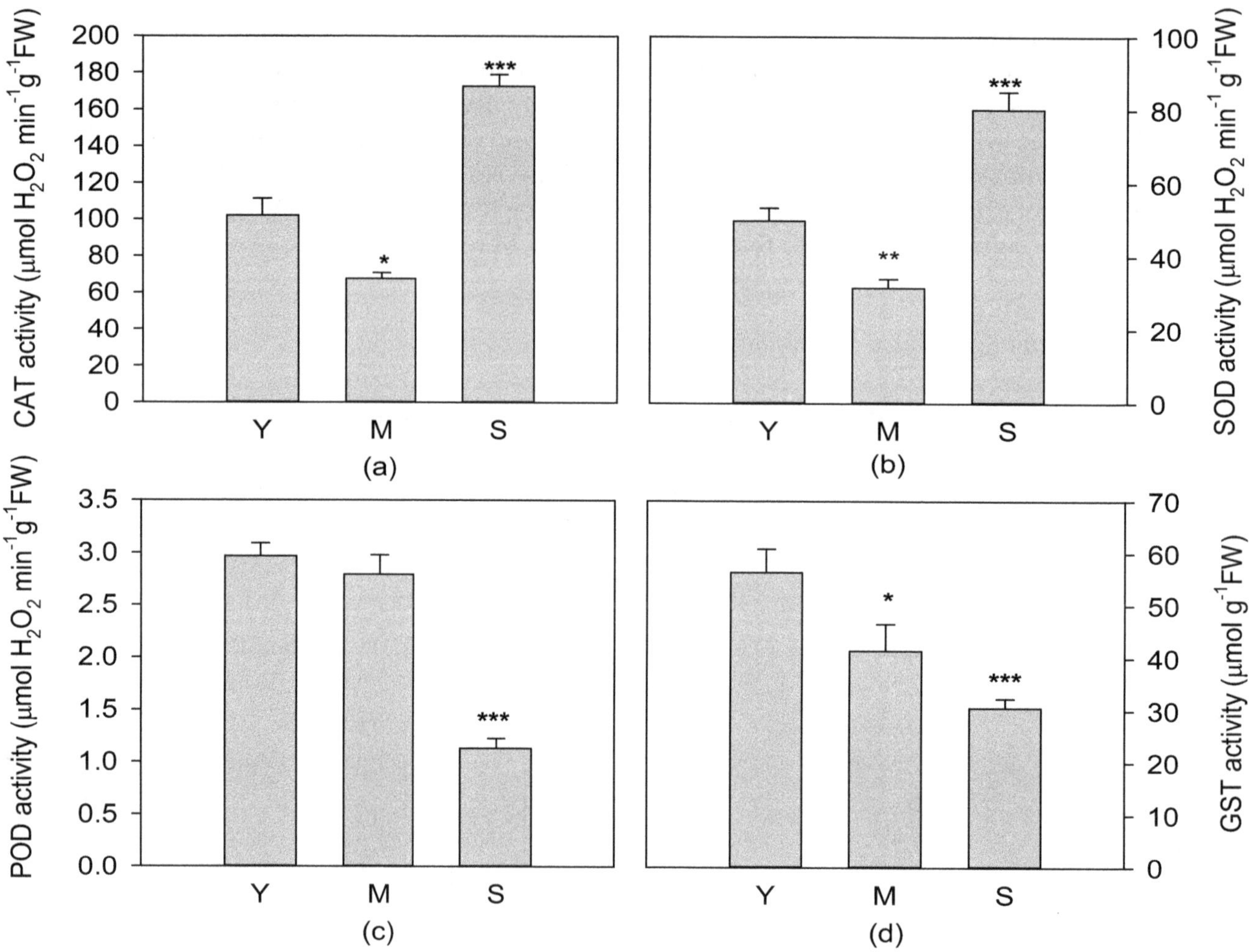

Figure 5. Activity of antioxidant enzymes: catalase, CAT (**a**), superoxide dismutase, SOD (**b**), peroxidase, POD (**c**), and glutathione S-transferase, GST (**d**) in young (Y), mature (M) and senescent (S) leaves *of Cistus incanus* L. Data are mean ± SD ($n = 5$). Results were analysed by one-way ANOVA with Holm–Sidak post hoc test. Asterisks represent different levels of significance (*** $0 < p \leq 0.001$, ** $0.001 < p \leq 0.01$, * $0.01 < p \leq 0.05$) compared to Y leaves.

4. Discussion

The ageing process has been associated with an increase of reactive oxygen species (ROS), responsible for cumulative damage to biological macromolecules, including lipids, proteins, and nucleic acids [1–4]. In the chloroplast, an impairment of photosynthetic membrane functionality leads to the decrease of photosynthetic activity. In fact, with increasing age, the PSII photochemical activity declines indicating a limited capacity of photosynthetic apparatus in the light-harvesting and conversion at reaction centres. Contextually, the increase of PARP activity at the nuclear level may be considered a rapid response of cells to DNA injuries. It has been demonstrated that poly(ADP-ribosyl)ation is one of the most immediate cellular response to genotoxic insults prompted by ionising radiation, alkylating agents, and heavy metals [4,59,60].

Numerous studies suggest a link between poly(ADP-ribosyl)ation induced by the DNA damage and ageing/longevity [60,61]; an evident correlation between PARP-1 deficiency and accelerated aging in mammalian cells was also demonstrated [62].

In the present work, we demonstrate for the first time that the PARP activity and consequent poly(ADPR) production are strongly influenced by the leaf aging in plants. In plants, three different nuclear PARPs have been previously characterized [38–40], and in particular in *Cistus incanus* species a PARP of 80 kDa has already identified [41].

The young, mature and senescent leaves of *C. incanus* express only the protein of 80 kDa, recognized by anti-PARP antibodies, specific in the identification of the highly conserved catalytic domain of all three nuclear PARPs. The *Cistus* species used in our experiment express only one of three PARPs documented in plants. We observe a single acceptor of poly(ADPR), corresponding to the same PARP modified with a polymer of about 20 units of ADPR, as indicated by the shift of the molecular weight from 80 kDa to 90 kDa. Therefore, at different leaf growth stages, we do not observe any difference in the length of the synthesized polymer, but only in the amount of poly(ADPR). In particular, the highest levels of poly(ADPR) are found in the Y and S leaves, according to the PARP activity.

In senescent (S) leaves, the increase of PARP activity as well as of CAT and SOD enzymes indicates that all these scavenger activities co-work against the oxidative stress occurring in cells during ageing. The reduction of the total antioxidant capacity together with the decrease of POD and GST activity suggests the inability of cells to counteract definitely ROS production [63]. However, it is likely that in our study, the oxidative damage to DNA does not lead to cell death because PARP enzymes are hyperactivated guaranteeing a continuous repair of the genomic material. It has been demonstrated in plants that, during senescence, cells are subjected to a degradation of cellular structures, and only the nucleolus and mitochondria remain intact until final stages of leaf senescence [64]; therefore, it is reasonable to suppose that mitochondria could be still functioning ensuring an adequate intracellular NAD^+ supply for PARP activity.

At the functional level, ageing determines a reduction of photosynthetic capacity in senescent leaves (F_v/F_m ratio and Φ_{PSII} decreasing), suggesting a loss of functionality of PSII reaction centres. Although it is likely that the decline of photochemistry may be due to a direct effect of ROS on photosynthetic membranes [5], it cannot be excluded that in senescent leaves the photosynthetic electron transport is diverted in the process other than carbon fixation to provide the reductive power for the scavenger enzymes.

By contrast with S leaves, the high PARP activity found in young (Y) leaves, responsible for the high poly(ADPR) production, seems to be not due to the oxidative DNA damage but rather to other cellular events such as gene expression, transcriptional activity, cell growth, and tissue proliferation [65]. The active growth phase and the high photosynthetic efficiency would guarantee in Y leaves an adequate production of ATP for the synthesis of NAD^+ for PARP [41]. Moreover, the high total antioxidant capacity, together with the elevated POD and GST activities might assure the apoplastic ROS homeostasis, that control the cell expansion [66]. It is well known that, during the leaf development, a wave of ROS-dependent cell growth sweeps through the leaves [67,68]. Our data suggest that with the progression of ageing, leaves have to carry out a fine-tuning of ROS production and removal adapting to continuous changes of cell metabolism and modulating the antioxidant defences differently [69,70].

5. Conclusions

Our work provides the novel insight that the PARP activity and the consequent poly(ADPR) production are strongly influenced by the leaf aging in plants. Leaf aging is associated to oxidative stress due to the ROS increase, responsible for DNA damage and consequent PARP activation. In the senescent leaves, the hyperactivation of PARP activity guarantees a continuous repair of the genomic material avoiding cell death. Overall, this study contributes to a further understanding of physiological, molecular, and biochemical changes occurring in cells during leaf aging and might act as "trailblazer study" for additional research, such as the study of the impact of environmental stressors (e.g., water shortage) on leaf aging.

Author Contributions: Conceptualization, C.A., L.V., A.D.M.; methodology, C.A., A.D.M., C.P.; investigation, C.A., A.D.M., A.R.B., C.M., L.V., G.G.; resources, C.A., A.D.M., V.M.; data curation, E.V.; writing—original draft preparation, C.A., A.D.M; writing—review and editing, C.A., A.D.M., L.V.; supervision, C.A., V.M, G.G.

References

1. Noodén, L.D.; Guiamet, J.J. Genetic control of senescence and aging in plants. In *Handbook of the Biology of Aging*, 4th ed.; Schneider, E.L., Rowe, J.W., Eds.; Academic Press: San Diego, CA, USA, 1996; pp. 94–118.
2. Dangl, J.L.; Dietrich, R.A.; Thomas, H. Cell death and senescence. In *Biochemistry and Molecular Biology of Plants*; Buchanan, B., Gruissem, W., Jones, R., Eds.; ASPP Press: Rockville, MD, USA, 2000; pp. 1044–1100.
3. Troen, B.R. The biology of aging. *Mt. Sinai J. Med.* **2003**, *70*, 3–22. [PubMed]
4. Day, R.E.; Rogers, P.J.; Dawes, I.W.; Higgins, V.J. Molecular analysis of maltotriose transport and utilization by *Saccharomyces cerevisiae*. *Appl. Environ. Microbiol.* **2002**, *68*, 5326–5335. [CrossRef] [PubMed]
5. Woo, H.R.; Kim, H.J.; Nam, H.G.; Lim, P.O. Plant leaf senescence and death–regulation by multiple layers of control and implications for aging in general. *J. Cell Sci.* **2013**, *126*, 4823–4833. [CrossRef] [PubMed]
6. Watanabe, M.; Balazadeh, S.; Tohge, T.; Erban, A.; Giavalisco, P.; Kopka, J.; Mueller-Roeber, B.; Fernie, A.R.; Hoefgen, R. Comprehensive dissection of spatiotemporal metabolic shifts in primary, secondary, and lipid metabolism during developmental senescence in Arabidopsis. *Plant Physiol.* **2013**, *162*, 1290–1310. [CrossRef] [PubMed]
7. Livingstone, D.R.; Garcia Martinez, P.; Michel, X.; Narbonne, J.F.; O'Hara, S.; Ribera, D.; Winston, G.W. Oxyradical production as a pollution mediated mechanism of toxicity in the common mussel *Mytilusedulis* L. and other mollusks. *Funct. Ecol.* **1990**, *4*, 415–424. [CrossRef]
8. Arena, C.; De Micco, V.; De Maio, A. Growth alteration and leaf biochemical responses in *Phaseolus vulgaris* exposed to different doses of ionising radiation. *Plant Biol.* **2014**, *16*, 194–202. [CrossRef]
9. Svilar, D.; Goellner, E.M.; Almeida, K.H.; Sobol, R.W. Base excision repair and lesion-dependent sub pathways for repair of oxidative DNA damage. *Antioxid. Redox Signal.* **2011**, *14*, 2491–2507. [CrossRef]
10. Barnes, D.E.; Lindahl, T. Repair and genetic consequences of endogenous DNA base damage in mammalian cells. *Annu. Rev. Genet.* **2004**, *38*, 445–476. [CrossRef]
11. Kannan, K.; Jain, S.K. Oxidative stress and apoptosis. *Pathophysiology* **2000**, *7*, 153–163. [CrossRef]
12. Choi, K.; Kim, J.; Kim, G.W.; Choi, C. Oxidative stress-induced necrotic cell death via mitochondira-dependent burst of reactive oxygen species. *Curr. Neurovasc. Res.* **2009**, *6*, 213–222. [CrossRef]
13. Filomeni, G.; De Zio, D.; Cecconi, F. Oxidative stress and autophagy: The clash between damage and metabolic needs. *Cell Death Differ.* **2015**, *22*, 377–388. [CrossRef] [PubMed]
14. Ziegler, D.V.; Wiley, C.D.; Velarde, M.C. Mitochondrial effectors of cellular senescence: Beyond the free radical theory of aging. *Aging Cell* **2015**, *14*, 1–7. [CrossRef] [PubMed]
15. Van der Oost, R.; Beyer, J.; Vermeulen, N.P.E. Fish bioaccumulation and biomarkers in environmental risk assessment: A review. *Environ. Toxicol. Pharmacol.* **2003**, *13*, 57–149. [CrossRef]
16. Gruber, J.; Fong, S.; Chen, C.B.; Yoong, S.; Pastorin, G.; Schaffer, S.; Cheah, I.; Halliwell, B. Mitochondria-Targeted antioxidants and metabolic modulators as pharmacological interventions to slow ageing. *Biotechnol. Adv.* **2013**, *31*, 563–592. [CrossRef]
17. Bodnar, A.G. Marine invertebrates as models for aging research. *Exp. Gerontol.* **2009**, *44*, 477–484. [CrossRef]
18. Lobo, V.; Patil, A.; Phatak, A.; Chandra, N. Free radicals, antioxidants and functional foods: Impact on human health. *Pharmacogn. Rev.* **2010**, *4*, 118–126. [CrossRef]
19. Dizdaroglu, M.; Jaruga, P.; Birincioglu, M.; Rodriguez, H. Free radical-induced damage to DNA: Mechanisms and measurement. *Free Radic. Biol. Med.* **2002**, *32*, 1102–1115. [CrossRef]
20. Bjelland, S.; Seeberg, E. Mutagenicity, toxicity and repair of DNA base damage induced by oxidation. *Mutat. Res.* **2003**, *531*, 37–80. [CrossRef]
21. Maslov, A.Y.; Vijg, J. Genome instability, cancer and aging. *Biochim. Biophysic. Acta* **2009**, *1790*, 963–969. [CrossRef]
22. David, S.S.; O'Shea, V.L.; Kundu, S. Base-excision repair of oxidative DNA damage. *Nature* **2007**, *447*, 941–950. [CrossRef]
23. Berquist, B.R.; Wilson, D.M. Pathways for repairing and tolerating the spectrum of oxidative DNA lesions. *Cancer Lett.* **2012**, *327*, 61–72. [CrossRef] [PubMed]
24. Sancar, A.; Lindsey-Boltz, L.A.; Ünsal-Kaçmaz, K.; Linn, S. Molecular mechanisms of mammalian DNA repair and the DNA damage checkpoints. *Annu. Rev. Biochem.* **2004**, *73*, 39–85. [CrossRef] [PubMed]

25. Schreiber, V.; Amé, J.C.; Dollé, P.; Schultz, I.; Rinaldi, B.; Fraulob, V.; Ménissier-de Murcia, J.; de Murcia, G. Poly (ADP-ribose) polymerase-2 (PARP-2) is required for efficient base excision DNA repair in association with PARP-1 and XRCC1. *J. Biol. Chem.* **2002**, *277*, 23028–23036. [CrossRef] [PubMed]
26. Bürkle, A. Poly(ADP-ribosyl)ation: A post translation protein modification linked with genome protection and mammalian longevity. *Biogerontology* **2000**, *1*, 41–46. [CrossRef]
27. D'Amours, D.; Desnoyers, S.; D'Silva, I.; Poirier, G.G. Poly (ADP-ribosyl)ation reactions in the regulation of nuclear functions. *Biochem. J.* **1999**, *342*, 249–268. [CrossRef]
28. Kim, M.Y.; Zhang, T.; Kraus, W.L. Poly (ADP-ribosyl)ation by PARP-1: 'PAR-laying' NAD+ into a nuclear signal. *Genes Dev.* **2005**, *19*, 1951–1967. [CrossRef]
29. Amé, J.C.; Rolli, V.; Schreiber, V.; Niedergang, C.; Apiou, F.; Decker, P.; Muller, S.; Höger, T.; Ménissier-de Murcia, J.; de Murcia, G. PARP-2, A novel mammalian DNA damage-dependent poly (ADP-ribose) polymerase. *J. Biol. Chem.* **1999**, *274*, 17860–17868. [CrossRef]
30. Shieh, W.M.; Amé, J.C.; Wilson, M.V.; Wang, Z.Q.; Koh, D.W.; Jacobson, M.K.; Jacobson, E.L. Poly (ADP-ribose) polymerase null mouse cells synthesize ADP-ribose polymers. *J. Biol. Chem.* **1998**, *273*, 30069–30072. [CrossRef]
31. Hooten, N.; Fitzpatrick, M.; Kompaniez, K.; Jacob, K.D.; Moore, B.R.; Nagle, J.; Barnes, J.; Lohani, A.; Evans, M.K. Coordination of DNA repair by NEIL1 and PARP-1: A possible link to aging. *Aging* **2012**, *4*, 674–685. [CrossRef]
32. Grube, K.; Bürkle, A. Poly (ADP-ribose) polymerase activity in mononuclear leukocytes of 13 mammalian species correlates with species-specific life span. *Proc. Natl. Acad. Sci. USA* **1992**, *89*, 11759–11763. [CrossRef]
33. Kapahi, P.; Boulton, M.E.; Kirkwood, T.B. Positive correlation between mammalian life span and cellular resistance to stress. *Free Radic. Biol. Med.* **1999**, *26*, 495–500. [CrossRef]
34. Beneke, S.; Alvarez-Gonzalez, R.; Bürkle, A. Comparative characterisation of poly (ADP-ribose) polymerase-1 from two mammalian species with different life span. *Exp. Gerontol.* **2000**, *35*, 989–1002. [CrossRef]
35. Doucet-Chabeaud, G.; Godon, C.; Brutesco, C.; De Murcia, G.; Kazmaier, M. Ionising radiation induces the expression of PARP-1 and PARP-2 genes in Arabidopsis. *Mol. Genet. Genomics* **2001**, *265*, 954–963. [CrossRef]
36. Babiychuk, E.; Cottrill, P.B.; Storozhenko, S.; Fuangthong, M.; Chen, Y.; O'Farrell, M.K.; Van Montagu, M.; Inze, D.; Kushnir, S. Higher plants possess two structurally different poly (ADP-ribose) polymerases. *Plant J.* **1998**, *15*, 635–645. [CrossRef]
37. Briggs, A.G.; Bent, A.F. Poly (ADP-ribosyl) ation in plants. *Trends Plant Sci.* **2011**, *16*, 372–380. [CrossRef]
38. Bianchi, A.R.; De Maio, A. Synthesis and degradation of poly (ADP-ribose) in plants. *Front. Biosci.* **2014**, *19*, 1436–1444. [CrossRef]
39. Lamb, R.S.; Citarelli, M.; Teotia, S. Functions of the poly (ADP-ribose) polymerase superfamily in plants. *Cell. Mol. Life Sci.* **2012**, *69*, 175–189. [CrossRef]
40. Chen, I.; Haehnel, U.; Altschmied, L.; Schubert, I.; Puchta, H. The transcriptional response of *Arabidopsis* to genotoxic stress—A high-density colony array study (HDCA). *Plant J.* **2003**, *35*, 771–786. [CrossRef]
41. Arena, C.; Mistretta, C.; Di Natale, E.; FaraoneMennella, M.R.; Virzo De Santo, A.; De Maio, A. Characterization and role of poly(ADP-ribosyl)ation in the Mediterranean species *Cistus incanus* L. under different temperature conditions. *Plant Physiol. Biochem.* **2011**, *49*, 435–440. [CrossRef]
42. Adams-Phillips, L.; Briggs, A.G.; Bent, A.F. Disruption of poly (ADP-ribosyl)ation mechanisms alters responses of Arabidopsis to biotic stress. *Plant Physiol.* **2010**, *152*, 267–280. [CrossRef]
43. Arena, C.; De Micco, V.; De Maio, A.; Mistretta, C.; Aronne, G.; Vitale, L. Winter and summer leaves of *Cistus incanus*: Differences in leaf morphofunctional traits, photosynthetic energy partitioning, and poly(ADP-ribose) polymerase (PARP) activity. *Botany* **2013**, *91*, 805–813. [CrossRef]
44. Hunt, L.; Holdsworth, M.J.; Gray, J.E. Nicotinamidase activity is important for germination. *Plant J.* **2007**, *51*, 341–351. [CrossRef]
45. De Micco, V.; Aronne, G. Seasonal dimorphism in wood anatomy of the Mediterranean *Cistus incanus* L. subsp. *incanus*. *Trees* **2009**, *23*, 981–989. [CrossRef]
46. De Micco, V.; Arena, C.; Vitale, L.; Aronne, G.; Virzo De Santo, A. Mediterranean *Cistus incanus* winter leaves under natural outdoor and warmer indoor conditions. *Botany* **2011**, *89*, 677–688. [CrossRef]
47. Vitale, L.; Magliulo, V.; Arena, C. Morphological and physiological modifications of *Cistus salvifolius* L. winter leaves to rise of winter temperatures. *Plant Biosyst.* **2014**, *148*, 1093–1101. [CrossRef]
48. Van Kooten, O.; Snell, J.F.H. The use of chlorophyll fluorescence nomenclature in plant stress physiology. *Photosynth. Res.* **1990**, *25*, 147–150. [CrossRef]

49. Genty, B.; Briantais, J.M.; Baker, N.R. The relationship between the quantum yield of photosynthetic electron transport and quenching of chlorophyll fluorescence. *Biochim. Biophys. Acta* **1989**, *90*, 87–92. [CrossRef]
50. De Maio, A.; Natale, E.; Rotondo, S.; Di Cosmo, A.; Faraone Mennella, M.R. Vault-Poly-ADP-ribose polymerase in the *Octopus vulgaris* brain: A regulatory factor of actin polymerization dynamic. *Comp. Biochem. Phys. B Biochem. Mol. Biol.* **2013**, *166*, 40–47. [CrossRef]
51. Prieto, P.; Pineda, M.; Aguilar, M. Spectrophotometric quantitation of antioxidant capacity through the formation of a phosphomolybdenum complex: Specific application to the determination of vitamin E. *Anal. Biochem.* **1999**, *269*, 337–341. [CrossRef]
52. Wong, D.W.S.; Whitaker, J.R. Catalase. In *Handbook of Food Enzymology*; Whitaker, J.R., Voragen, A.G.J., Wong, D.W.S., Eds.; Marcel Dekker Inc.: New York, NY, USA, 2003; pp. 389–401. [CrossRef]
53. Chance, B.; Maehly, A.C. Assay of catalases and peroxidases. *Methods Enzymol.* **1955**, *2*, 764–817. [CrossRef]
54. Verma, S.; Dubey, R.S. Lead toxicity induces lipid peroxidation and alters the activities of antioxidant enzymes in growing rice plants. *Plant Sci.* **2003**, *164*, 645–655. [CrossRef]
55. Yuan, Z.W.; Jiang, T.J. Peroxidase. In *Handbook of Food Enzymology*; Whitaker, J.R., Voragen, A.G.J., Wong, D.W.S., Eds.; Marcel Dekker Inc.: New York, NY, USA, 2003; pp. 389–401. [CrossRef]
56. Sun, M.; Zigman, S. An improved spectrophotometric assay for superoxide dismutase based on epinephrine autoxidation. *Anal. Biochem.* **1977**, *90*, 81–89. [CrossRef]
57. Khopde, S.M.; Priyadarsini, K.I.; Mohan, H.; Gawandi, V.B.; Satav, J.G.; Yakhmi, J.V.; Banavaliker, M.M.; Biyani, M.K.; Mittal, J.P. Characterizing the antioxidant activity of amla (*Phyllanthusemblica*) extract. *Curr. Sci.* **2001**, *81*, 185–190.
58. Habig, W.H.; Pabst, M.J.; Jakoby, W.B. Glutathione S-transferases the first enzymatic step in mercapturic acid formation. *J. Biol. Chem.* **1974**, *249*, 7130–7139.
59. Arena, C.; De Maio, A.; De Nicola, F.; Santorufo, L.; Vitale, L.; Maisto, G. Assessment of eco-physiological performance of *Quercus ilex* L. leaves in urban area by an integrated approach. *Water Air Soil Pollut.* **2014**, *225*, 1824. [CrossRef]
60. Burkle, A. DNA repair and PARP in aging. *Free Radic. Res.* **2006**, *40*, 1295–1302. [CrossRef]
61. Burkle, A.; Diefenbach, J.; Brabeck, C.; Beneke, S. Ageing and PARP. *Pharmacol. Res.* **2005**, *52*, 93–99. [CrossRef]
62. Piskunova, T.S.; Yurova, M.N.; Ovsyannikov, A.I.; Semenchenko, A.V.; Zabezhinski, M.A.; Popovich, I.G.; Wang, Z.Q.; Anisimov, V.N. Deficiency in poly(ADP-ribose) polymerase-1 (PARP-1) accelerates aging and spontaneous carcinogenesis in mice. *Curr. Gerontol. Geriatr. Res.* **2008**, 754190. [CrossRef]
63. Ding, S.; Wang, L.; Yang, Z.; Lu, Q.; Wen, X.; Lu, C. Decreased glutathione reductase 2 leads to early leaf senescence in *Arabidopsis*. *J. Integr. Plant Biol.* **2016**, *58*, 29–47. [CrossRef]
64. Xie, X.; He, Z.; Chen, N.; Tang, Z.; Wang, Q.; Cai, Y. The roles of environmental factors in regulation of oxidative stress in plant. *Hindaw iBioMed. Res. Int.* **2019**, *19*, 9732325. [CrossRef]
65. Munne'-Bosch, S.; Alegre, L. The function of tocopherols and tocotrienols in plants. *Crit. Rev. Plant Sci.* **2002**, *21*, 31–57. [CrossRef]
66. Krishnakumar, R.; Gamble, M.J.; Frizzell, K.M.; Berrocal, J.G.; Kininis, M.; Kraus, W.L. Reciprocal binding of PARP-1 and histone H1 at promoters specifies transcriptional outcomes. *Science* **2008**, *319*, 819–821. [CrossRef] [PubMed]
67. Schmidt, R.; Kunkowska, A.B.; Schippers, J.H.M. Role of reactive oxygen species during cell expansion in leaves. *Plant Physiol.* **2016**, *172*, 2098–2106. [CrossRef]
68. Rodrıguez, A.A.; Grunberg, K.A.; Taleisnik, E.L. Reactive oxygen species in the elongation zone of maize leaves are necessary for leaf extension. *Plant Physiol.* **2002**, *129*, 1627–1632. [CrossRef]
69. Gapper, C.; Dolan, L. Control of plant development by reactive oxygen species. *Plant Physiol.* **2006**, *141*, 341–345. [CrossRef]
70. Kunert, K.J.; Ederer, M. Leaf aging and lipid peroxidation: The role of the antioxidants vitamin C and E. *Physiol. Plant.* **1985**, *65*, 85–88. [CrossRef]

4

Chitosan Upregulates the Genes of the ROS Pathway and Enhances the Antioxidant Potential of Grape (*Vitis vinifera* L. 'Touriga Franca' and 'Tinto Cão') Tissues

Rupesh K. Singh [1,2,*], Bruno Soares [2,3], Piebiep Goufo [2,4], Isaura Castro [4], Fernanda Cosme [1], Ana L. Pinto-Sintra [4], António Inês [1], Ana A. Oliveira [2,4] and Virgílio Falco [1,2,*]

1 Centro de Química de Vila Real (CQ-VR), Universidade de Trás-os-Montes e Alto Douro (UTAD), Quinta de Prados, 5000-801 Vila Real, Portugal; fcosme@utad.pt (F.C.); aines@utad.pt (A.I.)
2 Departamento de Agronomia, Universidade de Trás-os-Montes e Alto Douro (UTAD), Quinta de Prados, 5000-801 Vila Real, Portugal; bmgoncalves90@gmail.com (B.S.); pgoufo@utad.pt (P.G.); anaolive@utad.pt (A.A.O.)
3 CoLAB Vines&Wines, Associação para o Desenvolvimento da Viticultura Duriense (ADVID), Régia Douro Park, 5000-033 Vila Real, Portugal
4 Centro de Investigação e Tecnologias Agroambientais e Biológicas (CITAB), Universidade de Trás-os-Montes e Alto Douro (UTAD), Quinta de Prados, 5000-801 Vila Real, Portugal; icastro@utad.pt (I.C.); asintra@utad.pt (A.L.P.-S.)
* Correspondence: rupesh@utad.pt (R.K.S.); vfalco@utad.pt (V.F.)

Abstract: Chitosan is an environmentally-friendly active molecule that has been explored for numerous agricultural uses. Its use in crop protection is well-known, however, other properties, such as bioactivity, deserve attention. Moreover, the modes of actions of chitosan remain to be elucidated. The present study assessed the levels of total phenolic compounds, the antioxidant potential, and the expression of reactive oxygen species (ROS) scavenging genes in the berries (skins and seeds), leaves, cluster stems, and shoots upon chitosan application on two red grapevine varieties (Touriga Franca and Tinto Cão). The application of chitosan on the whole vine before and after veraison led to the increased levels of polyphenols, anthocyanins, and tannins in Tinto Cão berries, and polyphenols and tannins in Touriga Franca berries, respectively. CUPric Reducing Antioxidant Capacity (CUPRAC) and Ferric Reducing Antioxidant Power (FRAP) assays indicated an increase in the antioxidant potential of berries. With the exception of ascorbate peroxidase (APX), all the ROS pathway genes tested, i.e., iron-superoxide dismutase (Fe-SOD), copper-zinc-superoxide dismutase (Cu/Zn-SOD), catalase (CAT), glutathione reductase (GR), glutaredoxin (Grx), respiratory burst oxidase (Rboh), amine oxidase (AO), peroxidase (POD) and polyphenol oxidase (PPO), were found up-regulated in chitosan-treated berries. Results from the analyses of leaves, stems, and shoots revealed that chitosan not only induced the synthesis of phenolic compounds but also acted as a facilitator for the transfer of polyphenols from the leaves to the berries.

Keywords: chitosan; elicitor; *Vitis vinifera* L.; antioxidant activity; secondary metabolites; anthocyanins; ROS pathway; polyphenols; tannins

---

1. Introduction

Grapes are non-climacteric berry fruits produced by deciduous and perennial woody vines. For centuries, grapes have had a huge impact on the world's economy due to their use in winemaking and in a vast variety of food products [1,2]. The quality of grape-derived products depends on the levels of secondary metabolites in the berries, including anthocyanins, tannins, stilbenes, flavonoids, and other phenolic compounds. Although these compounds are known for their antioxidant properties, they are particularly important for the flavor, color, and taste of wines as well as other industrial interests like neutraceuticals and pharmaceutical products [3–5].

For years now, increasing the levels of secondary metabolites in grape berries has been a major subject of interest worldwide. Several cultural practices have been studied to improve the accumulation of secondary metabolites in grape and wine [1,6–12]. It was found that vine 'cordon training' and pruning led to the increased levels of tannins and anthocyanins in red wine [9]. The total anthocyanin and the total phenolic contents were also higher in Tannat, Malbec, and Xinomavro grapes obtained from vines subjected to cluster thinning, when compared with "not thinned" controls [6–8]. Koundouras et al. reported that deficit irrigation caused a substantial increase of the level of anthocyanins in the skins of Cabernet Sauvignon grapes [10]. Subsequent studies in the following years confirmed this finding [11,12].

Besides the above-mentioned traditional cultural practices, the use of electrical [1] and chemical [13] elicitors have also been attempted. Elicitor-mediated induction of plant defenses is one of the most effective strategies in enhancing the production of secondary metabolic compounds, and the approach has been tested in various crops [13]. An elicitor is defined as "a compound introduced in small concentration to a living system to promote specific responses" [13]. In plants, several molecules, such as phenylalanine, urea, benzothiadiazole, methyl jasmonate, yeast extract, abscisic acid, and chitosan have been used as elicitors [14–18]. Chitosan, in particular, has received a scrutinized interest giving its efficacy in the crop protection [18–22]. Chitosan is a cheap linear polysaccharide derived from the chitin of fungal cell walls and exoskeleton of arthropods [23,24]. Chitosan is safe, biodegradable, non-allergic, bio-compatible, and the second most abundant renewable carbon source after cellulose [25–28]. These properties have ensured their wide industrial use [23]. Chitosan was introduced for agricultural purposes in the 1980s. The first report of chitosan as an elicitor was in pea (*Pisum sativum* L.) and tomato (*Solanum lycopersicum* L.) [19]. Its application resulted in an increased accumulation of the phytoalexin pisatin in pea, and anti-nutrient proteinase inhibitors in tomato. Chitosan was later used to induce the accumulation of secondary metabolites in many other plant species. For example, it was demonstrated that foliar application of chitosan enhances tomato fruit weight, productivity, and resistance against fungal pathogens [20,21], improves the yield and vegetative growth of strawberries [29], and the quality and self-life of kiwifruit [30].

There is a worldwide trend to explore the properties of chitosan for sustainable agriculture [25,27,28]. There are, however, only sporadic reports on in-field applications of chitosan on grapevine plants. Spraying of chitosan combined with yeast on vines before harvest reduced the decay of berries during cold storage [31]. In vitro cultured grape plantlets supplemented with a chitosan gel demonstrated improved growth and resistance towards the fungal pathogen *Botrytis cinerea* [22]. Vitalini et al. reported increased levels of aromatic compounds in wines obtained from chitosan-treated vines, as compared to those vines which were treated with conventional fungicides [15]. There was also an improvement in the overall sensory acceptance of wines [15]. Although Portu et al. did not find a substantial impact on the phenolic composition of grapes and wines following chitosan treatment [16]. The above-mentioned studies highlight the necessity of conducting more research on the response of vines concerning the application of chitosan.

Moreover, the mode of action of chitosan is still unknown for several aspects of plant protection and productivity. Chitosan is known to induce plant defense mechanisms by indirectly stimulating the synthesis of secondary metabolites [16,23]. The metabolic pathways regulated by chitosan, which ultimately lead to drought resistance in white clover, have been reported recently [26]. In particular, it was found that chitosan-induced drought resistance is associated with the accumulation of endogenous chitosan, and the enhancement of the ascorbate–glutathione and tricarboxylic acid cycle pathways [26]. In the present study, an additional insight in the mode of action of chitosan is provided by analyzing vines in a tissue-specific manner, after field application of chitosan during and after veraison on *Vitis vinifera* L. 'Touriga Franca' and 'Tinto Cão'. 'Touriga Franca' and 'Tinto Cão' are the most widely grown varieties of grapevines in the Douro Demarcated Region of Portugal and are very important varieties for Port wine production. Berries' seeds and skins, stems, leaves, and shoots were analyzed for the total phenolic content (TPC), total anthocyanin content (TAC), total tannin content (TTC), antioxidant potential, and the expression of the reactive oxygen species (ROS) pathway genes.

2. Materials and Methods

2.1. Plant Material, Treatments, and Sample Collection

Vitis vinifera L. 'Touriga Franca' and *Vitis vinifera* L. 'Tinto Cão' (red varieties), grown in the Quinta de Nossa Senhora de Lourdes vineyard (41°19′ N, 7°44′ W, 500 m above mean sea level), were used in this study. The vineyard is in Vila Real, which is situated in the lower Corgo sub-region of the Douro Demarcated Region of northern Portugal. The vines (420A rootstock) were trained with the Guyot system, spaced 2.3 × 0.9 m, on a grass-covered morainic soil, with 15% of gravel and loamy sand. The trial was set up as a complete randomized block design in three replications (lines), with 12 vines per line (chitosan-treated vines and control vines) in each line. Spraying of crop protectants, weed control, and shoot guiding were the same for all vines.

Chitosan (molecular weight of 76 kDa and deacetylation degree of 85%) was obtained from Sigma-Aldrich (St. Louis MO) and dissolved in 0.01 M aqueous acetic acid solution to a concentration of 0.1% (w/v). The whole vine, including the fruits, was sprayed with 200 mL of the chitosan solution, using a manual spray lance. Chitosan was applied on the same vines twice, first at the beginning of veraison at the onset of berry ripening, and second at the completion of veraison with all the berries colored red. The meteorological conditions from the beginning of veraison until berry harvesting were as follows: average temperature, humidity, and wind velocity of 23.4 °C, 57.4%, and 6.5 km/h, respectively. Berries and leaves were collected during veraison, after complete veraison, and at complete maturation. Stems and shoots were collected at the complete maturation of the berries, as explained diagrammatically in Figure 1. The same tissues from all 12 vines from a line were pooled to constitute a sample.

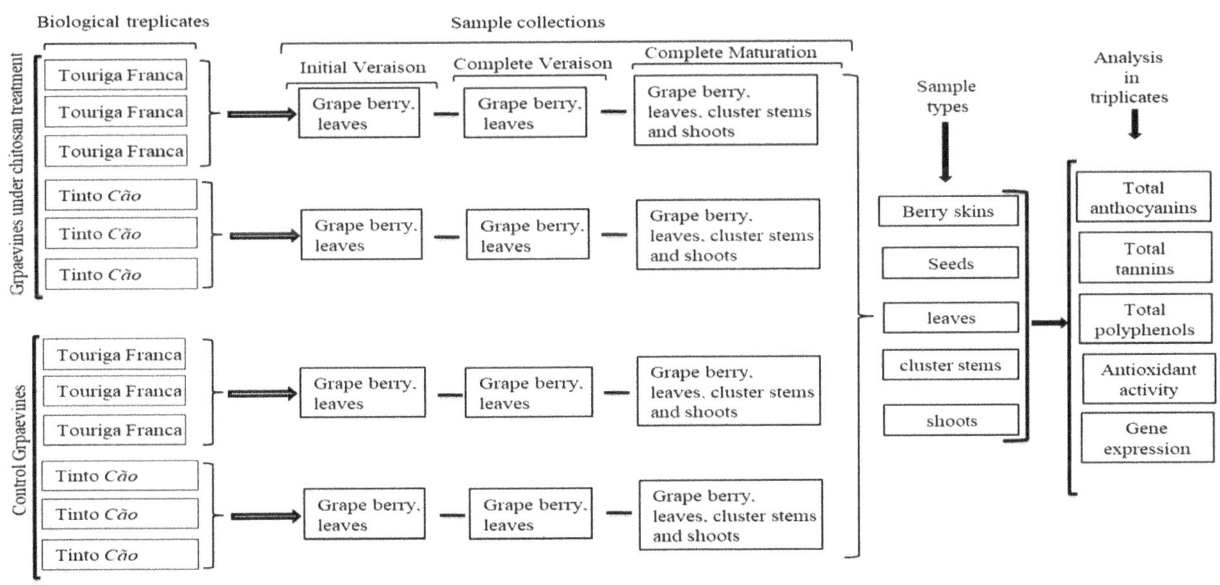

Figure 1. Diagrammatic representation of the complete experimental setup, including sample collection and sample analyses.

2.2. *Sample Preparation*

The leaves were removed from the vines using a scissor. The seeds were manually separated from the berries' skins. The stems and shoots were chopped in small pieces using a scissor. All processed samples were immediately frozen in liquid nitrogen and stored in –80 °C. Before the analyses, all the samples were freeze-dried and grounded into a fine powder (ca. 0.5 mm size) using a coffee grinder (Q.5321, Qilive, China).

Aqueous ethanol (1.5 mL, 50% *v*/*v*) was added to 100 mg of powdered sample and the contents were mixed in an agitator for 1 h at room temperature (ca. 25 °C). The mixture was then centrifuged at 10,000 rpm for 20 min at 4 °C. The supernatant was transferred to a tube and the residue was re-extracted as described above. Both the supernatants were combined, made up to 3 mL with 50% aqueous ethanol, and stored at −20 °C until further analyses.

2.3. *Determination of the Total Phenolic Content (TPC) and the Total Anthocyanin Content (TAC)*

The TPC and TAC were estimated using a protocol adapted from two previous publications [5,32]. An Aliquot (200 µL) of the supernatant extract obtained in 2.2 was acidified with 3.8 mL of 1.0 M aqueous HCl. The two solutions were mixed by gentle inversion, followed by incubation at 28 °C for 3 h. The absorbances at 520 nm and 280 nm were recorded using a spectrophotometer (Evolution 201, Thermo Scientific). A520 values were converted into malvidin-3-*O*-glucoside contents based on a generic extinction coefficient of malvidin-3-*O*-glucoside [5], and the TAC was expressed as µg of malvidin-3-*O*-glucoside equivalents per mg dry weight. The TPC was calculated and expressed as µg of epicatechin equivalents per mg dry weight, using a calibration curve for epicatechin.

2.4. *Determination of the Total Tannin Content (TTC)*

The TTC was determined using the methyl cellulose-mediated precipitation method [33,34]. A mixture was prepared by adding 600 µL of 0.04% methyl cellulose to 200 µL of the supernatant extract obtained in Section 2.2. After incubation for 3 min at room temperature, 400 µL of a saturated solution of ammonium sulphate and 0.8 mL of deionized water were added and the contents mixed by inversion. A control mixture was also prepared, but with deionized water (600 µL) instead of 0.04% methyl cellulose. After incubation for 10 min at room temperature, the mixtures were centrifuged for 5 min at 4,000 rpm (4 °C). The resulting supernatant was read at 280 nm. The TTC was obtained by subtracting the A280 nm extract and the A280 nm control. The tannin value was converted into epicatechin equivalents (µg per mg dry weight) with the aid of a calibration curve.

2.5. *Determination of the Antioxidant Activity Using the DPPH, CUPRAC, and FRAP Assays*

2.5.1. Sample Preparation

Powdered lyophilized samples (40 mg) were extracted with 1 mL of 70% methanol. After incubation at 70 °C for 30 min, the solution was cooled at room temperature and centrifuged at 11,000 rpm for 20 min in order to obtain the supernatant used for the analyses.

2.5.2. DPPH (2,2-diphenyl-1-picrylhydrazyl) Assay

The DPPH assay used in this study was a modification of previous methods [35–37]. A 4% DPPH solution was prepared in 95% ethanol. The supernatant obtained in 2.5.1 (15 µL) was added to the DPPH solution in a 96-wells microplate to a final volume of 300 µL. The decrease in absorbance at 517 nm (*ABS sample*) was measured with a microplate reader, against a blank prepared with 70% methanol instead of the supernatant (*ABS blank*). The DPPH antioxidant activity was expressed as the percentage of the remaining DPPH in the reaction medium (% AA) using Equation (1).

$$\%AA = \left(ABS\ blank - \left(\frac{ABS\ sample}{ABS\ blank}\right)\right) \times 100 \qquad (1)$$

2.5.3. FRAP (Ferric Reducing Antioxidant Power) Assay

The FRAP assay was modified from [38,39]. A portion of an aqueous 10 mM solution of 2,4,6-Tri(2-pyridyl)-s-triazine in 40 mM HCl was mixed with the same volume of 20 mM $FeCl_3.6H_2O$ and 10 times higher volume of acetate buffer of pH 3.6 (3.1 g sodium acetate and 16 mL acetic acid per L). The mixture was incubated at 37 °C for 20 min. A portion (900 μL) of the ferric tripyridyltriazine (Fe^{3+}-TPTZ) mixture and 20 μL of supernatant were diluted up to 1000 μL with deionized water and incubated for 10 min, and the absorbance was measured at 593 nm. A standard curve was prepared using fresh $FeSO_4$ solutions, and the ferric reducing power was expressed as mg $FeSO_4$ equivalents per mg dry weight.

2.5.4. CUPRAC (CUPric Reducing Antioxidant Capacity) Assay

The CUPRAC assay was carried out as described by [40]. Briefly, 0.2 mL of supernatant was added with 0.5 mL of 2×10^{-3} M Cu(II) chloride, 0.5 mL ethanol, 1 mL of 7.5×10^{-3} M neocuproine (Nc), 1 mL of 1 M ammonium acetate (NH_4/Ac) buffer pH 7.0, and 0.8 mL ethanol. After 30 min of incubation at room temperature, the absorbance at 450 nm was recorded against a water blank. The standard calibration curve of trolox was constructed as absorbance vs. concentration, and the antioxidant activity expressed as mg trolox equivalents per mg dry weight.

2.6. RNA Extraction, cDNA Synthesis, and qRT-PCR Analysis

Total RNA was extracted from frozen tissues harvested after complete maturation of the vines, using the CTAB (cetyl trimethyl ammonium bromide) buffer as described in [41]. Total RNA was extracted by using the RNeasy Mini Kit (Qiagen, Germany) and a DNase treatment was performed respectively. The integrity of the total RNA was verified by running samples on 1.0% denaturing agarose gels, followed by quantification with a ND-1000 Spectrophotometer (NanoDrop Technologies, Wilmington, DE) at wavelengths of 230 and 280 nm. A total of 1 μg of DNase-treated RNA was used for cDNA synthesis, using the First Strand cDNA Synthesis Kit (Solis Biodyne, Tartu, Estonia). Specific internal primers (Table 1) were designed using Primer3 online tool (http://primer3.ut.ee/). Quantitative real-time PCR (qRT-PCR) was performed to estimate the relative expression of 10 selected genes of the ROS pathway. Each reaction contained 5 × HOT FIREPol® EvaGreen® qPCR Mix Plus (Solis Biodyne, Tartu, Estonia), 200 nM of each primer, 1 μL 10 × diluted cDNA sample, and nuclease-free water to a final volume of 20 μl. Reactions were carried out on PicoReal-Time Thermal Cycler (PRO 96-1500-512, Thermo Scientific, Ratastie, Finland). The PCR program was: 95 °C for 12 min (initial denaturation), followed by 40 cycles of heating at 95 °C for 15 s (denaturation), 52 °C for 20 s (annealing), and 72 °C for 20 s (extension). No final extension was carried out. The software PicoReal 2.0 was used to analyze the data. The relative mRNA level for each gene was calculated as RQ values (relative quantification by using PicoReal 2.0 tool), using actin as the internal reference gene.

Table 1. Primer sequences for the genes of the reactive oxygen species pathway analyzed in this study.

	Gene	Primer Sequence
Fe-SOD	Superoxide dismutase	F 5′ CCTTTGTGAACCTAGGCGAACC 3′ R 5′ TGGCCGGGTTAGCTTGAACTC 3′
Cu/Zn-SOD	Superoxide dismutase	F 5′ AGATTGGCATGTGGTGTTGTTG 3′ R 5′ ACTCCCACATTACCCAACAACA 3′
CAT	Catalase	F 5′ GGTGTTCACACCTTCACTCT 3′ R 5′ GAGATCCTGAGTAGCATGACTG 3′
APX	Ascorbate peroxidase	F 5′ ATCTGGTGGTCATACTCTGG 3′ R 5′ TCTAGGAGAGCCTTGTCTGA 3′
GR	Glutathione reductase	F 5′ AATACTAGGGGGAGGGTACA 3′ R 5′ GTTGTCCTTGGATGCAGA 3′

Table 1. *Cont.*

	Gene	Primer Sequence
Grx	Glutaredoxin	F 5′ CTGCTCTCACAGCTAAAAGC 3′ R 5′ CATCACAGAGTCACATCCAC 3′
Rboh	Respiratory burst oxidase	F 5′ CTGATTCTCAGTAGGAACTGGT 3′ R 5′ GAGGTTTGGTCATGTATAGTGC 3′
AO	Amine oxidase	F 5′ GTGAATCCGAACAAGAGAAC 3′ R 5′ AGACTTATTGTAGGGTGTGACC 3′
POD	Peroxidase	F 5′ CCTGATCCAACACTAGATGC 3′ R 5′ GTCTGACTGAAGAAGTCCAGAG 3′
PPO	Polyphenol oxidase	F 5′ CTAGAACTCCAGGTTCATGC 3′ R 5′ GCTTCTCGTCGTAGAGTGAT 3′

2.7. Statistical Analyses

Statistical analyses were carried out using JMP 11.0 (SAS Institute Inc, Cary, NC). For each vine tissue, the effects of chitosan application on the TPC, TAC, and TTC during veraison, after veraison and at maturity were assessed using a two-way analysis of variance (ANOVA). The comparison among means was determined according to Turkey's test. When only two means were available, an independent *t*-test was used. Antioxidant activities (DPPH, FRAP, CUPRAC) and gene expression data for each tissue were subjected to an independent *t*-test analysis. All analyses were performed in triplicates (technical replicates) and results are expressed as mean ± standard deviation of data recorded from three independent samples (biological replicates). Differences between means were considered meaningful at $p < 0.05$ and represented by different letters for ANOVA and asterisks for the *t*-test.

3. Results

3.1. Effect of Chitosan Application on Polyphenols in Grapevine

Grape seeds, skins, and juice contain a substantial amount of polyphenolic compounds which explain the health benefits of grape-based nutraceutical products [23]. In this study, polyphenols accumulated more in the seeds than in the skins of 'Tinto Cão', and more in the skins than in the seeds of 'Touriga Franca' (Table 2). At the first and second harvesting times, the TPC increased in leaves, skins, and seeds collected from vines treated with chitosan, in comparison to samples from control vines: increases of 24.18%, 29.51%, and 8.96% in 'Touriga Franca' during veraison; 31.80%, 22.83%, and 13.31% in 'Touriga Franca' after veraison; 10.53%, 23.01%, and 3.40% in 'Tinto Cão' during veraison; 2.09%, 20.40%, and 18.17% in 'Tinto Cão' after veraison (Table 2). At maturity, the TPC was also higher in the skins of chitosan-treated vines than in those of control vines, with 8.96% and 25.71% increases in 'Touriga Franca' and 'Tinto Cão', respectively. An increase was also recorded in the leaves of 'Touriga Franca' (41.44%), with no change in the leaves of 'Tinto Cão'. At complete maturation, the TFC was higher in the stems from chitosan-treated vines than those of control vines, with increases of 28.73% and 31.75% for 'Touriga Franca' and 'Tinto Cão', respectively. The application of chitosan had no effect in the TPC in the seeds and shoots of the two varieties (Table 2).

Table 2. The total phenolic content.

Variety/Tissue	Harvesting Time/Treatment					
	During Veraison		After Veraison		At Complete Maturation	
	Control	Chitosan	Control	Chitosan	Control	Chitosan
'Touriga Franca'						
berry seeds	109.69 ± 5.04 b	119.27 ± 6.58 a	89.95 ± 5.13 c	101.92 ± 2.47 b	94.67 ± 2.26 c	93.65 ± 5.10 c
berry skins	105.54 ± 2.73 c	136.69 ± 5.40 a	110.61 ± 4.67 c	135.86 ± 5.37 a	112.70 ± 4.23 c	122.80 ± 1.92 ab
cluster stems	NA	NA	NA	NA	88.90 ± 5.55 b	114.44 ± 5.78 a
leaves	70.28 ± 6.27 c	87.27 ± 1.70 a	63.11 ± 0.56 d	83.18 ± 0.86 a	55.65 ± 1.09 e	78.71 ± 1.48 b
shoots	NA	NA	NA	NA	37.23 ± 2.36 a	34.09 ± 0.64 a
'Tinto Cão'						
berry seeds	192.64 ± 5.34 b	199.19 ± 2.64 a	154.15 ± 1.86 d	182.14 ± 5.54 c	146.32 ± 4.25 e	144.05 ± 1.23 e
berry skins	80.50 ± 3.61 f	99.03 ± 2.14 e	112.24 ± 5.07 d	135.13 ± 0.60 b	129.95 ± 4.17 c	163.36 ± 9.57 a
cluster stems	NA	NA	NA	NA	158.61 ± 5.28 b	208.97 ± 8.37 a
leaves	223.28 ± 10.10 b	246.79 ± 1.92 a	191.77 ± 3.86 d	195.77 ± 3.86 d	204.76 ± 8.89 c	201.16 ± 3.16 cd
shoots	NA	NA	NA	NA	49.59 ± 2.44 a	52.65 ± 1.51 a

(TPC; μg epicatechin equivalents per mg dry weight) in grapevine tissues after chitosan application before and after veraison. All data were obtained from three biological replicates; means ± standard deviations within a row followed by different letters (a–f) are statistically different ($p < 0.05$; ANOVA Turkey's test). NA = not analyzed.

3.2. Effect of Chitosan Application on Anthocyanins in Grapevine

Anthocyanins are a special class of polyphenols involved in resistance against diseases in the leaves and shoots, and responsible for the color of berries and wines [1]. Table 3 shows that anthocyanins accumulated mainly in the skins of the fruits, as compared to the seeds where they were usually not detected. After application of chitosan, the TAC increased in the skins of 'Touriga Franca' and 'Tinto Cão' during veraison by 35.27% and 56.02%, respectively and after veraison by 17.42% and 10.51%, respectively. At complete maturation, a 22.33% increase in TAC was also recorded in the skins of 'Tinto Cão', in comparison with the skins from control vines. On the contrary, the TAC decreased by 10.55% in the skins of 'Touriga Franca' at maturity (Table 3). Before maturation, the TAC also increased in the leaves of chitosan-treated vines, 28.21% in 'Touriga Franca' and 22.37% in 'Tinto Cão' during veraison, and 37.50% and 14.93% after veraison. At complete maturation, the TAC was also higher in 'Touriga Franca' leaves (35.90% increases) treated with chitosan, while no change was recorded for 'Tinto Cão' (Table 3). At complete maturation, more anthocyanins accumulated in the stems of chitosan-treated vines than in those of control vines, with 113.76% and 16.04% increases of TAC in 'Touriga Franca' and 'Tinto Cão', respectively. The TAC of the shoots was not affected by the application of chitosan (Table 3).

Table 3. The total anthocyanin content.

Variety/Tissue	Harvesting time/Treatment					
	During Veraison		After Veraison		At Complete Maturation	
	Control	Chitosan	Control	Chitosan	Control	Chitosan
'Touriga Franca'						
berry seeds	0.00 ± 0.00 a	0.00 ± 0.00 a	0.00 ± 0.00 a	0.00 ± 0.00 a	0.00 ± 0.00 a	0.00 ± 0.00 a
berry skins	16.52 ± 0.44 e	22.35 ± 0.91 cd	20.86 ± 0.87 d	24.49 ± 0.34 ab	26.02 ± 1.02 a	23.28 ± 0.55 bc
cluster stems	NA	NA	NA	NA	8.26 ± 0.02 b	17.66 ± 0.01 a
leaves	0.39 ± 0.02 b	0.50 ± 0.01 a	0.32 ± 0.02 c	0.44 ± 0.01 ab	0.39 ± 0.01 b	0.53 ± 0.01 a
shoots	NA	NA	NA	NA	1.55 ± 0.11 a	1.12 ± 0.02 a
'Tinto Cão'						
berry seeds	0.00 ± 0.00 a	0.00 ± 0.00 a	0.00 ± 0.00 a	0.00 ± 0.00 a	0.00 ± 0.00 a	0.00 ± 0.00 a
berry skins	5.32 ± 0.19 f	8.30 ± 0.40 e	21.16 ± 1.32 d	23.39 ± 0.25 c	28.74 ± 0.97 b	35.16 ± 1.16 a
cluster stems	NA	NA	NA	NA	2.25 ± 0.01 b	2.68 ± 0.02 a
leaves	0.76 ± 0.06 b	0.93 ± 0.04 a	0.67 ± 0.01 c	0.77 ± 0.03 b	0.78 ± 0.03 b	0.77 ± 0.03 b
shoots	NA	NA	NA	NA	0.14 ± 0.02 a	0.15 ± 0.01 a

(TAC; μg malvidin-3-*O*-glucoside equivalents per mg dry weight) in grapevine tissues after chitosan application before and after veraison. All data were obtained from three biological replicates; means ± standard deviations within a row followed by different letters (a–f) are statistically different ($p < 0.05$; ANOVA Turkey's test). NA = not analyzed.

3.3. *Effect of Chitosan Application on Tannins in Grapevine*

Tannins contribute to the astringency of grapes and, in combination with anthocyanins, define the stability of wine color [34]. In this study, tannins tended to accumulate more in the seeds than the skins, although this depended on the growth stage of the vines (Table 4). At all harvesting times, the TTC increased in the skins and seeds collected from vines treated with chitosan, as compared to control vines: 14.18% and 325.15% in 'Touriga Franca', 9.61% and 9.23% in 'Tinto Cão' during veraison; 37.48% and 159.02% in 'Touriga Franca', 52.62% and 21.32% in 'Tinto Cão' after veraison; 20.46% and 32.12% in 'Touriga Franca', 48.46% and 6.01% in 'Tinto Cão' at complete maturation. The effect of chitosan on the TTC in the leaves was varied among varieties. A decrease in the leaves of 'Touriga Franca' during (11.62%) and after (9.73%) veraison and an increase at complete maturation (90.33%); and the exact opposite for 'Tinto Cão' i.e., an increase during (36.14%) and after (23.71%) veraison, and a decrease after complete maturation (11.98%) (Table 4), was observed. A variety effect was also observed at maturity for the stems, with chitosan application leading to a lower TTC in 'Touriga Franca' (35.12 decreases) and a higher TTC in 'Tinto Cão' (37.01% increase), in comparison of the stems of control vines. Treatment of vines with chitosan led to an increased TTC in the shoots of both varieties, with 114.48% and 23.48% increases in 'Touriga Franca' and 'Tinto Cão', respectively (Table 4).

Table 4. The total tannin content.

Variety/Tissue	Harvesting time/Treatment					
	During Veraison		After Veraison		At Complete Maturation	
	Control	Chitosan	Control	Chitosan	Control	Chitosan
'Touriga Franca'						
berry seeds	33.19±4.57 e	141.11 ± 5.95 a	48.23 ± 5.28 d	124.93 ± 14.32 b	95.35 ± 5.32 c	125.98 ± 4.54 b
berry skins	74.24 ± 3.68 b	84.78 ± 4.78 a	59.72 ± 3.78 c	82.09 ± 6.41 a	47.41 ± 3.02 d	57.11 ± 0.76 c
cluster stems	NA	NA	NA	NA	31.63 ± 3.11 a	20.52 ± 2.28 b
leaves	41.75 ± 5.42 a	36.90 ± 1.11 b	35.97 ± 3.45 b	32.47 ± 2.33 c	17.06 ± 1.15 d	32.47 ± 3.37 c
shoots	NA	NA	NA	NA	17.26 ± 1.62 b	37.02 ± 0.66 a
'Tinto Cão'						
berry seeds	129.04 ± 5.34 b	140.95 ± 2.64 a	108.50 ± 1.86 c	131.63 ± 5.54 b	106.81 ± 1.23 c	113.64 ± 4.25 c
berry skins	43.69 ± 3.61 b	47.89 ± 2.14 a	26.47 ± 5.07 d	40.40 ± 0.60 b	32.03 ± 4.17 c	47.55 ± 9.57 a
cluster stems	NA	NA	NA	NA	102.66 ± 5.28 b	140.65 ± 8.37 a
leaves	64.78 ± 10.10 c	88.19 ± 1.92 a	62.67 ± 3.86 c	77.53 ± 3.86 b	89.43 ± 8.89 a	78.72 ± 3.16 b
shoots	NA	NA	NA	NA	30.75 ± 2.44 b	37.97 ± 1.51 a

(TTC; μg epicatechin equivalents per mg dry weight) in grapevine tissues after chitosan application before and after veraison. All data were obtained from three biological replicates; means ± standard deviations within a row followed by different letters (a–e) are statistically different ($p < 0.05$; ANOVA Turkey's test). NA = not analyzed.

3.4. *Effect of Chitosan Application on the Antioxidant Potential of Grapevine*

In this study, three assays were used for assessing the antioxidant potential of grapevine samples, namely DPPH, FRAP, and CUPRAC [35,38,40]. Data in Figure 2 shows that chitosan application did not affect the antioxidant potential of the seeds of the two varieties studied. Among skins, the antioxidant potential remained unchanged for 'Touriga Franca' (Figure 2A,C,E), while it increased for 'Tinto Cão': 11.54%, 27.06%, and 32.51% for the DPPH (Figure 2B), FRAP (Figure 2D) and CUPRAC (Figure 2F), respectively. With respect to the control, FRAP and CUPRAC increased in the stems of 'Touriga Franca' (5.30% and 6.10%, respectively; Figure 2C,E) and 'Tinto Cão' (25.74% and 31.53%, respectively; Figure 2D,F) following chitosan application, while no change was recorded for the DPPH (Figure 2A,B). Chitosan-treated leaves exhibited more antioxidant activities than control leaves: for 'Touriga Franca', 7.72%, 13.60%, 10.32% increases for the DPPH (Figure 2A), FRAP (Figure 2C) and CUPRAC (Figure 2E), respectively, were recorded. For 'Tinto Cão', 1.23% ($p > 0.05$), 9.70%, 8.70% increases for the DPPH (Figure 2B), FRAP (Figure 2D) and CUPRAC (Figure 2F), respectively, were recorded; however, these increases were non-significant. Chitosan application had no effect on the FRAP and the CUPRAC of the shoots of both varieties (Figure 2C–F). In terms of DPPH, a decreased activity (8.33%) was recorded in the shoots of chitosan-treated 'Touriga Franca' (Figure 2A), while an increased activity (11.80%) was recorded in the shoots of chitosan-treated 'Tinto Cão' (Figure 2B).

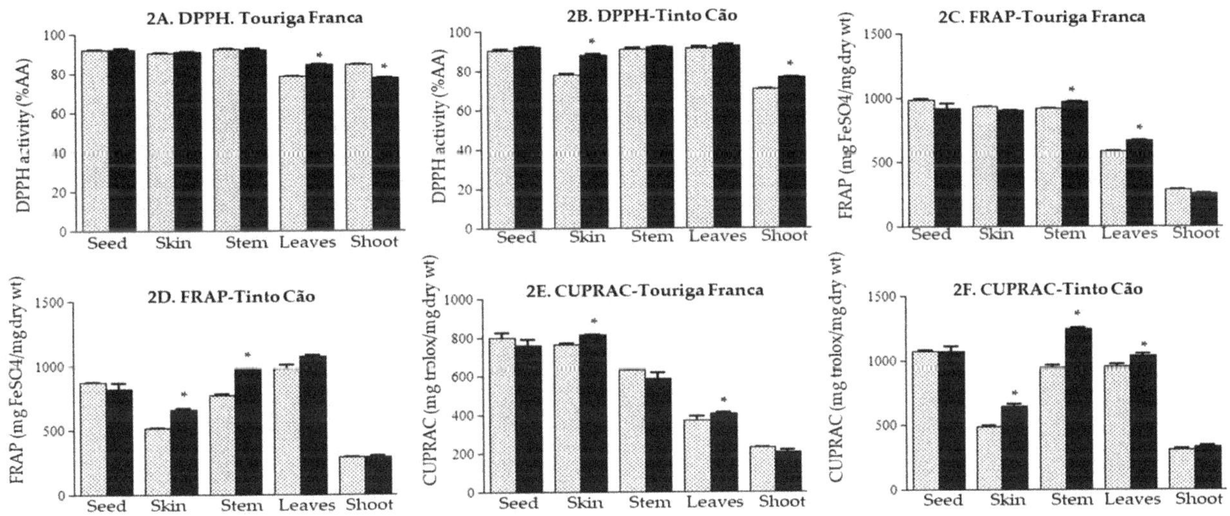

Figure 2. (**A**,**B**) 2,2-diphenyl-1-picrylhydrazyl activity (DPPH), (**C**,**D**) Ferric Reducing Antioxidant Power (FRAP), (**E**,**F**), Copper Reducing Antioxidant Capacity (CUPRAC) in the tissues of grapevine after application of chitosan before and after veraison. Tissues were collected at complete maturation of the berries. Gray column = control samples; black column = chitosan-treated samples. All data were obtained from three biological replicates; standard deviations and the levels of significance ($p < 0.05$; *t*-test) are represented by error bars and asterisks above the columns, respectively.

3.5. Effect of Chitosan Application of the Expression of Genes of the ROS Producing and Scavenging Pathway

A qRT-PCR was used for assessing the effect of chitosan on 10 genes of the ROS pathway in grapevine tissues collected at complete berry maturation.

With one exception (*Fe-SOD* remained unaffected in the shoots of 'Tinto Cão'), chitosan upregulated or tended to upregulate the expression of the genes *Cu/Zn-SOD* and *Fe-SOD* in all tissues of 'Touriga Franca' and 'Tinto Cão' (Figure 3). The most important changes were observed for *Fe-SOD* in the stems of 'Touriga Franca' (2.0-fold increase; Figure 3A); and for *Cu/Zn-SOD* in the leaves of 'Touriga Franca' (3.3-fold increase), leaves of 'Tinto Cão' (3.0-fold increase), stems of 'Touriga Franca' (7.4-fold increase), and stems of 'Tinto Cão' (7.1-fold increase) (Figure 3C,D).

CAT expression was also upregulated in all the tissues of grapevine after application of chitosan, with 1.3 to 7.6-fold increases recorded. The magnitude of the change was particularly important in the skins (3.2 and 4.3-fold for 'Touriga Franca' and 'Tinto Cão', respectively) and the shoots (7.6 and 5.0-fold for 'Touriga Franca' and 'Tinto Cão', respectively) (Figure 4A,B). A varietal effect was observed for *GR* expression, which following chitosan application, was upregulated in the tissues of 'Tinto Cão' (with the exception of a non-significant effect in the shoots), and remained unchanged in the tissues of 'Touriga Franca' (with the exception of a 1.3-fold increase in the skins) (Figure 4E,F). For both varieties, Grx expression was upregulated in the seeds, skins, and leaves, downregulated in the stems, and remained unaffected in the shoots. The highest changes were recorded in the berries, with 5.1-fold and 5.0-fold increases in the skins and seeds of 'Touriga Franca' (Figure 4G), and 3.0-fold and 2.7-fold increases in the skins and seeds of 'Tinto Cão' (Figure 4H). In contrast to *CAT*, *GR*, and *Grx*, the expression of *APX* was downregulated in all the tissues of both varieties upon chitosan treatment, although a non-significant effect was observed for the shoots (Figure 4C,D). The highest decreases were recorded in the skins and leaves (4.0–4.5-fold changes) and in the stems (5.2–8.1-fold changes).

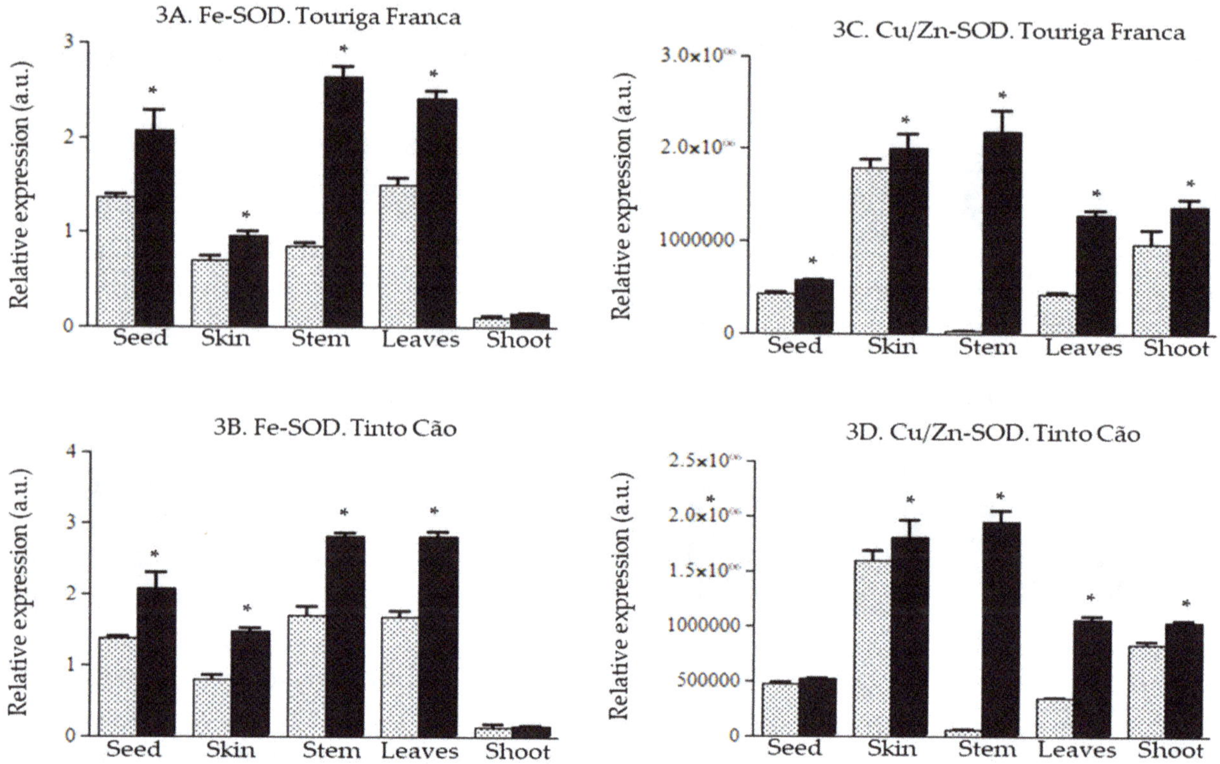

Figure 3. Effect of chitosan on *Fe-SOD* (Fe superoxide dismutase) and *Cu/Zn-SOD* (Cu/Zn superoxide dismutase) genes in the tissues of grapevine (**A**). relative expression of *Fe-SOD* gene in Touriga Franca, (**B**). relative expression of *Fe-SOD* gene in Tinto Cão, (**C**). relative expression of *Cu/Zn-SOD* gene in Touriga Franca, (**D**). relative expression of *Cu/Zn-SOD* gene in Tinto Cão. Tissues were collected at complete maturation of the berries. Gray column = control samples; black column = chitosan-treated samples. All data were obtained from three biological replicates; standard deviations and the levels of significance ($p < 0.05$; *t*-test) are represented by error bars and asterisks above the columns, respectively.

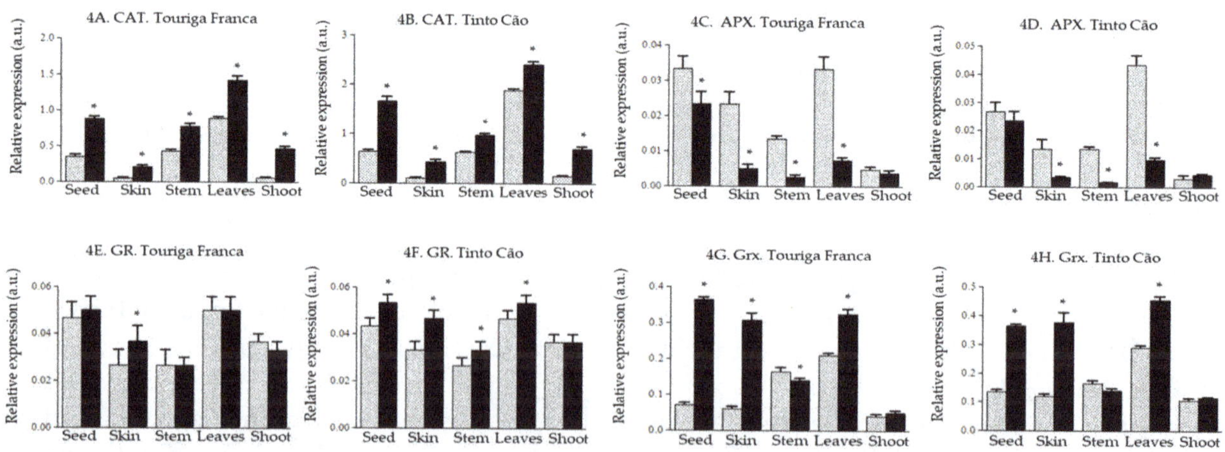

Figure 4. Effect of chitosan on *CAT* (catalase), *APX* (ascorbate peroxidase), *GR* (gluthione reductase), and *Grx* (glutaredoxin) genes in the tissues of grapevine (**A**). relative expression of *CAT* gene in Touriga Franca, (**B**). relative expression of *CAT* gene in Tinto Cão, (**C**). relative expression of *APX* gene in Touriga Franca, (**D**). relative expression of *APX* gene in Tinto Cão, (**E**). relative expression of *GR* gene in Touriga Franca, (**F**). relative expression of *GR* gene in Tinto Cão, (**G**). relative expression of *Grx* gene in Touriga Franca, (**H**). relative expression of *Grx* gene in Tinto Cão. Tissues were collected at complete maturation of the berries. Gray column = control samples; black column = chitosan-treated samples. All data were obtained from three biological replicates; standard deviations and the levels of significance ($p < 0.05$; *t*-test) are represented by error bars and asterisks above the columns, respectively.

Following chitosan application, there was an upregulation of the expression of *AO* (1.4-3.0-fold increases) and *POD* (1.4–31.5-fold increases) genes in all tissues of 'Touriga Franca' and 'Tinto Cão' (Figure 5C–F). In particular, the expression of *POD* was 31.5 and 11.0 times higher in the leaves of chitosan-treated 'Touriga Franca' and 'Tinto Cão', respectively, compared to that in the leaves of control vines (Figure 5E,F). The genes *Rboh* (Figure 5A,B) and *PPO* (Figure 5G,H) were also studied upon chitosan application and, with few exceptions of non-significance (*Rboh* in the stems and shoots of both varieties and POP in the stems and seeds of 'Tinto Cão'), all showed an upregulation in the tissues of 'Touriga Franca' and 'Tinto Cão' i.e., 1.3–1.4-fold increase for *Rboh* and 1.5–5.0-fold increase for *PPO*, with the highest changes noticed in the shoots.

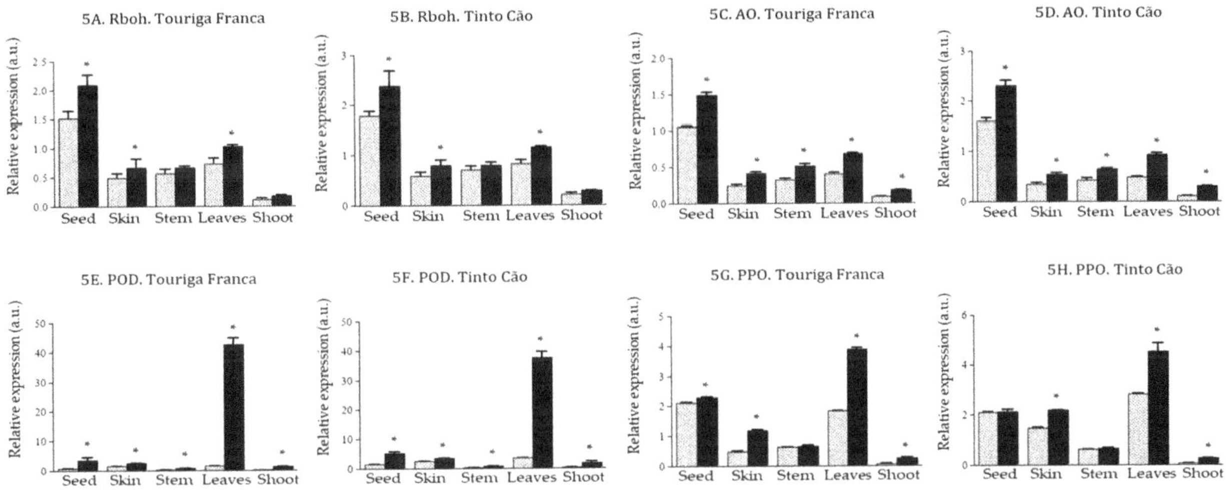

Figure 5. Effect of chitosan on *Rboh* (respiratory burst oxidase), *AO* (amine oxidase), *POD* (peroxidase), and *PPO* (polyphenol oxidase) genes in the tissues of grapevine (**A**). relative expression of Rboh gene in Touriga Franca, (**B**). relative expression of Rboh gene in Tinto Cão, (**C**). relative expression of *AO* gene in Touriga Franca, (**D**). relative expression of *AO* gene in Tinto Cão, (**E**). relative expression of *POD* gene in Touriga Franca, (**F**). relative expression of *POD* gene in Tinto Cão, (**G**). relative expression of *PPO* gene in Touriga Franca, (**H**). relative expression of *PPO* gene in Tinto Cão). Tissues were collected at complete maturation of the berries. Gray column = control samples; black column = chitosan-treated samples. All data were obtained from three biological replicates; standard deviations and the levels of significance ($p < 0.05$; *t*-test) are represented by error bars and asterisks above the columns, respectively.

4. Discussion

Improved fruit yield and quality is a high priority in vitiviniculture, and the use of elicitors holds the potential to help achieve this aim, while decreasing the reliance on pesticides and fertilizers. Previous experiments have instructively illustrated the feasibility of using elicitors to raise the levels of secondary metabolites in grapes. In the previous studies by Portu et al. solutions of methyl jasmonate (10 mM) and chitosan (0.03% in acetic acid 0.01 M) were sprayed over the leaves of *Vitis vinifera* L. ´Tempranillo´ (200–400 mL per vine) at veraison, and a week later, methyl jamsoate was applied which led to an increased accumulation of anthocyanins and stilbenes in the grape and wine, while chitosan treatment did not have a substantial impact on phenolic compounds [16,42]. There is, however, a report that the synthesis of anthocyanins and stilbenes is promoted by chitosan (50 μg/mL in 0.1% v/v acetic acid) in *Vitis vinifera* L.´Barbera´ grape cell suspensions [43]. For *Vitis vinifera* L.´Cabernet Sauvignon´, application of chitosan in three weekly intervals from fruit set to harvest had no effect on the TPC of the grapes, and caused a reduction in extraction of phenolics into the wine during vinification [44]. Thus, chitosan efficacy needs to be validated for each variety/clone in consideration with the dosage, timing, and mode of application. In this study, these variables were studied with a view to establishing the most appropriate conditions for ensuring high levels of phenolic compounds in all tissues. Thus, the whole vine was sprayed with a solution of 0.1% (w/v) chitosan in 0.01 M aqueous acetic acid before and after veraison.

The data in Tables 2 and 3 show that during veraison, after veraison, and at complete maturation, chitosan raised the TPC and the TAC of the leaves and berry's skins of ´Touriga Franca´ and ´Tinto Cão´, with one exception of slight decrease in TAC in the skins of ´Touriga Franca´ at the maturity. The TTC increased in the seeds of chitosan-treated ´Touriga Franca´ and ´Tinto Cão´ at all harvesting times, while an opposite response was observed for the leaves (Table 4), suggesting a varietal dependency. Chitosan is known to stimulate the activity of phenylalanine ammonia lyase, suggesting an induction of the phenylpropanoid pathway [45,46], which would explain the TPC and TAC in this study. The varietal difference observed for the TTC in the leaves might be explained by a preferential activation of the enzymes responsible for anthocyanin synthesis when chitosan is applied on ´Touriga Franca´ leaves, limiting the substrates for the production of tannins [47]. The positive effects of chitosan on the TPC, TAC, and TTC indicate that the treatment could also be regarded as a useful methodology in increasing resistance to some pathogens, as observed with grey mould [45] and powdery mildew infections [18].

Despite increased TPC, TAC and TTC in the berry's skins at maturity following chitosan application, the antioxidant potential measured by DPPH, CUPRAC and FRAP increased in ´Tinto Cão´, and remained unchanged in ´Touriga Franca´ (Figure 2), which again indicates a varietal dependency. The synthesis and accumulation of phenolic compounds in plants in response to the environmental conditions and cultural practices, are dependent on various factors including the cultivars [2,4,48,49]. This may contribute towards an increased or decreased antioxidant potential [50]. Indeed, chitosan and its formulations are reported to improve and/or reduce the antioxidant potential depending on the crop studied e.g., maize, soybean, jaborandi, Thymus and tobacco [51–54]. Overall, there was only a loose correlation between TPC/TAC/TTC and the DPPH. On the other hand, a good correlation was found between TPC/TAC/TTC and FRAP/CUPRAC in the stems and the leaves (data not shown).

ROS, such as $O_2{\cdot}^-$ (superoxide), are free radicals derived from oxygen that is continuously produced during normal aerobic respiration in plants or in response to injury and can be unfavorable to plant fitness [55–59]. Several endogenous antioxidants are produced by plants to slow or stop the production of ROS. SOD (including *Cu/Zn-SOD* and *Fe-SOD*) plays a major role in scavenging ROS in several plants [60–63], including grapevine [64,65]. The present study demonstrated an upregulation of the SOD gene upon chitosan application. Overall, there was an upregulation of genes of the ROS pathway (*Fe-SOD, Cu/Zn-SOD, CAT, GR, Rboh, AO, PPD,* and *PPO*) in all the tissues of both varieties, with the exception of *APX* in all tissues and *Grx* in the stems of both varieties (Figures 3–5). Notably, *POD* increased following application of chitosan, 51.5 and 11.0-fold in the leaves of ´Touriga Franca´ and ´Tinto Cão´, respectively (Figure 5E), suggesting a crucial role for *POD* in the ROS pathway. Recently, it was reported that maize plants are able to cope with drought, following application of chitosan because of the potentiation of the activities of the antioxidant enzymes SOD, *CAT, APX, GR* and guaiacol peroxidase [46]. This shows that chitosan may induce defense responses in grapevine by activating the ROS pathway, as suggested in some earlier studies [66–69]. Exogenous chitosan reportedly interacts with chitin synthase, and chitin deacetylase in plants, to produce chitosan oligomers that are responsible for signal perception in cells, leading to the activation of various genes, including genes of the ROS pathway [50,70,71]. Elicitor molecules have been used to stimulate a broad range of responses including defense mechanism [72] in field conditions as well as in vitro cultures for elicitation of secondary metabolites [73–76]. However, it was reported earlier that acetic acid alone itself may exert some effect [77]. This aspect will be included in the next study as an extra control sample.

A tissue-specific analysis of the response of grapevine to chitosan in this study shows that chitosan might also act by influencing the transport of secondary metabolites in the vine. At maturity, the TPC, and the TAC remained unchanged in the shoots (Tables 2 and 3), suggesting that shoots are not implicated in the transport of phenolic acids and anthocyanins to the berries. Instead, chitosan seems to stimulate the transfer of leaf anthocyanins to the berries via the stems. At maturity, that transfer seems complete in the case of ´Tinto Cão´, since the TAC and TPC increased in the stems and skins, but remained unchanged in the leaves; and incomplete in the case of ´Touriga Franca´ since a high TAC and TPC is still measured in the leaves. The response of the TTC in the leaves depended on the variety and the harvesting time (Table 4), suggesting that leaf tannins are not transported to the berries. Instead, shoot tannins seem to be transferred to the berries. Finally, the data in Tables 2–4 and Figure 2 shows that there is a prospect for adding values to grapevine stems and leaves via the application of chitosan. In general, the levels of phenolics and the antioxidant potential were constantly high in the stems and leaves of both varieties at all harvesting times.

5. Conclusions

Present study steps ahead towards a better understanding of chitosan–*Vitis vinifera* L. interaction and elicitation mechanism at the molecular level and suggests its application for value addition in viticulture practices.

Author Contributions: "Conceptualization, R.K.S. and V.F.; methodology and software, P.G.; validation, R.K.S., V.F., F.C. and I.C.; formal analysis, R.K.S. and B.S.; investigation, R.K.S.; resources, V.F. and R.K.S.; data curation, P.G. and R.K.S.; writing—original draft preparation, R.K.S. and P.G.; writing—review and editing, A.L.P.-S., A.A.O., A.I., P.G. and R.K.S.; supervision, V.F.; project administration, V.F.; funding acquisition."

Acknowledgments: The Centro de Química de Vila Real (CQ-VR) is gratefully acknowledged for providing research facilities and services to perform the analyses.

References

1. Saw, N.; Riedel, H.; Cai, Z.; Kütük, O.; Smetanska, I. Stimulation of anthocyanin synthesis in grape (*Vitis vinifera*) cell cultures by pulsed electric fields and ethephon. *Plant Cell Tiss. Organ. Cult.* **2012**, *108*, 47–54. [CrossRef]
2. Mattivi, F.; Zulian, C.; Nicolini, G.; Valenti, L. Wine, biodiversity, technology, and antioxidants. *Ann. N.Y. Acad. Sci.* **2002**, *957*, 37–56. [CrossRef]
3. Lingua, M.S.; Fabani, M.P.; Wunderlin, D.A.; Baroni, M.V. From grape to wine: Changes in phenolic composition and its influence on antioxidant activity. *Food Chem.* **2016**, *208*, 228–238. [CrossRef]
4. Mazza, G. Anthocyanins in grapes and grape products. *Crit. Rev. Food Sci. Nutr.* **1995**, *35*, 341–371. [CrossRef]
5. Tudo, J.A.; Buica, A.; Nieuwoudt, H.; Aleixandre, J.L.; du Toit, W. Spectrophotometric analysis of phenolic compounds in grapes and wines. *J. Agric. Food Chem.* **2017**, *65*, 4009–4026. [CrossRef] [PubMed]
6. Neves, G.G.; Gil, G.; Ferrer, M. Effect of different vineyard treatments on the phenolic contents in tannat (*Vitis vinifera* L.) grapes and their respective wines. *Food Sci. Technol. Int.* **2002**, *8*, 315–321. [CrossRef]
7. Fanzone, M.; Zamora, F.; Jofre, V.; Assof, M.; Neira, P.A. Phenolic composition of malbec grape skins and seeds from Valle de Uco (Mendoza, Argentina) during Ripening: Effect of cluster thinning. *J. Agric. Food Chem.* **2011**, *59*, 6120–6136. [CrossRef] [PubMed]
8. Soufleros, E.H.; Stavridou, K.; Dagkli, V. The effect of cluster thinning on phenolic maturity of *Vitis vinifera* cv. Xinomavro grapes. *Int. J. Vine Wine Sci.* **2011**, *45*, 171–179. [CrossRef]
9. Lamela, P.C.; Falcon, G.M.S.; Gándara, S.J.; Fernández, O.I. Influence of grape variety, vine system and enological treatments on the colour stability of young red wines. *Food Chem.* **2007**, *101*, 601–606. [CrossRef]
10. Koundouras, S.; Hatzidimitriou, E.; Karamolegkou, M.; Dimopoulou, E.; Kallithraka, S.; Tsialtas, J.T.; Zioziou, E.; Nikolaou, N.; Kotseridis, Y. Irrigation and rootstock effects on the phenolic concentration and aroma potential of *Vitis vinifera* L. cv. Cabernet Sauvignon grapes. *J. Agric. Food Chem.* **2009**, *57*, 7805–7813. [CrossRef]
11. Romero, P.; Fernandez, F.J.I.; Cutillas, M.A. Physiological thresholds for efficient regulated deficit-irrigation management in wine grapes grown under semiarid conditions. *Am. J. Enol. Vitic.* **2010**, *61*, 300–312.

12. Basile, B.; Marsal, J.; Mata, M.; Vallverdú, X.; Bellvert, J.; Girona, J. Phenological sensitivity of Cabernet Sauvignon to water stress: Vine physiology and berry composition. *Am. J. Enol. Vitic.* **2011**, *62*, 452–461. [CrossRef]
13. Karla, R.E.; Limon, H.V.; Hidalgo, D.; Moyano, E.; Golenioswki, M.; Cusido, R.M.; Palazon, J. Elicitation, an effective strategy for the biotechnological production of bioactive high added value compounds in plant cell factories. *Molecules* **2016**, *21*, 182.
14. Arenzana, G.L.; Portu, J.; López, R.; Garijo, P.; Cerdán, G.T.; Alfaro, L.I. Phenylalanine and urea foliar application: Effect on grape and must microbiota. *Int. J. Food Microbiol.* **2017**, *245*, 88–97. [CrossRef] [PubMed]
15. Vitalini, S.; Ruggiero, A.; Rapparini, F.; Neri, L.; Tonni, M.; Iriti, M. The application of chitosan and benzothiadiazole in vineyard (*Vitis vinifera* L. cv Groppello Gentile) changes the aromatic profile and sensory attributes of wine. *Food Chem.* **2014**, *162*, 192–205. [CrossRef] [PubMed]
16. Portu, J.; López, R.; Baroja, E.; Santamaría, P.; Cerdán, G.T. Improvement of grape and wine phenolic content by foliar application to grapevine of three different elicitors: Methyl jasmonate, chitosan, and yeast extract. *Food Chem.* **2016**, *15*, 213–221. [CrossRef]
17. Gonzalez, L.V.; Neira, P.A.; Ibanez, F.; Pastenes, C. Long term effect of abscisic acid (ABA) on the grape berry phenylpropanoids pathway: Gene expression and metabolic content. *Plant Physiol. Biochem.* **2016**, *105*, 213–223. [CrossRef]
18. Iriti, M.; Vitalini, S.; Tommaso, G.D.; Amico, S.D.; Borgo, M.; Faoro, F. New chitosan formulation prevents grapevine powdery mildew infection and improves polyphenol content and free radical scavenging activity of grape and wine. *Aust. J. Grape Wine Res.* **2011**, *17*, 263–269. [CrossRef]
19. Simmons, W.M.; Hadwiger, L.; Ryan, C.A. Chitosans and pectic polysaccharides both induce the accumulation of the antifungal phytoalexin pisatin in pea pods and antinutrient proteinase inhibitors in tomato leaves. *Biochem. Biophys. Res. Commun.* **1983**, *110*, 194–199. [CrossRef]
20. Sathiyabama, M.; Akila, G.; Einstein, C.R. Chitosan-induced defence responses in tomato plants against early blight disease caused by *Alternaria solani* (Ellis and Martin) Sorauer. *Arch. Phytopathol. Plant Protect.* **2014**, *47*, 1777–1787. [CrossRef]
21. Sathiyabama, M.; Charles, R.E. Fungal cell wall polymer based nanoparticles in protection of tomato plants from wilt disease caused by *Fusarium oxysporumf*.sp. lycopersici. *Carbohydr. Polym.* **2015**, *133*, 400–407. [CrossRef] [PubMed]
22. Barka, E.A.; Eullaffroy, P.; Clement, C.; Vernet, G. Chitosan improves development, and protects *Vitis vinifera* L. against *Botrytis cinerea*. *Plant Cell Rep.* **2004**, *22*, 608–614. [CrossRef] [PubMed]
23. Malerba, M.; Cerana, R. Chitosan effects on plant systems. *Int. J. Mol. Sci.* **2016**, *17*, 996. [CrossRef] [PubMed]
24. Muxika, A.; Etxabide, A.; Uranga, J.; Guerrero, P.; Caba, K. Chitosan as a bioactive polymer: Processing, properties and applications. *Int. J. Biol. Macromol.* **2017**, *105*, 1358–1368. [CrossRef] [PubMed]
25. Deepmala, K.; Hemantranjan, A.; Singh, B.; Bhanu, A.N. A future perspective in crop protection: Chitosan and its oligosaccharides. *Adv. Plants Agric. Res.* **2014**, *1*, 23–30.
26. Li, Z.; Zhang, Y.; Zhang, X.; Merewitz, E.; Peng, Y.; Ma, X.; Huang, L.; Yan, Y. Metabolic pathways regulated chitosan contributing to drought resistance in white clover. *J. Proteome Res.* **2017**, *16*, 3039–3052. [CrossRef]
27. Hidangmayum, A.; Dwivedi, P.; Katiyar, D.; Hemantranjan, A. Application of chitosan on plant responses with special reference to abiotic stress. *Physiol. Mol. Biol. Plants* **2018**, *25*, 313–326. [CrossRef]
28. Sharif, R.; Mujtaba, M.; Rahman, M.; Shalmani, A.; Ahmad, H.; Anwar, T.; Tianchan, D.; Wang, X. The Multifunctional Role of Chitosan in Horticultural Crops; A Review. *Molecules* **2018**, *23*, 872. [CrossRef]
29. Miniawy, S.M.; Ragab, M.; Youssef, S.; Metwally, A. Response of strawberry plants to foliar spraying of chitosan. *Res. J. Agric. Biol. Sci.* **2013**, *9*, 366–372.
30. Kaya, M.; Cesoniene, L.; Daubaras, R.; Leskauskaite, D.; Zabulione, D. Chitosan coating of red kiwifruit (*Actinidia melanandra*) for extending of the shelf life. *Int. J. Biol. Macromol.* **2016**, *85*, 355–360. [CrossRef]
31. Meng, X.; Tian, S. Effects of preharvest application of antagonistic yeast combined with chitosan on decay and quality of harvested table grape fruit. *J. Sci. Food Agric.* **2009**, *89*, 1838–1842. [CrossRef]
32. Cynkar, W.U.; Cozzolino, D.; Dambergs, R.G.; Janik, L.; Gishen, M. The effects of homogenisation method and freezing on the determination of quality parameters in red grape berries of *Vitis vinifera*. *Aust. J. Grape Wine Res.* **2004**, *10*, 236–242. [CrossRef]

33. Mercurio, M.D.; Dambergs, R.G.; Herderich, M.J.; Smith, P.A. High throughput analysis of red wine and grape phenolics—adaptation and validation of methyl cellulose precipitable tannin assay and modified Somers color assay to a rapid 96 well plate format. *J. Agric. Food Chem.* **2007**, *55*, 4651–4657. [CrossRef]
34. Sarneckis, C.J.; Dambergs, R.G.; Jones, P.; Mercurio, M.; Herderich, M.J.; Smith, P.A. Quantification of condensed tannins by precipitation with methyl cellulose: Development and validation of an optimised tool for grape and wine analysis. *Aust. J. Grape Wine Res.* **2006**, *12*, 39–49. [CrossRef]
35. Brand-Williams, W.; Cuvelier, M.E.; Berset, C. Use of a free radical method to evaluate antioxidant activity. *LWT Food Sci. Technol.* **1995**, *28*, 25–30. [CrossRef]
36. Moreno, C.S.; Larrauri, J.A.; Calixto, F.S. Procedure to measure the antiradical efficiency of polyphenols. *J. Sci. Food Agric.* **1998**, *270*, 270–276. [CrossRef]
37. Siddhuraju, P.; Becker, K. Antioxidant properties of various solvent extracts of total phenolic constituents from three different agroclimatic origins of drumstick tree (*Moringa oleifera* Lam.) leaves. *J. Agric. Food Chem.* **2003**, *9*, 2144–2155. [CrossRef]
38. Benzie, I.F.F.; Strain, J.J. The ferric reducing ability of plasma (FRAP) as a measure of "antioxidant power": The FRAP assay. *Anal. Biochem.* **1996**, *76*, 70–76. [CrossRef]
39. Stratil, P.; Klejdus, B.; Kuban, V. Determination of total content of phenolic compounds and their antioxidant activity in vegetables: Evaluation of spectrophotometric methods. *J. Agric. Food Chem.* **2006**, *8*, 607–616. [CrossRef]
40. Apak, R.; Guçlu, K.; Ozyurek, M.; Karademir, S.E. Novel total antioxidant capacity index for dietary polyphenols and Vitamins C and E, using their cupric ion reducing capability in the presence of neocuproine: CUPRAC Method. *J. Agric. Food Chem.* **2004**, *29*, 7970–7981. [CrossRef]
41. Gasic, K.; Hernandez, A.; Korban, S.S. RNA extraction from different apple tissues rich in polyphenols and polysaccharides for cDNA library construction. *Plant Mol. Biol. Rep.* **2004**, *22*, 437–438. [CrossRef]
42. Portu, J.; Santamaría, P.; Alfaro, L.I.; López, R.; Cerdán, G.T. Methyl jasmonate foliar application to Tempranillo vineyard improved grape and wine phenolic content. *J. Agric. Food Chem.* **2015**, *63*, 2328–2337. [CrossRef] [PubMed]
43. Ferri, M.; Tassoni, A.; Franceschetti, M.; Righetti, L.; Naldrett, M.J.; Bagni, N. Chitosan treatment induces changes of protein expression profile and stilbene distribution in *Vitis vinifera* cell suspensions. *Proteomics* **2009**, *9*, 610–624. [CrossRef]
44. Duxbury, M.; Hotter, G.; Reglinski, T.; Sharpe, N. Effect of chitosan and 5-chlorosalicylic acid on total phenolic content of grapes and wine. *Am. J. Enol. Viticult.* **2004**, *55*, 191–194.
45. Reglinski, T.; Elmer, G.; Taylor, J.T.; Wood, N.; Hoyte, S.M. Inhibition of *Botrytis cinerea* growth and suppression of botrytis bunch rot in grapes using chitosan. *Plant Pathol.* **2010**, *59*, 882–890. [CrossRef]
46. Rabêlo, V.M.; Magalhães, P.C.; Bressanin, L.A.; Carvalho, D.T.; Reis, C.O.; Karam, D.; Doriguetto, A.C.; Santos, M.H.; Filho, P.P.S.S.; Souza, T.C. The foliar application of a mixture of semisynthetic chitosan derivatives induces tolerance to water deficit in maize, improving the antioxidant system and increasing photosynthesis and grain yield. *Sci. Rep.* **2019**, *9*, 8164. [CrossRef]
47. García, Y.R.; Cascales, I.R.; Ortín, A.B.B.; Muñoz, R.G.; Cutillas, A.M.; Plaza, E.G. Increasing bioactive phenolic compounds in grapes: Response of six monastrell grape clones to benzothiadiazole and methyl jasmonate treatments. *Am. J. Enol. Vitic.* **2013**, *64*, 459–465. [CrossRef]
48. Espada, M.A.C.; Wood, K.V.; Bordelon, B.; Watkins, B.A. Anthocyanin quantification and radical scavenging capacity of Concord, Norton, and Marechal Foch grapes and wines. *J. Agric. Food Chem.* **2004**, *52*, 6779–6786. [CrossRef]
49. Yang, J.; Martinson, T.E.; Liu, R.H. Phytochemical profiles and antioxidant activities of wine grape. *Food Chem.* **2009**, *116*, 332–339. [CrossRef]
50. Kapłan, M.; Najda, A.; Klimek, K.; Borowy, A. Effect of gibberellic acid (GA_3) inflorescence application on content of bioactive compounds and antioxidant potential of grape (*Vitis* L.) 'Einset Seedless' berries. *S. Afr. J. Enol. Vitic.* **2019**, *40*, 1–10. [CrossRef]
51. Khan, W.M.; Prithiviraj, B.; Smith, D.L. Effect of foliar application of chitin and chitosan oligosaccharides on photosynthesis of maize and soybean. *Photosynthetica* **2002**, *40*, 621–624. [CrossRef]
52. Zhang, H.; Zhao, X.; Yang, J.; Yin, H.; Wang, W.; Lu, H.; Du, Y. Nitric oxide production and its functional link with OIPK in tobacco defense response elicited by chitooligosaccharide. *Plant Cell Rep.* **2011**, *30*, 1153–1162. [CrossRef] [PubMed]

53. Dousseau, S.; Rodrigues, A.C.; Sousa Lira, J.M.; Ribeiro Junior, P.M.; Pacheco, F.V.; de Alvarenga, A.A.; Vilela Resende, M.L.; Filha Ferreira de Paula, A.C.C. Exogenous chitosan application on antioxidant systems of Jaborandi. *Cienc. Rural* **2016**, *46*, 191–197. [CrossRef]
54. Bistgani, Z.E.; Siadat, S.A.; Bakhshandeh, A.; Pirbalouti, A.G.; Hashemi, M. Interactive effects of drought stress and chitosan application on physiological characteristics and essential oil yield of *Thymus daenensis* Celak. *Crop J.* **2017**, *5*, 407–415. [CrossRef]
55. Laxa, M.; Liebthal, M.; Telman, W.; Chibani, K.; Dietz, K.J. The role of the plant antioxidant system in drought tolerance. *Antioxidants* **2019**, *8*, 94. [CrossRef]
56. Avramova, V.; AbdElgawad, H.; Vasileva, I.; Petrova, A.S.; Holek, A.; Mariën, J.; Asard, H.; Beemster, G.T. High antioxidant activity facilitates maintenance of cell division in leaves of drought tolerant maize hybrids. *Front. Plant Sci.* **2017**, *8*, 84. [CrossRef]
57. Ahmad, N.; Malagoli, M.; Wirtz, M.; Hell, R. Drought stress in maize causes differential acclimation responses of glutathione and sulfur metabolism in leaves and roots. *BMC Plant Biol.* **2016**, *16*, 247. [CrossRef]
58. Anjum, S.A.; Tanveer, M.; Ashraf, U.; Hussain, S.; Shahzad, B.; Khan, I.; Wang, L. Effect of progressive drought stress on growth, leaf gas exchange, and antioxidant production in two maize cultivars. *Environ. Sci. Pollut. Res.* **2016**, *23*, 17132–17141. [CrossRef]
59. Goufo, P.; Trindade, H. Factors influencing antioxidant compounds in rice. *Cri. Rev. Food Sci. Nutr.* **2017**, *57*, 893–922. [CrossRef]
60. Tepperman, J.M.; Dunsmuir, P. Transformed plants with elevated levels of chloroplastic SOD are not more resistant to superoxide toxicity. *Plant Mol. Biol.* **1990**, *14*, 501–511. [CrossRef]
61. Alscher, R.G.; Erturk, N.; Heath, L.S. Role of superoxide dismutases (SODs) in controlling oxidative stress in plants. *J. Exp. Bot.* **2002**, *53*, 1331–1341. [CrossRef] [PubMed]
62. Feng, K.; Yu, J.; Cheng, Y.; Ruan, M.; Wang, R.; Ye, Q.; Zhou, G.; Li, Z.; Yao, Z.; Yang, Y.; et al. The SOD gene family in tomato: Identification, phylogenetic relationships, and expression patterns. *Front. Plant Sci.* **2016**, *7*, 1279. [CrossRef] [PubMed]
63. Puig, S.; Thiele, D.J. Molecular mechanisms of copper uptake and distribution. *Curr. Opin. Plant Biol.* **2002**, *6*, 171–180. [CrossRef]
64. Hu, X.; Hao, C.; Cheng, Z.M.; Zhong, Y. Genome-wide identification, characterization, and expression analysis of the grapevine superoxide dismutase (SOD) family. *Int. J. Genomics* **2019**, 7350414. [CrossRef]
65. Leng, X.; Jia, H.; Sun, X.; Shangguan, L.; Mu, Q.; Wang, B.; Fang, J. Comparative transcriptome analysis of grapevine in response to copper stress. *Sci. Rep.* **2015**, *17*, 17749. [CrossRef]
66. Chamnanmanoontham, N.; Pongprayoon, W.; Pichayangkura, R.; Roytrakul, S.; Chadchawan, S. Chitosan enhances rice seedling growth via gene expression network between nucleus and chloroplast. *Plant Growth Regul.* **2015**, *75*, 101–114. [CrossRef]
67. Choudhary, R.C.; Kumarswamy, R.V.; Kumari, S.; Sharma, S.S.; Pal, A.; Raliya, R.; Biswas, P.; Saharan, V. Cu-chitosan nanoparticle boost defense responses and plant growth in maize (*Zea mays* L.). *Sci. Rep.* **2017**, *7*, 9754. [CrossRef]
68. Hadwiger, L.A. Anatomy of a nonhost disease resistance response of pea to *Fusarium solani*: PR gene elicitation via DNase, chitosanand chromatin alterations. *Front. Plant Sci.* **2015**, *6*, 373. [CrossRef]
69. Zeng, K.; Deng, Y.; Ming, J.; Deng, L. Induction of disease resistance and ROS metabolism in navel oranges by chitosan. *Sci. Hortic.* **2010**, *126*, 223–228. [CrossRef]
70. Hadwiger, L.A. Multiple effects of chitosan on plant systems: Solid science or hype. *Plant Sci.* **2013**, *208*, 42–49. [CrossRef]
71. Pichyangkura, R.; Chadchawan, S. Biostimulant activity of chitosan in horticulture. *Sci. Hortic.* **2015**, *196*, 49–65. [CrossRef]
72. Rudolf, J.R.; Resurreccion, A.V.A. Elicitation of Resveratrol in Peanut Kernels by Application of Abiotic Stresses. *J. Agric. Food Chem.* **2005**, *53*, 10186–10192. [CrossRef] [PubMed]
73. Xu, A.; Zhan, J.-C.; Huang, W.-D. Combined elicitation of chitosan and ultraviolet C enhanced stilbene production and expression of chitinase and β-1, 3-glucanase in Vitis vinifera cell suspension cultures. *Plant Cell Tissue Organ Cult.* **2016**, *124*, 105–117. [CrossRef]
74. Brasili, E.; Miccheli, A.; Marini, F.; Praticò, G.; Sciubba, F.; Di Cocco, M.E.; Cechinel, V.F.; Tocci, N.; Valletta, A.; Pasqua, G. Metabolic Profile and Root Development of Hypericum perforatum L. In vitro Roots under Stress Conditions Due to Chitosan Treatment and Culture Time. *Front. Plant Sci.* **2016** *7*, 507. [CrossRef]

75. Badiali, C.; De Angelis, G.; Simonetti, G.; Elisa, B.; Tobaruela, E.C.; Purgatto, E.; Yin, H.; Valletta, A.; Pasqua, G. Chitosan oligosaccharides affect xanthone and VOC biosynthesis in Hypericum perforatum root cultures and enhance the antifungal activity of root extracts. *Plant Cell Rep.* **2018**, *37*, 1471–1484. [CrossRef] [PubMed]
76. Jiao, J.; Qing-Yan, G.; Xin, W.; Qi-Ping, Q.; Zi-Ying, W.; Jing, L.; Fu, Y.-J. Chitosan elicitation of Isatis tinctoria L. hairy root cultures for enhancing flavonoid productivity and gene expression and related antioxidant activity. *Ind. Crop. Prod.* **2018**, *124*, 28–35. [CrossRef]
77. Valletta, A.; De Angelis, G.; Badiali, C.; Brasili, E.; Miccheli, A.; Di Cocco, M.E.; Pasqua, G. Acetic acid acts as an elicitor exerting a chitosan-like effect on xanthone biosynthesis in *Hypericum perforatum* L. root cultures. *Plant Cell Rep.* **2016**, *35*, 1009–1020. [CrossRef]

5

Molecular Mechanism of Oxidation of P700 and Suppression of ROS Production in Photosystem I in Response to Electron-Sink Limitations in C3 Plants

Chikahiro Miyake [1,2]

[1] Department of Applied Biological Science, Faculty of Agriculture, Graduate School for Agricultural Science, Kobe University, 1-1, Rokkodai, Nada, Kobe 657-8501, Japan; cmiyake@hawk.kobe-u.ac.jp

[2] Core Research for Environmental Science and Technology (CREST), Japan Science and Technology Agency, 7 Goban, Chiyoda, Tokyo 102-0076, Japan

Abstract: Photosynthesis fixes CO_2 and converts it to sugar, using chemical-energy compounds of both NADPH and ATP, which are produced in the photosynthetic electron transport system. The photosynthetic electron transport system absorbs photon energy to drive electron flow from Photosystem II (PSII) to Photosystem I (PSI). That is, both PSII and PSI are full of electrons. O_2 is easily reduced to a superoxide radical (O_2^-) at the reducing side, i.e., the acceptor side, of PSI, which is the main production site of reactive oxygen species (ROS) in photosynthetic organisms. ROS-dependent inactivation of PSI in vivo has been reported, where the electrons are accumulated at the acceptor side of PSI by artificial treatments: exposure to low temperature and repetitive short-pulse (rSP) illumination treatment, and the accumulated electrons flow to O_2, producing ROS. Recently, my group found that the redox state of the reaction center of chlorophyll P700 in PSI regulates the production of ROS: P700 oxidation suppresses the production of O_2^- and prevents PSI inactivation. This is why P700 in PSI is oxidized upon the exposure of photosynthesis organisms to higher light intensity and/or low CO_2 conditions, where photosynthesis efficiency decreases. In this study, I introduce a new molecular mechanism for the oxidation of P700 in PSI and suppression of ROS production from the robust relationship between the light and dark reactions of photosynthesis. The accumulated protons in the lumenal space of the thylakoid membrane and the accumulated electrons in the plastoquinone (PQ) pool drive the rate-determining step of the P700 photo-oxidation reduction cycle in PSI from the photo-excited P700 oxidation to the reduction of the oxidized P700, thereby enhancing P700 oxidation.

Keywords: P700; P700 oxidation system; photorespiration; photosynthesis; Photosystem I (PSI); reactive oxygen species (ROS); reduction-induced suppression of electron flow (RISE); repetitive short-pulse (SP) illumination (rSP illumination treatment)

---

1. Introduction

Numerous researchers have shown that, in photosynthetic organisms, oxidative damage due to enhanced production of reactive oxygen species (ROS) occurs when environmental stress (e.g., extreme low/high temperatures, high salinity, and oligotrophic inorganic components) decreases the photosynthetic efficiency [1–21]. Much of this research has focused on the Mehler reaction, whereby a superoxide radical (O_2^-), one of the most important ROS species, is formed via a one-electron reduction of O_2 in Photosystem I (PSI) of the chloroplast thylakoid membranes. In addition to PSI, ROS production in Photosystem II (PSII) has long been studied, and numerous studies have been published regarding its relationships to oxidative damage in PSII across oxygenic photosynthetic organisms [22–29]. This review focuses on our recent findings regarding the molecular mechanisms of the production and suppression of ROS in PSI.

We try to answer the following questions in this review: "Is the excess accumulation of electrons in PSI truly harmful to photosynthetic organisms?" and "Does ROS production occur in vivo?" Driever and Baker [30,31] and Ruuska et al. [30,31] reported that the in vivo activity of the Mehler reaction is too small compared to the electron flux in photosynthesis. Additionally, considering the photosynthetic capacity of many photosynthetic organisms, the photosynthetic rate is saturated at a light intensity level corresponding to approximately 25% of natural sunlight or lower [2,3,31–37]. This indicates that photosynthesis regularly proceeds under conditions of an excess photon supply, which can cause the photosynthetic electron transport system to be full of electrons. To address this issue, we asked "How do photosynthetic organisms manage to escape from the accumulation of electrons in PSI?" We then developed a method for suppressing photosynthesis and imitating the accumulation of electrons in PSI. Using this approach, we succeeded in showing how ROS are produced and suppressed in PSI [38]. In what follows, the robust and universal characteristics of the suppression of ROS production in photosynthetic organisms are discussed.

2. The Accumulation of Electrons in the PSI Acceptor Side Promotes ROS Production and Oxidative Damage

In the photosynthetic electron transport system, the reaction center chlorophyll P700 in PSI functions in the photo-oxidation reduction cycle (Figure 1). Ground state P700 is photoexcited to produce the excited P700 (P700*), which then donates an electron to the primary electron carrier (A_0) to produce the oxidized P700 ($P700^+$). Subsequently, $P700^+$ accepts an electron from the reduced plastocyanin (PC) to regenerate P700. The oxidized PC in turn is reduced by the electrons coming from PSII. The electron in the reduced A_0 flows to ferredoxin (Fd) through the electron carriers: A_1, F_X, and F_A/F_B [39]. If a light pulse (light intensity: 20,000 μmol photons m^{-2} s^{-1}; pulse duration: 400 ms) is applied to C3-sunflower intact leaves in the dark under atmospheric conditions, all P700 molecules are oxidized to $P700^+$ (Figure 2A). $P700^+$ then rapidly decreases during the light pulse, which is due to the occupation of P700* in the photo-oxidation reduction cycle of P700, indicating the accumulation of electrons at the acceptor side of PSI. If the light pulse is applied every 10 s on C3-sunflower intact leaves (i.e., repetitive short-pulse (rSP) illumination treatment), PSI is deactivated (Figure 3) [38].

Conversely, PSII is not deactivated at all. PSI deactivation is all suppressed when atmospheric O_2 partial pressure drops to 2 kPa. The atmospheric rSP illumination treatment then produces a superoxide radical (O_2^-) i n P S I [32,40–43]. These results suggest that the electrons accumulated on the PSI acceptor side during one light pulse are used for O_2 reduction, which triggers the O_2^- production reflected as P700* accumulation. O_2^- accumulation occurring over time with rSP illumination treatment indicates progressive PSI oxidative damage to the photosynthetic organism. PSI inactivation due to rSP illumination treatment represents a case of photosynthetically induced CO_2 fixation inactivation [9,35,38,44]. Unlike PSII, PSI requires approximately one week for functional recovery following inactivation [7,9,35,36,45,46]. Therefore, PSI inactivation can be fatal to plant growth [7,35,36,42]. Exposure of plants to low temperatures also inactivates PSI and the recovery of PSI function can take several weeks.

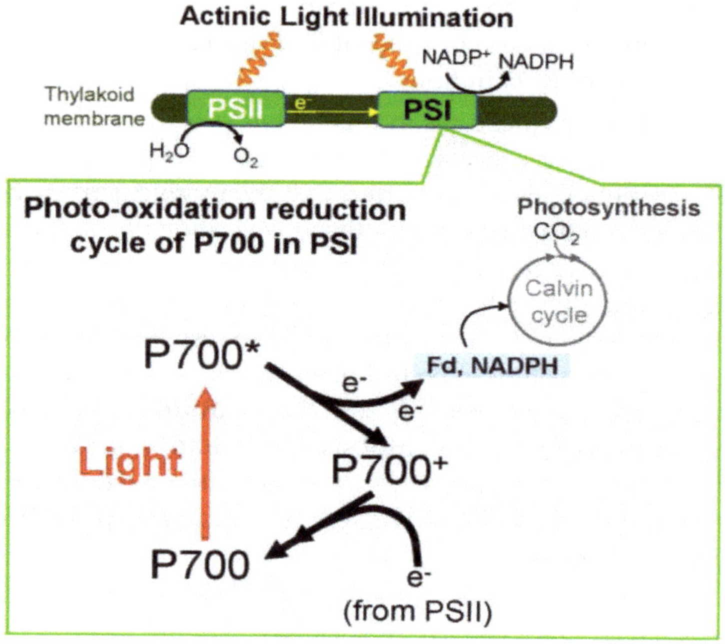

Figure 1. The photo-oxidation reduction cycle of P700 in Photosystem I (PSI). The reaction center chlorophyll P700 in PSI absorbs the light energy, and the P700 is excited to P700*. The P700* donates electrons to the electron carrier, A_0, and concomitantly produces the oxidized P700, $P700^+$. The $P700^+$ is reduced by electrons from Photosystem II (PSII) through plastoquinone (PQ), the cytochrome (Cyt) *b6/f*-complex, and plastocyanin (PC). The reduced PC directly donates electrons to P700. There are then P700 turnovers in the photo-oxidation reduction cycle of P700 in PSI. The electron on A_0 flows to $NADP^+$ to produce NADPH through the electron carriers A_1, F_X, F_A/F_B, and Fd. In the photosynthetic linear electron flow, electrons extracted from H_2O oxidation in PSII flow to $NADP^+$ for the production of NADPH.

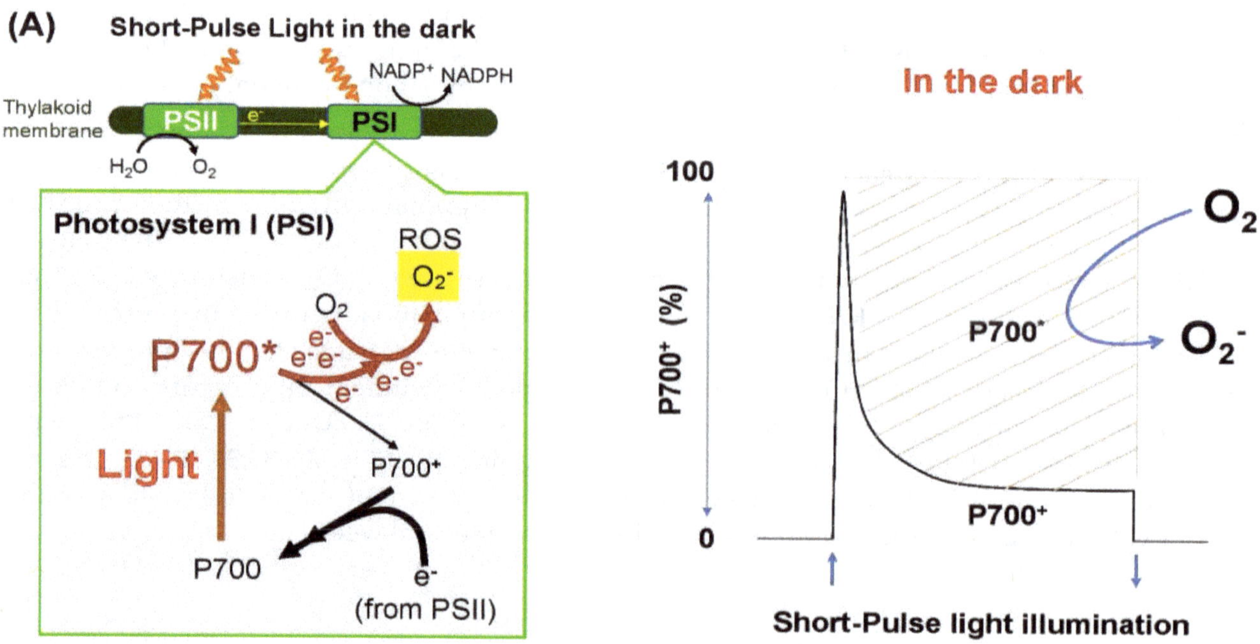

Figure 2. *Cont.*

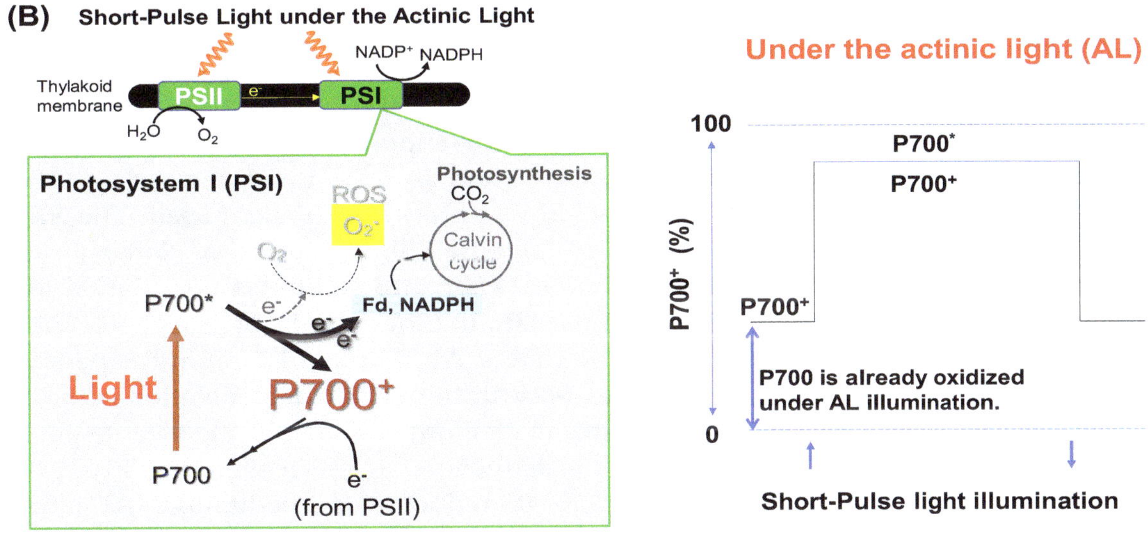

Figure 2. The kinetics of oxidized P700 during short-pulse (SP) illumination in the dark and under actinic light. (**A**) An SP illumination (20,000 μmol photons m^{-2} s^{-1}, 400 ms) to intact leaves in the dark rapidly oxidizes the reaction center chlorophyll P700 in PSI from the fully reduced state to the maximally oxidized state. Afterwards, oxidized P700 ($P700^+$) is reduced by electrons from PSII, with the accumulation of the excited state of P700 (P700*), as shown by the increase in shaded area. P700* can donate an electron to O_2, thereby producing a superoxide radical (O_2^-) through the one-electron reduction by the electron carriers, A_0/A_1, F_X, and F_A/F_B, localized at the acceptor side of PSI. (**B**) In actinic light (AL) illumination, before SP illumination, P700 is already oxidized to $P700^+$. SP illumination further oxidizes P700 to its maximum, which is lower than the oxidation obtained in the dark, because P700* accumulates under the steady-state AL illumination. $P700^+$ produced by SP illumination is not reduced, because P700 in PSI experiences turnover with the rate-determining step of $P700^+$ reduction in the P700 photo-oxidation reduction cycle. The Y-axis shows the oxidation ratio of P700 from 0% to 100%, and the X-axis shows the time for SP illumination: upward arrow: SP illumination starts; downward arrow: SP illumination stops.

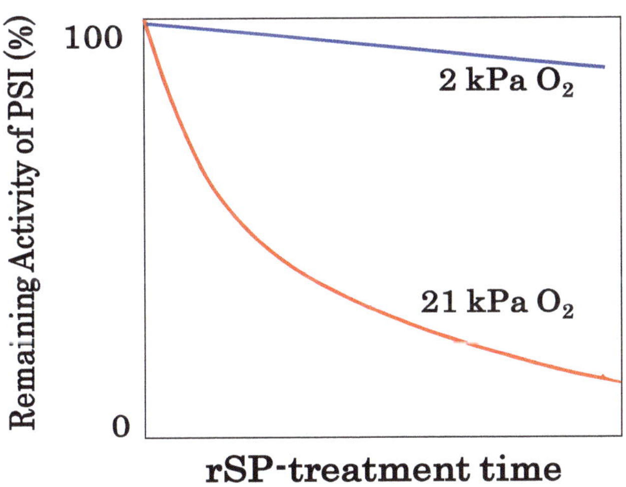

Figure 3. Repetitive short-pulse (rSP) illumination treatment inactivates PSI. In angiosperms, rSP illumination treatment of the intact leaves inactivates PSI activity [38]. The inactivation is suppressed under low O_2 conditions (2 kPa O_2). Short-pulse illumination reduces O_2 to a superoxide radical, O_2^-, and the accumulated O_2^- from rSP illumination treatment oxidatively attacks PSI components. Red line: the PSI activity gradually decreases as rSP illumination treatment proceeds under atmospheric O_2 (21 kPa) condition; blue line: PSI inactivation by rSP illumination treatment is suppressed under low O_2 (2 kPa) condition.

3. ROS Production Is Suppressed by Oxidation of P700

As described above, ROS-dependent oxidative damage in PSI was shown in vivo. Next, a new question arose: essentially, how do plants survive under natural sunlight? The answer was quickly obtained. PSI inactivation by rSP illumination treatment was suppressed under illumination by actinic light (AL) [38]. Furthermore, the extent of PSI inactivation decreased with increasing AL intensity. In other words, the light-induced inactivation of PSI was prevented by light itself. The effect of AL illumination was that P700 was oxidized to $P700^+$ (Figure 2B). Increasing AL intensity increased $P700^+$ in the P700 photo-oxidation reduction cycle. Afterwards, $P700^+$ accumulation prevented P700* accumulation during SP illumination (Figure 2B). In other words, high levels of $P700^+$ were maintained, even during light pulses. Thus, the role of AL illumination from the perspective of the P700 photo-oxidation reduction cycle was made clear: both P700* accumulation in the dark and $P700^+$ accumulation under AL illumination during an SP illumination, indicate the rate-determining step (RdS) transition in the P700 photo-oxidation reduction cycle. In the dark, P700* oxidation is rate-limited, whereas under AL illumination, $P700^+$ reduction is rate-limited. AL illumination suppresses rSP-illumination-treatment-induced PSI inactivation through the rate-determining step transition mechanism. That is, AL illumination suppresses the accumulation of P700*, which can donate electrons to O_2 to produce O_2^- in PSI. Subsequently, we found the robust effect of AL illumination, which was that P700 was oxidized and suppressed PSI inactivation by rSP illumination treatment [42].

Numerous studies have reported the oxidation of P700 in PSI under high intensity AL or low CO_2 conditions [33,34,47–54]. Accumulated $P700^+$ during the P700 photo-oxidation reduction cycle absorbs excess photon energy and dissipates it as heat [55–57]. Concomitantly, $P700^+$ accumulation inhibits O_2 reduction, producing O_2^- by suppressing P700* accumulation [38]. This is the molecular mechanism that reduces the rate of the Mehler reaction in intact leaves under AL illumination. If P700 is not subsequently oxidized, ROS production will increase, due to the promotion of the Mehler reaction, and PSI will stop functioning [38,42,58].

4. Molecular Mechanism of Oxidation of P700 in Photosystem I under AL Illumination

In photosynthesis, a light reaction functions with a dark reaction in a tightly coupled state. For example, the light reaction proceeds in the thylakoid membranes of chloroplasts in C_3-plants, where photosynthetic linear electron flow produces NADPH and ATP. The dark reaction proceeds in both photosynthesis and photorespiration, where both NADPH and ATP are used for the regeneration of ribulose 1,5-bisphosphate (RuBP) (one of the substrates for RuBP carboxylase/oxygenase which is consumed in the dark reaction). In photorespiration, reduced Fd is used in chloroplasts. Electron flow, which produces both NADPH and the reduced Fd, and proton flow, which produces ATP in the light reaction, are coupled, with both e^- flow and H^+ flow required for the consumption of NADPH, the reduced Fd, and ATP in the dark reaction. This tight coupling of the light reaction with the dark reaction can cause a dangerous situation in which ROS are produced in the photosynthetic electron transport system. The limitation of the dark reaction is the limitation of the light reaction, which can lead to the accumulation of electrons in the photosynthetic electron transport system. For example, in the photo-oxidation reduction cycle of P700 in PSI, electrons would accumulate at the acceptor side of PSI, where O_2 can be reduced to O_2^-, a ROS. As described above, unless P700 in PSI is oxidized, PSI suffers from oxidative damage.

We show below how the light reaction tightly couples with the dark reaction in photosynthesis. We later show how P700 is oxidized in response to the limitation of the dark reaction, for example, in drought, low temperature, and high light situations.

4.1. Robust Relationships Support Tight Coupling between the Light and Dark Reactions in Photosynthesis

Much evidence has accumulated that clarifies a robust tight coupling between the light and dark reactions in C3 photosynthesis (Figure 4). For instance, Genty et al. [59] reported a positive linear relationship with an origin of zero between the photosynthetic linear electron flow rate (J_f) in the light reaction and the photosynthetic electron consumption rate (J_g) based on NADPH consumption—in the dark reaction [59]. Consistently, a positive linear relationship with an origin of zero was also found between the photosynthetic linear electron flow rate (J_f) in the light reaction and the electron consumption rate (J_g)—based on NADPH consumption and reduced Fd—in the dark reaction, where photosynthesis and photorespiration both occur (Equation (1)) [30,31,60]. These results indicate that photosynthetic linear electron flow is the sole driver of both photosynthesis and photorespiration. In other words,

$$Jf = Jg \tag{1}$$

Recently, a positive linear relationship with an origin of zero between the reduced Fd oxidation rate (vFd) and the photosynthetic linear electron flow rate (J_f) was found (Equation (2) [61]). These results indicate that electron flow via Fd is driven by both photosynthesis and photorespiration. In other words,

$$vFd = k_{fd} \times Fd^- = Jg \tag{2}$$

The constant k_{fd} is the apparent rate constant of the reduced Fd oxidation rate, while Fd^- is the amount of reduced ferredoxin.

A positive linear relationship with an origin of zero was reported between Jf and the pmf production rate (($vH^+ = gH^+ \times pmf$ (gH^+, proton conductance) (Equation (3)) [62–65]). These results indicate that photosynthetic linear electron flow provides the formation of pmf. In other words,

$$k_H{}^+ \times Jf = vH^+ \tag{3}$$

The rate constant k_{H+} reflects lumen H^+ accumulation efficiency, which depends on H_2O oxidation in PSII and on the Q-cycle in the Cyt b_6/f complex. Furthermore, a positive linear relationship with an origin of zero was found between vH^+ and the proton usage rate (JgH^+), due to the regeneration of ATP required for both photosynthesis and photorespiration (Equation (4)) [66]. These results indicate that vH^+ also reflects the pmf usage rate during photosynthesis and photorespiration. In other words,

$$vH^+ = JgH^+ \tag{4}$$

Furthermore, a positive linear relationship with an origin of zero was found between JgH^+ and J_g (=J_f) (Equation (5)) [66]. These results indicate that JgH^+ was determined by the electron consumption rate (J_g), based on the consumption of both reduced Fd and NADPH. In other words,

$$JgH^+ = k_{jg} \times Jg = k_H{}^+ \times Jf \tag{5}$$

Equation (5) indicates that there is a tight coupling between the light and dark reactions during photosynthesis and photorespiration. Constant k_{jg} is the coefficient that links e^- flow with H^+ flow during photosynthesis and photorespiration. These findings indicate that the operation of the light reaction is driven by the dark reaction and vice versa, and provide a concrete explanation for the observed decrease in J_f under environmental stress, when the photosynthetic rate decreases due to a dark reaction rate limitation (i.e., when JgH^+ and J_g decrease).

Tight coupling between the light and dark reactions, as shown in Equation (5), can cause the accumulation of electrons at the acceptor side of PSI and the accumulation of protons in the lumenal space of thylakoid membranes under stress conditions, e.g., drought, low temperatures, and high intensity of light, which decrease the photosynthesis efficiency [2,3,8,67–69]. For example, under constant light intensity, decreases in both Jg and JgH^+ decrease Jf. These situations cause a reduction in PQ and the non-photochemical quenching of chlorophyll fluorescence, both of which reflect the accumulations of electrons and protons in the photosynthetic electron transport system [32,34,42,58,70–72].

However, the electrons do not accumulate at the acceptor side of PSI; P700 is oxidized under stress conditions [32,34,49,50,53,72]. From the model of the photo-oxidation reduction cycle of P700, the reduction of $P700^+$ must be the rate-determining step to accumulate the oxidized P700 in PSI. The suppression of electron fluxes in both light and dark reactions does not limit the reduction of $P700^+$ in the cycle. That is, the electron flux to the oxidized P700 should be suppressed such that P700 is oxidized.

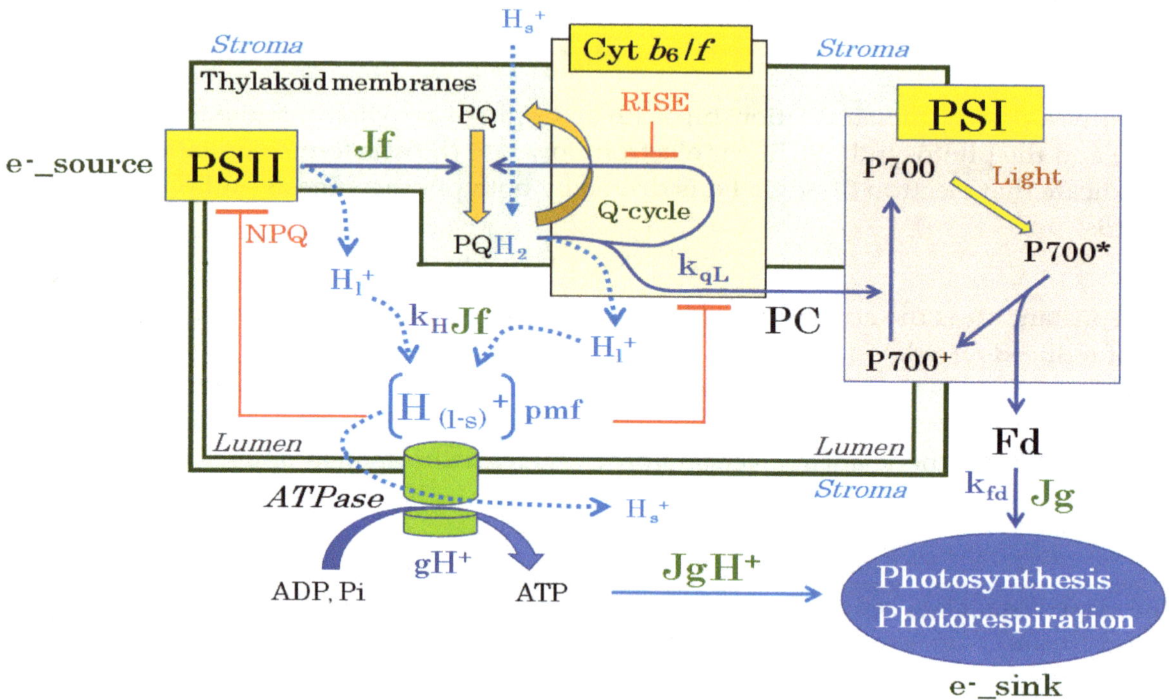

Figure 4. Tight coupling between the light and the dark reactions in photosynthesis of C3 angiosperms. Blue arrows: electron flow; blue dotted arrows: H^+ flow; Hs^+: stroma H^+; H_l^+: lumen H^+; $[H_{(l-s)}^+]$: difference in H^+ concentration (ΔpH) between the stroma and the lumen, i.e., the proton motive force (pmf). Photosystem II (PSII): electron source. P700 functions in the photo-oxidation reduction cycle to catalyze the electron flow from plastocyanin (PC) to Fd in Photosystem I (PSI). P700 is photoexcited to P700*, which donates electrons to Fd through electron carriers (A_0, A_1, F_A/F_B, and F_X) to produce $P700^+$. In turn, $P700^+$ is reduced by electrons from PSII through plastoquinone (PQ), the Cyt b_6/f complex, and PC. The Q-cycle in the Cyt b_6/f complex transfers electrons from reduced PQ along two routes: first, PC; second, oxidized PQ. Enhanced reduction of PQ suppresses the reduction of oxidized PQ in the Cyt b_6/f complex, which slows down the activity of the Q-cycle, and the electron flux from the Cyt b_6/f complex to $P700^+$ in PSI. H^+ conductance (gH^+) is an apparent rate constant that depends on the concentrations of ADP, Pi, ATPase, and the catalytic constant ($k_{(ATPase)}$). NPQ, non-photochemical quenching of chlorophyll fluorescence, is activated by the acidification of the lumenal space of the thylakoid membrane, which is reflected by the increase in pmf. NPQ suppresses the quantum yield of PSII, which is reflected by the electron flux in photosynthetic linear electron flow, Jf; the pmf also suppresses the oxidation activity of reduced PQ by the Cyt b_6/f complex. Tight coupling of the light reactions (the production of electron and proton flows) to the dark reaction (the consumption of electrons and protons in both photosynthesis and photorespiration) yields a robust relationship whereby electron flux (Jf) is equal to the electron flux in electron sinks (photosynthesis and photorespiration) expressed as Jg. The pmf is the driving force to synthesize ATP by ATP synthase using ADP and Pi, and the pmf-dissipation rate (vH^+) is expressed as $gH^+ \times$ pmf. As a robust model, $J_f = vFd = J_g$ [61]; vFd is the oxidation rate of reduced Fd; $k_{Jf} \times Jf = gH^+ \times$ pmf [64,65]. $vH^+ = gH^+ \times pmf = J_{gH+}$ [66]. J_g and J_{gH+} mutually determine each other. Meanwhile, pH decreases in the lumenal space of the thylakoid membrane during the photo-oxidation of water in PSII and the Q-cycle in the Cyt b_6/f complex. The decrease in pH (i.e., increased pmf) acts as the driving force for ATP production by ATPase.

4.2. How Do Protons and Electrons Accumulate in Both the Lumenal Space of Thylakoid Membranes and the PQ Pool in Response to the Suppression of the Dark Reaction?

Generally, it is proposed that the electron flux from PQ to P700 in PSI is regulated by the oxidation activity of the reduced PQ (PQH_2) by the cytochrome (Cyt) b_6/f-complex [48,67,70,71,73–81]. The PQH_2 oxidation activity by the Cyt b_6/f-complex decreases with a lowering pH. Therefore, the accumulation of protons in the lumenal space of thylakoid membranes suppresses the electron flow to the oxidized P700 in PSI, which is known as photosynthesis control [67,73,74]. Furthermore, the overreduction of PQ inhibits the Q-cycle in the Cyt b_6/f-complex, and contributes to the oxidation of P700 in PSI, which is known as the reduction-induced suppression of electron flow (RISE) [70,71]. We propose that the RISE is also involved in photosynthesis control. To clarify the molecular mechanism of oxidation of P700 in PSI upon exposure to the limitation of photosynthesis, we must understand how protons and electrons accumulate in both the lumenal space and the PQ pool in the photosynthetic electron transport system in response to the activities of both photosynthesis and photorespiration.

The induction mechanism of proton gradient (ΔpH) formation across the thylakoid membrane, the accumulation of protons in the lumenal space of thylakoid membranes, is shown as below: ΔpH formation is observed as a pmf increase [65,82,83]. In general, ΔpH mainly occupies the pmf at low photosynthetic efficiency, e.g., high light and low CO_2 conditions. The velocity of change in pmf [d(pmf)/dt] is determined by the difference between the pmf generation rate and the pmf decay rate. The pmf generation rate depends on the photosynthetic linear electron flow rate (Equation (3)), and the pmf decay rate depends on the ATP production rate, which is driven by both photosynthesis and photorespiration (Equation (4)): thus, the velocity of change in pmf is as shown (Equations (6) and (7)).

$$d(pmf)/dt = k_H{}^+ \times Jf - vH^+ \tag{6}$$

$$= k_H{}^+ \times Jf - gH^+ \times (pmf) \tag{7}$$

The rate constant k_{H+} reflects the H^+ accumulation in the luminal space of thylakoid membranes, which is driven by photosynthetic linear electron flow. At the steady state, d(pmf)/dt = 0, and the pmf production rate is $k_H{}^+ \times Jf$. This depends on H_2O oxidation in PSII and on Q-cycle rotation in the Cyt b_6/f complex. Furthermore, the pmf decay rate, vH^+, is expressed as $gH^+ \times (pmf)$. The gH^+, H^+ conductance, is a rate constant that reflects the apparent rate constant of pmf decay. The vH^+ also reflects the pmf dissipation rate, and vH^+ can be replaced with JgH^+ as follows (Equation (8)):

$$d(pmf)/dt = k_H{}^+ \times Jf - JgH^+ \tag{8}$$

We reveal that, in a steady state where d(pmf)/dt = 0, based on Equations (7) and (8), Equation (9) was obtained.

$$k_{H+} \times Jf = vH^+ \ (= gH^+ \times pmf) = JgH^+ \tag{9}$$

Equation (9) shows that the photosynthetic linear electron flow activity in the light reaction links photosynthesis and photorespiration activity in the dark reaction through pmf formation and dissipation. The dependence of pmf on Jf, gH^+, and JgH is shown as follows (Equations (10) and (11)).

$$pmf = (k_H{}^+ \times Jf)/gH^+ \tag{10}$$

$$= JgH^+/gH^+ \tag{11}$$

Based on this model, we will discuss the molecular mechanism of proton accumulation in the lumenal space of thylakoid membranes in response to the activities of photosynthesis and photorespiration. For example, if the extent of the decrease in the dark reaction rate, shown as the decrease in both Jg and JgH^+, was smaller than that in gH^+, then pmf increased. This is how protons accumulate in the lumenal space of thylakoid membranes.

We consider the accumulation of electrons in the photosynthetic electron transport system, reflected as the redox states of both PQ and Fd. The reduced state of PQ is evaluated by a parameter of chlorophyll fluorescence, 1 − qL [72,84]. The qL shows the oxidation state of PQ in thylakoid membranes. A higher value of 1 − qL shows the high accumulation of electrons in the PQ pool. The oxidation rate of the reduced PQ can be expressed as kqL × (1 − qL). Additionally, the oxidation rate of the reduced Fd (Fd^-) is expressed as $k_{Fd} \times Fd^-$. Both kqL and kFd are the apparent rate constants. In the light reaction, the electron fluxes for the oxidations of both the reduced PQ and the reduced Fd are equal to Jf, as shown in the equation (12).

$$Jf = kqL \times (1 - qL) = k_{Fd} \times Fd^- = Jg \tag{12}$$

The kqL is given a positive value and is expected to decrease by two factors: first, kqL decreases with a decrease in pH in the lumenal space of thylakoid membranes (photosynthesis control) [67,85,86]; second, kqL decreases by the RISE [70,71]. Generally, a decrease in Jg increases the value of 1 − qL [34,72]. These facts indicate, in response to the decrease in electron sink activity (both photosynthesis and photorespiration), that the extent of the decrease in kqL is larger than that in Jg. That is, from Equation (12), we can clearly understand why 1 − qL increases. This is how electrons accumulate in the PQ pool of thylakoid membranes.

On the other hand, a decrease in Jg does not induce the reduction of Fd [61]. For example, lowering the intercellular partial pressure of CO_2 decreases the photosynthesis rate, and Jg decreases; however, the reduced level of Fd either does not change or decrease. These facts show that the extent of the decrease in k_{Fd} was almost the same as that in Jg. The k_{Fd} depends on both the activity of Fd^-NADP^+ oxidoreductase (FNR) and $NADP^+$ regeneration efficiency. The decrease in the photosynthesis rate lowers the $NADP^+$ regeneration rate, which decreases k_{Fd}.

4.3. How Do the Accumulated Protons in the Lumenal Space and Electrons in the PQ Pool Oxidize P700 in PSI?

The decrease in Jf (=Jg) shows the negative linear relationship with $P700^+$ (Figure 5) [42,87]. The decrease in photosynthesis efficiency induces the oxidation of P700 in PSI. As described above, the suppression of photosynthesis enhances both the pmf and the RISE, which would cause the reduction of $P700^+$ in the photo-oxidation reduction cycle of P700 in PSI to be the rate-determining step. The decrease in Jg lowers Jf in the tight coupling of the light reaction with the dark reaction. Here, we have to pay an attention to the magnitude of the decrease in Jf. The extent of the decreases in gH^+ and kqL are larger than that in Jg, so the pmf and (1 − qL) increase. We also must consider that the magnitude of Jf would be downregulated by both the pmf and the RISE, such that the reduction of $P700^+$ should be the rate-determining step in the photo-oxidation reduction cycle of P700 in PSI. Jf must be lower than the theoretical value of Jg obtained by the suppression of photosynthesis. For example, if Jf decreases to the same extent as the decrease in Jg in the response to the decrease in Ci, the photo-excited P700 (P700*) would accumulate, because the efficiency of photo-excitation of P700 to P700* does not change at a constant intensity of AL. This deduction contradicts the experimental facts; that is, that suppressed photosynthesis induces the oxidation of P700.

The RISE was evidenced to have the potential to stop the electron flow in the light reaction and to oxidize P700 in PSI of cyanobacteria [70,71]. Flavodiiron protein (FLV) catalyzes the reduction of O_2 to H_2O using NADPH in the cyanobacteria [88,89]. The electron flux in the FLV reaction is so high that it alternates the electron flux in photosynthesis under CO_2-deficient conditions [88,90]. The cyanobacteria deficient in FLV show suppressed photosynthesis, with the accumulation of electrons in the PQ pool [70]. Furthermore, illumination by SP light reduces PQ almost transiently with the suppression of O_2 evolution under steady-state conditions [70]. FLV can oxidize PQ even though photosynthesis can function at a higher activity. FLV keeps PQ in an oxidized state so as not to induce the RISE for the full activity of photosynthesis, if enough CO_2 is supplied to cyanobacteria.

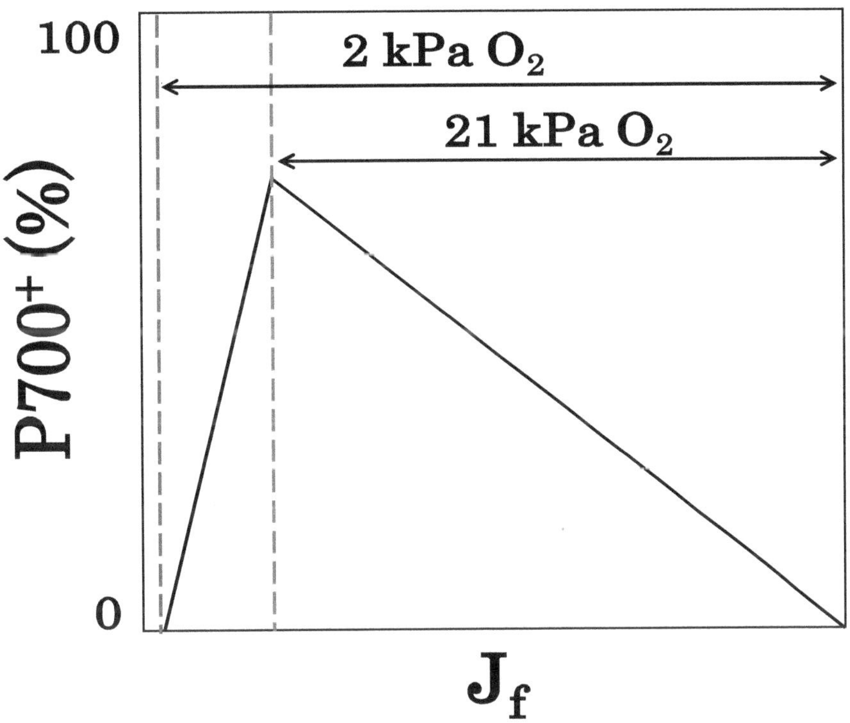

Figure 5. Dependence of oxidized P700 ($P700^+$) induction on electron flux in photosynthetic linear electron flow in intact leaves. Generally, P700 is oxidized with a decreasing electron flux of photosynthetic linear electron flow (J_f) by lowering CO_2 partial pressure. At 21 kPa O_2, when both photorespiration and photosynthetic CO_2 fixation are functioning, $P700^+$ shows the maximum value at the CO_2 compensation point, while J_f shows the minimum value (blue dotted line). The horizontal arrow at 21 kPa O_2 shows the range over which J_f is driven by both photosynthetic CO_2 fixation and photorespiration. At 2 kPa O_2, when no detectable photorespiration occurs and photosynthetic CO_2 fixation functions normally, $P700^+$ starts to decrease to the minimum at the CO_2 compensation point after reaching its maximum value, while J_f reaches its minimum value (red dotted line). The horizontal arrow at 2 kPa O_2 shows the range over which J_f is driven by photosynthetic CO_2 fixation.

Furthermore, we found that inductions of the pmf and RISE oxidize P700 in PSI of the C3-plant wheat leaves [42,87]. A decrease in the intercellular partial pressure of CO_2 lowers the photosynthetic rate and increases the photorespiration rate. However, the photorespiration rate cannot reach the value theoretically derived from Rubisco kinetics at lower CO_2 [37,91]. The induction of the pmf and RISE lowers the activity of the light reaction, which causes the activity of the dark reaction, the photorespiration rate, to be suppressed, even though the potential activity of the dark reaction exceeds the suppressed activity of the light reaction. These results show that the oxidation of P700 requires the suppression of electron flux in the light reaction much more so than in the dark reaction; thus, the suppressed activity in the dark reaction is lower than the theoretical potential activity. As a result, the reduction of $P700^+$ becomes the rate-determining step in the P700 photo-oxidation reduction cycle. This is the essence of the molecular mechanism of P700 oxidation.

On the other hand, the induction of the pmf and RISE does not always oxidize P700 in response to the limitation of photosynthesis. Unless the electron flux of the light reaction dips below the potential electron flux of the dark reaction, which is, theoretically, expected from the limited activity of photosynthesis, the limitation of the PSI acceptor side is so strong that P700 is not oxidized (Figure 5). When photosynthesis is severely limited, where both photosynthesis and photorespiration functions are suppressed, P700 cannot be oxidized because the oxidation of P700* becomes the rate-determining step in the P700 photo-oxidation reduction cycle.

4.4. Suppression of the Production of ROS in PSI by the P700 Oxidation System

We summarize the molecular mechanism of P700 oxidation in PSI of chloroplasts (Figure 6). Both PSI and PSII absorb photon energy to excite the reaction center chlorophylls P680 (PSII) and P700 (PSI). Similarly to P700 in PSI, P680 undergoes a photo-oxidation reduction cycle. The ground state of P680 is photo-excited to P680*. The P680* donates electrons to PQ through electron carriers in PSII. Additionally, then, the electrons start to flow from the reduced PQ to PSI, through the Cyt b_6/f-complex, and plastocyanin (PC). During the electron flow, H^+ from both H_2O oxidation in PSII and the reduced PQ (PQH_2, plastoquinol) oxidation of the Cyt b_6/f-complex accumulate in the lumenal space of thylakoid membranes. The accumulated protons are the pmf to drive ATP synthase to produce ATP in the stroma. On the other hand, NADPH is produced by ferredoxin-NADP oxidoreductase (FNR) at the acceptor side of PSI. ATP, NADPH, and the reduced Fd are used for the regeneration of RuBP in both photosynthesis and photorespiration. The regeneration rates of $NADP^+$, the oxidized Fd, and ADP are determined by the activities of both photosynthesis and photorespiration.

As shown in Equations (5), (9), and (12), the light reaction tightly couples with the dark reaction. The suppression of photosynthesis decreases Jg or JgH^+. Afterwards, the regeneration rates of ADP, Pi, and $NADP^+$ decrease. That is, gH^+ decreases to increase the pmf, and kqL decreases to increase $(1 - qL)$, which induces the RISE. Both the pmf and the RISE lower kqL, as reflected in the suppressed activity of the reduced PQ oxidation by the Cyt b_6/f-complex. NPQ induced by the pmf also downregulates the quantum yield of PSII. The suppressed electron flow from PSII to the oxidized P700 also causes the reduction of $P700^+$ in the photo-oxidation reduction cycle to be the rate-determining step, which leads to the oxidation of P700.

We compared the rate-determining step in the photo-oxidation reduction cycle of P700 in PSI in response to the limitation of electron sink activities and the photosynthesis and photorespiration activities at high light conditions (Figure 6). In Case (I), we set the electron sink to a large size. The black arrows show the electron flow in the photosynthetic electron transport system from the redox reaction of PQ to the electron sink, including both photosynthesis and photorespiration, through the Cyt b_6/f-complex, PC, and P700 in PSI and Fd. The width of the black arrows reflects the magnitude of the apparent rate constant in each elementary reaction. The rate-determining step of the P700 photo-oxidation reduction cycle in PSI is the reduction of $P700^+$, leading toward to the oxidation of $P700^+$. Both the pmf and RISE contribute to the suppression of the Q-cycle activity of the Cyt b_6/f-complex, which causes the reduction of $P700^+$ to be the rate-determining step.

In Case (II), the electron sink is set to a medium size. Environmental stress, e.g., drought and/or low temperatures, can suppress photosynthetic activities. That is, both the pmf and RISE increase in response to the decrease in both gH^+ and kqL, as determined by Equations (11) and (12). The apparent rate constant for the reduction of $P700^+$ further decreases, compared to Case (I). The magnitude of the apparent rate constant for $P700^+$ reduction is lower than the apparent constant for P700* oxidation. The increase in the suppression of the Q-cycle in the Cyt b_6/f-complex enhances the rate-determining step, the reduction of $P700^+$, in the P700 photo-oxidation reduction cycle of PSI.

In Case (III), the electron sink is smaller than the medium size in Case (II). The activities of both photosynthesis and photorespiration are further suppressed, compared to Case (II). However, the rate-determining step leads to the elementary reaction of the oxidation of the photo-excited P700, P700*, and P700* then accumulates, although both the pmf and RISE are kept to the same or a slightly greater extent or as in Case (II). That is, the suppression of the apparent rate constant of the oxidation of P700* is larger than that of the oxidation activity of the reduced PQ in the Cyt b_6/f-complex in response to the limitation of the electron sink. Electrons accumulate at the acceptor side of PSI, and the electrons flow to O_2 to produce O_2^-, the production of ROS. The transition of the rate-determining step from the reduction of $P700^+$ to the oxidation of P700* is a dangerous condition that induces oxidative damage to PSI.

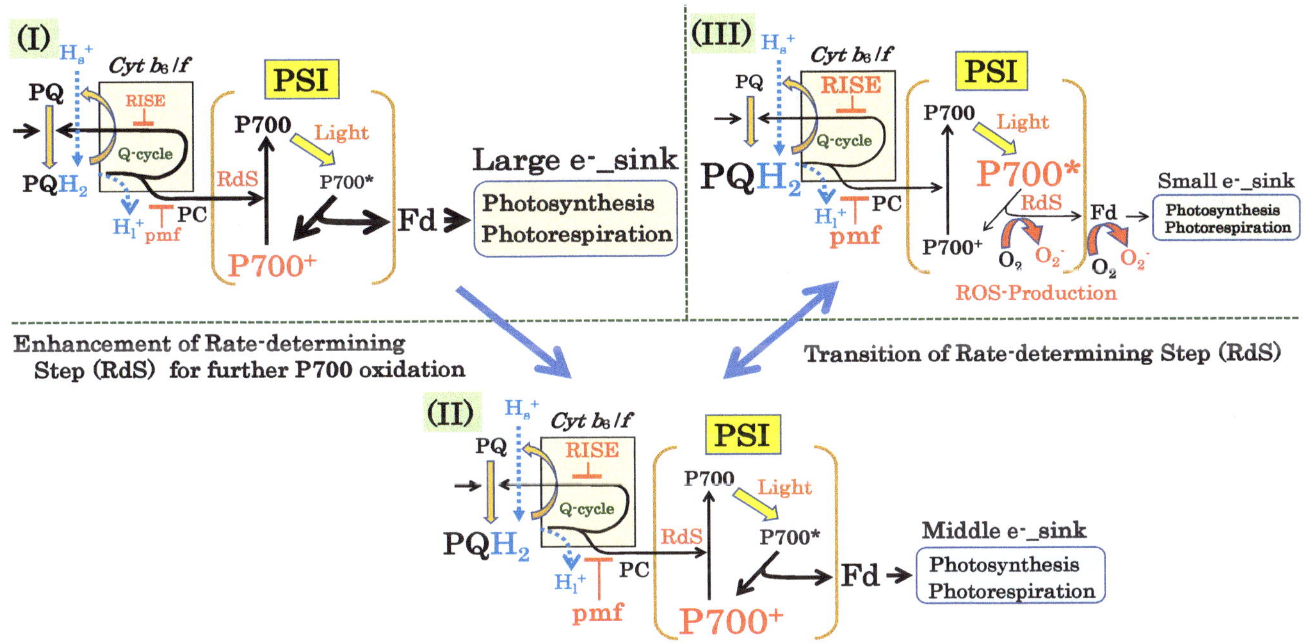

Figure 6. The P700 oxidation system and ROS production mechanism. The photosynthetic linear electron flow in the light reaction is tightly coupled to the dark reaction, photosynthesis and photorespiration (electron sinks). The P700 oxidation system enhances the rate-determining step of $P700^+$ reduction to stimulate the oxidation of P700 in PSI, in response to the decrease in electron sinks. Once exposed to a further decrease in electron sinks, $P700^+$ reduction cannot remain the rate-determining step, and the rate-determining step transitions from $P700^+$ reduction to the oxidation of P700*. Three distinct scenarios of the rate-determining step can take place under saturated light intensity: In Case (I) with the large electron sink, both the pmf and RISE are induced by the two electron sinks: photosynthesis and photorespiration. Afterwards, the oxidation activity of the reduced PQ by the Cyt b_6/f-complex is suppressed to decrease the reduction rate of $P700^+$. As a result, the photo-oxidation reduction cycle of P700 in PSI turns over with the rate-determining step of $P700^+$ reduction. In Case (II) with the medium-sized electron sink, PQ is further reduced, and the pmf is further accumulated, compared to Case (I), by the decreased gH^+ and k_{qL}. The electron flux from the reduced PQ to the oxidized P700 is then further suppressed. The oxidation of P700 becomes the rate-determining step in the photo-oxidation reduction cycle of P700 in PSI. In Case (III) with the small electron sink, the further induced pmf and RISE cannot decrease the reduction rate more than the suppression of the oxidation of P700* due to the smaller electron sink, compared to Case (II). Afterwards, P700* accumulates. As a result, the photo-oxidation reduction cycle of P700 in PSI turns over with the rate-determining step of P700* oxidation. From Case (II) to Case (III), the rate-determining step transitions from the reduction of $P700^+$ to the oxidation of P700* in the photo-oxidation reduction cycle of P700 in PSI. The accumulated P700* increases the chance of the reduction of O_2 to O_2^-, the production of ROS at the acceptor side of PSI. The reduced electron carriers in PSI—A_0, A_1, F_X, F_A/F_B, and the reduced Fd—can react with O_2 to produce O_2^- [2,39–41,92,93].

4.5. Relationship between Photorespiration- and Flavodiiron Protein (FLV)-Dependent Electron Flows and Their Contribution to P700 Oxidation in PSI

Following the model of the P700 oxidation system, photosynthetic linear electron flow contributes to the induction of either the pmf or RISE, which suppresses the electron flow from the Cyt b_6/f-complex to the oxidized P700 through PC. Afterwards, P700 is oxidized, as shown in Figures 5 and 6. From cyanobacteria to gymnosperms, in the evolution of green lineage oxygenic photosynthetic organisms, flavodiiron protein (FLV) drives O_2-dependent electron flow in the photosynthetic electron transport system, in addition to the photorespiration pathway [88–90,94–96]. FLV reduces O_2 to H_2O using NADPH as the electron donor [88]. Different from photorespiration, FLV-dependent electron flow does not consume ATP; i.e., it only induces the pmf. Therefore, FLV-dependent electron flow has a higher capacity to oxidize P700. In fact, in the cyanobacteria *Synechocystis* sp. PCC6803 and *Synechococcus* sp. PCC7002, the deletion of *flv* genes caused the photoinhibition of PSI, which was enhanced by the reduction of P700 in PSI [58]. Furthermore, the liverwort *Marchantia polymorpha*, which was deficient in flv genes, suffered from the photoinhibition of PSI due to the reduction of P700 [58]. Angiosperms, however, do not have FLV genes in their genomes [87,97]. Among land plants, the electron flux of photorespiration-dependent photosynthetic linear electron flow was compared to FLV-dependent electron flux at the CO_2 compensation point, where photosynthetic activity was suppressed to zero. The maximum activities of both FLV-dependent and photorespiration-dependent photosynthetic linear electron flow were then evaluated among land plants [98].

Photorespiration activities at the CO_2 compensation point increased with the evolution of photosynthetic organisms from liverworts to C3 angiosperms to ferns and gymnosperms (Figure 7A). C3-C4 and C4 angiosperms show lower activities of photorespiration, compared to C3 angiosperms. C3 angiosperms show the maximum activities of photorespiration [98]. In C3 angiosperms, photosynthetic linear electron flow was mainly driven by photorespiration, which was the main electron sink in photosynthesis, at the CO_2 compensation point (Figure 7B). An electron sink other than photorespiration was found in liverworts, ferns, and angiosperms. FLV could be an electron sink at the CO_2 compensation point. C3-C4 and C4 angiosperms also had an electron sink other than photorespiration and FLV-dependent photosynthetic linear electron flows, but its mechanism is unknown.

What do C3-angiosperms achieve by discarding FLV genes in their genomes? Different from photorespiration, FLV-dependent electron flow does not consume ATP; it produces it, as described above. FLV-dependent electron flow induces the pmf, which can be a self-regulating electron flow and suppress photosynthesis, because an enhanced pmf suppresses the activity of the Cyt b_6/f-complex, as shown in Figures 5 and 6. In liverworts, ferns, and gymnosperms, FLV-dependent photosynthetic linear electron flow functions immediately after the actinic light illumination starts, and its flux is suppressed at the steady state of photosynthesis [98]. This indicates that FLV-dependent photosynthetic linear electron flow is generally not essential to photosynthesis organisms, and that there is a molecular mechanism to reduce FLV-dependent photosynthetic linear electron flow under a steady state of photosynthesis. On the other hand, photorespiration drives both the light reaction and the dark reaction, and inevitably functions with photosynthesis. This situation implies that a tight coupling of the light reaction with the dark reaction driven by both photosynthesis and photorespiration can regulate both the pmf and RISE in response to their activities, because the dark reaction induces and dissipates the pmf and RISE simultaneously.

This could be a sensitive trigger for the regulation of P700 oxidation. If FLV-dependent photosynthetic linear electron flow participates in the photosynthetic linear electron flow driven by both photosynthesis and photorespiration, the automatic regulation of photosynthetic linear electron flow will collapse. Therefore, C3 angiosperms do not require FLV-dependent photosynthetic linear electron flow, and discard FLV genes in their genomes. These conclusions explain why angiosperms can have a high activity of photosynthesis, if photosynthetic organisms having FLV genes in their genomes have lower photosynthetic activity.

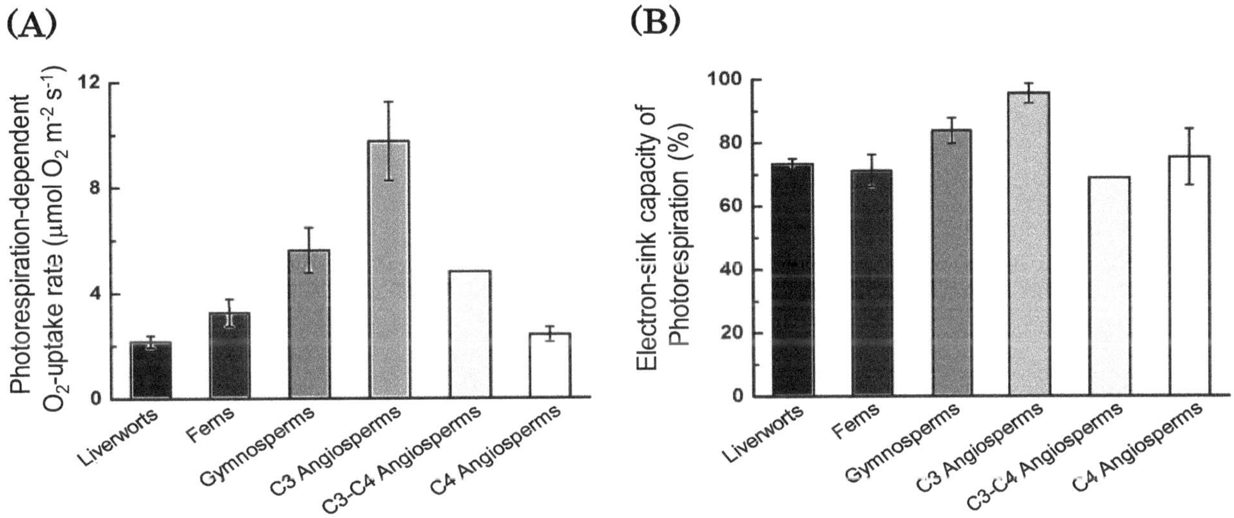

Figure 7. Evolution of tight coupling between the light and the dark reactions in photosynthesis. Upon spreading on the surface of the earth, oxygenic photosynthetic organisms inevitably metabolized 2-phosphoglycolate to regenerate 3-phosphoglycerate via the photorespiration pathway, because ribulose 1,5-bisphophate (RuBP) carboxylase/oxygenase (Rubisco) catalyzes both RuBP carboxylation and oxygenation under atmospheric conditions. Consequently, photorespiration inevitably accompanies photosynthetic CO_2 fixation. In C3 angiosperms, both photosynthetic CO_2 fixation and photorespiration have become the main electron sinks in the photosynthetic process. Hanawa et al. [98] found that photorespiration activity increased with the evolution of oxygenic photosynthetic organisms. Photorespiration activity is evaluated as a photorespiration-dependent O_2-uptake at the CO_2 compensation point, at which point its activity peaks [66]. (**A**) Photorespiration activity increases with the evolution of oxygenic photosynthetic organisms from liverworts to C3 angiosperms through ferns and gymnosperms. (**B**) The electron sink capacity of photorespiration reaches its maximum in C3 angiosperms. The activities of flavodiiron (FLV)-dependent photosynthetic linear electron flow and other alternative photosynthetic linear electron flow (AEF) decrease upon the evolution of oxygenic photosynthetic organisms from liverworts to C3 angiosperms through ferns and gymnosperms. On the other hand, C3-C4 and C4 plants show lower photorespiration activity, and other electron sinks function at the CO_2 compensation point. Data are from Hanawa et al. [98].

5. Conclusions

Oxygenic photosynthetic organisms have a robust, common molecular mechanism to suppress the production of reactive oxygen species (ROS), the P700 oxidation system, because the ROS-dependent oxidative inactivation of PSI is lethal to these organisms. When cyanobacteria first appeared on the Earth about 2–3 billion years ago, they already had a P700 oxidation system [58]. Photosynthetic organisms regulate the activity of plastoquinol oxidation by the cytochrome (Cyt) b_6/f-complex using two strategies: pmf formation and the reduction-induced suppression of electron flow (RISE), both of which are used to suppress the electron flux from the Cyt b_6/f-complex to the reaction center chlorophyll P700 in PSI. The induction of both the pmf and RISE is sensitive to changes in the activities of both photosynthesis and photorespiration in land plants, which drive photosynthetic linear electron flow. The magnitudes of both the pmf and RISE are determined by the activity of photosynthetic linear electron flow. Upon exposure to environmental stress, e.g., low temperature, and/or drought, suppressed activities of both photosynthesis and photorespiration cause both the pmf and RISE to oxidize P700 in the manner described in the text for the suppression of ROS production in PSI.

Acknowledgments: The author is thankful to MDPI for English language editing.

References

1. Aro, E.M.; Virgin, I.; Andersson, B. Photoinhibition of photosystem II. Inactivation, protein damage and turnover. *Biochim. Biophys. Acta* **1993**, *1143*, 113–134. [CrossRef]
2. Asada, K. The Water-Water Cycle IN Chloroplasts: Scavenging of Active Oxygens and Dissipation of Excess Photons. *Annu. Rev. Plant. Physiol. Plant. Mol. Biol.* **1999**, *50*, 601–639. [CrossRef]
3. Asada, K. The water-water cycle as alternative photon and electron sinks. *Philos. Trans. R. Soc. Lond. B Biol. Sci.* **2000**, *355*, 1419–1431. [CrossRef] [PubMed]
4. Allakhverdiev, S.I.; Tsvetkova, N.; Mohanty, P.; Szalontai, B.; Moon, B.Y.; Debreczeny, M.; Murata, N. Irreversible photoinhibition of photosystem II is caused by exposure of Synechocystis cells to strong light for a prolonged period. *Biochim. Biophys. Acta* **2005**, *1708*, 342–351. [CrossRef] [PubMed]
5. Powles, S.B.; Osmond, C.B. Photoinhibition of intact attached leaves of C3 plants illuminated in the absence of both carbon dioxide and of photorespiration. *Plant. Physiol.* **1979**, *64*, 982–988. [CrossRef]
6. Roberty, S.; Bailleul, B.; Berne, N.; Franck, F.; Cardol, P. PSI Mehler reaction is the main alternative photosynthetic electron pathway in Symbiodinium sp., symbiotic dinoflagellates of cnidarians. *New Phytol.* **2014**, *204*, 81–91. [CrossRef]
7. Sonoike, K. Photoinhibition of photosystem I. *Physiol. Plant.* **2011**, *142*, 56–64. [CrossRef]
8. Tikkanen, M.; Mekala, N.R.; Aro, E.M. Photosystem II photoinhibition-repair cycle protects Photosystem I from irreversible damage. *Biochim. Biophys. Acta* **2014**, *1837*, 210–215. [CrossRef]
9. Tjus, S.E.; Moller, B.L.; Scheller, H.V. Photosystem I is an early target of photoinhibition in barley illuminated at chilling temperatures. *Plant. Physiol.* **1998**, *116*, 755–764. [CrossRef]
10. Tyystjarvi, E. Photoinhibition of Photosystem II. *Int. Rev. Cell. Mol. Biol.* **2013**, *300*, 243–303. [CrossRef]
11. Tyystjarvi, E.; Maenpaa, P.; Aro, E.M. Mathematical modelling of photoinhibition and Photosystem II repair cycle. I. Photoinhibition and D1 protein degradation in vitro and in the absence of chloroplast protein synthesis in vivo. *Photosynth. Res.* **1994**, *41*, 439–449. [CrossRef] [PubMed]
12. van Wijk, K.J.; van Hasselt, P.R. Photoinhibition of photosystem II in vivo is preceded by down-regulation through light-induced acidification of the lumen: Consequences for the mechanism of photoinhibition in vivo. *Planta* **1993**, *189*, 359–368. [CrossRef]
13. Weisz, D.A.; Gross, M.L.; Pakrasi, H.B. Reactive oxygen species leave a damage trail that reveals water channels in Photosystem II. *Sci. Adv.* **2017**, *3*, eaao3013. [CrossRef] [PubMed]
14. Yamamoto, Y.; Nishi, Y.; Yamasaki, H.; Uchida, S.; Ohira, S. Assay of photoinhibition of photosystem II and protease activity. *Methods Mol. Biol.* **2004**, *274*, 217–227. [CrossRef] [PubMed]
15. Tjus, S.E.; Scheller, H.V.; Andersson, B.; Moller, B.L. Active oxygen produced during selective excitation of photosystem I is damaging not only to photosystem I, but also to photosystem II. *Plant. Physiol.* **2001**, *125*, 2007–2015. [CrossRef] [PubMed]
16. Egneus, H.; Heber, U.; Matthiesen, U.; Kirk, M. Reduction of oxygen by the electron transport chain of chloroplasts during assimilation of carbon dioxide. *Biochim. Biophys. Acta* **1975**, *408*, 252–268. [CrossRef]
17. Fryer, M.J.; Oxborough, K.; Mullineaux, P.M.; Baker, N.R. Imaging of photo-oxidative stress responses in leaves. *J. Exp. Bot.* **2002**, *53*, 1249–1254.
18. Marsho, T.V.; Kok, B. Interaction between electron transport components in chloroplasts. *Biochim. Biophys. Acta* **1970**, *223*, 240–250. [CrossRef]
19. Ort, D.R.; Baker, N.R. A photoprotective role for O_2 as an alternative electron sink in photosynthesis? *Curr. Opin. Plant. Biol.* **2002**, *5*, 193–198. [CrossRef]
20. Furbank, R.T.; Badger, M.R.; Osmond, C.B. Photosynthetic oxygen exchange in isolated cells and chloroplasts of C3 plants. *Plant. Physiol.* **1982**, *70*, 927–931. [CrossRef]
21. Heber, U. Irrungen, Wirrungen? The Mehler reaction in relation to cyclic electron transport in C3 plants. *Photosynth Res.* **2002**, *73*, 223–231. [CrossRef] [PubMed]

22. Jimbo, H.; Izuhara, T.; Hihara, Y.; Hisabori, T.; Nishiyama, Y. Light-inducible expression of translation factor EF-Tu during acclimation to strong light enhances the repair of photosystem II. *Proc. Natl. Acad. Sci. USA* **2019**. [CrossRef] [PubMed]
23. Oxborough, K.; Baker, N.R. An evaluation of the potential triggers of photoinactivation of photosystem II in the context of a Stern-Volmer model for downregulation and the reversible radical pair equilibrium model. *Philos. Trans. R Soc. Lond. B Biol. Sci.* **2000**, *355*, 1489–1498. [CrossRef]
24. Wilson, S.; Ruban, A.V. Quantitative assessment of the high-light tolerance in plants with an impaired photosystem II donor side. *Biochem. J.* **2019**, *476*, 1377–1386. [CrossRef]
25. Wilson, S.; Ruban, A.V. Enhanced NPQ affects long-term acclimation in the spring ephemeral Berteroa incana. *Biochim. Biophys. Acta Bioenerg.* **2019**. [CrossRef] [PubMed]
26. Krieger-Liszkay, A.; Trosch, M.; Krupinska, K. Generation of reactive oxygen species in thylakoids from senescing flag leaves of the barley varieties Lomerit and Carina. *Planta* **2015**, *241*, 1497–1508. [CrossRef] [PubMed]
27. Krieger-Liszkay, A.; Kienzler, K.; Johnson, G.N. Inhibition of electron transport at the cytochrome *b6/f* complex protects photosystem II from photoinhibition. *FEBS Lett.* **2000**, *486*, 191–194. [CrossRef]
28. Krieger, A.; Rutherford, A.W.; Vass, I.; Hideg, E. Relationship between activity, D1 loss, and Mn binding in photoinhibition of photosystem II. *Biochemistry* **1998**, *37*, 16262–16269. [CrossRef]
29. Aro, E.M.; Suorsa, M.; Rokka, A.; Allahverdiyeva, Y.; Paakkarinen, V.; Saleem, A.; Battchikova, N.; Rintamaki, E. Dynamics of photosystem II: A proteomic approach to thylakoid protein complexes. *J. Exp. Bot.* **2005**, *56*, 347–356. [CrossRef]
30. Driever, S.M.; Baker, N.R. The water-water cycle in leaves is not a major alternative electron sink for dissipation of excess excitation energy when CO_2) assimilation is restricted. *Plant. Cell Environ.* **2011**, *34*, 837–846. [CrossRef]
31. Ruuska, S.A.; Badger, M.R.; Andrews, T.J.; von Caemmerer, S. Photosynthetic electron sinks in transgenic tobacco with reduced amounts of Rubisco: Little evidence for significant Mehler reaction. *J. Exp. Bot.* **2000**, *51*, 357–368. [CrossRef] [PubMed]
32. Miyake, C. Alternative electron flows (water-water cycle and cyclic electron flow around PSI) in photosynthesis: Molecular mechanisms and physiological functions. *Plant. Cell Physiol.* **2010**, *51*, 1951–1963. [CrossRef] [PubMed]
33. Miyake, C.; Horiguchi, S.; Makino, A.; Shinzaki, Y.; Yamamoto, H.; Tomizawa, K. Effects of light intensity on cyclic electron flow around PSI and its relationship to non-photochemical quenching of Chl fluorescence in tobacco leaves. *Plant. Cell Physiol.* **2005**, *46*, 1819–1830. [CrossRef] [PubMed]
34. Miyake, C.; Miyata, M.; Shinzaki, Y.; Tomizawa, K. CO_2 response of cyclic electron flow around PSI (CEF-PSI) in tobacco leaves–relative electron fluxes through PSI and PSII determine the magnitude of non-photochemical quenching (NPQ) of Chl fluorescence. *Plant. Cell Physiol.* **2005**, *46*, 629–637. [CrossRef] [PubMed]
35. Zivcak, M.; Brestic, M.; Kunderlikova, K.; Sytar, O.; Allakhverdiev, S.I. Repetitive light pulse-induced photoinhibition of photosystem I severely affects CO_2 assimilation and photoprotection in wheat leaves. *Photosynth. Res.* **2015**, *126*, 449–463. [CrossRef]
36. Zivcak, M.; Kalaji, H.M.; Shao, H.B.; Olsovska, K.; Brestic, M. Photosynthetic proton and electron transport in wheat leaves under prolonged moderate drought stress. *J. Photochem. Photobiol. B* **2014**, *137*, 107–115. [CrossRef]
37. Miyake, C.; Yokota, A. Determination of the rate of photoreduction of O_2 in the water-water cycle in watermelon leaves and enhancement of the rate by limitation of photosynthesis. *Plant. Cell Physiol.* **2000**, *41*, 335–343. [CrossRef]
38. Sejima, T.; Takagi, D.; Fukayama, H.; Makino, A.; Miyake, C. Repetitive short-pulse light mainly inactivates photosystem I in sunflower leaves. *Plant. Cell Physiol.* **2014**, *55*, 1184–1193. [CrossRef]
39. Charepanov, D.A.; Milanovsky, G.E.; Petrova, A.A.; Tikhonov, A.N.; Semenov, A.Y. Electron transfer through the acceptor side of photosystem I: Interaction with exogenous acceptors and molecular oxygen. *Biochemistry* **2017**, *82*, 1249–1268. [CrossRef]
40. Takahashi, M.; Asada, K. Dependence of oxygen affinity for Mehler reaction on photochemical activity of chloroplast thylakoids. *Plant. Cell Physiol.* **1982**, *23*, 1457–1461.

41. Takahashi, M.A.; Asada, K. Superoxide anion permeability of phospholipid membranes and chloroplast thylakoids. *Arch. Biochem. Biophys.* **1983**, *226*, 558–566. [CrossRef]
42. Shimakawa, G.; Miyake, C. Oxidation of P700 Ensures Robust Photosynthesis. *Front. Plant. Sci.* **2018**, *9*, 1617. [CrossRef] [PubMed]
43. Khorobrykh, S.; Havurinne, V.; Mattila, H.; Tyystjarvi, E. Oxygen and ROS in Photosynthesis. *Plants* **2020**, *9*. [CrossRef] [PubMed]
44. Tikkanen, M.; Grebe, S. Switching off photoprotection of photosystem I-a novel tool for gradual PSI photoinhibition. *Physiol. Plant.* **2018**, *162*, 156–161. [CrossRef]
45. Scheller, H.V.; Haldrup, A. Photoinhibition of photosystem I. *Planta* **2005**, *221*, 5–8. [CrossRef]
46. Havaux, M.; Davaud, A. Photoinhibition of photosynthesis in chilled potato leaves is not correlated with a loss of Photosystem-II activity: Preferential inactivation of Photosystem I. *Photosynth. Res.* **1994**, *40*, 75–92. [CrossRef]
47. Golding, A.J.; Johnson, G.N. Down-regulation of linear and activation of cyclic electron transport during drought. *Planta* **2003**, *218*, 107–114. [CrossRef]
48. Golding, A.J.; Joliot, P.; Johnson, G.N. Equilibration between cytochrome f and P700 in intact leaves. *Biochim. Biophys. Acta* **2005**, *1706*, 105–109. [CrossRef]
49. Harbinson, J.; Foyer, C.H. Relationships between the Efficiencies of Photosystems I and II and Stromal Redox State in CO_2-Free Air: Evidence for Cyclic Electron Flow in Vivo. *Plant. Physiol.* **1991**, *97*, 41–49. [CrossRef]
50. Harbinson, J.; Genty, B.; Baker, N.R. Relationship between the Quantum Efficiencies of Photosystems I and II in Pea Leaves. *Plant. Physiol.* **1989**, *90*, 1029–1034. [CrossRef]
51. Harbinson, J.; Genty, B.; Baker, N.R. The relationship between CO_2 assimilation and electron transport in leaves. *Photosynth. Res.* **1990**, *25*, 213–224. [CrossRef]
52. Harbinson, J.; Genty, B.; Foyer, C.H. Relationship between Photosynthetic Electron Transport and Stromal Enzyme Activity in Pea Leaves: Toward an Understanding of the Nature of Photosynthetic Control. *Plant. Physiol.* **1990**, *94*, 545–553. [CrossRef] [PubMed]
53. Harbinson, J.; Hedley, C.L. Changes in P-700 Oxidation during the Early Stages of the Induction of Photosynthesis. *Plant. Physiol.* **1993**, *103*, 649–660. [CrossRef] [PubMed]
54. Miyake, C.; Shinzaki, Y.; Miyata, M.; Tomizawa, K. Enhancement of cyclic electron flow around PSI at high light and its contribution to the induction of non-photochemical quenching of chl fluorescence in intact leaves of tobacco plants. *Plant. Cell Physiol.* **2004**, *45*, 1426–1433. [CrossRef] [PubMed]
55. Klughammer, C.; Schreiber, U. An improved method, using saturating light pulses, for the determination of photosystem I quantum yield via $P700^+$-absorbance changes at 830 nm. *Planta* **1994**, *192*, 261–268. [CrossRef]
56. Nuijs, A.M.; Shuvalov, V.A.; van Gorkom, H.J.; Plijter, J.J.; Duysens, L.N. Picosecond absorbance difference spectroscopy on the primary reactions and the antenna-excited states in Photosystem I particle. *Biochim. Biophys. Acta Bioenerg.* **1986**, *850*, 310–318. [CrossRef]
57. Bukhov, N.G.; Carpentier, R. Measurement of photochemical quenching of absorbed quanta in photosystem I of intact leaves using simultaneous measurements of absorbance changes at 830 nm and thermal dissipation. *Planta* **2003**, *216*, 630–638. [CrossRef]
58. Shimakawa, G.; Shaku, K.; Miyake, C. Oxidation of P700 in Photosystem I Is Essential for the Growth of Cyanobacteria. *Plant. Physiol.* **2016**, *172*, 1443–1450. [CrossRef]
59. Genty, B.; Briantais, J.; Baker, N.R. The relationship between the quantum yield of photosynthetic electron transport and quenching of chlorophyll fluorescence. *Biochim. Biophys. Acta* **1989**, *990*, 87–92. [CrossRef]
60. Ghashghaie, J.; Cornic, G. Effect of temperature on partioning of photosynthetic electron flow between CO_2 assimilation and O_2 reduction and on the CO_2/O_2 specificity of Rubisco. *J. Plant. Physiol.* **1994**, *143*, 642–650. [CrossRef]
61. Kadota, K.; Furutani, R.; Makino, A.; Suzuki, Y.; Wada, S.; Miyake, C. Oxidation of P700 Induces Alternative Electron Flow in Photosystem I in Wheat Leaves. *Plants* **2019**, *8*. [CrossRef] [PubMed]
62. Kanazawa, A.; Kramer, D.M. In vivo modulation of nonphotochemical exciton quenching (NPQ) by regulation of the chloroplast ATP synthase. *Proc. Natl. Acad. Sci. USA* **2002**, *99*, 12789–12794. [CrossRef] [PubMed]
63. Kanazawa, A.; Ostendorf, E.; Kohzuma, K.; Hoh, D.; Strand, D.D.; Sato-Cruz, M.; Savage, L.; Cruz, J.A.; Fisher, N.; Froehlich, J.E.; et al. Chloroplast ATP Synthase Modulation of the Thylakoid Proton Motive Force: Implications for Photosystem I and Photosystem II Photoprotection. *Front. Plant. Sci.* **2017**, *8*, 719. [CrossRef] [PubMed]

64. Avenson, T.J.; Cruz, J.A.; Kanazawa, A.; Kramer, D.M. Regulating the proton budget of higher plant photosynthesis. *Proc. Natl. Acad. Sci. USA* **2005**, *102*, 9709–9713. [CrossRef]
65. Avenson, T.J.; Cruz, J.A.; Kramer, D.M. Modulation of energy-dependent quenching of excitons in antennae of higher plants. *Proc. Natl. Acad. Sci. USA* **2004**, *101*, 5530–5535. [CrossRef]
66. Sejima, T.; Hanawa, H.; Shimakawa, G.; Takagi, D.; Suzuki, Y.; Fukayama, H.; Makino, A.; Miyake, C. Post-illumination transient O2 -uptake is driven by photorespiration in tobacco leaves. *Physiol. Plant.* **2016**, *156*, 227–238. [CrossRef]
67. Tikhonov, A.N. The cytochrome *b6/f* complex at the crossroad of photosynthetic electron transport pathways. *Plant. Physiol. Biochem.* **2014**, *81*, 163–183. [CrossRef]
68. Tikkanen, M.; Gollan, P.J.; Mekala, N.R.; Isojarvi, J.; Aro, E.M. Light-harvesting mutants show differential gene expression upon shift to high light as a consequence of photosynthetic redox and reactive oxygen species metabolism. *Philos. Trans. R. Soc. Lond. B Biol. Sci.* **2014**, *369*, 20130229. [CrossRef]
69. Tikkanen, M.; Rantala, S.; Aro, E.M. Electron flow from PSII to PSI under high light is controlled by PGR5 but not by PSBS. *Front. Plant. Sci.* **2015**, *6*, 521. [CrossRef]
70. Shaku, K.; Shimakawa, G.; Hashiguchi, M.; Miyake, C. Reduction-Induced Suppression of Electron Flow (RISE) in the Photosynthetic Electron Transport System of Synechococcus elongatus PCC 7942. *Plant. Cell Physiol.* **2016**, *57*, 1443–1453. [CrossRef]
71. Shimakawa, G.; Shaku, K.; Miyake, C. Reduction-Induced Suppression of Electron Flow (RISE) Is Relieved by Non-ATP-Consuming Electron Flow in Synechococcus elongatus PCC 7942. *Front. Microbiol.* **2018**, *9*, 886. [CrossRef] [PubMed]
72. Miyake, C.; Amako, K.; Shiraishi, N.; Sugimoto, T. Acclimation of tobacco leaves to high light intensity drives the plastoquinone oxidation system-relationship among the fraction of open PSII centers, non-photochemical quenching of Chl fluorescence and the maximum quantum yield of PSII in the dark. *Plant. Cell Physiol.* **2009**, *50*, 730–743. [CrossRef] [PubMed]
73. Baker, N.R.; Harbinson, J.; Kramer, D.M. Determining the limitations and regulation of photosynthetic energy transduction in leaves. *Plant. Cell Environ.* **2007**, *30*, 1107–1125. [CrossRef] [PubMed]
74. Rantala, S.; Lempiainen, T.; Gerotto, C.; Tiwari, A.; Aro, E.M.; Tikkanen, M. PGR5 and NDH-1 systems do not function as protective electron acceptors but mitigate the consequences of PSI inhibition. *Biochim. Biophys. Acta Bioenerg.* **2020**, *1861*, 148154. [CrossRef] [PubMed]
75. Zhang, M.M.; Fan, D.Y.; Murakami, K.; Badger, M.R.; Sun, G.Y.; Chow, W.S. Partially dissecting electron fluxes in both photosystems in spinach leaf discs during photosynthetic induction. *Plant. Cell Physiol.* **2019**. [CrossRef]
76. Tikhonov, A.N. The Cytochrome b 6 f Complex: Biophysical Aspects of Its Functioning in Chloroplasts. *Subcell. Biochem.* **2018**, *87*, 287–328. [CrossRef]
77. Rott, M.; Martins, N.F.; Thiele, W.; Lein, W.; Bock, R.; Kramer, D.M.; Schottler, M.A. ATP synthase repression in tobacco restricts photosynthetic electron transport, CO2 assimilation, and plant growth by overacidification of the thylakoid lumen. *Plant. Cell* **2011**, *23*, 304–321. [CrossRef]
78. Anderson, J.M. Cytochrome *b6/f* complex: Dynamic molecular organization, function and acclimation. *Photosynth. Res.* **1992**, *34*, 341–357. [CrossRef]
79. Hope, A.B.; Valente, P.; Matthews, D.B. Effects of pH on the kinetics of redox reactions in and around the cytochrome *b6/f* complex in an isolated system. *Photosynth. Res.* **1994**, *42*, 111–120. [CrossRef]
80. Hope, A.B.; Matthews, D.B.; Valente, P. The kinetics of reactions around the cytochrome *b6/f* complex studied in an isolated system. *Photosynth. Res.* **1994**, *40*, 199–206. [CrossRef]
81. Hope, A.B. Electron transfers amongst cytochrome f, plastocyanin and photosystem I: Kinetics and mechanisms. *Biochim. Biophys. Acta* **2000**, *1456*, 5–26. [CrossRef]
82. Kramer, D.M.; Avenson, T.J.; Edwards, G.E. Dynamic flexibility in the light reactions of photosynthesis governed by both electron and proton transfer reactions. *Trends Plant. Sci.* **2004**, *9*, 349–357. [CrossRef] [PubMed]
83. Johnson, M.P.; Ruban, A.V. Rethinking the existence of a steady-state $\Delta\psi$ component of the proton motive force across plant thylakoid membranes. *Photosynth. Res.* **2014**, *119*, 233–242. [CrossRef]
84. Kramer, D.M.; Johnson, G.; Kiirats, O.; Edwards, G.E. New Fluorescence Parameters for the Determination of QA Redox State and Excitation Energy Fluxes. *Photosynth. Res.* **2004**, *79*, 209. [CrossRef] [PubMed]

85. Foyer, C.; Furbank, R.; Harbinson, J.; Horton, P. The mechanisms contributing to photosynthetic control of electron transport by carbon assimilation in leaves. *Photosynth. Res.* **1990**, *25*, 83–100. [CrossRef] [PubMed]
86. Schottler, M.A.; Toth, S.Z. Photosynthetic complex stoichiometry dynamics in higher plants: Environmental acclimation and photosynthetic flux control. *Front. Plant. Sci.* **2014**, *5*, 188. [CrossRef] [PubMed]
87. Shimakawa, G.; Murakami, A.; Niwa, K.; Matsuda, Y.; Wada, A.; Miyake, C. Comparative analysis of strategies to prepare electron sinks in aquatic photoautotrophs. *Photosynth. Res.* **2019**, *139*, 401–411. [CrossRef]
88. Shimakawa, G.; Shaku, K.; Nishi, A.; Hayashi, R.; Yamamoto, H.; Sakamoto, K.; Makino, A.; Miyake, C. FLAVODIIRON2 and FLAVODIIRON4 proteins mediate an oxygen-dependent alternative electron flow in Synechocystis sp. PCC 6803 under CO_2-limited conditions. *Plant. Physiol.* **2015**, *167*, 472–480. [CrossRef]
89. Helman, Y.; Tchernov, D.; Reinhold, L.; Shibata, M.; Ogawa, T.; Schwarz, R.; Ohad, I.; Kaplan, A. Genes encoding A-type flavoproteins are essential for photoreduction of O2 in cyanobacteria. *Curr. Biol.* **2003**, *13*, 230–235. [CrossRef]
90. Hayashi, R.; Shimakawa, G.; Shaku, K.; Shimizu, S.; Akimoto, S.; Yamamoto, H.; Amako, K.; Sugimoto, T.; Tamoi, M.; Makino, A.; et al. O_2-dependent large electron flow functioned as an electron sink, replacing the steady-state electron flux in photosynthesis in the cyanobacterium Synechocystis sp. PCC 6803, but not in the cyanobacterium Synechococcus sp. PCC 7942. *Biosci. Biotechnol. Biochem.* **2014**, *78*, 384–393. [CrossRef]
91. von Caemmerer, S. *Biochemical Models of Leaf Photosynthesis*; CSIRO Publishing: Collingwood, Australia, 2000.
92. Asada, K.; Kiso, K.; Yoshikawa, K. Univalent reduction of molecular oxygen by spinach chloroplasts on illumination. *J. Biol. Chem.* **1974**, *249*, 2175–2181.
93. Asada, K.; Takahashi, M. Production and scavenging of active oxygen in photosynthesis. In *Photoinhibition*; Kyle, D.J., Osmond, C.B., Arntzen, C.J., Eds.; Elsevier: Amsterdam, The Netherland, 1987; pp. 227–287.
94. Bersanini, L.; Battchikova, N.; Jokel, M.; Rehman, A.; Vass, I.; Allahverdiyeva, Y.; Aro, E.M. Flavodiiron protein Flv2/Flv4-related photoprotective mechanism dissipates excitation pressure of PSII in cooperation with phycobilisomes in Cyanobacteria. *Plant. Physiol.* **2014**, *164*, 805–818. [CrossRef] [PubMed]
95. Allahverdiyeva, Y.; Mustila, H.; Ermakova, M.; Bersanini, L.; Richaud, P.; Ajlani, G.; Battchikova, N.; Cournac, L.; Aro, E.M. Flavodiiron proteins Flv1 and Flv3 enable cyanobacterial growth and photosynthesis under fluctuating light. *Proc. Natl. Acad. Sci. USA* **2013**, *110*, 4111–4116. [CrossRef] [PubMed]
96. Burlacot, A.; Sawyer, A.; Cuine, S.; Auroy-Tarrago, P.; Blangy, S.; Happe, T.; Peltier, G. Flavodiiron-Mediated O_2 Photoreduction Links H2 Production with CO_2 Fixation during the Anaerobic Induction of Photosynthesis. *Plant. Physiol.* **2018**, *177*, 1639–1649. [CrossRef] [PubMed]
97. Shimakawa, G.; Matsuda, Y.; Nakajima, K.; Tamoi, M.; Shigeoka, S.; Miyake, C. Diverse strategies of O_2 usage for preventing photo-oxidative damage under CO_2 limitation during algal photosynthesis. *Sci. Rep.* **2017**, *7*, 41022. [CrossRef] [PubMed]
98. Hanawa, H.; Ishizaki, K.; Nohira, K.; Takagi, D.; Shimakawa, G.; Sejima, T.; Shaku, K.; Makino, A.; Miyake, C. Land plants drive photorespiration as higher electron-sink: Comparative study of post-illumination transient O_2 -uptake rates from liverworts to angiosperms through ferns and gymnosperms. *Physiol. Plant.* **2017**, *161*, 138–149. [CrossRef] [PubMed]

6

Inactivation of Carbonyl-Detoxifying Enzymes by H_2O_2 is a Trigger to Increase Carbonyl Load for Initiating Programmed Cell Death in Plants

Md. Sanaullah Biswas [1,*], Ryota Terada [2] and Jun'ichi Mano [3,4,*]

1 Department of Horticulture, Bangabandhu Sheikh Mujibur Rahman Agricultural University, Gazipur 1706, Bangladesh
2 Faculty of Agriculture, Yamaguchi University, Yoshida 1677-1, Yamaguchi 753-8515, Japan; y.rt.univ028.ge78@gmail.com
3 Science Research Center, Organization of Research Initiatives, Yamaguchi University, Yamaguchi 753-8511, Japan
4 Graduate School of Science and Technology for Innovation, Yamaguchi University, Yamaguchi 753-8511, Japan
* Correspondence: sanaullah@bsmrau.edu.bd (M.S.B.); mano@yamaguchi-u.ac.jp (J.M.)

Abstract: H_2O_2-induced programmed cell death (PCD) of tobacco Bright Yellow-2 (BY-2) cells is mediated by reactive carbonyl species (RCS), degradation products of lipid peroxides, which activate caspase-3-like protease (C3LP). Here, we investigated the mechanism of RCS accumulation in the H_2O_2-induced PCD of BY-2 cells. The following biochemical changes were observed in 10-min response to a lethal dose (1.0 mM) of H_2O_2, but they did not occur in a sublethal dose (0.5 mM) of H_2O_2. (1) The C3LP activity was increased twofold. (2) The intracellular levels of RCS, i.e., 4-hydroxy-(*E*)-hexenal and 4-hydroxy-(*E*)-nonenal (HNE), were increased 1.2–1.5-fold. (3) The activity of a reduced nicotinamide adenine dinucleotide phosphate (NADPH)-dependent carbonyl reductase, scavenging HNE, and *n*-hexanal was decreased. Specifically, these are the earliest events leading to PCD. The proteasome inhibitor MG132 suppressed the H_2O_2-induced PCD, indicating that the C3LP activity of the β1 subunit of the 20S proteasome was responsible for PCD. The addition of H_2O_2 to cell-free protein extract inactivated the carbonyl reductase. Taken together, these results suggest a PCD-triggering mechanism in which H_2O_2 first inactivates a carbonyl reductase(s), allowing RCS levels to rise, and eventually leads to the activation of the C3LP activity of 20S proteasome. The carbonyl reductase thus acts as an ROS sensor for triggering PCD.

Keywords: lipid peroxide; oxidative stress; oxylipin; reactive carbonyl species (RCS); reactive electrophile species (RES); redox signal; ROS

---

1. Introduction

Reactive oxygen species (ROS), such as superoxide radical and hydrogen peroxide (H_2O_2), are metabolic outcomes of various cellular processes in plants. Versatile biological actions of ROS, collectively termed as oxidative signals, largely depends on the ROS concentration [1]. Recent evidence shows that ROS are involved in diverse signaling functions in plants, from cell cycle regulation to whole plant senescence [2]. The transition from cellular proliferation to cell elongation, which is necessary for plant growth in the early stages of cell differentiation, is regulated by ROS signals [3]. The levels of superoxide radical and H_2O_2 are high in the elongation and meristematic zone of Arabidopsis root [4]. ROS also play an important role in seed germination; superoxide and H_2O_2 are increased in the radicle of Arabidopsis during germination [5]. Thus, ROS signals are necessary in various stages of plant growth and development.

Programmed cell death (PCD) is an essential phenomenon in plants for maintaining their growth and development. In many cases, PCD is started by the rise of ROS levels, which is stimulated by various endogenous or exogenous stimuli. Aleurone layer of cereal grains and the lateral root cap cells are eliminated as a result of PCD with a rise of H_2O_2 level [6,7]. Under water deficit condition rice developing anthers started PCD due to enhancement of H_2O_2 level and down-regulation of antioxidant transcripts [8]. In response to salinity and sorbitol stress, tobacco BY-2 cells increase the level of superoxide radical, which triggers PCD [9]. High temperature causes a rise in the H_2O_2 level in tobacco BY-2 cells before initiation of PCD [10]. H_2O_2 is also involved in the hypersensitive response (HR)-like cell death in tobacco leaves [11]. However, the mechanism of ROS action to initiate PCD was largely unclear.

Recently, we found that, in the mechanism by which ROS initiates PCD, the lipid peroxide-derived reactive carbonyl species (RCS) play critical roles in conveying ROS signals [12]. RCS is a group name of the α,β-unsaturated aldehydes and ketones, such as acrolein and 4-hydroxy-(*E*)-2-nonenal (HNE) [13,14]. RCS are generated from lipid peroxides (LOOH) and act as agents to mediate ROS signal to target proteins in the heat shock-responsive gene regulation [15], senescence [16], abscisic acid (ABA) signaling for stomatal closure [17,18], and auxin signaling for lateral root formation [19] in plants. We reported that tobacco BY-2 cells exposed to H_2O_2 generated acrolein and HNE, and a chemical carbonyl scavenger prevented the initiation of PCD, without affecting the increase in the intracellular ROS level [12].

We further revealed that the targets of RCS were two caspase-like proteases [20]. Caspases are cysteine-containing aspartate-specific proteases that are involved in the initiation of PCD in animals. Plants do not have proteins homologous to animal caspases but have caspase-like proteases that show the same substrate specificity for the artificial peptides to caspase-1 and caspase-3. Vacuolar processing enzyme (VPE), an ortholog of asparaginyl endopeptidase (legumain), shows the caspase-1-like protease (C1LP) activity and is responsible for processing vacuolar proteins [21]. The functions of C1LP in plants and caspase-1 in animals to determine cell fate are comparable [22]. Activation of C1LP is associated with the aluminum-induced PCD in the root tip of a peanut [23]. The caspase-3-like protease (C3LP) is also involved in the initiation of PCD in cadmium-induced PCD [24] and hypersensitive response-induced PCD [25] in *Arabidopsis thaliana*. High temperature treatment to tobacco BY-2 cells increases the H_2O_2 level, followed by the activation of C3LP before PCD occurs [10]. In our previous study [20], we found that H_2O_2 addition to BY-2 cells activated both C3LP and C1LP before PCD symptoms become apparent. The addition of acrolein and HNE to the cell-free extract rapidly activated both C3LP and C1LP, but the addition of H_2O_2 to the extract failed to activate the proteases. These results indicated that, in BY-2 cells, the RCS accumulating due to the H_2O_2 stimulus activated C3LP and C1LP and started PCD. Thus, the activation of C3LP and C1LP was ascribed to RCS that was generated by the oxidative stimulus.

We tentatively concluded that C3LP was more relevant than C1LP to the initiation of PCD because, on addition of lethal dose of acrolein, the activation of the former was more rapid and larger than that of the latter [20]. However, the contribution of C1LP to the H_2O_2-induced PCD remains open. It is also unclear which protein is responsible for the RCS-stimulated activity of C3LP because two distinct proteins have been known to show C3LP activity: the $\beta 1$ subunit of the 20S proteasome (PBA1) [25,26] and cathepsin B [27].

Another critical question is about the type of RCS that activated C3LP. In our previous study, we analyzed carbonyls in the cells 2 h after H_2O_2 addition and found that at least eight types of oxylipin carbonyls were increased [12]. The two potent RCS acrolein and HNE, found as the increasing carbonyls, showed the ability to activate C3LP and C1LP in vitro [20]. On the other hand, C3LP was activated within 30 min after H_2O_2 addition. It is therefore expected that certain types of RCS should be increased prior to the C3LP activation on the addition of H_2O_2 to BY-2 cells.

In this study, attempting to elucidate the detailed mechanism of the H_2O_2-induced PCD of BY-2 cells, we monitored early biochemical events in response to oxidative stimulus. We confirmed that

C3LP is critical for the PCD, and found its activity is attributable to PBA1. In response to H_2O_2, BY-2 cells accumulated HNE and 4-hydroxy-(*E*)-hexenal (HHE) in 10 min. Interestingly, at the same time, an NADPH-dependent carbonyl reductase(s), showing HNE-reducing activity, was largely inactivated. The same activity in cell-free extract was inactivated by the addition of H_2O_2, indicating that H_2O_2 attenuated the carbonyl reductase activity to enhance the RCS level. We propose a new mechanism of ROS-induced PCD, where the carbonyl reductase(s) acts as an ROS sensor for initiating the PCD process.

2. Materials and Methods

2.1. Culture of Cells

Cell suspension of tobacco BY-2 (*Nicotiana tabacum* L. cv. Bright Yellow-2) (RIKEN BioResource Research Center, Tsukuba, Japan) were cultured in Murashige and Skoog medium (MP Bio, Tokyo, Japan) supplemented with sucrose (30 g L^{-1}), myo-inositol (100 mg L^{-1}), KH_2PO_4 (200 mg L^{-1}), thiamine HCl (0.5 mg L^{-1}), and 2,4-dichlorophenoxyacetic acid (0.2 mg L^{-1}) (all these chemicals were purchased from FUJIFILM Wako Pure Chemical, Osaka, Japan) at pH 5.6. Cells were propagated with continuous rotation at 120 rpm in darkness at 25 °C. Every seven days, 0.5 mL dense cells were sub-cultured to a fresh medium (50 mL). The exponential growth phase cells at the fourth day were selected for experiment [28]. H_2O_2 was added to the culture medium (H_2O_2 treatment). After an incubation, the cells were harvested and washed with distilled water for analyses.

2.2. Cell Viability Assay

Cell viability was determined by fluorescein diacetate (FDA) staining. Harvested cells were incubated in FDA solution (1 µg mL^{-1}) for 5 min, washed twice with distilled water, and then observed under a fluorescence microscope (Leica AF6000, Wetzlar, Germany). Viable cells cleave FDA to form fluorescein, which fluoresce with excitation at 485 nm and emission at 515 nm. Dead cells do not fluoresce.

2.3. Determination of C1LP and C3LP Activities

Protein extract from BY-2 cells was prepared as described previously [20,21]. The harvested cells were frozen with liquid nitrogen and ground to powder with a mortar and pestle. Then, the ground powder of cells was transferred to a centrifuge tube and immediately added the protease extraction medium containing 50 mM sodium acetate, pH 5.5, 50 mM NaCl, 1 mM EDTA, 1 mM phenylmethylsulfonyl fluoride (PMSF), and 0.1 mM E64-d (a thiol protease inhibitor). After centrifugation at 14,000× *g* for 30 min at 4 °C, supernatant was collected as the protein extract.

The tetrapeptide fluorogenic substrates Ac-YVAD-AMC for C1LP and Ac-DEVD-AMC for C3LP and the respective inhibitors (Ac-YVAD-CHO and Ac-DEVD-CHO) were purchased from Peptide Institute (Osaka, Japan). For measuring a protease activity, two tubes were prepared. One had a reaction mixture (0.2 mL) containing the protease sample (50 µL of protein extract or 50 µg of protein) in 20 mM Na-acetate, pH 5.5, 0.1 M dithiothreitol, 0.1 mM EDTA, and 1 mM PMSF with an inhibitor at 0.1 mM, and the other had the same reaction mixture without the inhibitor. They were preincubated at 37 °C for 1 h. Then, the substrate peptide was added to each tube at 0.1 mM. After an incubation at 37 °C for 1 h, the AMC released from the substrate was determined with a fluorescence microplate reader (Twincle LB970, Berthold Japan, Tokyo, Japan) (excitation 380 nm; emission 445 nm). The specific activity of proteases was calculated from the difference between the absence and the presence of an inhibitor. A standard curve was developed using a series of AMC (Peptide Institute) solutions in the 0 nM to 200 nM range.

In vitro activation of C1LP and C3LP activity was done by the addition of either acrolein or H_2O_2 to protein extract. After incubation, the protein extract was allowed passage through a PD MiniTrap G-25 column (GE Healthcare, Tokyo, Japan), equilibrated with protease extraction medium, to remove small molecules. A 0.2 mL of reaction mixture with the eluted protein extract was prepared similarly as described above.

2.4. Analysis of Glutathione

Total glutathione (GSH + oxidized form (GSSG)) was determined as described previously [20,29] with a minor modification. Briefly, harvested cells (about 0.4 g) were immediately ground in liquid nitrogen with mortar and pestle and suspended in two volumes of cold 5% sulphosalicylic acid in a 1.5 mL centrifuge tube. For glutathione analysis, the supernatant was collected after centrifugation at 20,000× *g* for 15 min at 4 °C. The cell extract (a 0.1 mL aliquot) was neutralized by mixing with 0.9 mL of 0.1 M HEPES-KOH, pH 7.4, containing 5 mM EDTA. The mixture was divided into two fractions for determining total GSH (GSH + oxidized form (GSSG)) and GSSG. A 1.0 mL reaction mixture was prepared with 0.3 mL of neutralized cell extract containing 0.1 M HEPES-KOH, pH 7.4, 5 mM EDTA, 10 mM 5,5′-dithiobis-2-nitrobenzoic acid (DTNB), 0.5 unit glutathione reductase. The reaction was started with the addition of 0.2 mM NADPH, and the rate of absorbance increase at 412 nm was recorded for 1 min. The GSSG content was determined after masking GSH by 20 μL of 2-vinylpyridine to the neutralized cell extract. The emulsified residual 2-vinylpyridine was removed through a brief centrifuge; then, GSSG was determined, as described above, for total GSH. The GSH content was estimated after the subtraction of GSSG to the total GSH. A GSH standard curve was prepared with authentic GSH (Wako, Japan) in the range of 0–0.1 mM.

2.5. Analysis of Ascorbate

Asc and dehydroascorbic acid (DHA) were determined as described previously [20,30] with some modifications. The cells were extracted with the same protocol of GSH analysis. The total ascorbate (Asc + DHA) content was estimated in a 1 mL reaction mixture containing 0.1 mL aliquot of cell extract, 0.2 mL of 150 mM phosphate, pH 7.4, containing 5 mM EDTA, and 0.05 mL of 10 mM dithiothreitol (DTT). To remove excess DTT, 0.05 mL of 0.5% *N*-ethylmaleimide was added after 10-min incubation at room temperature. To determine Asc, 0.1 mL of water was added in the reaction mixture instead of DTT and *N*-ethylmaleimide. Then, the reaction mixture was incubated at 40 °C for 40 min with 0.2 mL of 10% trichloroacetic acid, 0.2 mL of 44% ortho-phosphoric acid, 0.2 mL of 4% α,α-dipyridyl in 70% ethanol, and 0.3% (*w*/*v*) $FeCl_3$ for color development. The absorbance at 525 nm was measured, and the concentration of DHA was calculated from the difference of total ascorbate (Asc + DHA) and Asc. A standard curve was prepared with Asc in the 0–250 μM range.

2.6. Detection and Quantification of Carbonyls Using HPLC

The fresh cells (0.4 g) were filtered and washed twice with distilled water, then immediately transferred into a screw-capped glass tube containing 2.5 mL of acetonitrile containing internal standard (25 nmol of 2-ethylhexanal) and 0.005% (*w*/*v*) butylhydroxytoluene to prevent autoxidation of unsaturated organic compounds. Then, the tube was closed and incubated in a water bath at 60 °C for 30 min. Avoiding cell debris, only clear solution was collected by Pasteur pipette to another glass tube, and the carbonyls were derivatized with 2,4-dinitrophenylhydrazine (0.5 mM) and formic acid (0.5 M) for 60 min at 25 °C. Then, the formic acid in the solution neutralized with 0.45 g of $NaHCO_3$ in 2.5 mL of saturated NaCl. After thorough mixing and subsequent centrifugation, the upper layer was collected, dried, and dissolved in a smaller volume of acetonitrile. The hydrazone-derivatized carbonyls were determined on a reverse phase (C30) HPLC [12,17,31,32].

2.7. Determination of Enzymatic Activity

Proteins were extracted as follows. The cells harvested and washed on a nylon mesh were frozen with liquid nitrogen and ground to powder with a mortar and pestle. About 0.4 g powdered cells were transferred to the reductase extraction medium containing 50 mM KH_2PO_4, pH 7.4, 1 mM EDTA, and 1×Protease Inhibitor Cocktail (Sigma-Aldrich Japan, Tokyo, Japan) and extensively mixed. After centrifugation with 10,000× *g* for 10 min at 4 °C, 0.5 mL cell extract was allowed passage of the PD MiniTrap G-25 column equilibrated with reductase extraction medium. Protein content was determined with the Protein Assay CBB solution (Nacalai Tesque, Kyoto, Japan) with bovine serum albumin as the standard.

The aldehyde reducing activity by the reductases of the protein extract was measured using 50 mM Mes-NaOH, pH 6.0, containing 0.1 mM NADPH. Various species of aldehydes (HNE, *n*-hexanal, acetaldehyde, butyraldehyde, (Z)-3-Hexenal) were used as the substrate of reductases and the rate of oxidation of NADPH was measured at 340 nm using a Shimadzu (Kyoto, Japan) MPS-2000 spectrophotometer.

3. Results

3.1. Threshold Level of Hydrogen Peroxide for Inducing PCD

Because addition of H_2O_2 to BY-2 cells caused a rapid increase (within 10 min) in C3LP and C1LP [20], it is expected that certain types RCS are generated in the cells. On the other hand, addition of H_2O_2 is generally oxidative stimulus, and it will cause lipid peroxidation, leading to the formation of various RCS. We attempted to distinguish the RCS that are responsible for the triggering of PCD from those generated as a consequence of the progress of oxidative stress. For this purpose, we have set up two H_2O_2 treatment conditions, i.e., lethal and sublethal. BY-2 cells were treated with H_2O_2 at various concentrations, and it was found that 1 mM H_2O_2 induced about 75% cell death in 5-h incubation, while 0.5 mM H_2O_2 did not increase cell death as compared with untreated control (Figure 1). Specifically, 1 mM H_2O_2 was a lethal dose to induce cell death and 0.5 mM was sublethal. The biochemical events that triggered PCD thus should have occurred earlier than 5 h only in the cells treated with 1 mM H_2O_2 but should not in those treated with 0.5 mM H_2O_2.

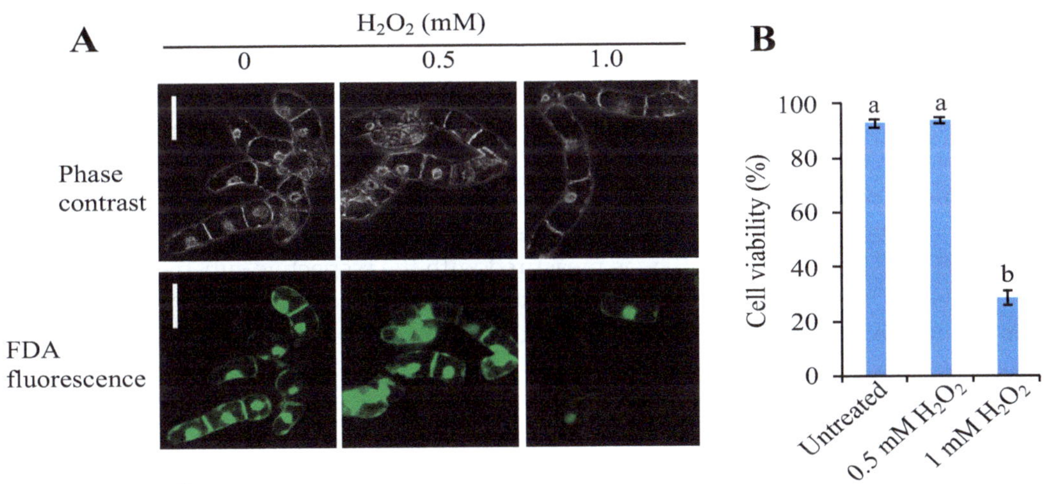

Figure 1. Effects of H_2O_2 concentration on the viability of Bright Yellow-2 (BY-2) cells. A 0.5 mL aliquot of cells from 7-d culture were sub-cultured in 50 mL of fresh culture medium and, after 4 d, the culture medium was supplemented with H_2O_2 to 0.5 mM or 1 mM. After 5 h incubation cells were harvested and stained with fluorescein diacetate (FDA). Living cells and dead cells were counted as described in the Materials and Methods section. Cells forming a single layer under microscopy were chosen for evaluation. (**A**) Typical phase-contrast (top) and FDA-fluorescence (bottom) images of the same field. Bar = 50 µm. (**B**) Percentage of cell viability. The total number of cells was counted under phase contrast observation, and the FDA-stained cells were counted under fluorescence observation. A total of 200 cells were observed in each treatment. Means ± SE of three independent experiments are shown. Differences among treatments were analyzed by Tukey's test: $p < 0.05$. Different letters among the treatments represent statistically significant differences.

3.2. Caspase-Like Proteases in Tobacco BY-2 Cells Are Activated Only When H_2O_2 Induces PCD

The addition of H_2O_2 to BY-2 cells at a lethal dose activated C3LP and C1LP within 30 min [20]. To compare the effects of sublethal and lethal doses of H_2O_2, we treated BY-2 cells with H_2O_2 at 0.5 and 1 mM and determined the caspase activities (Figure 2). Untreated cells exhibited constitutive levels of C3LP and C1LP activity. H_2O_2 at the lethal dose (1 mM) increased C3LP significantly in 10 min, and after 30 min, the activity reached 2.5-fold. A sublethal dose of H_2O_2 (0.5 mM) did not increase the C3LP activity (Figure 2A). By the lethal dose of H_2O_2, the C1LP activity was also increased to a significantly higher level in 60 min, while in 10 min, the increase was apparently smaller (Figure 2B). The sublethal dose of H_2O_2 caused a very slight increase in the C1LP activity. Thus, early activation of C3LP and C1LP was consistently associated with the later occurrence of PCD.

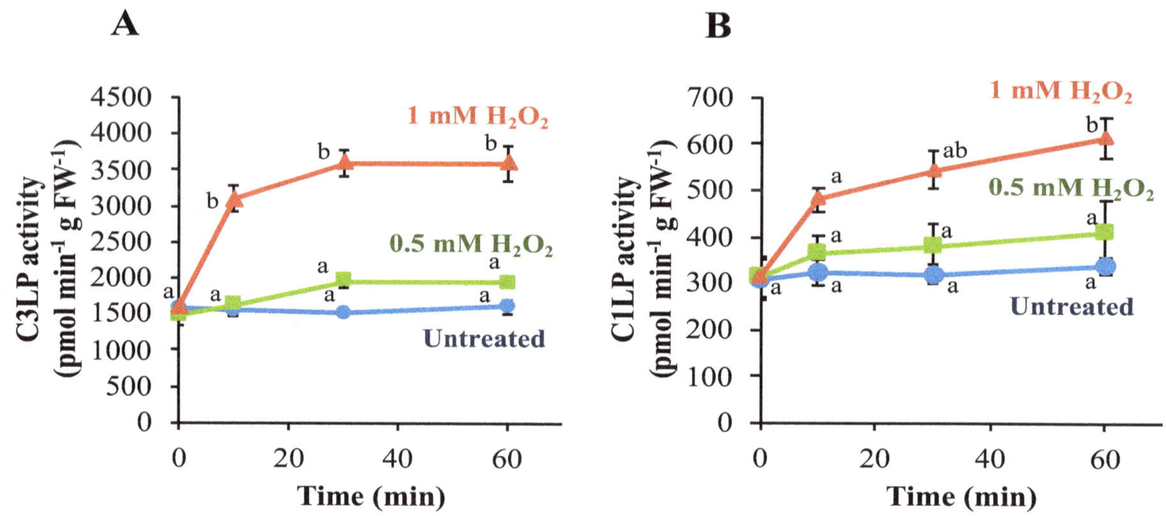

Figure 2. Activation of caspase-3-like protease (C3LP) (**A**) and caspase-1-like protease (C1LP) (**B**) in BY-2 cells after exposure to H_2O_2. Four-d cultured cells were treated with H_2O_2 to 0.5 mM or 1 mM, as in Figure 1. Proteins was extracted from the cells as described in the Material and Methods section. The activity of C3LP (**A**) and C1LP (**B**) was measured with the respective substrates Ac-DEVD-AMC and Ac-YVAD-AMC, as in the Materials and Methods section. The data are the mean ± SE. Different letters represent significantly different values ($p < 0.05$ on Tukey test).

3.3. Activation of C3LP, Rather Than C1LP, Triggers PCD

The activation of C3LP in response to a lethal dose of H_2O_2 was more rapid and to a larger extent than C1LP (Figure 2). To evaluate the contribution of each proteases to PCD initiation, we treated the cells with the inhibitors specific to C3LP or C1LP and examined whether one of them or both prevents H_2O_2-induced PCD or not (Figure 3). The H_2O_2 treatment caused PCD in 75% cells in 5 h. The addition of C3LP inhibitor efficiently suppressed H_2O_2-induced cell death; only 25% cells were dead. On the other side, C1LP inhibitor could not prevent H_2O_2-induced cell death effectively; 70% cells were dead. We also tested whether these caspase inhibitors prevent the PCD induced by 0.2 mM acrolein. The addition of acrolein at 0.2 mM induced 90% cell death after 5-h incubation (Figure S1). Addition of the C3LP inhibitor prior to the acrolein treatment suppressed the cell death to 30%, as compared with 80% death for the acrolein-treatment without protease inhibitors. On the other side, the C1LP inhibitor did not efficiently prevent acrolein-induced cell death. These results suggest that the activation of C3LP is critical for the PCD initiated by oxidative stimuli.

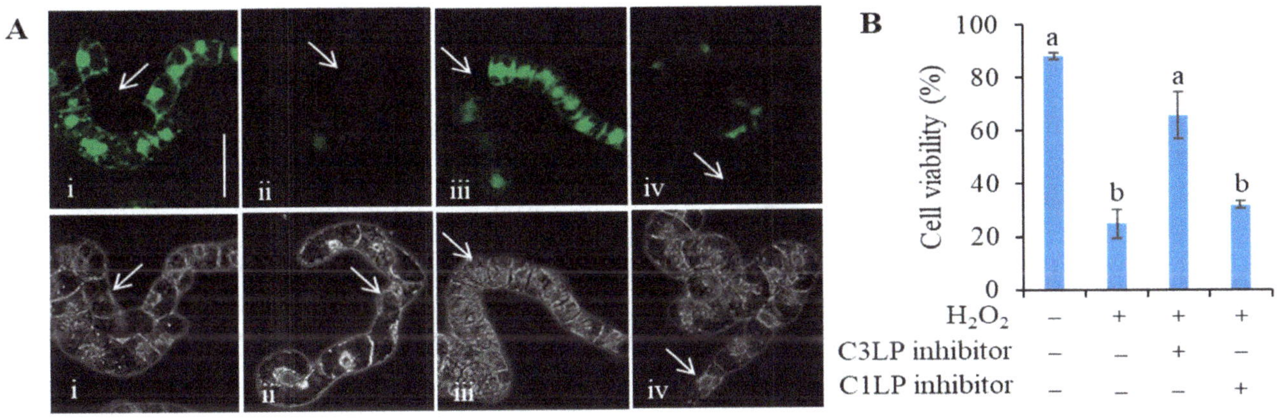

Figure 3. Effects of caspase inhibitors on the H_2O_2-induced cell death. Four-d cultured BY-2 cells were treated, as in Figure 1, with 1 mM H_2O_2 in the presence or absence of an inhibitor. The cells were harvested after 5-h incubation, and FDA-stained cells were counted as living cells. (**A**) Typical fluorescence images (top) and phase contrast images (bottom) of the same field: (i) untreated cells as control, (ii) 1 mM H_2O_2, (iii) 1 mM H_2O_2 + 20 µM C3LP inhibitor, and (iv) 1 mM H_2O_2 + 20 µM C1LP inhibitor (top). White arrows indicate dead cells. Bar = 50 µm. (**B**) Percentage of living cells. A total of 400 cells were observed in each treatment. Mean of 3 runs ± SEM. Different letters represent significantly different values ($p < 0.05$ on Tukey test).

3.4. The C3LP Activity of the 20S Proteasome Is Responsible for the H_2O_2-Initiated PCD

Two distinct proteins are known to show the C3LP activity, i.e., PBA1, a subunit of the 20S proteasome [25,26,33] and cathepsin B [27]. Vacca et al. [34] measured ROS accumulation and C3LP in the heat shocked tobacco BY-2 cells in absence and presence of the proteasome inhibitor MG132. They found that MG132 prevented PCD and specifically decreased ROS level and C3LP activity. To clarify which protein is really involved in the H_2O_2-induced PCD, we tested the effect of MG132. BY-2 cells were pre-incubated with MG132 at 0.1 mM for 30 min, and then H_2O_2 was added at 1 mM to induce PCD. H_2O_2 at 1 mM in 5-h incubation induced about 75% cell death, and proteasome inhibitor MG132 significantly reduced cell death to 40% (Figure 4). Thus, the C3LP activity responsible for PCD was attributed to the ß1 subunit of the 20S proteasome.

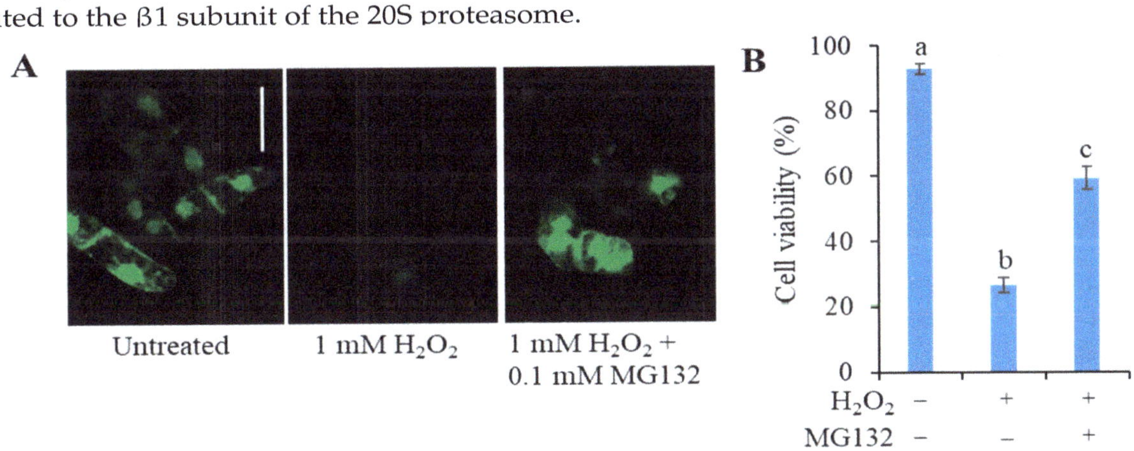

Figure 4. Effect of MG132 on the cell survival in H_2O_2-treated tobacco BY-2 cells. Four-d cultured cells were treated with either 1 mM H_2O_2 or 1 mM H_2O_2 plus 0.1 mM MG132. MG132 was added 30 min before addition of H_2O_2. After 5-h incubation, cell death was determined with FDA staining, as in Figure 1. (**A**) Typical photographs are shown: untreated cells as control (left), treated with H_2O_2 at 1 mM (middle), and H_2O_2 at 0.2 mM plus 0.1 mM MG132 (right). Bar = 50 µm. (**B**) Cell viability (%). All values are mean ± SE, and the data represent three independent experiments. Different letters represent significantly different values ($p < 0.05$ on Tukey test).

3.5. Limited Types of Oxylipin Carbonyls Are Increased in Very Early Stages of PCD

We presume that oxylipin carbonyls generated in the H_2O_2-stimulated cells activate C3LP and C1LP and thereby initiate PCD [20]. Because C3LP is activated rapidly on H_2O_2 addition (Figure 2), the oxylipin carbonyls to activate it should also be rapidly increased.

BY-2 cells were treated with the lethal (1 mM) and sublethal (0.5 mM) dose of H_2O_2, and after 10-min incubation, oxylipin carbonyls were extracted and quantified on HPLC (Figure S2). We found that the lethal dose of H_2O_2 increased several types of RCS and other carbonyls as compared with their basal levels in the untreated samples (Figure 5). Among the RCS, HNE increased significantly. HHE also showed an increasing trend although the difference was insignificant. The non-RCS carbonyls acetaldehyde, butyraldehyde, (Z)-3-hexenal, and *n*-hexanal were also increased significantly (Figure 5). In the cells treated with the sublethal dose of H_2O_2 (0.5 mM), HNE, HHE, acetaldehyde, and *n*-hexanal were not changed, but butyraldehyde and (Z)-3-hexenal were increased twofold compared with the untreated cells (Figure 5). When the lethal and sublethal conditions are compared, only limited types of oxylipin carbonyls, i.e., HNE (a RCS), acetaldehyde, and *n*-hexanal (non-RCS carbonyls), were increased. It is expected that these carbonyls are relevant to the C3LP activation and initiation of PCD.

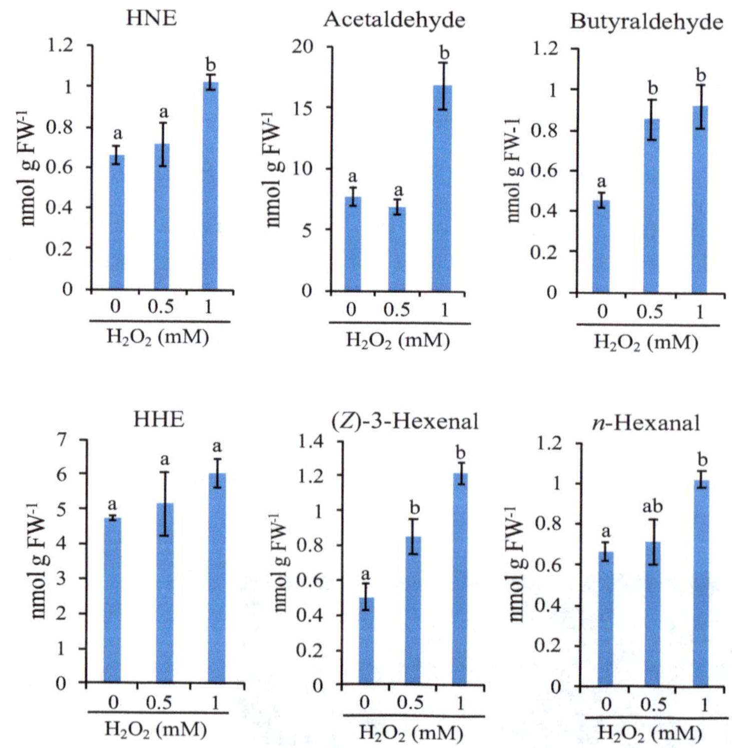

Figure 5. Effects of H_2O_2 on the carbonyl levels in BY-2 cells. To 4-d-cultured cells, H_2O_2 was added to the indicated concentration. After a 10-min incubation, cells were harvested, and the intracellular contents of carbonyl levels were determined as in the Materials and Methods section. Results of 4-hydroxy-(*E*)-2-nonenal (HNE), acetaldehyde, butyraldehyde, 4-hydroxy-(*E*)-hexenal (HHE), 3-(Z)-hexanal and *n*-hexanal are shown. Each point represents the mean of three independent experiments and the error bars the SEM. Different letters represent significantly different values ($p < 0.05$ on Tukey test).

HNE and *n*-hexanal have been known to induce PCD in tobacco BY-2 cells [12], and HNE can directly activate C3LP [20]. Here, we tested the ability of acetaldehyde for the activation of C3LP (Figure S3). Four-d cultured cells were treated with various concentrations of acetaldehyde, from 0.5 mM to 10 mM, and then measured for C3LP activity. Acetaldehyde up to 3 mM did not increase C3LP, but at 10 mM, it significantly increased the C3LP activity compared with that in the untreated cells. Thus, the three oxylipin carbonyls that were increased rapidly on the addition of lethal dose of H_2O_2 were capable of activating C3LP.

3.6. Effects of H_2O_2 on the Intracellular Glutathione and Ascorbate Levels

A possible cause of the RCS increases is the depletion of glutathione (GSH) in the cells because the thiol moiety of GSH provides a major defense against the various electrophiles, including RCS [35,36]. Consumption of ascorbate by H_2O_2 also may facilitate the loss of GSH pool and can be an indirect cause of RCS increase. To measure early changes in the GSH and ascorbate status, tobacco BY-2 cells were treated with a lethal and sublethal doses of H_2O_2 and sampled at different time points (Figure 6). Lethal level of H_2O_2 (1 mM) decreased total glutathione pool (GSH + GSSG) by 40% in 30 min. In 60-min incubation, this lowered level of GSH recovered by about 10% (Figure 6A). The GSH reduction ratio was not dropped by H_2O_2 at 1 mM in 30 min (Figure 6B). The sublethal level of H_2O_2 (0.5 mM) lowered the total glutathione pool by about 35% in 10 min, but afterwards, it recovered (Figure 6A). In contrast, 0.2 mM acrolein, a lethal level, consumed 90% glutathione in 10 min, as observed previously [20]. The lethal level of H_2O_2, although, induced 68% cell death, but only 10% total GSH pool was decreased in 10 min, and in 60-min incubation, the total GSH pool was not decreased, rather showing an increasing trend. The result suggests that glutathione pool size and GSH reduction ratio are not directly correlated to the initiation of PCD in H_2O_2-stimulated cells.

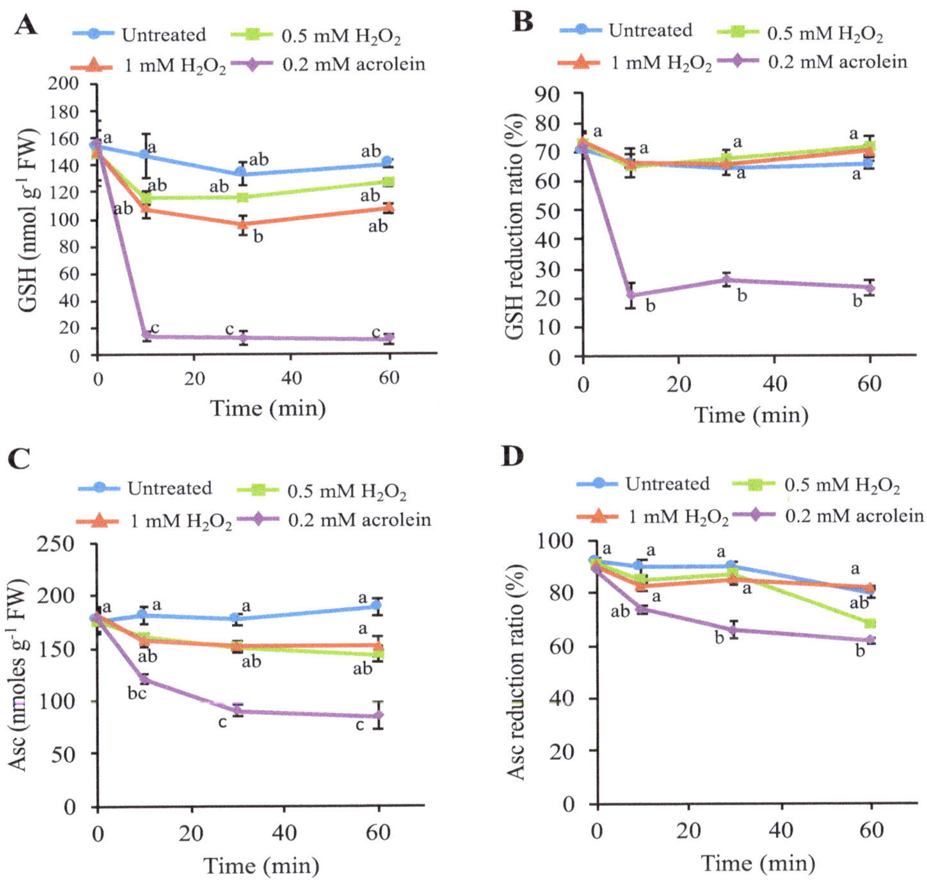

Figure 6. Changes in the contents and reduction ratios of ascorbate (Asc) and glutathione (GSH) in tobacco BY-2 cells treated with H_2O_2 or acrolein. H_2O_2 and acrolein at the indicated concentration was added to 4-d cultured cells, and then Asc and GSH contents were measured as described in the Materials and Methods section. (**A**) Reduced Asc, (**B**) Asc reduction ratio (%), (**C**) contents of reduced GSH, and (**D**) GSH reduction ratio (%). Each point represents the mean of three independent experiments and the error bars of the SEM. Different letters represent significant different data ($p < 0.05$ on Tukey test).

The changes of Asc content and Asc reduction ratio (Asc/Asc + dehydroascorbate (DHA)) was also measured after treatment with H_2O_2 and acrolein. The lethal level of H_2O_2 (1 mM) decreased an insignificant level of Asc ca. 12% in 10 min, and the sublethal level of H_2O_2 (0.5 mM) did not show any adverse effects on Asc level (Figure 6C). The Asc reduction ratio was also slightly decreased and recovered afterwards (Figure 6D). In contrast, the lethal level of acrolein, i.e., 0.2 mM consumed 70% Asc content in 10 min, as observed previously [20]. These results suggest that the lethal dose of H_2O_2 induced two-third percentage cells death without affecting cellular redox status. Thus, the mechanism of carbonyls accumulation is critical for cell viability.

3.7. H_2O_2 Stimulus Regulates a Carbonyl-Scavenging Capacity in Tobacco BY-2 Cells

In BY-2 cells stimulated with H_2O_2, only limited types of carbonyls were increased early (Figure 5). This was rather unexpected from our previous observation that, in oxidative stressed plant cells, many types of oxylipin carbonyl, including various RCS, are produced [12,31]. This contradiction lead us to conceive the presence of a regulation mechanism of carbonyl increase. We conceived a possibility that H_2O_2 attenuated certain types of carbonyl-scavenging enzyme and eventually increased the levels of limited types of carbonyls, especially HNE. Plant cells have NAD(P)H-dependent HNE-scavenging reductases, including 2-alkenal reductase (AER) [37,38] and aldo-keto reductase [13,14,39]. We treated BY-2 cells with lethal and sublethal doses of H_2O_2, and then extracted proteins and determined the NADPH-dependent HNE reductase activity (Figure 7A). It was found that the H_2O_2 treatment at a lethal dose (1 mM) decreased the reductase activity by 35% in 10 min and by 60% in 60 min. Sublethal H_2O_2 (0.5 mM) treatment caused a slight but insignificant decrease in the reductase activity. Thus, both the accumulation of HNE and the suppression of the HNE-reducing enzyme activity were observed only in the cells undergoing PCD. This result suggested that the inactivation of HNE-reducing enzyme by the lethal level of H_2O_2 was a cause of the enhancement of the intracellular HNE level, which eventually leads to PCD.

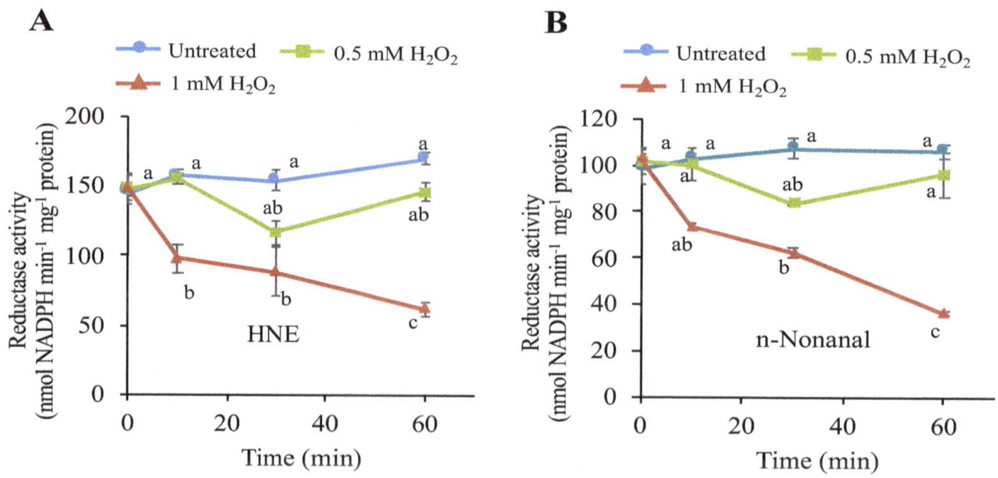

Figure 7. Effects of H_2O_2 on the NADPH-dependent HNE-reducing and *n*-nonanal-reducing activities in tobacco BY-2 cells. Four-d cultured cells were treated with H_2O_2 at 0.5 mM and 1 mM, as in Figure 1. Then, cells were harvested at the indicated time point, and proteins were extracted as in the Materials and Methods section. The reductase activities for (**A**) HNE and (**B**) *n*-nonanal were determined as in the Materials and Methods section. Each point represents the mean of three independent experiments and the error bars of the SEM. Different letters represent significantly different values ($p < 0.05$ on Tukey test).

To characterize the inactivated reductase, we also measured the NADPH-dependent activity for the saturated aldehyde *n*-nonanal. The *n*-nonanal-reducing activity responded to H_2O_2 treatment similar to the HNE-reducing activity; it was quickly decreased in the lethal level of H_2O_2, and in the sublethal level, it was decreased slightly and recovered afterward (Figure 7B). These results suggest that the H_2O_2-sensitive reductase had substrate specificity for both HNE and *n*-nonanal, i.e., it was a carbonyl reductase. Alternatively, there might be two different reductases for HNE and *n*-nonanal, and both showed similar sensitivity to H_2O_2.

3.8. H_2O_2 Directly Inactivates the Carbonyl Reductase

The addition of H_2O_2 to living cells will cause a broad range of biochemical changes, and the inactivation of the carbonyl reductase described above (Figure 7) could be a consequence of sequential reactions involving multiple proteins as observed in signal transduction. To verify the role of H_2O_2 in the inactivation of the carbonyl reductase, we prepared protein extract from untreated BY-2 cells and examined the effects of H_2O_2 addition on the NADPH-dependent HNE-reducing activity in it (Figure 8). In this experiment, we added 0.5 mM hydroxylamine to eliminate the catalase activity so that the added H_2O_2 was not immediately scavenged. It was found that H_2O_2 caused a rapid decrease in the HNE-reducing activity. In 5 min at 0.2 mM H_2O_2, the carbonyl reductase activity was lost by about half. Because the inactivation of HNE-reducing activity was observed in the cell-free protein extract, where intricate interactions among cellular components are very unlikely, we concluded that H_2O_2 directly acted on the carbonyl reductase and inactivated it.

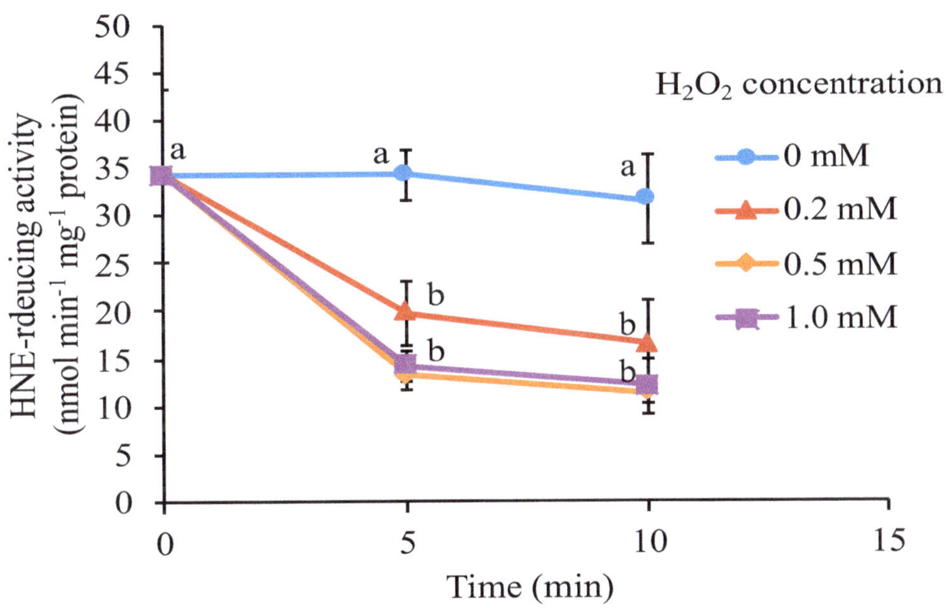

Figure 8. Effect of H_2O_2 on the NADPH-dependent HNE-reducing activity in protein extract from BY-2 cells. Proteins were extracted from untreated cells. H_2O_2 was added at the indicated concentration to the protein extract in the presence of 0.5 mM NH_2OH. The HNE-reducing activity was determined at the indicated time after addition of H_2O_2, as in the Materials and Methods section. Different letters among the treatments represent significantly different values ($p < 0.05$ on Tukey test).

Because H_2O_2 treatment of the cells caused inactivation of both HNE-reducing and *n*-nonanal-reducing activity, as shown in Figure 7, the H_2O_2-sensitive carbonyl reductase may show broad substrate specificity for various aldehydes. In the protein extract, we determined carbonyl reductase activity for five different aldehydes, i.e., HNE, *n*-hexanal, acetaldehyde, butyraldehyde, and (Z)-3-hexenal, and examined the effect of H_2O_2 treatment on distinct activities (Figure 9). After H_2O_2 treatment, the carbonyl reductase activity for all these substrates became significantly lower than those in the untreated control. These results suggest that the H_2O_2-sensitive NADPH-dependent carbonyl reductase possesses a broad substrate specificity.

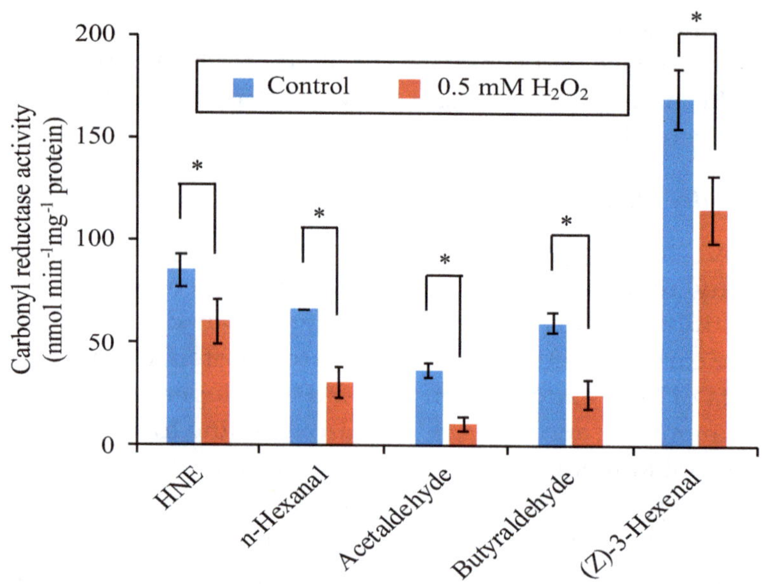

Figure 9. Effects of H_2O_2 on the NADPH-dependent carbonyl reductase activity for different substrates. Proteins were extracted from the untreated tobacco BY-2 cells. H_2O_2 at 0.5 mM was added to the cell extracts, and after 10-min incubation, the enzyme activity for the indicated substrate were determined as described in the Materials and Methods section. Each point represents the mean of three independent experiments and the error bars of the SEM. Asterisks indicate a statistical difference ($p < 0.05$, Student's t-test).

4. Discussion

4.1. H_2O_2-Induced Inactivation of Carbonyl-Detoxifying Enzyme(s) Enhances the HNE Level and Triggers PCD

We recently reported that, in response to oxidative stimulus, plant cells generated RCS and that the generated RCS triggered PCD via activation of caspase-like proteases [20]. In this study, we investigated the biochemical mechanism of RCS accumulation at the early stage of oxidative stimulus. We found a larger fraction (70%) of tobacco BY-2 cells were dying in 5 h after the treatment with 1 mM H_2O_2. In exploring the early events caused by the lethal dose of H_2O_2, we found a 2.5-fold increase of C3LP activity even in 10-min incubation, while the increase in the C1LP activity was insignificant. Suppression of C3LP activity by the specific C3LP inhibitor protected cell death, but the C1LP inhibitor did not. We found a 10-min exposure of the cells to a lethal level of H_2O_2 increased only limited species of carbonyls, including highly reactive HNE and HHE. Importantly, the addition of H_2O_2 at lethal level decreased a carbonyl reductase activity in 10-min incubation. These findings suggest that the inactivation of the carbonyl reductase by H_2O_2 is the very initial step of the ROS-mediated PCD in BY-2 cells.

As compared with the condition in living cells, where the HNE reductase was not inactivated by 0.5 mM H_2O_2 (Figure 7), in a cell-free solution, it was rather labile; H_2O_2 at 0.2 mM was high enough to affect the reductase activity (Figure 8). This difference is accounted for by the fact that living cells are equipped with the potent H_2O_2-scavenging system, including catalase and peroxidases. The addition of 0.5 mM H_2O_2 was not harmful to BY-2 cells because catalase and peroxidases with ascorbate scavenged H_2O_2 and thereby protected the carbonyl reductase. An excessive amount of H_2O_2 (1 mM<) could escape from the scavenging system, and the intracellular level of H_2O_2 would go over the threshold, to inactivate the carbonyl reductase. The inactivation profile of the carbonyl reductase in Figure 8, obtained in the presence of a strong inhibitor of heme-type enzymes, likely represented the intrinsic H_2O_2 sensitivity of the enzyme. Attenuation of the carbonyl reductase indirectly activates C3LP for the initiation of PCD. Therefore, the carbonyl reductase acts as a sensor of H_2O_2.

4.2. *Consumption of Antioxidants Is Not a Major Cause of PCD*

In our previous work, when PCD in BY-2 cells was started by the addition of acrolein, we observed a rapid drop of glutathione pool and significant decrease in the ascorbate pool, along with the activation of caspase-like proteases [20]. This posed a possibility that the loss of these antioxidants by RCS could also be a critical factor in causing PCD. In this work, we analyzed changes in the cellular redox status in response to H_2O_2 at different levels. We found that the effects of H_2O_2 of the lethal and sublethal doses were not different on the pool size and reduction level of both glutathione and ascorbate (Figure 6). Addition of H_2O_2 lowered the glutathione pool size slightly, but it was recovered in 60 min. Importantly, even though glutathione level was restored in 1 mM H_2O_2, PCD was not stopped (Figure 1). These results demonstrate that the main cause of PCD in H_2O_2-stimulated cells was not the redox status changes and supports the central role of RCS as an executor of PCD.

4.3. *Candidates of the H_2O_2-Sensitive Carbonyl Reductase*

The protein of the H_2O_2-sensing carbonyl reductase in BY-2 cells has not been identified, but several characters of it have been revealed in this study: (i) Its activity depends on NADPH. (ii) It recognizes both HNE and *n*-nonanal as a substrate, i.e., it reduces the carbonyl moiety. The enzyme that has these properties is aldo/keto reductase (AKR). In plants and animals, and in the microorganism, 16 families and above 190 members of reductases present in the AKR superfamily to catabolize RCS. The identified AKR in higher plants classified in four families depending on their functions; detoxify reactive carbonyls, osmoprotectants, secondary metabolism, and membrane transport. In addition to the four families, AKR have three subfamilies; 4A, 4B, and 4C [14,40,41]. There are only limited number of isozymes thus far characterized of the substrate specificity [40,41] and, to our knowledge, there are no reports of their sensitivity to H_2O_2.

In cell-free extract broad range of substrates from RCS, such as HNE, to less reactive carbonyls, such as *n*-hexanal, acetaldehyde, and butyraldehyde, (Z)-3-Hexenal showed H_2O_2 at 0.5 mM significantly reduced aldehyde reducing activity (Figure 7). If all these activities are ascribed to a single species of protein, the carbonyl reductase has a broad substrate specificity. Alternatively, it is possible that several different isozymes of carbonyl reductase participated, and they all were sensitive to H_2O_2. Purification of the carbonyl reductase will clarify this question.

4.4. *Mechanism of the Inactivation of the Carbonyl Reductase by H_2O_2*

ROS can oxidize various amino acid residues on a protein. Cys residue is especially sensitive to the oxidation and indeed such oxidative modification generally changes the normal activity or function of the protein [42]. For example, in response to H_2O_2, Cys199 moiety of OxyR, a six cysteine residues protein, oxidized to sulfenic acid and destabilized its activity. Cys199 modification accelerates to form a disulfide bond with Cys208, which enhances conformational changes of global OxyR and becomes susceptible to oxidation. A site-directed mutant C199S OxyR lost oxidative response and did not activate in the presence of H_2O_2 [43]. Peroxiredoxin (Prx) are ubiquitous enzymes appear in all types of organisms and metabolize peroxide to maintain cellular redox balance. The enzymatic activity of Prxs largely depends on the cysteine residues. The Prx family consists of three cysteine subfamilies, viz. 2-Cys, atypical 2-Cys, and 1-Cys. In human cell, 90% of the H_2O_2 generated in mitochondria reacts with Prx [44]. Thus, it is expected that the thiol moiety of the carbonyl reductase to be sensitive to oxidative signal. H_2O_2 sensors in the AKR family in plants have not been characterized yet. To elucidate the inactivation mechanism, we are trying to purify the H_2O_2-sensing carbonyl reductase.

5. Conclusions

In response to oxidative stimulus by H_2O_2, plant cells significantly increased C3LP activity and accumulate limited species of carbonyls, including RCS. The addition of H_2O_2 did not notably disturb redox balance of the cells but inactivated NADPH-dependent carbonyl reductase(s) in the cells and cell-free extract. Eventually, the carbonyls load increased and cells underwent PCD. Our results show a new mechanism of ROS sensing in plants: ROS is sensed by the carbonyl reductase(s), and its oxidative inactivation eventually increases carbonyls load and initiates the program of cell death.

Supplementary Materials

Figure S1. Acrolein-induced cell death was suppressed by the C3LP inhibitor. Four-d cultured BY-2 cells were treated with 0.2 mM acrolein or 0.2 mM acrolein + 20 μM of each inhibitor (C1LP or C3LP) and harvested after 5 h. (A) Typical fluorescence photographs (top row) and phase contrast microscopy of the same images (bottom): (i) untreated cells as control, (ii) 0.2 mM acrolein, (iii) 0.2 mM acrolein + 20 μM C3LP inhibitor, (iv) 0.2 mM acrolein + 20 μM C1LP inhibitor (top). White arrows indicate dead cells. (B) FDA-staining cells were counted for living cells and expressed as percentage. The data are the mean ± SE. Different letters represent significant different data ($p < 0.05$ on Tukey test). Figure S2: Typical chromatograms of DNP-carbonyls in BY-2 cells. The cells were treated with water as control (A) and H_2O_2 (B). Four-d-cultured cells treated with H_2O_2 at 1 mM and incubation for 10 min. The identified aldehydes are labelled at the top of each peak. Figure S3. Activation of C3LP in BY-2 cells after exposure to acetaldehyde. Four-d cultured cells were treated with various concentration of acetaldehyde as indicated. Protease inhibitors were added to cell extracts to a final concentration of 0.1 mM and caspase assay was performed as described in Materials and Methods. Statistical differences compared to controls are indicated by asterisks. ($p < 0.05$, Student's t-test).

Author Contributions: Conceptualization, M.S.B. and J.M.; methodology, M.S.B., R.T.; validation, M.S.B. and J.M.; formal analysis, M.S.B.; investigation, M.S.B., R.T.; resources, J.M.; data curation, M.S.B. and J.M.; writing—original draft preparation, M.S.B. and J.M.; writing—review and editing, J.M.; visualization, M.S.B.; supervision, J.M.; project administration, J.M.; funding acquisition, J.M. All authors have read and agreed to the published version of the manuscript.

References

1. Petrov, V.; Hille, J.; Mueller-Roeber, B.; Gechev, T.S. ROS-mediated abiotic stress-induced programmed cell death in plants. *Front. Plant Sci.* **2015**, *6*, 69. [CrossRef] [PubMed]
2. Mittler, R. ROS are good. *Trends Plant Sci.* **2017**, *22*, 11–19. [CrossRef] [PubMed]
3. Tsukagoshi, H.; Busch, W.; Benfey, P.N. Transcriptional regulation of ROS controls transition from proliferation to differentiation in the root. *Cell* **2010**, *143*, 606–616. [CrossRef] [PubMed]
4. Dunand, C.; Penel, C. Localization of superoxide in the root apex of Arabidopsis. *Plant Signal. Behav.* **2007**, *2*, 131–132. [CrossRef]
5. Leymarie, J.; Vitkauskaité, G.; Hoang, H.H.; Gendreau, E.; Chazoule, V.; Meimoun, P.; Corbineau, F.; El-Maarouf-Bouteau, H.; Bailly, C. Role of reactive oxygen species in the regulation of Arabidopsis seed dormancy. *Plant Cell Physiol.* **2012**, *53*, 96–106. [CrossRef]
6. Bethke, P.C.; Jones, R.L. Cell death of barley aleurone protoplasts is mediated by reactive oxygen species. *Plant J.* **2001**, *25*, 19–29. [CrossRef]
7. Van Breusegem, F.; Dat, J.F. Reactive oxygen species in plant cell death. *Plant Physiol.* **2006**, *141*, 384–390. [CrossRef]
8. Nguyen, G.N.; Hailstones, D.L.; Wilkes, M.; Sutton, B.G. Drought-induced oxidative conditions in rice anthers leading to a programmed cell death and pollen abortion. *J. Agron. Crop Sci.* **2009**, *195*, 157–164. [CrossRef]
9. Monetti, E.; Kadono, T.; Tran, D.; Azzarello, E.; Arbelet-Bonnin, D.; Biligui, B.; Briand, J.; Kawano, T.; Mancuso, S.; Bouteau, F. Deciphering early events involved in hyperosmotic stress-induced programmed cell death in tobacco BY-2 cells. *J. Exp. Bot.* **2014**, *65*, 1361–1375. [CrossRef]
10. Locato, V.; Gadaleta, C.; De Gara, L.; de Pinto, M.C. Production of reactive species and modulation of antioxidant network in response to heat shock: a critical balance for cell fate. *Plant Cell Environ.* **2008**, *31*, 1606–1619. [CrossRef]
11. Yoda, H.; Yamaguchi, Y.; Sano, H. Induction of hypersensitive cell death by hydrogen peroxide produced through polyamine degradation in tobacco plants. *Plant Physiol.* **2003**, *132*, 1973–1981. [CrossRef]
12. Biswas, M.S.; Mano, J. Lipid peroxide-derived short-chain carbonyls mediate hydrogen peroxide-induced and salt-induced programmed cell death in plants. *Plant Physiol.* **2015**, *168*, 885–898. [CrossRef] [PubMed]

13. Mano, J. Reactive carbonyl species: Their production from lipid peroxides, action in environmental stress, and the detoxification mechanism. *Plant Physiol. Biochem.* **2012**, *59*, 90–97. [CrossRef] [PubMed]
14. Mano, J.; Biswas, M.S.; Sugimoto, K. Reactive carbonyl species: a missing link in ROS signaling. *Plants* **2019**, *8*, 391. [CrossRef] [PubMed]
15. Yamauchi, Y.; Kunishima, M.; Mizutani, M.; Sugimoto, Y. Reactive short-chain leaf volatiles act as powerful inducers of abiotic stress-related gene expression. *Sci. Rep.* **2015**, *5*, 8030. [CrossRef] [PubMed]
16. Srivastava, S.; Brychkova, C.; Yarmolinsky, D.; Soltabayeva, A.; Samani, T.; Sagi, M. Aldehyde oxidase 4 plays a critical role in delaying silique senescence by catalyzing aldehyde detoxification. *Plant Physiol.* **2017**, *173*, 1977–1997. [CrossRef]
17. Islam, M.M.; Ye, W.; Matsushima, D.; Munemasa, S.; Okuma, E.; Nakamura, Y.; Biswas, M.S.; Mano, J.; Murata, Y. Reactive carbonyl species mediate ABA signaling in guard cells. *Plant Cell Physiol.* **2016**, *57*, 2552–2563. [CrossRef]
18. Islam, M.M.; Ye, W.; Matsushima, D.; Rhaman, M.S.; Munemasa, S.; Okuma, E.; Nakamura, Y.; Biswas, M.S.; Mano, J.; Murata, Y. Reactive carbonyl species function as signal mediators downstream of H_2O_2 production and regulate $[Ca^{2+}]_{cyt}$ elevation in ABA signal pathway in Arabidopsis guard cells. *Plant Cell Physiol.* **2019**, *60*, 1146–1159. [CrossRef]
19. Biswas, M.S.; Fukaki, H.; Mori, I.C.; Nakahara, K.; Mano, J. Reactive oxygen species and reactive carbonyl species constitute a feed-forward loop in auxin signaling for lateral root formation. *Plant J.* **2019**, *100*, 536–548. [CrossRef]
20. Biswas, M.S.; Mano, J. Reactive carbonyl species activate caspase-3-like protease to initiate programmed cell death in plants. *Plant Cell Physiol.* **2016**, *57*, 1432–1442. [CrossRef]
21. Hatsugai, N.; Kuroyanagi, M.; Yamada, K.; Meshi, T.; Tsuda, S.; Kondo, M.; Mikio Nishimura, M.; Hara-Nishimura, I. A plant vacuolar protease, VPE, mediates virus-induced hypersensitive cell death. *Science* **2004**, *305*, 855–858. [CrossRef] [PubMed]
22. Hatsugai, N.; Yamada, K.; Goto-Yamada, S.; Hara-Nishimura, I. Vacuolar processing enzyme in plant programmed cell death. *Front. Plant Sci.* **2015**, *6*, 234. [CrossRef] [PubMed]
23. Yao, S.; Huang, W.; Pan, C.; Zhan, J.; He, L.-F. Caspase-like proteases regulate aluminum-induced programmed cell death in peanut. *Plant Cell Tissue Organ Cult.* **2016**, *127*, 691–703. [CrossRef]
24. Ye, Y.; Li, Z.; Xing, D. Nitric oxide promotes MPK6-mediated caspase-3-like activation in cadmium-induced Arabidopsis thaliana programmed cell death. *Plant Cell Environ.* **2013**, *36*, 1–15. [CrossRef]
25. Hatsugai, N.; Iwasaki, S.; Tamura, K.; Kondo, M.; Fuji, K.; Ogasawara, K.; Nishimura, M.; Hara-Nishimura, I. A novel membrane fusion-mediated plant immunity against bacterial pathogens. *Genes Dev.* **2009**, *23*, 2496–2506. [CrossRef]
26. Han, J.J.; Lin, W.; Oda, Y.; Cui, K.M.; Fukuda, H. The proteasome is responsible for caspase-3-like activity during xylem development. *Plant J.* **2012**, *72*, 129–141. [CrossRef]
27. Ge, Y.; Cai, Y.; Bonneau, L.; Rotari, V.; Danon, A.; McKenzie, E.A.; McLellan, H.; Mach, L.; Gallois, P. Inhibition of cathepsin B by caspase-3 inhibitors blocks programmed cell death in Arabidopsis. *Cell Death Differ.* **2016**, *25*, 1532. [CrossRef]
28. Nagata, T.; Sakamoto, K.; Shimizu, T. Tobacco BY-2 cells: The present and beyond. *Vitr. Cell Dev. Biol. Plant* **2004**, *40*, 163–166. [CrossRef]
29. de Pinto, M.C.; Francis, D.; De Gara, L. The redox state of the ascorbate–dehydroascorbate pair as a specific sensor of cell division in tobacco BY-2 cells. *Protoplasma* **1999**, *209*, 90–97. [CrossRef]
30. Kampfenkel, K.; Montagu, M.V.; Inzé, D. Extraction and determination of ascorbate and dehydroascorbate from plant tissue. *Anal. Biochem.* **1995**, *225*, 165–167. [CrossRef]
31. Yin, L.; Mano, J.; Wang, S.; Tsuji, W.; Tanaka, K. The involvement of lipid peroxide-derived aldehydes in aluminum toxicity of tobacco roots. *Plant Physiol.* **2010**, *152*, 1406–1417. [CrossRef] [PubMed]
32. Mano, J.; Biswas, M.S. Analysis of reactive carbonyl species generated under oxidative stress. In *Plant Programmed Cell Death: Methods and Protocols*; De Gara, L., Locato, V., Eds.; Springer: New York, NY, USA, 2018; pp. 117–124.
33. Cai, Y.-M.; Yu, J.; Ge, Y.; Mironov, A.; Gallois, P. Two proteases with caspase-3-like activity, cathepsin B and proteasome, antagonistically control ER-stress-induced programmed cell death in Arabidopsis. *New Phytol.* **2018**, *218*, 1143–1155. [CrossRef] [PubMed]

34. Vacca, R.A.; Valenti, D.; Bobba, A.; de Pinto, M.C.; Merafina, R.S.; De Gara, L.; Passarella, S.; Marra, E. Proteasome function is required for activation of programmed cell death in heat shocked tobacco Bright-Yellow 2 cells. *FEBS Lett.* **2007**, *581*, 917–922. [CrossRef]
35. Esterbauer, H.; Zollner, H.; Scholz, N. Reaction of glutathione with conjugated carbonyls. *Z. Naturforsch. C* **1975**, *30*, 466–473. [CrossRef]
36. Mano, J.; Miyatake, F.; Hiraoka, E.; Tamoi, M. Evaluation of the toxicity of stress related aldehydes to photosynthesis in chloroplasts. *Planta* **2009**, *230*, 639–648. [CrossRef] [PubMed]
37. Mano, J.; Torii, Y.; Hayashi, S.; Takimoto, K.; Matsui, K.; Nakamura, K.; Inzé, D.; Babiychuk, E.; Kushnir, S.; Asada, K. The NADPH: quinone oxidoreductase P1-ζ-crystallin in Arabidopsis catalyzes the α,β-hydrogenation of 2-alkenals: Detoxication of the lipid peroxide-derived reactive aldehydes. *Plant Cell Physiol.* **2002**, *43*, 1445–1455. [CrossRef] [PubMed]
38. Mano, J.; Belles-Boix, E.; Babiychuk, E.; Inzé, D.; Torii, Y.; Hiraoka, H.; Takimoto, K.; Slooten, L.; Asada, K.; Kushnir, S. Protection against photooxidative injury of tobacco leaves by 2-alkenal reductase. Detoxication of lipid peroxide-derived reactive carbonyls. *Plant Physiol.* **2005**, *139*, 1773–1783. [CrossRef]
39. Yalcinkaya, T.; Uzilday, B.; Ozgur, R.; Turkan, I.; Mano, J. Lipid peroxidation-derived reactive carbonyl species (RCS): their interaction with ROS and cellular redox during environmental stresses. *Environ. Exp. Bot.* **2019**, *165*, 139–149. [CrossRef]
40. Yamauchi, Y.; Hasegawa, A.; Taninaka, A.; Mizutani, M.; Sugimoto, Y. NADPH-dependent reductases involved in the detoxification of reactive carbonyls in plants. *J. Biol. Chem.* **2011**, *286*, 6999–7009. [CrossRef]
41. Mano, J.; Kanameda, S.; Kuramitsu, R.; Matsuura, N.; Yamauchi, Y. Detoxification of Reactive Carbonyl Species by Glutathione Transferase Tau Isozymes. *Front. Plant Sci.* **2019**, *10*, 487. [CrossRef]
42. Wible, R.S.; Sutter, T.R. Soft cysteine signaling network: the functional significance of cysteine in protein function and the soft acids/bases thiol chemistry that facilitates cysteine modification. *Chem. Res. Toxicol.* **2017**, *30*, 729–762. [CrossRef] [PubMed]
43. Kullik, I.; Toledano, M.B.; Tartaglia, L.A.; Storz, G. Mutational analysis of the redox-sensitive transcriptional regulator OxyR: regions important for oxidation and transcriptional activation. *J. Bacteriol.* **1995**, *177*, 1275–1284. [CrossRef] [PubMed]
44. Karplus, P.A. A primer on peroxiredoxin biochemistry. *Free Radic. Biol. Med.* **2015**, *80*, 183–190. [CrossRef] [PubMed]

7

Serratia marcescens BM1 Enhances Cadmium Stress Tolerance and Phytoremediation Potential of Soybean through Modulation of Osmolytes, Leaf Gas Exchange, Antioxidant Machinery and Stress-Responsive Genes Expression

Mohamed A. El-Esawi [1,*], Amr Elkelish [2], Mona Soliman [3], Hosam O. Elansary [4,5], Abbu Zaid [6] and Shabir H. Wani [7]

1 Botany Department, Faculty of Science, Tanta University, Tanta 31527, Egypt
2 Botany Department, Faculty of Science, Suez Canal University, Ismailia 41522, Egypt; amr.elkelish@science.suez.edu.eg
3 Botany and Microbiology Department, Faculty of Science, Cairo University, Giza 12613, Egypt; monahsh1@gmail.com
4 Plant Production Department, College of Food and Agriculture Sciences, King Saud University, P.O. Box 2455, Riyadh 11451, Saudi Arabia; helansary@ksu.edu.sa
5 Floriculture, Ornamental Horticulture, and Garden Design Department, Faculty of Agriculture (El-Shatby), Alexandria University, Alexandria 21526, Egypt
6 Plant Physiology and Biochemistry Laboratory, Department of Botany Aligarh Muslim University, Aligarh 202002, India; zaidabbu19@gmail.com
7 Mountain Research Centre for Field Crops, Sher-e-Kashmir University of Agricultural Sciences and Technology of Kashmir, Khudwani Anantnag 192101, India; shabirhussainwani@gmail.com
* Correspondence: mohamed.elesawi@science.tanta.edu.eg

Abstract: The heavy metal contamination in plant-soil environment has increased manifold recently. In order to reduce the harmful effects of metal stress in plants, the application of beneficial soil microbes is gaining much attention. In the present research, the role of *Serratia marcescens* BM1 in enhancing cadmium (Cd) stress tolerance and phytoremediation potential of soybean plants, was investigated. Exposure of soybean plants to two Cd doses (150 and 300 μM) significantly reduced plant growth, biomass, gas exchange attributes, nutrients uptake, antioxidant capacity, and the contents of chlorophyll, total phenolics, flavonoids, soluble sugars, and proteins. Additionally, Cd induced the stress levels of Cd, proline, glycine betaine, hydrogen peroxide, malondialdehyde, antioxidant enzymes (i.e., catalase, CAT; ascorbate peroxidase, APX; superoxide dismutase, SOD; peroxidise, POD), and the expression of stress-related genes (i.e., *APX, CAT, Fe-SOD, POD, CHI, CHS, PHD2, VSO, NR,* and *P5CS*) in soybean leaves. On the other hand, inoculation of Cd-stressed soybean plants with *Serratia marcescens* BM1 significantly enhanced the plant growth, biomass, gas exchange attributes, nutrients uptake, antioxidant capacity, and the contents of chlorophyll, total phenolics, flavonoids, soluble sugars, and proteins. Moreover, *Serratia marcescens* BM1 inoculation reduced the levels of cadmium and oxidative stress markers, but significantly induced the activities of antioxidant enzymes and the levels of osmolytes and stress-related genes expression in Cd-stressed plants. The application of 300 μM $CdCl_2$ and *Serratia marcescens* triggered the highest expression levels of stress-related genes. Overall, this study suggests that inoculation of soybean plants with *Serratia marcescens* BM1 promotes phytoremediation potential and Cd stress tolerance by modulating the photosynthetic attributes, osmolytes biosynthesis, antioxidants machinery, and the expression of stress-related genes.

Keywords: *Serratia marcescens* BM1; cadmium; soybean; osmolytes; antioxidants; genes expression

1. Introduction

Over the last few decades, there has been a growing incidence of heavy metal (HM) circulation in plant-soil continuum owing to various natural and anthropogenic activities [1]. HMs, including cadmium (Cd), pose a serious danger to the growth of crop plants, as well as human and animal health [2,3]. Cadmium is a toxic HM with strong mobility in the soil-plant interface and biological toxicity [4]. Cadmium retards vital physio-biochemical activities in plants, which include photosynthesis, biosynthesis of chlorophyll and accessory pigments, and the uptake and assimilation of essential mineral nutrients by triggering the overproduction of reactive oxygen species (ROS), including singlet oxygen ($^1O_2$), hydrogen peroxide (H_2O_2), superoxide radical ($O_2\bullet^-$), and the hydroxyl radical (OH•) [5,6]. ROS accumulation causes high toxicity in plant cells [7–16]. Cadmium evokes peroxidation of membrane lipids and modulates the expression of genes of antioxidant defense systems in plants [17]. The presence of Cd inside plants thus causes stress by inducing ionic, osmotic, and oxidative stress [17,18]. Various studies also support the active involvement of antioxidant proteins in combating Cd-induced stress. Gene expression analysis further revealed the active participation of antioxidants genes (*Mn/SOD, FeSOD, POD, CAT, APX,* and *GR*) under Cd-induced stress [19,20].

During the course of evolution, plants have evolved intricate strategies to cope with the Cd-induced oxidative stress by activating various signalling networks, which include optimum nutrient homeostasis, enhanced accumulation of osmolytes, ROS detoxification by enhanced activities of antioxidants, and the production of thiol-related compounds [21–24]. Plant stress physiologists are engaged in devising potential sustainable strategies that can unravel mechanisms behind Cd-stress tolerance; however, the field is still emerging. Out of the various strategies that have been adopted to reverse HM-induced stress impacts in plants, the plant growth promoting rhizobacteria (PGPR) interaction is an emerging and effective sustainable way. Various recent studies in the discipline of ecological engineering and management have strongly advocated the use of PGPRs to alleviate HM-induced oxidative stress in plants, like energy crops [25], *Solanum nigrum* [26], *Spartina maritime* [27], and *Lycopersicon esculentum* [20,28,29]. Plant growth promoting rhizobacteria are known to induce tolerance against metal stress via modulating various intrinsic or underlying mechanisms [30]. These mechanisms primarily involve exclusion, extrusion, biotransformation, methylation/demethylation, and accommodation (complex formation) of metals [20,31–34]. Plant growth promoting rhizobacteria also induce the formation of different phytohormones, such as cytokinins, auxins, and gibberellins [35,36]. In addition, the synergistic interaction between microorganisms and plant roots under various abiotic pressures positively regulates the plant performance and edaphic factors [32,37,38]. *Serratia* species are potent PGPRs that are known to induce HM stress tolerance in crop plants [39]. Studies, undertaken so far, have established the role of *Serratia marcescens* in inducing HM stress tolerance in plant species. Cristani et al. [39] studied the possible action of *Serratia marcescens* in lead, Cd, and chromium metal biosorption. Khan et al. [40] analyzed the genome of *Serratia marcescens* RSC-14 and found that this species is an efficient PGPR that can alleviate Cd stress in host plants. By employing proteomic approach, Queiroz et al. [41] indicated that *Serratia marcescens* LG1 can be successfully used for adaptation and tolerance against manganese (Mn). However, the potential of the isolate *Serratia marcescens* BM1 in conferring tolerance against Cd stress in soybean (*Glycine max* L.) has not been studied yet.

Soybean is an economically important food crop worldwide and experiences various environmental stresses, including salinity and HMs, which limit the plant growth and productivity [42–44]. Therefore, enhancing the stress tolerance, productivity and phytoremediation potential of this crop is of utmost importance. Considering the importance of investigating these issues, the present research intends to elucidate the various underlying mechanisms that *Serratia marcescens* BM1 triggers in response to Cd stress in soybean plants. Growth traits of soybean plants have been investigated. Additionally, various physiological, biochemical and molecular approaches have been assayed.

2. Material and Methods

2.1. Investigation of Cadmium Tolerance of *Serratia marcescens* BM1 Strain

Serratia marcescens BM1 strain used in the present study was isolated from the maize rhizospheric soils of the Egyptian Suez Canal area and proved its capability for indole acetic acid production and inorganic phosphate solubilization [45,46]. In the present research, Cd tolerance of *Serratia marcescens* BM1 strain was investigated in nutrient broth media, containing 0, 150, 300, and 450 μM $CdCl_2$ (i.e., $CdCl_2$ was used to provide Cd stress). Bacterial growth was estimated by measuring the optical density (OD = 600 nm) following 36, 48, 72, and 96 h of incubation at 29 °C. This experiment was conducted in four replications.

2.2. Inoculation and Growth of Soybean Plants

Following culturing of *Serratia marcescens* BM1 in nutrient broth media for 4 days at 29 °C, bacterial cells were centrifuged for 4 min at 2000× *g* and then collected. Resultant pellets were re-suspended in sterilized water, and bacterial culture was then set to 10^8 colony-forming units (CFU) mL^{-1} and used to inoculate soybean plants. Soybean cultivar Giza 35 seeds were obtained from Legumes Institute of Kafrelsheikh in Egypt, and were then sterilized using sodium hypochlorite (8%, *v/v*) for 7 min, washed with sterilized water several times, and left to grow on a wet filter paper for 6 days at 24 °C. The 6-day old plants were then inoculated with *Serratia marcescens* BM1 suspension for 25 min and transferred into hydroponic plastic pots, containing Hoagland plant nutrient solution. Control plants were kept in fresh nutrient broth for 25 min. Experimental treatments were performed as follows: (i) control plants without $CdCl_2$ and bacterial inoculation (T1); (ii) plants inoculated with *Serratia marcescens* BM1 alone (T2); (iii) plants treated with 150 μM $CdCl_2$ alone (T3); (iv) plants treated with 150 μM $CdCl_2$ and *Serratia marcescens* BM1 (T4); (v) plants treated with 300 μM $CdCl_2$ alone (T5); and, (vi) plants treated with 300 μM $CdCl_2$ and *Serratia marcescens* BM1 (T6). Soybean plants were irrigated with a Hoagland nutrient solution supplemented with 0, 150, and 300 μM $CdCl_2$ three times a week. The pots were left in a completely randomized block design in growth chambers of a temperature of 27/19 °C (day/night) and a humidity of 76%. After 7 weeks, plants were collected for the subsequent experimental analyses.

2.3. Morphological Parameters of Plant Root and Shoot

Root and shoot lengths were determined using a measuring tape. Separated roots and shoots were washed with deionized water and weighed to measure their fresh weight. Separated roots and shoots were then oven-dried at 72 °C for 50 h to estimate their dry weights.

2.4. Measurement of Phosphorus, Nitrogen, and Cadmium Uptake

Oven-dried leaf samples were ground into a fine powder and then digested in H_2SO_4 at 190 °C for 5 h. H_2O_2 was then added to the samples and left for 60 min. Digested samples were filtered and then diluted with sterile distilled H_2O. Nitrogen concentration was calculated following Kjeldahl methodology as mentioned by Bremner [47]. Phosphorus concentration was calculated according to the protocol of Murphy and Riley [48]. To determine Cd content, dried leaf samples were grinded to fine powder and then digested in a mixture of HNO_3: $HClO_4$ (3:1, *v/v*) at 120 °C. Cadmium was then quantified using an atomic absorption spectrometer (AA6300C, Shimadzu, Kyoto, Japan).

2.5. Measurements of Chlorophyll Content, Leaf Relative Water Content, and Gas-Exchange Attributes

Total leaf chlorophyll content was determined by homogenizing 0.2 g fresh leaf samples in 50 mL of 80% acetone, followed by centrifugation at 14,000× *g* for 7 min and the absorbance was then spectrophotometrically recorded at 662 and 645 nm as reported by Lichtenthaler [49].

Net photosynthesis rate (P_n), transpiration rate (E), and stomatal conductance (g_s) were measured on expanded leaves of similar developmental stages with a portable gas-exchange system (LI-6400, LI-COR Inc., Lincoln, NE, USA) between 9:30 and 10:30 am, according to the methodology of Holá et al. [50]. Leaf relative water content (RWC) was determined as previously explained by El-Esawi and Alayafi [12].

2.6. Measurement of Total Soluble Sugars, Soluble Protein, Proline and Glycine Betaine Levels

Leaves were ground into a fine powder and homogenized in 100 mM Tris buffer (pH 8.0), followed by centrifugation at at 14,000× *g* for 14 min. Using the protocol of Dey [51], total soluble sugar content was determined by recording the absorbance at 485 nm. The Bradford method [52] was used to estimate the total protein content.

The protocol of Bates et al. [53] was used to estimate proline content. Leaf samples were digested in 5% (*w/v*) sulfosalicylic acid, and centrifuged at 10,000× *g* for 7.0 min. Supernatant was diluted with sterile distilled water and then mixed with 2% ninhydrin, followed by heating up at 96 °C for 30 min, then cooling. Toluene was then added to the mixture, and the absorbance of the upper aqueous phase formed was recorded at 520 nm. Following the protocol of Grieve and Grattan [54], glycine betaine content was estimated by extracting dry leafy samples in hot distilled water at 72 °C. To the extract formed, 2 N HCl and potassium tri-iodide solution were added, mixed and cooled on ice for 2 h. Cold 1,2-dichloromethane and distilled water were then added to the mixture where two layers were formed. Organic layer absorbance was recorded at 365 nm.

2.7. Determination of Total Flavonoids and Phenols Contents

The protocol of Zhishen et al. [55] was used to estimate the total flavonoid content by homogenizing oven-dried leaf powder (1.0 g) in distilled water (100 mL), followed by filtration and mixing with a solution composed of distilled H_2O, $AlCl_3$, and $NaNO_2$. Few drops of NaOH was then added to the mixed solution, which was then diluted with distilled water. The mixture absorbance was recorded at 510 nm. Catechin calibration curve was used.

Total phenolic content was measured by extracting leaf samples (2.0 g) in methanol solution (10 mL, 80%), followed by agitation at 72 °C for 18 min. Methanolic extract (2.0 mL) was diluted in 10 mL distilled H_2O comprising 1 N Folin–Ciocalteau reagent (500 μL), and then incubated at 30 °C. The mixture absorbance was recorded at 725 nm [56] using gallic acid as a standard.

2.8. Estimation of Hydrogen Peroxide and Malondialdehyde Levels

Following the protocol of Velikova et al. [57], hydrogen peroxide (H_2O_2) content was measured by extracting leaf samples in 0.1% trichloroacetic acid (TCA), followed by centrifugation at 12,000× *g* for 15 min. Potassium phosphate buffer (10 mM, pH 7.0) and potassium iodide (1 M) were added and well-mixed with the supernatant. The mixture absorbance was then read at 390 nm and H_2O_2 content was estimated following H_2O_2 standard curve. Using the methodology of Heath and Packer [58], malondialdehyde (MDA) level was determined by homogenizing leaf samples in 0.1% TCA followed by centrifugation at 14,000× *g* for 6 min. Thiobarbituric acid (0.5%) and 20% TCA were added to the supernatant. The mixture was then heated up at 96°C for 25 min, followed by cooling and centrifugation at 9,000× *g* for 12 min. The supernatant absorbance was recorded at 532 and 660 nm.

2.9. Determination of Leaf Antioxidant Capacity

Leaf antioxidant capacities were estimated by β-carotene-linoleic acid and 2,2′-diphenypicrylhydrazyl (DPPH) assays [59,60]. Leaf samples (2.0 g) were extracted in methanol solution (10 mL, 80%), and then agitated at 72°C for 18 min. For DPPH assay, 50 μL of methanolic extract (1 mg mL^{-1}) was added to 5 mL of methanolic DPPH solution (0.004%), and then mixed and left in darkness for 25 min.

The mixture absorbance was read at 517 nm. For β-carotene-linoleic acid assay, 50 μL of methanolic extract was transferred to β-carotene mixture, and then mixed and incubated for 46 h. The absorbance was read at 470 nm. The results were expressed as IC_{50} in mg mL^{-1}.

2.10. Estimation of Antioxidant Enzymes

Leaf samples were extracted in Tris-HCl (100 mM, pH 7.5) and then mixed with Dithiothreitol (5 mM), PVP-40 (1.5%), $MgCl_2$ (10 mM), EDTA (1 mM), magnesium acetate (5 mM), phenylmethanesulfonyl fluoride (1 mM), and aproptinin (1 μg mL^{-1}). The mixture was filtered and then centrifuged at 14,000× *g* for 8 min. The supernatants were utilized for estimating antioxidant enzymes activities. The protocol of Nakano and Asada [61] was used to estimate ascorbate peroxidase (APX) activity by extracting leaf samples in 2 mM AsA, followed by recording the absorbance at 265 nm. Moreover, the methodology of Aebi [62] was used to estimate catalase (CAT) activity, and the absorbance was read at 240 nm. Following the nitroblue tetrazolium photo-reduction method [63], superoxide dismutase (SOD) activity was measured, and the absorbance was read at 540 nm. Using the methodology of Putter and Becker [64], peroxidase (POD) activity was determined by recording the oxidized guaiacol production rate at 436 nm. Enzymes activities were expressed in enzyme unit per milligramme protein (EU mg^{-1} protein).

2.11. Expression Analysis of Stress-Related Genes

Quantitative real-time PCR (RT-qPCR) assay was applied to investigate the expression levels of 10 stress-related genes, including *APX, CAT, Fe-SOD, POD, CHI* (encoding chalcone isomerase in flavonoids biosynthesis pathways; [65]), *CHS* (encoding chalcone synthase in flavonoids biosynthesis pathways), *PHD2* (plant-homeo-domain gene of DNA binding ability and stress tolerance induction; [66]), *VSP* (encoding vegetative storage defense protein; [67]), *NR* (encoding nitrate reductase; [67]) and *P5CS* (encoding pyrroline-5 carboxylate synthetase; [67]). Total RNA was extracted from leaf tissue samples using RNeasy Plant Mini kits (Qiagen, Manchester, UK). Contaminating DNA was then removed and first-strand cDNAs were prepared using Reverse Transcription kits (Qiagen, Manchester, UK). RT-qPCR analysis was conducted as reported in the protocol of QuantiTect SYBR Green PCR kit (Qiagen, Manchester, UK). Reaction volume and PCR amplification conditions were adjusted as mentioned by El-Esawi et al. [43]. Gene-specific primers of Sirhindi et al. [68], Vaishnav et al. [67], El-Esawi et al. [69], and Kim et al. [70] (Table S1) were used for genes amplification. A housekeeping gene, *actin*, was used. The method of Livak and Schmittgen [71] was utilized to estimate relative expression levels.

2.12. Statistical Analysis

The experimental data are means ± standard errors (SEs) ($n = 4$). One-way analysis of variance (ANOVA) and Duncan's multiple range test were carried out using SPSS version 16 (Chicago, IL, USA). Values at $p \leq 0.05$ differ significantly.

3. Results and Discussion

3.1. Cadmium Tolerance of Serratia marcescens BM1

Serratia marcescens BM1 strain grew in nutrient broth medium having 0, 150 and 300 μM $CdCl_2$ at different incubation time intervals with OD = 0.49–0.61, OD = 0.29–0.49 and OD = 0.22–0.35, respectively (Figure 1). However, the bacterium could not grow in the medium containing 450 μM $CdCl_2$ at all the incubation time intervals (OD = 0). These findings reveal that BM1 was able to grow and tolerate Cd stress up to 300 μM $CdCl_2$, indicating that this bacterium could be exploited in the phytoremediation and Cd stress tolerance studies.

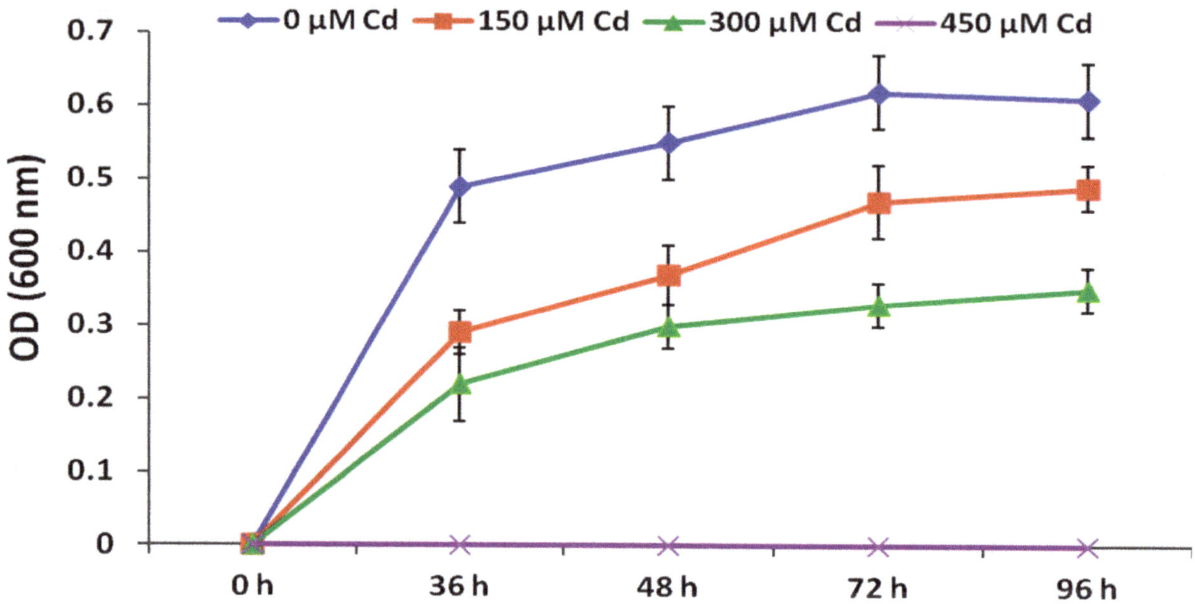

Figure 1. Growth curve of *Serratia marcescens* BM1 in nutrient broth media supplemented with 0, 150, 300, and 450 µM $CdCl_2$ after 0, 36, 48, 72, and 96 h of incubation at 29 °C.

3.2. *Serratia marcescens BM1 Enhances Growth and Biomass of Soybean Plants Under Cadmium Stress*

Soybean plants exhibited a significant decrease in growth and biomass traits under Cd stress at both doses (150 and 300 µM) with respect to control plants (Table 1). Cadmium stress at T3 (150 µM $CdCl_2$) and T5 (300 µM $CdCl_2$) treatments decreased root length by 27.56%, and 47.43% and shoot length by 30.95%, and 50.79%, respectively, with respect to the control plants. Inoculating Cd-stressed soybean with *Serratia marcescens* BM1 enhanced the length of root by 14.20% at T4 (150 µM $CdCl_2$ + *Serratia marcescens* BM1) and 17.10% at T6 (300 µM $CdCl_2$ + *Serratia marcescens* BM1) treatments, as compared to T3, and T5, respectively (Table 1). A similar trend also follows in case of shoot length, by which a significant increase of 18.40% at T4 and 27.40% at T6 was noticed with respect to T3, and T5, respectively (Table 1).

Table 1. Growth and biomass of un-inoculated and inoculated soybean plants with *Serratia marcescens* BM1 under Cd stress.

$CdCl_2$ (µM)	BM1	Root Length (cm)	Root Fresh Weight (g Plant^{-1})	Root Dry Weight (g Plant^{-1})	Shoot Length (cm)	Shoot Fresh Weight (g Plant^{-1})	Shoot Dry Weight (g Plant^{-1})	Root DW/Shoot DW
0	−BM1 (T1)	15.6 ± 1.11b	1.37 ± 0.07b	0.15 ± 0.03b	25.2 ± 1.4b	1.83 ± 0.15b	0.37 ± 0.14b	0.41 ± 0.05b
	+BM1 (T2)	17.3 ± 1.25a	1.51 ± 0.08a	0.18 ± 0.06a	26.9 ± 1.6a	1.98 ± 0.12a	0.42 ± 0.12a	0.43 ± 0.04a
150	−BM1 (T3)	11.3 ± 1.11d	0.98 ± 0.06d	0.10 ± 0.05d	17.4± 1.3d	1.24 ± 0.11d	0.25 ± 0.15d	0.40 ± 0.05b
	+BM1 (T4)	12.9 ± 1.21c	1.18 ± 0.09c	0.12 ± 0.04c	20.6 ± 1.5c	1.41 ± 0.12c	0.29 ± 0.11c	0.42 ± 0.03a
300	−BM1 (T5)	8.2 ± 1.07f	0.61 ± 0.08f	0.06 ± 0.05f	12.4 ± 1.6f	0.77 ± 0.15f	0.16 ± 0.13f	0.37 ± 0.05d
	+BM1 (T6)	9.6 ± 1.10e	0.77 ± 0.06e	0.07 ± 0.04e	15.8 ± 1.8e	0.92 ± 0.12e	0.18 ± 0.11e	0.39 ± 0.03c

Values represent the means ± SE (n = 4). Different letters (a–f) next to numbers in the same column indicate significant differences between treatments ($p \leq 0.05$). DW denotes dry weight.

Root and shoot fresh weights showed a significant ($p \leq 0.05$) decrease under Cd stress at T3 and T5 treatments compared to T1 unstressed plants (Table 1). Cadmium stress at T3 and T5 decreases the root fresh weight by 28.46% and 55.47% and shoot fresh weight by 32.24%, and 57.92%, respectively, with respect to unstressed T1 plants. On the other hand, *Serratia marcescens* BM1 application to Cd-stressed soybean plants enhanced the root fresh weight by 20.41% at T4 and 26.23% at T6 treatments with respect to T3, and T5 treatments, respectively. A similar trend was also noticed in shoot fresh weight, in which a significant increase of 13.71% at T4 and 19.48% at T6, was observed with respect to T3, and T5 treatments, respectively (Table 1).

In case of dry weights of root and shoot, Cd induces a significant ($p \leq 0.05$) decrease at both T3 and T5 treatments (Table 1). Cadmium induced significant ($p \leq 0.05$) reductions of 33.33% and 60.00% in root dry weight at T3, and T5 treatments, respectively, with respect to control plants. Cadmium stress also decreased the shoot dry weight by 43.24%, and 59.45% at T3, and T5 treatments, respectively, with respect to control plants. On the other hand, *Serratia marcescens* BM1 application to Cd-stressed plants showed a significant increase in root and shoot dry weight. An increase of 20.00% and 16.66% in root dry weight and 38.10% and 26.66% in shoot dry weight was noticed at T4 and T6 treatments in comparison to plants treated only with Cd at T3, and T5, respectively (Table 1). Furthermore, the root dry weight/shoot dry weight ratio was also higher for plants inoculated with *Serratia marcescens* BM1 in the presence or absence of cadmium. Therefore, *Serratia marcescens* BM1 effect was most profound on root development with a recorded increment in the root/shoot ratio in response to cadmium stress. This could be attributed to the modulation of antioxidant machinery and nutrients uptake in plant cells.

The aforementioned growth effects are consequences of the plant growth-promoting traits of *Serratia marcescens* BM1 strain, which promoted soybean growth and biomass under Cd stress conditions. Similar reports revealed the potential of *Serratia* spp. in enhancing plant growth and biomass under control and stress conditions. *Serratia marcescens* strain SRM significantly improved plant growth and biomass of wheat plants grown in cold temperature [72]. *Serratia marcescens* AL2-16 also improved the growth and biomass of the medicinal plant, *Achyranthes aspera* L. [73]. Furthermore, *Serratia liquefaciens* KM4 improved maize growth and biomass under normal and saline conditions [74]. *Serratia* marcescens CDP-13 promoted plant growth and alleviated salinity stress in wheat [75]. Moreover, Khan et al. [76] revealed that Cd stress significantly reduced plant growth and biomass of *Solanum nigrum* L. plants, whereas *Serratia* sp. RSC-14 inoculation alleviated Cd stress-induced toxic effects by significantly promoting plant growth and biomass.

3.3. *Serratia marcescens BM1 Modulates the Contents of Cadmium, Nitrogen and Phosphorous in Cadmium-Stressed Soybean Plants*

Cadmium stress at T3 (150 μM $CdCl_2$) and T5 (300 μM $CdCl_2$) treatments significantly induced the accumulation of Cd over the control plants (Table 2). On the other hand, application of *Serratia marcescens* BM1 to Cd-stressed soybean plants significantly reduced Cd uptake by 21.05% at T4 (150 μM $CdCl_2$ + *Serratia marcescens* BM1) and 17.86% at T6 (300 μM $CdCl_2$ + *Serratia marcescens* BM1) treatments with respect to T3, and T5, respectively (Table 2).

Table 2. Minerals uptake, photosynthesis rate (P_n), transpiration rate (E) and stomatal conductance (g_s) in leaves of un-inoculated and inoculated soybean plants with *Serratia marcescens* BM1 under Cd stress.

$CdCl_2$ (μM)	BM1	Cd Content (mg g^{-1} DW)	N Content (mg g^{-1} DW)	P Content (mg g^{-1} DW)	P_n (μmol m^2 s^{-1})	E (mmol m^2 s^{-1})	g_s (mol m^2 s^{-1})
0	−BM1 (T1)	0.01 ± 0.01e	0.88 ± 0.12b	0.40 ± 0.08b	17.03 ±1.24b	1.84 ± 0.07b	0.11 ± 0.03b
	+BM1 (T2)	0.01 ± 0.01e	0.93 ± 0.11a	0.44 ± 0.11a	18.92 ±1.15a	2.03 ± 0.05a	0.13 ± 0.02a
150	−BM1 (T3)	0.19 ± 0.09c	0.72 ± 0.09d	0.29 ± 0.09d	11.12 ±1.21d	1.49 ± 0.04d	0.06 ± 0.03d
	+BM1 (T4)	0.15 ± 0.05d	0.81 ± 0.07c	0.36 ± 0.10c	13.25 ±1.23c	1.58 ± 0.07c	0.08 ± 0.02c
300	−BM1 (T5)	0.2 8± 0.11a	0.58 ± 0.08f	0.22 ± 0.11e	8.91 ± 1.05f	1.33 ± 0.07f	0.03 ± 0.01f
	+BM1 (T6)	0.23 ± 0.10b	0.65 ± 0.07e	0.30 ± 0.12d	10.33 ±1.07e	1.42 ± 0.08e	0.05 ± 0.03e

Values represent the means ± SE (n = 4). Different letters (a–f) next to numbers in the same column indicate significant differences between treatments ($p \leq 0.05$).

Under both doses of Cd stress at T3 and T5 treatments, soybean plants exhibited a decrease in N and P contents (Table 2). Cadmium stress at T3 and T5 decreased the N level by 18.18%, and 34.09%, respectively, with respect to unstressed T1 plants. *Serratia marcescens* BM1 application significantly enhanced N content at both Cd treatments. An increase of 12.50% and 12.07% was noticed in N content at T4 and T6 treatments with respect to T3, and T5, respectively (Table 2). Low and high doses of Cd stress also significantly decreased P content at T3 and T5 treatments with respect to unstressed T1 plants. Phosphorous contents showed a significant decline of 27.50% and 45.00% at T3, and T5, respectively, in relation to control plants. Application of *Serratia marcescens* BM1 to Cd-stressed soybean plants induced phosphorus content by 24.14% and 36.36% at T4 and T6 treatments with respect to T3, and T5, respectively (Table 2). These results indicate that *Serratia marcescens* BM1 was able to alleviate Cd stress-induced oxidative damage by enhancing nutrients uptake in soybean plants. Our results are in agreement with the findings of previous reports which indicated the potential of other *Serratia* strains in mitigating abiotic stress-induced oxidative damage through enhancing nutrients uptake in plants. For example, *Serratia marcescens* strain SRM significantly induced nutrients uptake of wheat plants grown under cold conditions [72]. *Serratia marcescens* TRS-1 also enhanced leaf and root phosphate content of tea plants [77]. Furthermore, *Serratia nematodiphila* LRE07 improved plant growth, biomass and nutrient uptake of *Solanum nigrum* L. plants under Cd stress conditions [78]. *Serratia liquefaciens* KM4 also significantly induced nutrient uptake of maize plants grown under normal and saline stress conditions [74].

3.4. Serratia marcescens BM1 Induces Leaf Gas Exchange Attributes, Leaf Relative Water Content and Biosynthesis of Chlorophyll, Sugars, Proteins, Osmolytes, Flavonoids, and Phenolics Under Cadmium Stress

Serratia marcescens BM1 application improved *Pn, E* and *gs* under stress-free and Cd-stressed conditions (Table 2). Soybean exposure to 150 μM $CdCl_2$ (T3) and 300 μM $CdCl_2$ (T5) doses significantly decreased *Pn, E* and *gs* values with respect to unstressed T1 plants. *Serratia marcescens* BM1 inoculation nullified the Cd-induced repressions in gas-exchange characteristics and enhanced *Pn, E* and *gs* at T4 (150 μM $CdCl_2$ + *Serratia marcescens* BM1) and T6 (300 μM $CdCl_2$ + *Serratia marcescens* BM1) treatments as compared to Cd stressed plants at T3, and T5, respectively (Table 2).

Cadmium stress decreased the chlorophyll content by 21.28% and 30.19% at T3, and T5 treatments, respectively, with respect to unstressed T1 plants (Table 3). *Serratia marcescens* BM1 inoculation significantly ($p \leq 0.05$) induced the biosynthesis of chlorophyll by 5.66% at T4 and 5.67% at T6 treatments as compared to Cd stressed plants at T3, and T5, respectively. Furthermore, Cd stress significantly reduced the leaf relative water content (RWC) with respect to unstressed T1 plants (Table 3). On the other hand, *Serratia marcescens* BM1 inoculation significantly induced RWC in Cd-stressed plants as compared to untreated Cd stressed plants.

Table 3. chlorophyll content, leaf relative water content (RWC), total soluble sugars, protein, proline and glycine betaine (GB) contents in leaves of un-inoculated and inoculated soybean plants with *Serratia marcescens* BM1 under Cd stress.

$CdCl_2$ (μM)	BM1	Chlorophyll (mg g^{-1} FW)	RWC (%)	Sugars (μg g^{-1} FW)	Proteins (mg g^{-1} FW)	Proline (μg g^{-1} FW)	GB (μmol g^{-1} FW)
0	−BM1 (T1)	2.02 ± 0.08b	91 ± 0.53b	2.41 ± 0.14b	1.11 ± 0.14b	20.5 ± 0.29f	2.36 ± 0.04f
	+BM1 (T2)	2.14 ± 0.07a	93 ± 0.48a	2.63 ± 0.16a	1.23 ± 0.13a	35.1 ± 0.24e	3.22 ± 0.07e
150	−BM1 (T3)	1.59 ± 0.09d	55 ± 0.51d	2.17 ± 0.14d	0.88 ± 0.10d	55.2 ± 0.41d	4.17 ± 0.06d
	+BM1 (T4)	1.68 ± 0.08c	68 ± 0.46c	2.29 ± 0.15c	1.05 ± 0.14c	77.3 ± 0.40c	5.13 ± 0.05c
300	−BM1 (T5)	1.41 ± 0.09f	42 ± 0.61e	1.93 ± 0.17e	0.67 ± 0.11f	81.7 ± 0.52b	6.06 ± 0.07b
	+BM1 (T6)	1.49 ± 0.07e	54 ± 0.45d	2.12 ± 0.13d	0.79 ± 0.14e	97.4 ± 0.55a	7.88 ± 0.06a

Values represent the means ± SE (n = 4). Different letters (a–f) next to numbers in the same column indicate significant differences between treatments ($p \leq 0.05$). FW denotes the fresh weight of soybean plants.

Low and high doses of Cd stress at T3 (150 μM $CdCl_2$) and T5 (300 μM $CdCl_2$) treatments in soybean plants showed a decrease in total soluble sugars and protein contents (Table 3). Both low and high dose of Cd at T3 and T5 caused significant ($p \leq 0.05$) decreases of 9.95%, and 19.91%, respectively, in sugars levels with respect to T1 control plants. A similar trend also follows in these treatments in case of total protein content, whereas a decrease of 20.72% and 39.93% at T3, and T5 was noticed with respect to control plants. On the other hand, *Serratia marcescens* BM1 inoculation significantly enhanced sugars and protein contents at T4 (150 μM $CdCl_2$ + *Serratia marcescens* BM1) and T6 (300 μM $CdCl_2$ + *Serratia marcescens* BM1) treatments with respect to T3, and T5, respectively. In case of sugars, a significant increase of 5.53% and 9.84% was noticed at T4 and T6 treatments with respect to T3, and T5, respectively. Significant increases of 19.32% and 17.91% were also recorded in protein content at T4 and T6 with respect to T3, and T5, respectively (Table 3).

Serratia marcescens BM1 enhanced the biosynthesis of osmolytes, including proline and glycine betaine in control and Cd-stressed soybean plants (Table 3). Cadmium stress at T3 and T5 caused a marked increase in proline content by 2.69, and 3.98 fold, respectively, over control plants. A similar trend was also noticed in glycine betaine, whereas an increase of 1.76 and 2.56 fold at T3 and T5, respectively, was recorded over control plants. Inoculation of plants with *Serratia marcescens* BM1 at both Cd stress conditions had an additive effect on proline and glycine betaine biosynthesis and maximum values of these parameters were recorded at T6 treatments.

Serratia marcescens BM1 inoculation improved the levels of flavonoids and phenolics under control and Cd stress conditions (Table 4). Cadmium stress lowered flavonoid content by 29.95% at T3 (150 μM $CdCl_2$) and by 59.77% at T5 (300 μM $CdCl_2$) with respect to unstressed T1 plants. However, inoculating Cd-stressed soybean plants with *Serratia marcescens* BM1 induced flavonoids content by 10.69% and 41.37% at T4 (150 μM $CdCl_2$ + *Serratia marcescens* BM1) and T6 (300 μM $CdCl_2$ + *Serratia marcescens* BM1) treatments, with respect to T3, and T5, respectively (Table 4). The phenolic content also exhibited a significant decrease of 23.59% and 41.01% at T3, and T5 treatments, respectively, with respect to T1 control plants. *Serratia marcescens* BM1 application exhibited higher phenolic content of 9.55% and 24.76% at T4 and T6 treatments with respect to T3, and T5, respectively (Table 4).

Table 4. The hydrogen peroxide (H_2O_2) content, lipid peroxidation (MDA) level, total flavonoid content (mg catechin/g extract), total phenolic content (mg gallic acid/g extract) and antioxidant activity (DPPH and β-carotene-linoleic acid) of leaves of un-inoculated and inoculated soybean plants with *Serratia marcescens* BM1 under Cd stress.

$CdCl_2$ (μM)	BM1	Total Flavonoids	Total Phenolics	H_2O_2 (μmol g^{-1} FW)	MDA (μmol g^{-1} FW)	DPPH (IC_{50}, μg mL^{-1})	β-Carotene Linoleic Acid (IC_{50}, μg mL^{-1})
0	−BM1 (T1)	7.21 ± 0.28b	17.8 ± 0.24b	0.27 ± 0.05d	20 ± 1.71e	0.53 ± 0.03a	0.51 ± 0.05a
	+BM1 (T2)	9.11 ± 0.30a	19.1 ± 0.23a	0.18 ± 0.06e	16 ± 1.66f	0.48 ± 0.05b	0.47 ± 0.04b
150	−BM1 (T3)	5.05 ± 0.22d	13.6 ± 0.22d	0.72 ± 0.04b	51 ± 2.27b	0.46 ± 0.04c	0.44 ± 0.05c
	+BM1 (T4)	5.59 ± 0.27c	14.9 ± 0.24c	0.61 ± 0.07c	42 ± 2.24d	0.42 ± 0.03d	0.41 ± 0.06d
300	−BM1 (T5)	2.9 ± 0.23f	10.5 ± 0.31e	0.85 ± 0.04a	61 ± 2.43a	0.38 ± 0.05e	0.37 ± 0.05e
	+BM1 (T6)	4.1 ± 0.26e	13.1 ± 0.29d	0.60 ± 0.05c	47 ± 2.11c	0.35 ± 0.04f	0.34 ± 0.05f

Values represent the means ± SE (n = 4). Different letters (a–f) next to numbers in the same column indicate significant differences between treatments ($p \leq 0.05$).

The results of the above parameters demonstrate that *Serratia marcescens* BM1 also mitigated the negative impacts of Cd stress by modulating leaf gas exchange characteristics, photosynthesis, and osmolytes. These results are in agreement with the findings of previous studies which revealed the key roles of *Serratia* isolates in alleviating abiotic stress effects in plants by regulating photosynthesis, leaf gas exchange parameters, and osmolytes biosynthesis. For example, *Serratia nematodiphila* LRE07 enhanced photosynthetic pigments biosynthesis of *Solanum nigrum* L. plants under Cd stress conditions [78]. Khan et al. [76] revealed that *Serratia* sp. RSC-14 inoculation mitigated the negative impacts of Cd stress by significantly improving chlorophyll biosynthesis. Moreover, *Serratia liquefaciens* KM4 significantly induced leaf gas exchange attributes, photosynthesis process, and osmolytes biosynthesis of maize plants grown under saline stress conditions [74].

3.5. *Serratia marcescens BM1 Inoculation Reduces the Contents of H_2O_2 and MDA Under Cadmium Stress*

Cadmium stress at T3 (150 μM $CdCl_2$) and T5 (300 μM $CdCl_2$) treatments significantly increased the production of H_2O_2 over the control plants. On the other hand, the application of *Serratia marcescens* BM1 to Cd-stressed soybean plants significantly reduced H_2O_2 level by 15.27% at T4 (150 μM $CdCl_2$ + *Serratia marcescens* BM1) and 29.41% at T6 (300 μM $CdCl_2$ + *Serratia marcescens* BM1) treatments with respect to T3, and T5, respectively. MDA production exhibited significant increases at T3 and T5 treatments with respect to T1 unstressed plants. A significant decrease of 17.64% and 22.95% in

MDA content was noticed in plants inoculated with *Serratia marcescens* BM1 at T4 and T6 treatments with respect to T3, and T5, respectively (Table 4), indicating that *Serratia marcescens* BM1 regulates membrane functions through scavenging ROS accumulation. These results are in agreement with the findings of the previous reports, which indicated the potential of *Serratia* isolates in mitigating the Cd stress impacts by reducing the oxidative stress markers, such as MDA, electrolyte leakage, and hydrogen peroxide. Khan et al. [76] revealed that *Serratia* sp. RSC-14 relived the toxic effects of Cd stress by reducing MDA and electrolytes production. Furthermore, *Serratia liquefaciens* KM4 significantly alleviated the salt stress-induced oxidative damage by reducing MDA and H_2O_2 contents [74].

3.6. *Serratia marcescens* BM1 Inoculation Modulates Antioxidant Capacity and Antioxidative Enzyme Activities Under Cadmium Stress

Cd stress decreased the DPPH values by 13.20% and 28.30% at T3 (150 μM $CdCl_2$), and T5 (300 μM $CdCl_2$) treatments, respectively, with respect to control plants (Table 4). Nevertheless, *Serratia marcescens* BM1 inoculation further caused a decrease in DPPH values by 8.69% and 7.89% at T4 (150 μM $CdCl_2$ + *Serratia marcescens* BM1) and T6 (300 μM $CdCl_2$ + *Serratia marcescens* BM1) treatments with respect to respective T3, and T5 treatments, respectively. β-carotene-linoleic acid of soybean plants also follows a same decreasing trend when grown under Cd stress conditions at T3 and T5 treatments. Inoculating soybean plants with *Serratia marcescens* BM1 further caused a significant decrease in β-carotene-linoleic acid values by 6.81% and by 8.10% at T4 and T6 with respect to T3, and T5, respectively (Table 4).

To scavenge ROS accumulation and mitigate cellular toxicity, plants activate their defence strategies, including their non-enzymatic and enzymatic antioxidant systems, such as ascorbate, SOD, CAT, APX, and POD. In the present study, the activities of antioxidant enzymes in soybean plants, such as CAT, APX, POD, and SOD were elevated in response to Cd stress at T3 (150 μM $CdCl_2$) and T5 (300 μM $CdCl_2$) with respect to T1 control plants (Figure 2). Furthermore, the highest significant increase in CAT, APX, POD, and SOD activities was recorded when soybean Cd-stressed plants were inoculated with *Serratia marcescens* BM1 at T6 (300 μM $CdCl_2$ + *Serratia marcescens* BM1) treatments (Figure 2). These findings are consistent with the results of the previous reports which showed the key roles of *Serratia* strains in alleviating the abiotic stress-induced oxidative damage through regulating the key antioxidant enzymes. For example, *Serratia nematodiphila* LRE07 alleviated the negative impacts of Cd stress in *Solanum nigrum* L. plants by further inducing the activities of anti-oxidative enzymes [78]. *Serratia liquefaciens* KM4 also significantly alleviated the salt stress-induced oxidative damage by upregulating various antioxidative enzymes, such as CAT, APX, POD, and SOD [74].

3.7. *Serratia marcescens* BM1 Induces the Expression Levels of Stress-Related Genes Under Cadmium Stress

The application of Cd stress in both doses on soybean plants exhibited a dynamic increase in the expression levels of antioxidant genes (*APX, CAT, Fe-SOD* and *POD*; Figure 3) and stress-induced genes (*CHI, CHS, PHD2, VSP,* and *P5CS*; Figure 4). The results also revealed that the expression levels of these 10 genes were further increased upon inoculating soybean plants with *Serratia marcescens* BM1, whereas, the expression levels reached to the maximum extent at T6 (300μM $CdCl_2$ + *Serratia marcescens* BM1) treatments in case of *APX, CAT, Fe-SOD, POD, PHD2, VSP,* and *P5CS* genes. The expression profiling of *CHI, CHS,* and *NR* genes were induced to a maximum extent in non-stressed soybean plants, inoculated with *Serratia marcescens* BM1 (Figure 4). These results indicate the key role of *Serratia marcescens* BM1 in eliminating toxic free radicals and conferring Cd stress tolerance through upregulating the abiotic stress-related genes. Similar reports indicated the potential of *Serratia* and other PGPBs in improving abiotic stress tolerance by inducing abiotic stress-related genes [74,79].

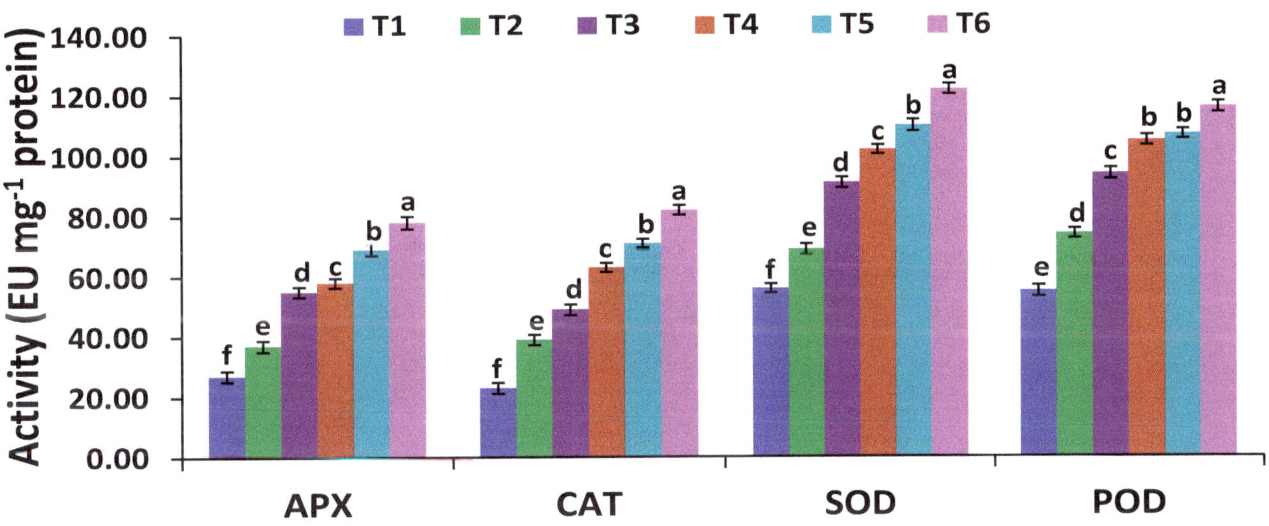

Figure 2. Ascorbate peroxidase (APX), catalase (CAT), superoxide dismutase (SOD), and peroxidase (POD) activities in leaves of un-inoculated and inoculated soybean plants with *Serratia marcescens* BM1 under Cd stress. Data are means ± SE ($n = 4$). Different letters (a–f) above the chart columns indicate significant differences between treatments ($p \leq 0.05$). T1, control with no Cd and BM1 treatment; T2, plant treated with *Serratia marcescens* BM1; T3, plant treated with 150 μM $CdCl_2$; T4, plant treated with 150 μM $CdCl_2$ and *Serratia marcescens* BM1; T5, plant treated with 300 μM $CdCl_2$; T6, plant treated with 300 μM $CdCl_2$ and *Serratia marcescens* BM1.

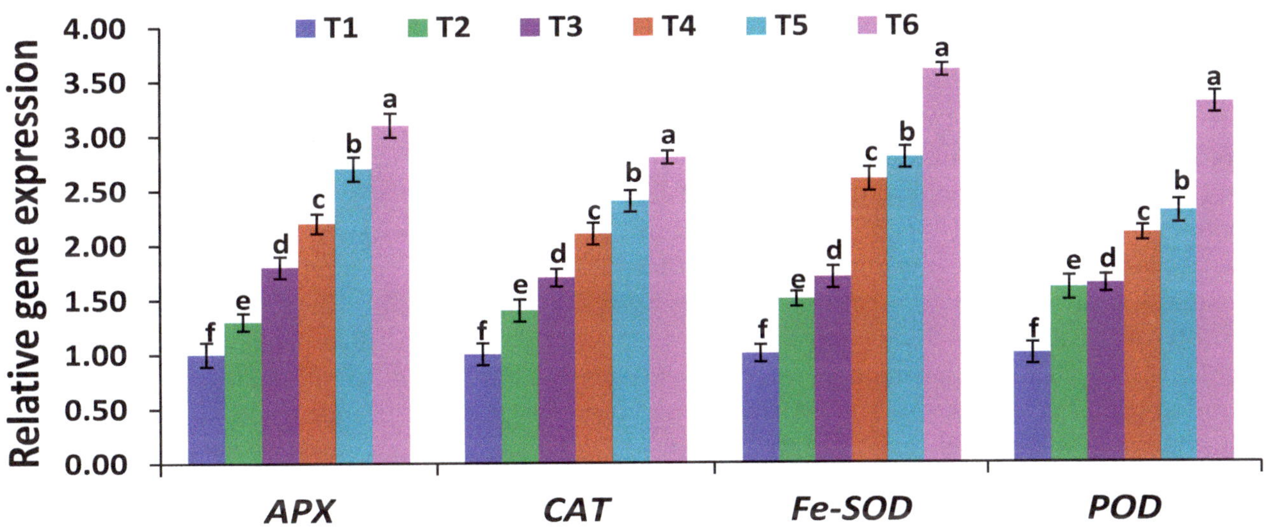

Figure 3. Expression levels of *APX, CAT, Fe-SOD,* and *POD* genes in leaves of un-inoculated and inoculated soybean plants with *Serratia marcescens* BM1 under Cd stress. Data are means ± SE ($n = 4$). Different letters (a–f) above the chart columns indicate significant differences between treatments ($p \leq 0.05$). T1, control with no Cd and BM1 treatment; T2, plant treated with *Serratia marcescens* BM1; T3, plant treated with 150 μM $CdCl_2$; T4, plant treated with 150 μM $CdCl_2$ and *Serratia marcescens* BM1; T5, plant treated with 300 μM $CdCl_2$; T6, plant treated with 300 μM $CdCl_2$ and *Serratia marcescens* BM1.

Overall, cadmium stress-induced oxidative effects have negative impacts on gas exchange attributes and photosynthetic activity, thereby, limiting plant growth and biomass. However, the present study revealed a key role of *Serratia marcescens* BM1 in improving phytoremediation and Cd stress tolerance in soybean plants by modulating the leaf gas exchange attributes, antioxidant machinery, osmolytes, and expression of genes associated with antioxidant defence systems and stress response.

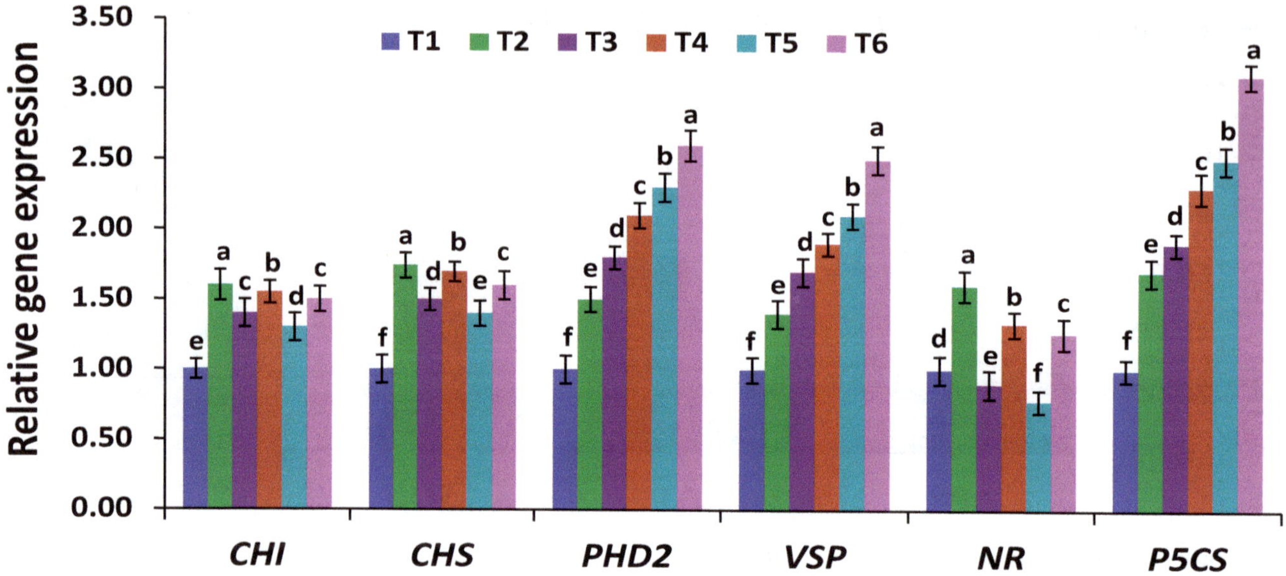

Figure 4. Expression levels of *CHI, CHS, PHD2, VSP, NR* and *P5CS* genes in leaves of un-inoculated and inoculated soybean plants with *Serratia marcescens* BM1 under Cd stress. Data are means ± SE ($n = 4$). Different letters (a–f) above the chart columns indicate significant differences between treatments ($p \leq 0.05$). T1, control with no Cd and BM1 treatment; T2, plant treated with *Serratia marcescens* BM1; T3, plant treated with 150 μM $CdCl_2$; T4, plant treated with 150 μM $CdCl_2$ and *Serratia marcescens* BM1; T5, plant treated with 300 μM $CdCl_2$; T6, plant treated with 300 μM $CdCl_2$ and *Serratia marcescens* BM1.

4. Conclusions

In the present study, Cd stress reduced the growth, biomass, photosynthetic attributes, mineral nutrients uptake, and contents of flavonoids, phenolics, soluble sugars, and proteins of soybean plants. Cadmium also increased the contents of cadmium, hydrogen peroxide, and malondialdehyde. Conversely, *Serratia marcescens* BM1 inoculation mitigated the deleterious impacts of cadmium stress by significantly inducing the plant growth, biomass, gas exchange attributes, nutrients uptake, antioxidant capacity, and the contents of chlorophyll, total phenolics, flavonoids, soluble sugars, and proteins in Cd-stressed soybean plants. *Serratia marcescens* BM1 also triggered the activities of antioxidant enzymes and the expression of stress-related genes. Considering these results, it could be concluded that *Serratia marcescens* BM1 might promote Cd stress tolerance through the modulation of photosynthesis, leaf gas exchange traits, antioxidant machinery, osmolyte biosynthesis, and the expression of stress-related genes.

Author Contributions: M.A.E.-E. performed the experiments, analyzed the data, and wrote the manuscript. A.E., M.S. and H.O.E. participated in experiments, data analysis and writing the manuscript. A.Z. and S.H.W. participated in writing the manuscript. All the authors revised and approved the final version of the manuscript. All authors have read and agreed to the published version of the manuscript.

Acknowledgments: The authors extend their appreciation to King Saud University's funding through Researchers Supporting Project number (RSP-2019/118). The authors also acknowledge the supporting facilities of Tanta University.

References

1. Wani, W.; Masoodi, K.Z.; Zaid, A.; Wani, S.H.; Shah, F.; Meena, V.S.; Wani, S.A.; Mosa, K.A. Engineering plants for heavy metal stress tolerance. *Rend. Lincei* **2018**, *29*, 709–723. [CrossRef]
2. Abou-Kassem, D.E.; Mahrose, K.M.; Alagawany, M. The role of vitamin E or clay in growing Japanese quail fed diets polluted by cadmium at various levels. *Animal* **2016**, *10*, 508–519. [CrossRef]

3. Ali, S.; Rizwan, M.; Zaid, A.; Arif, M.S.; Yasmeen, T.; Hussain, A.; Shahid, M.R.; Bukhari, S.A.H.; Hussain, S.; Abbasi, G.H. 5-Aminolevulinic Acid-Induced Heavy Metal Stress Tolerance and Underlying Mechanisms in Plants. *J. Plant Growth Regul.* **2018**, *37*, 1423–1436. [CrossRef]
4. Rizwan, M.; Ali, S.; Zia-ur-Rehman, M.; Maqbool, A. A critical review on the effects of zinc at toxic levels of cadmium in plants. *Environ. Sci. Pollut. Res.* **2019**, *26*, 6279–6289. [CrossRef]
5. Per, T.S.; Khan, N.A.; Masood, A.; Fatma, M. Methyl jasmonate alleviates cadmium-induced photosynthetic damages through increased S-assimilation and glutathione production in mustard. *Front. Plant Sci.* **2016**, *7*, 1933. [CrossRef]
6. Zaid, A.; Mohammad, F. Methyl Jasmonate and Nitrogen Interact to Alleviate Cadmium Stress in *Mentha arvensis* by Regulating Physio-Biochemical Damages and ROS Detoxification. *J. Plant Growth Regul.* **2018**, *37*, 1331–1348. [CrossRef]
7. Elkelish, A.A.; Alhaithloul, H.A.S.; Qari, S.H.; Soliman, M.H.; Hasanuzzaman, M. Pretreatment with *Trichoderma harzianum* alleviates waterlogging-induced growth alterations in tomato seedlings by modulating physiological, biochemical, and molecular mechanisms. *Environ. Exp. Bot.* **2019**, *171*, 103946. [CrossRef]
8. El-Esawi, M.; Arthaut, L.; Jourdan, N.; d'Harlingue, A.; Martino, C.; Ahmad, M. Blue-light induced biosynthesis of ROS contributes to the signaling mechanism of *Arabidopsis* cryptochrome. *Sci. Rep.* **2017**, *7*, 13875. [CrossRef]
9. Consentino, L.; Lambert, S.; Martino, C.; Jourdan, N.; Bouchet, P.-E.; Witczak, J.; Castello, P.; El-Esawi, M.; Corbineau, F.; d'Harlingue, A.; et al. Blue-light dependent reactive oxygen species formation by *Arabidopsis* cryptochrome may define a novel evolutionarily conserved signalling mechanism. *New Phytol.* **2015**, *206*, 1450–1462. [CrossRef]
10. El-Esawi, M.A.; Alaraidh, I.A.; Alsahli, A.A.; Ali, H.M.; Alayafi, A.A.; Witczak, J.; Ahmad, M. Genetic variation and alleviation of salinity stress in barley. *Molecules* **2018**, *23*, 2488. [CrossRef]
11. El-Esawi, M.A.; Alayafi, A.A. Overexpression of rice *Rab7* gene improves drought and heat tolerance and increases grain yield in rice (*Oryza sativa* L.). *Genes* **2019**, *10*, 56. [CrossRef] [PubMed]
12. El-Esawi, M.A.; Alayafi, A.A. Overexpression of *StDREB2* Transcription Factor Enhances Drought Stress Tolerance in Cotton (*Gossypium barbadense* L.). *Genes* **2019**, *10*, 142. [CrossRef] [PubMed]
13. Soliman, M.; Alhaithloul, H.A.; Hakeem, K.R.; Alharbi, B.M.; El-Esawi, M.; Elkelish, A. Exogenous Nitric Oxide Mitigates Nickel-Induced Oxidative Damage in Eggplant by Upregulating Antioxidants, Osmolyte Metabolism, and Glyoxalase Systems. *Plants* **2019**, *8*, 562. [CrossRef] [PubMed]
14. El-Esawi, M.A.; Al-Ghamdi, A.A.; Ali, H.M.; Ahmad, M. Overexpression of *AtWRKY30* Transcription Factor Enhances Heat and Drought Stress Tolerance in Wheat (*Triticum aestivum* L.). *Genes* **2019**, *10*, 163. [CrossRef] [PubMed]
15. El-Esawi, M.A.; Elkelish, A.; Elansary, H.O.; Ali, H.M.; Elshikh, M.; Witczak, J.; Ahmad, M. Genetic transformation and hairy root induction enhance the antioxidant potential of *Lactuca serriola* L. *Oxid. Med. Cell. Longev.* **2017**, *2017*, 5604746. [CrossRef]
16. Ahanger, M.A.; Qin, C.; Begum, N.; Maodong, Q.; Dong, X.X.; El-Esawi, M.; El-Sheikh, M.A.; Alatar, A.A.; Zhang, L. Nitrogen availability prevents oxidative effects of salinity on wheat growth and photosynthesis by up-regulating the antioxidants and osmolytes metabolism, and secondary metabolite accumulation. *BMC Plant Biol.* **2019**, *19*, 479. [CrossRef]
17. Wang, Y.Y.; Wang, Y.; Li, G.Z.; Hao, L. Salicylic acid-altering Arabidopsis plant response to cadmium exposure: Underlying mechanisms affecting antioxidation and photosynthesis-related processes. *Ecotoxicol. Environ. Saf.* **2019**, *169*, 645–653. [CrossRef]
18. Rizwan, M.; Ali, S.; Hussain, A.; Ali, Q.; Shakoor, M.B.; Zia-ur-Rehman, M.; Farid, M.; Asma, M. Effect of zinc-lysine on growth, yield and cadmium uptake in wheat (*Triticum aestivum* L.) and health risk assessment. *Chemosphere* **2017**, *187*, 35–42. [CrossRef]
19. Luo, H.; Li, H.; Zhang, X.; Fu, J. Antioxidant responses and gene expression in perennial ryegrass (*Lolium perenne* L.) under cadmium stress. *Ecotoxicology* **2011**, *20*, 770–778. [CrossRef]
20. Khanna, K.; Jamwal, V.L.; Kohli, S.K.; Gandhi, S.G.; Ohri, P.; Bhardwaj, R.; Abd-Allah, E.F.; Hashem, A.; Ahmad, P. Plant growth promoting rhizobacteria induced Cd tolerance in *Lycopersicon esculentum* through altered antioxidative defense expression. *Chemosphere* **2019**, *217*, 463–474. [CrossRef]

21. Mnasri, M.; Ghabriche, R.; Fourati, E.; Zaier, H.; Sabally, K.; Barrington, S.; Ghnaya, T. Cd and Ni transport and accumulation in the halophyte *Sesuvium portulacastrum*: Implication of organic acids in these processes. *Front. Plant Sci.* **2015**, *6*, 156. [CrossRef] [PubMed]
22. Silva, A.J.; Nascimento, C.W.A.; Gouveia-Neto, A.S. Assessment of cadmium phytotoxicity alleviation by silicon using chlorophyll a fluorescence. *Photosynthetica* **2016**, *55*, 648–654. [CrossRef]
23. Sakouhi, L.; Rahoui, S.; Gharsallah, C.; Munemasa, S.; El Ferjani, E.; Murata, Y.; Chaoui, A. Effects of calcium and EGTA on thiol homeostasis and defense-related enzymes in Cd-exposed chickpea roots. *Acta Physiol. Plant.* **2018**, *40*, 20. [CrossRef]
24. Rizwan, M.; Ali, S.; ur Rehman, M.Z.; Rinklebe, J.; Tsang, D.C.; Bashir, A.; Maqbool, A.; Tack, F.M.G.; Ok, Y.S. Cadmium phytoremediation potential of *Brassica* crop species: A review. *Sci. Total Environ.* **2018**, *631*, 1175–1191. [CrossRef]
25. Ma, Y.; Rajkumar, M.; Vicente, J.A.F.; Freitas, H. Inoculation of Ni-resistant plant growth promoting bacterium *Psychrobacter* sp. strain SRS8 for the improvement of nickel phytoextraction by energy crops. *Int. J. Phytoremed.* **2010**, *13*, 126–139. [CrossRef]
26. Shi, P.; Zhu, K.; Zhang, Y.; Chai, T. Growth and Cadmium Accumulation of *Solanum nigrum* L. Seedling were Enhanced by Heavy Metal-Tolerant Strains of *Pseudomonas aeruginosa*. *Water Air Soil Pollut.* **2016**, *227*, 459. [CrossRef]
27. Mesa-Marín, J.; Del-Saz, N.F.; Rodríguez-Llorente, I.D.; Redondo-Gómez, S.; Pajuelo, E.; Ribas-Carbó, M.; Mateos-Naranjo, E. PGPR Reduce Root Respiration and Oxidative Stress Enhancing *Spartina maritima* Root Growth and Heavy Metal Rhizoaccumulation. *Front. Plant Sci.* **2018**, *9*, 1500. [CrossRef]
28. Madhaiyan, M.; Poonguzhali, S.; Sa, T. Metal tolerating methylotrophic bacteria reduces nickel and cadmium toxicity and promotes plant growth of tomato (*Lycopersicon esculentum* L.). *Chemosphere* **2007**, *69*, 220–228. [CrossRef]
29. Wei, T.; Lv, X.; Jia, H.; Hua, L.; Xu, H.; Zhou, R.; Zhao, J.; Ren, X.; Guo, J. Effects of salicylic acid, Fe (II) and plant growth-promoting bacteria on Cd accumulation and toxicity alleviation of Cd tolerant and sensitive tomato genotypes. *J. Environ. Manag.* **2018**, *214*, 164–171. [CrossRef]
30. Sessitsch, A.; Kuffner, M.; Kidd, P.; Vangronsveld, J.; Wenzel, W.W.; Fallmann, K.; Puschenreiter, M. The role of plant-associated bacteria in the mobilization and phytoextraction of trace elements in contaminated soils. *Soil Biol. Biochem.* **2013**, *60*, 182–194. [CrossRef]
31. Kao, P.H.; Huang, C.C.; Hseu, Z.Y. Response of microbial activities to heavy metals in a neutral loamy soil treated with biosolid. *Chemosphere* **2006**, *64*, 63–70. [CrossRef] [PubMed]
32. Brunetti, G.; Farrag, K.; Soler-Rovira, P.; Ferrara, M.; Nigro, F.; Senesi, N. The effect of compost and *Bacillus licheniformis* on the phytoextraction of Cr, Cu, Pb and Zn by three brassicaceae species from contaminated soils in the Apulia region, Southern Italy. *Geoderma* **2012**, *170*, 322–330. [CrossRef]
33. Malekzadeh, E.; Alikhani, H.A.; Savaghebi-Firoozabadi, G.R.; Zarei, M. Bioremediation of cadmium-contaminated soil through cultivation of maize inoculated with plant growth–promoting rhizobacteria. *Bioremed. J.* **2012**, *16*, 204–211. [CrossRef]
34. Radhakrishnan, R.; Hashem, A.; Abd_Allah, E.F. *Bacillus*: A biological tool for crop improvement through bio-molecular changes in adverse environments. *Front. Physiol.* **2017**, *8*, 667. [CrossRef] [PubMed]
35. Gupta, C.; Dubey, R.; Maheshwari, D. Plant growth enhancement and suppression of *Macrophomina phaseolina* causing charcoal rot of peanut by fluorescent *Pseudomonas*. *Biol. Fertil. Soils* **2002**, *35*, 399–405.
36. Arkhipova, T.N.; Veselov, S.U.; Melentiev, A.I.; Martynenko, E.V.; Kudoyarova, G.R. Ability of bacterium *Bacillus subtilis* to produce cytokinins and to influence the growth and endogenous hormone content of lettuce plants. *Plant Soil* **2005**, *272*, 201–209. [CrossRef]
37. Gururani, M.A.; Upadhyaya, C.P.; Baskar, V.; Venkatesh, J.; Nookaraju, A.; Park, S.W. Plant growth-promoting rhizobacteria enhance abiotic stress tolerance in *Solanum tuberosum* through inducing changes in the expression of ROS-scavenging enzymes and improved photosynthetic performance. *J. Plant Growth Regul.* **2013**, *32*, 245–258. [CrossRef]
38. Chandra, P.; Tripathi, P.; Chandra, A. Isolation and molecular characterization of plant growth-promoting *Bacillus* spp. and their impact on sugarcane (*Saccharum* spp. hybrids) growth and tolerance towards drought stress. *Acta Physiol. Plant.* **2018**, *40*, 199. [CrossRef]
39. Cristani, M.; Naccari, C.; Nostro, A.; Pizzimenti, A.; Trombetta, D.; Pizzimenti, F. Possible use of *Serratia marcescens* in toxic metal biosorption (removal). *Environ. Sci. Pollut. Res.* **2012**, *19*, 161–168. [CrossRef]

40. Khan, A.R.; Park, G.S.; Asaf, S.; Hong, S.J.; Jung, B.K.; Shin, J.H. Complete genome analysis of *Serratia marcescens* RSC-14: A plant growth-promoting bacterium that alleviates cadmium stress in host plants. *PLoS ONE* **2017**, *12*, e0171534. [CrossRef]
41. Queiroz, P.S.; Ruas, F.A.D.; Barboza, N.R.; de Castro Borges, W.; Guerra-Sá, R. Alterations in the proteomic composition of *Serratia marcescens* in response to manganese (II). *BMC Biotechnol.* **2018**, *18*, 83. [CrossRef] [PubMed]
42. Egamberdieva, D.; Wirth, S.; Jabborova, D.; Räsänen, L.A.; Liao, H. Coordination between *Bradyrhizobium* and *Pseudomonas* alleviates salt stress in soybean through altering root system architecture. *J. Plant Interact.* **2017**, *12*, 100–107. [CrossRef]
43. El-Esawi, M.A.; Alaraidh, I.A.; Alsahli, A.A.; Alamri, S.A.; Ali, H.M.; Alayafi, A.A. *Bacillus firmus* (SW5) augments salt tolerance in soybean (*Glycine max* L.) by modulating root system architecture, antioxidant defense systems and stress-responsive genes expression. *Plant Physiol. Biochem.* **2018**, *132*, 375–384. [CrossRef] [PubMed]
44. Elkelish, A.A.; Alnusaire, T.S.; Soliman, M.H. Calcium availability regulates antioxidant system, physio-biochemical activities and alleviates salinity stress mediated oxidative damage in soybean seedlings. *J. Appl. Bot. Food Qual.* **2019**, *92*, 258–266.
45. Abd El-Azeem, S.A.M. Studies on Plant Growth Promoting Rhizosphere Microorganisms. Ph.D. Thesis, Faculty of Agriculture Suez Canal University, Ismailia, Egypt, 2006.
46. Abd El-Azeem, S.A.M.; Mehana, T.A.; Shabayek, A.A. Some plant-growth-promoting traits of rhizobacteria isolated from Suez Canal region, Egypt. *Afr. Crop Sci. Conf. Proc.* **2007**, *8*, 1517–1525.
47. Bremner, J.M. Total Nitrogen. In *Methods of Soil Analysis. Part 2. Chemical and Microbiological Properties (Methods of Soil)*; American Society of Agronomy, Soil Science Society of America: Madison, WI, USA, 1965; pp. 1149–1178.
48. Murphy, J.; Riley, J. A modified single solution method for the determination of phosphate in natural waters. *Anal. Chem. Acta* **1962**, *27*, 31–36. [CrossRef]
49. Lichtenthaler, H.K. Chlorophylls and carotenoids: Pigments of photosynthetic biomembranes. *Methods Enzym.* **1987**, *148*, 350–382.
50. Holá, D.; Benešová, M.; Honnerová, J.; Hnilička, F.; Rothová, O.; Kočová, M.; Hniličková, H. The evaluation of photosynthetic parameters in maize inbred lines subjected to water deficiency: Can these parameters be used for the prediction of performance of hybrid progeny? *Photosynthetica* **2010**, *48*, 545–558. [CrossRef]
51. Dey, P.M. Oligosaccharides. In *Methods in Plant Biochemistry, Carbohydrates*; Dey, P.M., Ed.; Academic Press: London, UK, 1990; Volume 2, pp. 189–218.
52. Bradford, M.M. A rapid and sensitive method for the quantification of microgram quantities of protein utilizing the principle of protein dye binding. *Anal. Biochem.* **1976**, *72*, 248–254. [CrossRef]
53. Bates, L.; Waldren, P.P.; Teare, J.D. Rapid determination of free proline of water stress studies. *Plant Soil* **1973**, *39*, 205–207. [CrossRef]
54. Grieve, C.; Grattan, S. Rapid assay for determination of water-soluble quaternary ammonium compounds. *Plant Soil* **1983**, *70*, 303–307. [CrossRef]
55. Zhishen, J.; Mengcheng, T.; Jianming, W. The determination of flavonoid contents in mulberry and their scavenging effects on superoxide radicals. *Food Chem.* **1999**, *64*, 555–559. [CrossRef]
56. Zieslin, N.; Ben-Zaken, R. Peroxidase activity and presence of phenolic substances in peduncles of rose flowers. *Plant Physiol. Biochem.* **1993**, *31*, 333–339.
57. Velikova, V.; Yordanov, I.; Edreva, A. Oxidative stress and some antioxidant systems in acid rain-treated bean plants: Protective role of exogenous polyamines. *Plant Sci.* **2000**, *151*, 59–66. [CrossRef]
58. Heath, R.L.; Packer, L. Photoperoxidation in isolated chloroplasts. I. Kinetics and stoichiometry of fatty acid peroxidation. *Arch. Biochem. Biophys.* **1968**, *125*, 189–198. [CrossRef]
59. Ueno, H.; Yamakura, S.; Arastoo, R.S.; Oshima, T.; Kokubo, K. Systematic evaluation and mechanistic investigation of antioxidant activity of fullerenols using β carotene bleaching assay. *Nanomaterials* **2014**, *7*, 802596.
60. Pyrzynska, K.; Pękal, A. Application of free radical diphenylpicrylhydrazyl (DPPH) to estimate the antioxidant capacity of food samples. *Anal. Methods* **2013**, *5*, 4288–4295. [CrossRef]
61. Nakano, Y.; Asada, K. Hydrogen peroxide is scavenged by ascorbate-specific peroxidase in spinach chloroplasts. *Plant Cell Physiol.* **1981**, *22*, 867–880.

62. Aebi, H. Catalase in vitro. *Methods Enzym.* **1984**, *105*, 121–126.
63. Kono, Y. Generation of superoxide radical during autoxidation of hydroxylamine and an assay for superoxide dismutase. *Arch. Biochem. Biophys.* **1978**, *186*, 189–195. [CrossRef]
64. Putter, J.; Becker, R. *Methods of Enzymatic Analysis*; Academic Press: New York, NY, USA, 1974; p. 685.
65. Zhou, Y.; Huang, J.L.; Zhang, X.L.; Zhu, L.M.; Wang, X.F.; Guo, N.; Zhao, J.M.; Xing, H. Overexpression of Chalcone Isomerase (CHI) Increases Resistance Against *Phytophthora sojae* in Soybean. *J. Plant Biol.* **2018**, *61*, 309–319. [CrossRef]
66. Wei, W.; Huang, J.; Hao, Y.J.; Zou, H.F.; Wang, H.W.; Zhao, J.Y.; Liu, X.Y.; Zhang, W.K.; Ma, B.; Zhang, J.S.; et al. Soybean GmPHD-type transcription regulators improve stress tolerance in transgenic *Arabidopsis* plants. *PLoS ONE* **2009**, *4*, e7209. [CrossRef] [PubMed]
67. Vaishnav, A.; Kumari, S.; Jain, S.; Varma, A.; Tuteja, N.; Choudhary, D.K. PGPR-mediated expression of salt tolerance gene in soybean through volatiles under sodium nitroprusside. *J. Basic Microbiol.* **2016**, *56*, 1274–1288. [CrossRef] [PubMed]
68. Sirhindi, G.; Mir, M.A.; Abd-Allah, E.F.; Ahmad, P.; Gucel, S. Jasmonic Acid Modulates the Physio-Biochemical Attributes, Antioxidant Enzyme Activity, and Gene Expression in *Glycine max* under Nickel Toxicity. *Front. Plant Sci.* **2016**, *7*, 591. [CrossRef]
69. El-Esawi, M.A.; Elansary, H.O.; El-Shanhorey, N.A.; Abdel-Hamid, A.M.E.; Ali, H.M.; Elshikh, M.S. Salicylic Acid-Regulated Antioxidant Mechanisms and Gene Expression Enhance Rosemary Performance under Saline Conditions. *Front. Physiol.* **2017**, *8*, 716. [CrossRef]
70. Kim, M.J.; Kim, H.J.; Pak, J.H.; Cho, H.S.; Choi, H.K.; Jung, H.W.; Lee, D.H.; Chung, Y.-S. Overexpression of *AtSZF2* from *Arabidopsis* showed enhanced tolerance to salt stress in soybean. *Plant Breed. Biotechnol.* **2017**, *5*, 1–15. [CrossRef]
71. Livak, K.J.; Schmittgen, T.D. Analysis of relative gene expression data using real-time quantitative PCR and the 2(−Delta Delta C(T)) method. *Methods* **2001**, *25*, 402–408. [CrossRef]
72. Selvakumar, G.; Mohan, M.; Kundu, S.; Gupta, A.D.; Joshi, P.; Nazim, S.; Gupta, H.S. Cold tolerance and plant growth promotion potential of *Serratia marcescens* strain SRM (MTCC 8708) isolated from flowers of summer squash (*Cucurbita pepo*). *Lett. Appl. Microbiol.* **2008**, *46*, 171–175. [CrossRef]
73. Devi, K.A.; Pandey, P.; Sharma, G.D. Plant Growth-Promoting Endophyte Serratia marcescens AL2-16 Enhances the Growth of *Achyranthes aspera* L., a Medicinal Plant. *HAYATI J. Biosci.* **2016**, *23*, 173–180. [CrossRef]
74. El-Esawi, M.A.; Alaraidh, I.A.; Alsahli, A.A.; Alzahrani, S.M.; Ali, H.M.; Alayafi, A.A.; Ahmad, M. *Serratia liquefaciens* KM4 Improves Salt Stress Tolerance in Maize by Regulating Redox Potential, Ion Homeostasis, Leaf Gas Exchange and Stress-Related Gene Expression. *Int. J. Mol. Sci.* **2018**, *19*, 3310. [CrossRef]
75. Singh, R.P.; Jha, P.N. The multifarious PGPR *Serratia marcescens* CDP-13 augments induced systemic resistance and enhanced salinity tolerance of wheat (*Triticum aestivum* L.). *PLoS ONE* **2016**, *11*, e0155026. [CrossRef] [PubMed]
76. Khan, A.R.; Ullah, I.; Khan, A.L.; Park, G.S.; Waqas, M.; Hong, S.J.; Jung, B.K.; Kwak, Y.; Lee, I.J.; Shin, J.H. Improvement in phytoremediation potential of *Solanum nigrum* under cadmium contamination through endophytic-assisted *Serratia* sp. RSC-14 inoculation. *Environ. Sci. Pollut. Res.* **2015**, *22*, 14032–14042. [CrossRef] [PubMed]
77. Chakraborty, U.; Chakraborty, B.N.; Chakraborty, A.P. Influence of Serratia marcescens TRS-1 on growth promotion and induction of resistance in *Camellia sinensis* against *Fomes lamaoensis*. *J. Plant Interact.* **2010**, *5*, 261–272. [CrossRef]
78. Wan, Y.; Luo, S.; Chen, J.; Xiao, X.; Chen, L.; Zeng, G.; Liu, C.; He, Y. Effect of endophyte-infection on growth parameters and Cd-induced phytotoxicity of Cd-hyperaccumulator *Solanum nigrum* L. *Chemosphere* **2012**, *89*, 743–750. [CrossRef] [PubMed]
79. El-Esawi, M.A.; Al-Ghamdi, A.A.; Ali, H.M.; Alayafi, A.A. *Azospirillum lipoferum* FK1 confers improved salt tolerance in chickpea (*Cicer arietinum* L.) by modulating osmolytes, antioxidant machinery and stress-related genes expression. *Environ. Exp. Bot.* **2019**, *159*, 55–56. [CrossRef]

8

On the Origin and Fate of Reactive Oxygen Species in Plant Cell Compartments

Martina Janků, Lenka Luhová and Marek Petřivalský *

Department of Biochemistry, Palacký University in Olomouc, Šlechtitelů 27, CZ-78371 Olomouc, Czech Republic; janku.martina@seznam.cz (M.J.); lenka.luhova@upol.cz (L.L.)

* Correspondence: marek.petrivalsky@upol.cz

Abstract: Reactive oxygen species (ROS) have been recognized as important signaling compounds of major importance in a number of developmental and physiological processes in plants. The existence of cellular compartments enables efficient redox compartmentalization and ensures proper functioning of ROS-dependent signaling pathways. Similar to other organisms, the production of individual ROS in plant cells is highly localized and regulated by compartment-specific enzyme pathways on transcriptional and post-translational level. ROS metabolism and signaling in specific compartments are greatly affected by their chemical interactions with other reactive radical species, ROS scavengers and antioxidant enzymes. A dysregulation of the redox status, as a consequence of induced ROS generation or decreased capacity of their removal, occurs in plants exposed to diverse stress conditions. During stress condition, strong induction of ROS-generating systems or attenuated ROS scavenging can lead to oxidative or nitrosative stress conditions, associated with potential damaging modifications of cell biomolecules. Here, we present an overview of compartment-specific pathways of ROS production and degradation and mechanisms of ROS homeostasis control within plant cell compartments.

Keywords: cell wall; chloroplasts; cytoplasmic membrane; cytosol; glyoxysomes; mitochondria; peroxisomes; plant cell; reactive oxygen species

---

1. Introduction

Subcellular compartmentalization in eukaryotic cells forms the basis for highly selective separation of biochemical reactions and metabolic pathways, and delimit their potential mutual interferences [1]. Current knowledge indicates that within each compartment of eukaryotic cells, specific redox characteristics have evolved [2,3]. Unique redox aspects of different cell compartments are integrated at the molecular level by a group of small diffusible and reactive molecules, reactive oxygen species (ROS), which provide communication between intracellular compartments [4]. ROS are produced in all forms of aerobic life by divergent reactions including a partial reduction of the molecular oxygen [5,6]. Previously, ROS were described as toxic by-products of the aerobic metabolism and their increased levels were connected with multiple forms of cellular damage mediated by oxidative modifications of cell biomolecules. Nowadays, the accumulated evidence suggest that ROS function as key redox signaling and effector molecules in vital biological processes such as cell growth, differentiation, proliferation and responses to a wide spectrum of external stimuli [7,8]. In plant cells, signaling networks of ROS are closely connected to central physiological processes of energy generation and consumption such as respiration, photosynthesis and photorespiration and in plant responses to abiotic and biotic stress conditions [9–11].

In general, signaling functions of ROS are related to tightly regulated site- and time-specific modulations of ROS levels, whereas uncontrolled ROS accumulation as a consequence of either induced production or defective catabolism, or their combination, is associated with damaging oxidative effect leading to cellular damages or even to the cell death [12]. ROS can modify proteins structures and functions through oxidative reactions namely with protein thiol groups or iron-containing clusters. On the other hand, toxic effects of accumulated ROS are also exploited as chemical defense within immune repertoires of diverse organisms against invading pathogens, such as the well-described ROS burst leading to the plant cell death within the plant hypersensitive response to biotrophic pathogens [13].

Individual ROS show a high variability in their chemical properties, reactivity and involvement in redox signaling pathways [14]. As a general rule, radical forms of ROS are more reactive compared to non-radical ROS. Moreover, ROS reactivity is significantly affected by presence of free Fe^{2+} ions involved in formation of highly reactive hydroxyl radical in a pathway termed the Fenton reaction, which results in production of hydroxyl radicals as the most reactive ROS form with a single unpaired electron, capable to react with virtually all biomolecules. High reactivity of hydroxyl radicals leads to subsequent cellular damages, including changes in protein structures, lipid peroxidation and membrane destruction [15]. However, recent findings support the suggested role of the hydroxyl radical as being more than a destructive agent, as its oxidative properties can facilitate seed germination, growth, stomatal closure, reproduction, immune responses, plant cell death and adaptations to stress conditions [16].

Singlet oxygen ($^1O_2$) is formed in photosynthetic membranes in reactions between triplet-state chlorophyll molecules and the molecular oxygen [17]. Observed half-life of $^1O_2$ is approximately 3 μs and it has been shown that a minor portion of singlet oxygen is able to diffuse to a distance of several nanometers, where it can react with diverse biomolecules and potentially mediate its signaling functions [18,19]. Enzyme activity of lipoxygenase represents an additional source of $^1O_2$ in plants [20]. Highly reactive $^1O_2$ can be efficiently removed by quenching with low-molecular lipophilic compounds like β-carotene and tocopherol, or by a scavenging action of protein D1 in photosystem II [21].

Superoxide anion radical (O_2^-) is a highly reactive nucleophilic species with a half-life about 1 μs, which often serves as an initiator of reaction cascades generating other ROS, namely hydrogen peroxide. Superoxide occurs with a similar half-life as singlet oxygen, but strongly differs in its target reaction partners. Polyunsaturated fatty acid within thylakoid membranes are considered the main reaction partners of $^1O_2$, whereas superoxide preferentially reacts with other radical compound including nitric oxide and components of protein hem-containing and non-hem iron centers [22].

Hydrogen peroxide (H_2O_2) is the least reactive non-radical ROS. Its signaling pathways are mediated mainly through interactions with proteins containing redox-sensitive moieties, such as metal centers or cysteine thiols, whose oxidation controls protein biological activity [23,24]. H_2O_2 can diffuse across lipid membranes through the aquaporin channels, which is a crucial feature for H_2O_2 intracellular signaling [25]. In plants, it was demonstrated that H_2O_2 transport could also be mediated by aquaporin homologues from the plasma membrane intrinsic factor (PIP) family and by the tonoplast intrinsic protein (TIP) [26]. Excess H_2O_2 is known to trigger chloroplast and peroxisome autophagy and programmed cell death in plants [27].

2. Subcellular Localization and Functions of ROS Production in Plant Cells

On a quantitative scale, in contrast to animals where mitochondria play a predominant role in the ROS biogenesis in most cell types, plant cells generate ROS in a highly variable manner by multiple pathways depending on the plant tissue, developmental stage and external conditions [9,28]. ROS metabolism has been extensively studied in all plant cell compartments including cell wall, apoplast, plasma membrane, cytosol, mitochondria, chloroplasts, peroxisomes and glyoxysomes [29]. More stable ROS species like hydrogen peroxide can even diffuse, cross cell membranes and transport their signaling effects into other compartments (Figure 1). Moreover, ROS cooperate and cross-talk with signaling pathways of plant hormones such as abscisic, jasmonic and salicylic acid or ethylene [30,31].

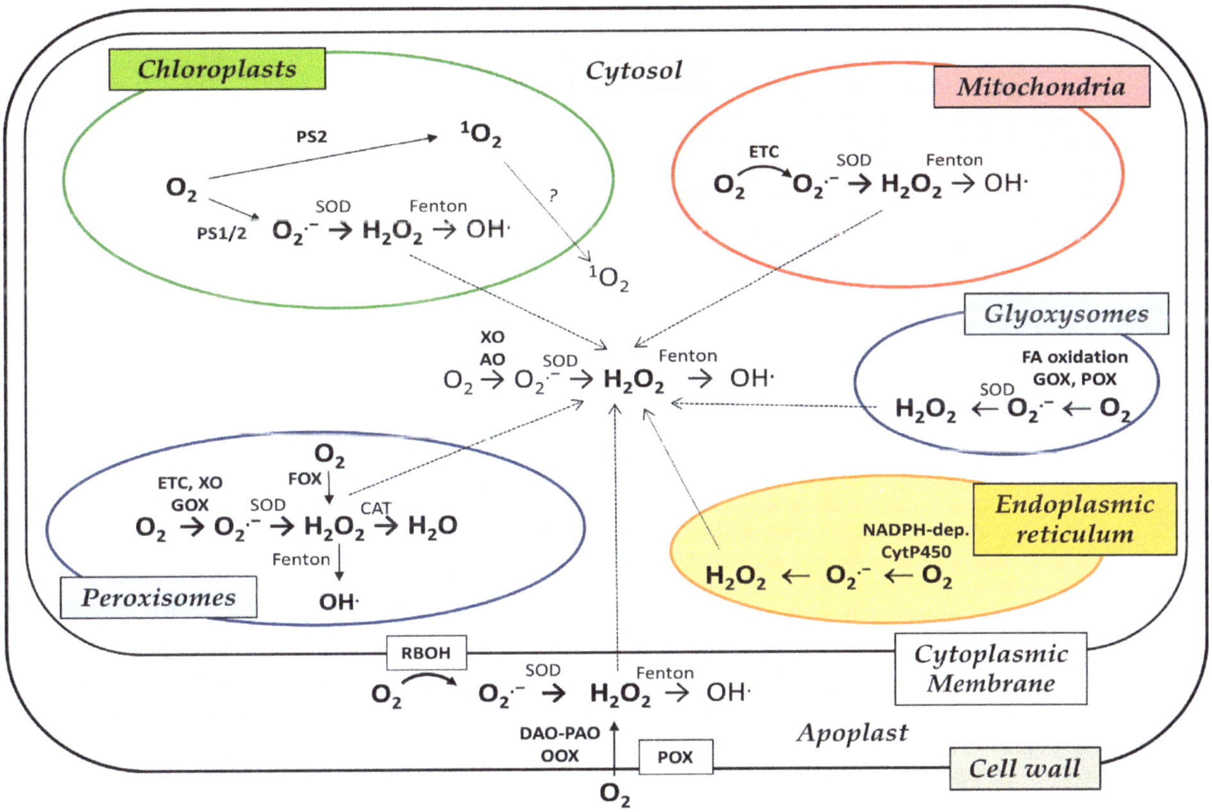

Figure 1. Schematic overview of reactive oxygen species (ROS) sources in plant cell compartments. AO, aldehyde oxidase; CAT, catalase; DAO, diamine oxidase; GOX, glycolate oxidase; ETC, electron transport chain; FOX, flavin oxidases; OOX, oxalate oxidase; PAO, polyamine oxidase; POX, peroxidases; PS, photosystem; RBOH, NADPH oxidase; SOD, superoxide dismutase; XO, xanthine oxidase/dehydrogenase.

2.1. Cell Wall, Apoplast and Cytoplasmic Membrane

The cell wall, apoplast and cytoplasmic membrane form crucial boundary compartments of plant cells securing their integrity vital for intracellular processes and communication with external environment including interactions with microbial symbionts or pathogens [32,33]. An important portion of ROS produced under physiological conditions and in plant responses to stress stimuli is formed within these compartments, which represent the major site of ROS production in plant-pathogen interactions [34]. Intensive ROS production termed as the oxidative burst is activated upon the recognition of pathogens or other biotic stress stimuli by immune receptors localized in plasma membrane, which triggers signaling pathways leading to transient localized increase of ROS [35]. In plants, receptor-like kinases (RLKs) represent key elements of the communication between the external environment and the cellular interior. Apoplastic ROS production occurs often following activation of RLK signaling in a wide array of cellular processes [36]. Thus, intricate interconnections exist between RLKs, extracellular ROS generation and ROS signal transduction and perception within the plant ROS sensing machinery. ROS can further diffuse as H_2O_2 inside the plant cells via membrane aquaporin channels [25]. Furthermore, apoplastic ROS are key players in physiological processes such as cell growth, which is particularly dependent on ROS involvement in increasing extensibility of the cell wall during root hair and pollen tube elongation, as well as in leaf growth [37,38]. It is known that hydroxyl radical promotes cell elongation by loosening the cell wall by oxidative cleavage of pectins and xyloglucans. On the other hand, H_2O_2 can initiate polysaccharide cross-linking and protein disulfide bond formation which restricts elongation growth. Several apoplastic and plasma-membrane proteins are involved in the regulation of apoplastic balance between hydroxyl radical and H_2O_2, which regulates cell expansion by direct interactions with cell wall components and also ROS interactions with intracellular transcriptional networks and cytoskeleton [37].

Plant-specific class III peroxidases, as members of a large multigene family of peroxidases (POXs, EC 1.11.1.7) are localized in the plant cell wall and comprise a significant source of apoplastic ROS [39]. Peroxidases contain eight conserved cysteine residues, one or more glycosylation sites, and a signal peptide which divert them into cell wall, apoplast or vacuole [40]. Peroxidases show a broad substrate specificity and can operate either in ROS-consuming or ROS-generating mode through reaction mechanisms based on peroxidation or hydroxylation cycle, respectively [41–44]. In ROS-consuming peroxidase mode, one oxygen atom from H_2O_2 is transferred to the hem group of the ferric enzyme in the ground state and the first intermediate is formed. Then an electron from a reducing substrate generates the second intermediate and the cycle is finished by regeneration of ferric enzyme [41]. H_2O_2 reduction catalyzed by peroxidases occurs with simultaneous oxidation of electron donors such as phenolic compounds [45]. Yet, the presence of certain reductants is not the only condition for peroxidases functions; the second one is the apoplast alkalinization. In ROS-producing mode, peroxidases produce ROS due to the oxidase cycle forming the third intermediate. Plant class III peroxidases are responsible for production of apoplastic ROS and involved in stress signaling after pathogen attack, wounding and under abiotic stress stimuli [46]. Beside their function in signaling pathways, POXs are also involved in polymerization of suberin and lignin, important compounds of passive plant defense barriers [42] in a cross-talk with signaling pathways of jasmonic acid [47].

The apoplastic oxidative burst occurs as a result of ROS-producing activities of several groups of enzymes including cell wall peroxidases, plasma membrane NADPH oxidases, amine oxidases, lipoxygenases, oxalate oxidases and quinone reductases [48,49]. Lower apoplastic pH compared to the cytosolic pH has a crucial control effect on redox properties of protein cysteine thiols and overall redox conditions. Another important feature of the apoplastic space is decreased antioxidant capacity as a consequence of low abundance of low-molecular weight antioxidants such as glutathione and ascorbate. Decreased antioxidant capacity in the apoplast is critical for activation of ROS signaling pathways. Thus, the apoplastic oxidative burst results in enhanced apoplastic oxidation state, while cytoplasm stays reduced. Redox gradients formed across the cellular membrane allow for differential regulation of proteins on cell surface such as receptors important for stress signal perception [50]. Oxalate oxidase and amine oxidases belong to ROS sources found in the apoplast. Oxalate oxidase (EC 1.2.3.4) catalyzes oxidation of oxalate to H_2O_2 and CO_2, and its gene expression is strongly induced by increased levels of salts and salicylic acid [51]. Copper-containing amine oxidases (CuAOs, EC 1.4.3.6) together with polyamine oxidases (PAOs, EC 1.5.3.3) are abundant plant proteins involved in the catabolism of polyamines, important plant growth regulators [52,53]. CuAOs are homodimer proteins contain copper ion and 2,4,5-trihydroxyphenylalanine quinone cofactor in both subunits. PAOs are monomeric and require flavin adenine nucleotide for their enzymatic activity. Enzymes of these classes are responsible for di- and polyamine catabolism, catalyzing oxidative deamination of their substrates, producing appropriate aldehydes, ammonia and H_2O_2 [54]. Produced apoplastic H_2O_2 is then used in modifications of the cell wall under physiological or stress conditions [55] or can function as a signal molecule, e.g., in the opening of Ca^{2+} channels [56,57].

Germins, a group of proteins with oxalate oxidases activity, represents another contribution to apoplastic ROS levels. Germins are apoplastic proteins with an oligomeric structure appearing at the onset of germination in plants, and are involved in plant development and defense responses to pathogen attack [58]. Oxalate oxidases contain manganese in their structure and are responsible for oxidation of oxalate. Moreover, plants with higher oxalate oxidase activity were found more resistant to plant pathogens [59]. Apart from germins, the germin protein family includes germin-like proteins (GLPs) which lack the oxalate oxidase activity but were shown to possess superoxide dismutase (SOD) activity [60].

NADPH oxidases (termed NOXs in animals) are family of proteins catalyzing production of superoxide in cytoplasmic membrane, termed as respiratory burst oxidase homologues (Rbohs) in plants [61,62]. Rbohs are considered the major ROS sources within the oxidative burst orchestrated in plant responses to pathogen attacks [63]. Within molecular events following the recognition of microbe- and damage-associated molecular patterns (MAMPs and DAMPs, respectively), the ROS is often observed as a biphasic ROS increase [64]. Rbohs were believed to specifically localize in plasma membrane [62]; however, recently, NtRbohD was observed localized in lipid rafts, membrane micro-domains in tobacco cells that can be coupled to other membrane components. It seems that different sub-membrane distribution of Rbohs is critical for their functions [65,66].

Protein structures of plant NADPH oxidases contain six conserved transmembrane helixes, where the 3rd and 5th helix bind two hem groups through four histidine residues. The hem groups are essential for transfer of electrons across the membrane to the molecular oxygen as the acceptor in the extracellular space [67]. N-terminal Ca^{2+} binding motifs and cytosolic FAD- and NADPH-binding domains in C-terminal regions are another typical structural feature of plant Rbohs. Plant Rbohs share a high similarity in their amino acid sequences of regions spanning membrane C-terminal cytosolic regions, whereas N-terminal cytosolic regions of NADPH oxidases show higher variability. NADPH oxidases require Ca^{2+} binding and phosphorylation to become fully activated; however, the exact activation mechanism has not been yet deciphered. These two activation events were suggested to occur synergistically [68], whereas other studies indicated the phosphorylation occurs as the first step followed by Ca^{2+} binding [69]. Both Rbohs activity and localization are controlled through phosphorylation by protein kinases from several families, including calcium-dependent and calmodulin-like domain protein kinases (CDPKs) [66,70]. The product of NADPH oxidase catalyzed reaction, superoxide anion radical is a membrane impermeable species due to its negative charge; however, under decreased pH conditions the superoxide is protonated and can pass throw the membranes. In this way, the pH status can affect compartmentalization of created product: in plants, where the physiological range is found around pH 5, approximately 16% of superoxide produced by Rbohs is transformed to a membrane permeable hydroperoxyl form [62].

Superoxide dismutases (SODs) constitute the first line of ROS catabolism with a crucial function in all compartments where superoxide radicals are produced [51]. Specifically in the apoplast, Cu-Zn SOD were reported to have a physiologically role in the lignification process in the vascular tissue of spinach hypocotyls [71]. It is noteworthy that concentrations of the key redox regulators glutathione (GSH) in the apoplast are orders of magnitude lower compared to the cytosol, where GSH is found in the milimolar range [72]. Furthermore, most apoplastic ascorbic acid (AA) in physiological conditions is oxidized, while over 90% of total leaf tissue AA is found in the reduced form [73], and oxidized AA derivatives, such as oxalic acid, were suggested to function as signaling molecules [74]. On the other hand, AA has been suggested to be the most important antioxidant for the detoxification of ROS in the apoplast in stress conditions [75]. Therefore, it seems that the low abundance of low molecular antioxidants in the apoplast might extend the half-life of H_2O_2, and thus enable H_2O_2 diffusion and propagation of ROS signals within and from the apoplast compartment.

2.2. Cytosol

The cytosol has not been regarded as an important compartment in plant ROS production. Cytoplasmic NADPH is a central component of redox maintenance in the cytosol as it controls the thiol/disulfide status and participates in the production of reducing substrates for antioxidant enzymes involved in ROS catabolism. On the other hand, NADPH also supplies electrons to plasma membrane NADPH oxidases as ROS producing enzymes [76]. Xanthine oxidase/dehydrogenase (XO; EC 1.2.1.37) and aldehyde oxidase (AO; EC 1.2.3.1) are molybdenum- and flavin-containing enzymes, which play important roles in plant purine metabolism and phytohormone biosynthesis, respectively. Both XO and AO contribute to increased ROS production in multiple animal pathologies and were suggested as potential sources of ROS also in plants. Interestingly, tomato and *Arabidopsis thaliana* XO was shown to produce superoxide but not H_2O_2 in vitro, whereas its animal counterpart can produce both superoxide and H_2O_2 [77]. Superoxide-generating activity was lost in *A. thaliana* Atxdh1 T-DNA insertion mutant and RNA interference lines, providing molecular evidence that plant XO generates superoxide. Unlike XO, AO produces only H_2O_2, which was surprisingly found not sensitive to inhibition by diphenyl iodonium, a known inhibitor of flavin-containing enzymes. Application of abscisic acid or water-stressed induced ROS production and AO and XO upregulation both in leaves and roots, indicating that plant AO and XO might be significant sources for ROS accumulation under stress condition [77].

Recently, AO isoforms AO2 and AO3 and their superoxide-producing activity were observed in *Nicotiana benthamiana* plants inoculated with the tomato bushy stunt virus, suggesting involvement of plant AO in defense mechanisms against viral infection [78]. It seems that another specific AO isoforms AsAO4 has a critical function in delaying senescence in siliques by catalyzing aldehyde detoxification under stress condition [79]. AAO4 enzyme oxidizes an array of aromatic and aliphatic aldehydes and generates superoxide and H_2O_2 in an aldehyde-dependent manner. Interestingly, XDH1 can play spatially specific dual and opposing roles in ROS metabolism in *A. thaliana* defense responses to powdery mildew [80]. In leaf epidermal cells, XDH1 functions as an oxidase, together with the NADPH oxidases RbohD and RbohF, to generate superoxide. Superoxide dismutation to H_2O_2 and subsequent accumulation of H_2O_2 in the fungal haustorial structures within infected epidermal cells contributes to constrain the haustorium, thereby contributing to powdery mildew resistance. In contrast, leaf XDH1 functions in the xanthine dehydrogenase mode producing uric acid in local and systemic tissues to scavenge H_2O_2 from stressed chloroplasts, thus contributing to plant protection from biotic stress-induced oxidative damage.

Besides its role in ROS production, the cytosol is supposed to play a key role in redox signal integration. ROS signaling from the apoplast and organelles needs to pass through the cytoplasm to achieve its effects and modulate gene expression in the cell nucleus [81]. The signal can be transmitted directly from H_2O_2 or via redox sensitive proteins to the MAPK cascade and to redox-dependent transcription factors such as heat stress transcription factors [82]. A detailed discussion of ROS signaling mechanisms in plant cells is not within the scope of this paper; nevertheless, multiple aspects of plant ROS signaling have been recently covered elsewhere [9,11,12,29,36,61,83].

2.3. Mitochondria

Mitochondrial respiration, encompassed by a directed electron flow from reduced organic substrates to the molecular oxygen through components of respiratory chain in the inner mitochondrial membrane, is inheritable associated with ROS generation [84]. Therefore, superoxide production occurs during normal operation of the respiratory chain, but its rate is highly increased in conditions of decelerated respiratory rates, e.g., by respiratory chain inhibition or limited ADP availability, resulting in a highly reduced state of mitochondrial electron transport chain. Superoxide disproportionation to H_2O_2 and O_2 is strongly accelerated by superoxide dismutase present in mitochondrial matrix [85].

In animals, respiratory complexes I and III are considered main sources of mitochondrial ROS, namely of superoxide anion radical, formed by a partial reduction of the molecular oxygen by highly reactive redox intermediates [86]. In plants, succinate-dependent H_2O_2 production in mitochondria was reported to be faster than pyruvate/malate-dependent H_2O_2 production, indicating a larger role for Complex II compared to Complex I [87]. Moreover, the ubiquinone pool might serve as another site of ROS production in plant mitochondria [88]. Using a *A. thaliana* mutant in the Complex II subunit succinate dehydrogenase 1-1, Complex II was shown to contribute to localized mitochondrial ROS production that regulates plant stress and defense responses [89].

ROS originating from plant mitochondria are known to impact both mitochondrial respiratory and cellular functions, including a variety of signaling cascades in the cell, comprising retrograde signaling, plant hormone action, programmed cell death and defense against pathogens [90]. Observed mitochondrial steady-state levels of H_2O_2 are two orders of magnitude lowers compared to peroxisomes and chloroplasts. However, in specific plant tissue or developmental stages, major production and accumulation of mitochondrial ROS can occur. During seed imbibition and hydration, high rates of mitochondrial respiration contribute as the major source of superoxide, which need to be properly balanced by the antioxidant machinery to avoid oxidative damage of biomolecules potentially resulting in defective seed germination [91]. Superoxide production in plant mitochondria can be minimized by several pathways that enable bypassing the electron transport chain. Proton leak across the membrane is facilitated by uncoupling proteins, whereas alternative oxidase (AOX) bypasses proton pumping on Complex III and IV [92]. Interestingly, overexpression of the *A. thaliana* uncoupling protein 1 in tobacco reduced ROS generation, induced the antioxidant system and increased resistance to diverse abiotic stress conditions [93,94].

2.4. Chloroplasts

Within photosynthetically active cells, chloroplasts act as another site of the major ROS production, connected with multiple redox reactions and electron transport chains localized in thylakoid membranes. Singlet oxygen ($^1O_2$) is a unique ROS species produced constitutively in plant leaves in light. Chlorophylls pigments in the antenna system and in the reaction center of photosystem II are primary sources of $^1O_2$ in plants, which is formed in a reaction between the molecular oxygen and chlorophyll molecules in the triplet state [17,95,96]. The highly reactive nature of $^1O_2$ is characterized by its half-life of about 3 μs in water solutions, but about 100 μs in hydrophobic environment of lipid membranes. A small part of singlet oxygen is able to diffuse to distances of several nanometers and it has been proposed that $^1O_2$ can be involved in the signaling of programmed cell death or light acclimation processes [18]. Production of $^1O_2$ is enhanced under light stress conditions, and lipoxygenase activity can provide an additional source of $^1O_2$ [19]. Singlet oxygen can be detoxified namely by quenching by small lipophilic antioxidants like tocopherol or β-carotene. Carotenoids are capable of effective $^1O_2$ scavenging resulting in formation of excited triplet states which return efficiently to the ground state through a heat release mechanism. Among carotenoids occurring in plant cells, β-carotene is functionally the most important [97]. Plastoquinone and D1 protein of photosystem II are known as other potential $^1O_2$ scavengers [21]. Recently, a role of $^1O_2$ produced by the chloroplast lipoxygenase was reported in leaf wounding using *A. thaliana lox2* mutants where $^1O_2$ produced in wounded plants is involved in lipid or protein oxidation which can act as signaling mechanisms [20].

Chloroplast superoxide anion radical is known to occur both in photosystem I and II. In photosystem I, superoxide is generated by the Mehler reaction during low $NADP^+$ concentrations and in presence of iron-sulfur proteins, reduced thioredoxin and ferredoxin [98]. Electron leakage to molecular oxygen on the acceptor side of photosystem II produces superoxide anion radical which after dismutation to H_2O_2 is reduced to hydroxyl radical by the non-hem iron. H_2O_2 formation by incomplete water oxidation occurs at the donor side of photosystem II, where it can be reduced to hydroxyl radical by manganese [99]. Another sources of chloroplastic ROS generation represent the oxygenase activity of ribulose-1,5-bisphosphate carboxylase/oxygenase (RuBisCO, EC 4.1.1.39) producing phosphoglycerate and phosphoglycolate that are metabolized by ongoing reactions in peroxisomes where H_2O_2 is formed [100].

Chloroplastic H_2O_2 is metabolized within glutathione-ascorbate cycle, called Foyer-Asada-Halliwell pathway, important cycle within the antioxidative mechanisms of plant cells [101–103]. Ascorbate peroxidase (APX, EC 1.11.1.11) scavenges H_2O_2 using ascorbate as an electron donor to form monodehydroascorbate (MDHA) as a first step of the pathway. Four APX isoforms are found within plant cells: thylakoid tAPX, chloroplast stromatal soluble sAPX, cytosol cAPX and glyoxysome membrane form mAPX. Then monodehydroascorbate reductase (MDHAR, EC 1.6.5.4) transforms MDHA to dehydroascorbate (DHA), which utilizes NADPH produced during photosynthesis to regenerate ascorbate. MDHAR also catalyzes the elimination of H_2O_2 from mitochondria and peroxisomes [104]. Ascorbate regeneration from DHA is mediated by dehydroascorbate reductase (DHAR, EC 1.8.5.1). Oxidized thiol groups of DHAR are then returned to a reduced state by reaction with reduced glutathione, which is in turn regenerated from the oxidized glutathione by glutathione reductase (GR, EC 1.6.4.2), enzyme with a critical role in redox pathways and essential for maintaining reduced pool of GSH [105]. Its localization is predominantly chloroplastic; however, minor GR activity is also found in cytosol and mitochondria [106].

2.5. *Peroxisomes and Glyoxysomes*

Plant peroxisomes accommodate multiple key metabolic pathways such as β-oxidation of fatty acids, photorespiration, ureides metabolism and xenobiotic detoxifications. Considerable amounts of ROS are produced in peroxisomal reaction pathways, peroxisomes are therefore included among major intracellular sources of plant ROS [107]. A short electron transport chain comprising cytochrome b and NADH-dependent ferric-chelate is located in the peroxisomal membrane (EC 1.16.1.7). Superoxide is formed on cytochrome b within the peroxisomal respiration and then released into the cytosol. Xanthine oxidase (XO, EC 1.17.3.2) contributes the peroxisomal ROS pool with superoxide formed during oxidation of xanthine or hypoxanthine to uric acid [108]. During photorespiration, H_2O_2 is produced in chloroplasts as a result of RuBisCO oxygenase activity followed by reaction of glycolate oxidase in peroxisomes [100]. Cu, Zn-SOD catalyzes dismutation of superoxide produced by matrix XO as another source of peroxisomal H_2O_2, which can be also produced by activities of several flavin oxidases. Furthermore, enzymes participating in ROS catabolism such as monodehydroascorbate reductase mediating superoxide detoxification are also present in peroxisomes [109].

Catalases (CAT, EC 1.11.1.6) are tetrameric hemoproteins catalyzing H_2O_2 decomposition to water and oxygen. CAT are primarily active in peroxisomes and glyoxysomes, the sites of high H_2O_2 generation and turnover [101]. According their catalytic mechanisms, CAT enzymes can be categorized into two groups: monofunctional with dismutating activity and bifunctional with dismutating/peroxidatic activities. By the current enzyme nomenclature, plant CAT isoforms comprise class I present in leaves and participating in H_2O_2 scavenging during photorespiration, class II are in vascular bundles, and CAT class III as key enzymes of H_2O_2 removal in the glyoxylate cycle [110].

Glyoxysomes were considered for long as a rather specialized organelle among plant microbodies derived from a single ancestral peroxisome, are important for specific plant aerobic pathways, especially glyoxylate cycle and β-oxidation of fatty acids [111]. However, recent studies accentuate unifying features of how these dynamic organelles contribute to energy metabolism, development and responses to environmental challenges. On the functional level, the major part of peroxisomal and glyoxysomal proteins are related to fatty acid oxidation. Analysis of triglyceride metabolism in *A. thaliana* seedlings showed that only two enzymes, isocitrate lyase and malate synthase, distinguish glyoxysomes from other peroxisomes [112]. Importantly, glyoxysomes as well as all peroxisome-like organelles share a number of characteristic enzymes related to ROS metabolism [107,113,114]. Nevertheless, glyoxysomes represent essential compartments of plant cells during seed germination [111,115]. Glyoxysomes are involved in mobilization of storage lipids in germinating seeds and represent an important source of both superoxide and H_2O_2 resulting from the β-oxidation of fatty acids and the activity of enzymes such as glycolate and urate oxidases [116,117]. Similarly to peroxisomes, glyoxysomes also contain catalase and H_2O_2-forming peroxidase.

2.6. *Endoplasmic Reticulum*

The endoplasmic reticulum (ER) is a dynamic organelle that fulfills many cellular functions including calcium storage, protein production and lipid metabolism. Metabolic pathways of protein biosynthesis, folding and post-translational modifications such as glycosylation, disulfide bond formation and chaperone-mediated protein folding processes are harbored in ER [118]. ER is also an essential cell compartment where oxidation, hydroxylation and deamination of cellular components or xenobiotics occur [119]. ROS generated in ER and moving the cytoplasm are partly produced in the electron transport chain located in membranes associated with ER; moreover, H_2O_2 is produced in ER for proper oxidative protein folding [120]. Stress-triggered H_2O_2 production can signal outside ER to induce components of antioxidant defense and balance the redox status of the cell [121,122]. In *A. thaliana*, ROS produced within the ER stress response pathway are involved the water stress-induced programed cell death [123]. Superoxide radical is a by-product of oxidation and hydroxylation processes involving cytochrome P450 and cytochrome P540 reductase in a presence of reduced NADPH [124].

3. Conclusions

In recent years, we have witnessed a substantial progress in our understanding of ROS-dependent redox signaling in plants involved in plant growth and developmental and responses to abiotic and biotic stress stimuli. It has become increasingly evident that cellular ROS signaling pathways are clearly confined in a spatio-temporal manner, similarly to that observed for other second messengers, and that ROS redox signaling is tightly connected to cellular compartmentalization, which allows organelle-specific signaling responses. Roles of low molecular antioxidants and ROS-catabolizing enzymes as important regulators of ROS levels in cellular compartments have been uncovered.

Within the plant ROS landscape, a simultaneous operation of both highly conserved and highly specific mechanisms of ROS production and degradation has been recognized in different plant cell organelles. It has become evident that light-driven reactions in thylakoid membranes can be considered the major ROS source in photosynthetically active tissues, whereas mitochondrial ROS production is of major importance in the dark or in non-green tissues. Furthermore, ROS production in the apoplastic space was uncovered to play a crucial role in plant interactions with external environment, including microbial organisms or root growth and development. However, some well-accepted concepts are continuously questioned by advances in the plant ROS field; e.g., singlet oxygen, generally thought to be produced in photosynthetic center, was reported to be involved in osmotic stress-induced cell death in *A. thaliana* roots, suggesting a novel light-independent mechanism of its generation [125].

Recent advances in the development and application of redox-sensitive probes enable specific measurements of individual ROS in different cellular compartments of intact plant cells. The application of these probes permits to analyze simultaneously spatio-temporal ROS modulations and functional associations between ROS metabolism and signaling and organelle functions. It is noteworthy that the current knowledge on the quantitative aspects of ROS metabolism in plant cells in vivo is still quite limited. This is mainly associated with current protocols used to detect ROS which are in many case not suitable for accurate and reliable ROS quantification [9,126]. Recent development of quantitative ROS transcription-based bioreporters seems a promising new strategy to achieve quantitative cellular mapping of ROS changes in plant responses to stress stimuli [127].

Moreover, actual advances of omics technologies provide new strategies to study in detail the site-specific functions of different components of ROS metabolism in plant cell signaling. With advances from single-cell to single-organelle transcriptomics, proteomics and metabolomics, specific roles of individual metabolites and enzymes in the regulation of ROS metabolism can be achieved. The recently introduced "ramanomics" approach might provide an efficient tool of non-invasive quantitative profiling of cellular compartments and monitoring of molecular interactions and transformations in live cells and their subcellular structures. Advancing field of the high-resolution microscopy combined with mentioned genetically-encoded redox sensors proteins is expected to provide new insights into the compartment-specific landscapes of short-lived reactive oxygen species in plant cell compartments, and to further elucidate their role in plant physiology and stress responses.

Author Contributions: Conceptualization, L.L. and M.P.; writing—original draft preparation, M.J.; writing—review and editing, M.J., L.L. and M.P.

References

1. Gabaldón, T.; Pittis, A.A. Origin and evolution of metabolic sub-cellular compartmentalization in eukaryotes. *Biochimie* **2015**, *119*, 262–268. [CrossRef] [PubMed]

2. Go, Y.M.; Jones, D.P. Redox compartmentalization in eukaryotic cells. *Biochim. Biophys. Acta* **2008**, *1780*, 1273–1290. [CrossRef]
3. Dey, S.; Sidor, A.; O'Rourke, B. Compartment-specific Control of Reactive Oxygen Species Scavenging by Antioxidant Pathway Enzymes. *J. Biol. Chem.* **2016**, *291*, 11185–11197. [CrossRef] [PubMed]
4. Kaludercic, N.; Deshwal, S.; Di Lisa, F. Reactive oxygen species and redox compartmentalization. *Front. Physiol.* **2014**, *5*, 285. [CrossRef] [PubMed]
5. D'Autréaux, B.; Toledano, M.B. ROS as signalling molecules: Mechanisms that generate specificity in ROS homeostasis. *Nat. Rev. Mol. Cell Biol.* **2007**, *8*, 813–824. [CrossRef] [PubMed]
6. Ray, P.D.; Huang, B.W.; Tsuji, Y. Reactive oxygen species (ROS) homeostasis and redox regulation in cellular signaling. *Cell Signal.* **2012**, *24*, 981–990. [CrossRef] [PubMed]
7. Gill, S.S.; Tuteja, N. Reactive oxygen species and antioxidant machinery in abiotic stress tolerance in crop plants. *Plant Physiol. Biochem.* **2010**, *48*, 909–930. [CrossRef]
8. Schieber, M.; Chandel, N.S. ROS function in redox signalling and oxidative stress. *Curr. Biol.* **2014**, *24*, R453–R462. [CrossRef] [PubMed]
9. Mhamdi, A.; Van Breusegem, F. Reactive oxygen species in plant development. *Development* **2018**, *145*, dev164376. [CrossRef] [PubMed]
10. Choudhury, F.K.; Rivero, R.M.; Blumwald, E.; Mittler, R. Reactive oxygen species, abiotic stress and stress combination. *Plant J.* **2017**, *90*, 856–867. [CrossRef] [PubMed]
11. Dreyer, A.; Dietz, K.J. Reactive Oxygen Species and the Redox-Regulatory Network in Cold Stress Acclimation. *Antioxidants* **2018**, *7*, 169. [CrossRef] [PubMed]
12. Mittler, R. ROS are good. *Trends Plant Sci.* **2017**, *22*, 11–19. [CrossRef] [PubMed]
13. Eckardt, N.A. The Plant Cell Reviews Plant Immunity: Receptor-Like Kinases, ROS-RLK Crosstalk, Quantitative Resistance, and the Growth/Defense Trade-Off. *Plant Cell* **2017**, *29*, 601–602. [CrossRef]
14. Vaahtera, L.; Brosché, M.; Wrzaczek, M.; Kangasjärvi, J. Specificity in ROS signalling and transcripts signatures. *Antioxid. Redox Signal.* **2014**, *21*, 1422–1441. [CrossRef]
15. Demidchik, V. Mechanisms of oxidative stress in plants: From classical chemistry to cell biology. *Environ. Exp. Bot.* **2015**, *109*, 212–228. [CrossRef]
16. Richards, S.L.; Wilkins, K.A.; Swarbreck, S.M.; Anderson, A.A.; Habib, N.; Smith, A.G.; McAinsh, M.; Davies, J.M. The hydroxyl radical in plants: From seed to seed. *J. Exp. Bot.* **2015**, *66*, 37–46. [CrossRef] [PubMed]
17. Mittler, R. Oxidative stress, antioxidants and stress tolerance. *Trends Plan Sci.* **2002**, *7*, 405–410. [CrossRef]
18. Skovsen, E.; Snyder, J.W.; Lambert, J.D.C.; Ogilby, P.R. Lifetime and diffusion of singlet oxygen in a cell. *J. Phys. Chem. Lett.* **2005**, *109*, 8570–8573. [CrossRef] [PubMed]
19. Dogra, V.; Rochaix, J.D.; Kim, C. Singlet oxygen-triggered chloroplast-to-nucleus retrograde signalling pathways: An emerging perspective. *Plant Cell Environ.* **2018**, *41*, 1727–1738. [CrossRef] [PubMed]
20. Prasad, A.; Sedlářová, M.; Kale, R.S.; Pospíšil, P. Lipoxygenase in singlet oxygen generation as a response to wounding: In vivo imaging in Arabidopsis thaliana. *Sci. Rep.* **2017**, *7*, 9831. [CrossRef]
21. Krieger-Liszkay, A.; Fufezan, C.; Trebst, A. Singlet oxygen production in photosystem II and related protection mechanism. *Photosynth. Res.* **2008**, *98*, 551–564. [CrossRef] [PubMed]
22. Halliwell, B. Reactive species and antioxidants. Redox biology is fundamental theme of aerobic life. *Plant Physiol.* **2006**, *141*, 312–322. [CrossRef]
23. Antunes, F.; Brito, P.M. Quantitative biology of hydrogen peroxide signaling. *Redox Biol.* **2017**, *13*, 1–7. [CrossRef]
24. Rampon, C.; Volovitch, M.; Joliot, A.; Vriz, S. Hydrogen Peroxide and Redox Regulation of Developments. *Antioxidants* **2018**, *7*, 159. [CrossRef]
25. Tamma, G.; Valenti, G.; Grossini, E.; Donnini, S.; Marino, A.; Marinelli, R.A.; Calamita, G. Aquaporin Membrane Channels in Oxidative Stress, Cell Signaling, and Aging: Recent Advances and Research Trends. *Oxidative Med. Cell. Longev.* **2018**, *2018*, 1501847. [CrossRef]
26. Nordzieke, D.E.; Medraño-Fernandez, I. The Plasma Membrane: A Platform for Intra- and Intercellular Redox Signaling. *Antioxidants* **2018**, *7*, 168. [CrossRef]
27. Smirnoff, N.; Arnaud, D. Hydrogen peroxide metabolism and functions in plants. *New Phytol.* **2019**, *221*, 1197–1214. [CrossRef]

28. Jacoby, R.P.; Millar, A.H.; Taylor, N.L. Mitochondrial Biochemistry: Stress Responses and Roles in Stress Alleviation. In *Annual Plant Reviews Online*; Roberts, J.A., Ed.; Wiley Online Library: New York, NY, USA, 2018.
29. Mignolet-Spruyt, L.; Xu, E.; Idänheimo, N.; Hoeberichts, F.A.; Mühlenbock, P.; Brosché, M.; Van Breusegem, F.; Kangasjärvi, J. Spreading the news: Subcellular and organellar reactive oxygen species production and signalling. *J. Exp. Bot.* **2016**, *67*, 3831–3844. [CrossRef]
30. Srivastava, A.; Redij, T.; Sharma, B.; Suprasanna, P. Interaction between Hormone and Redox Signaling in Plants: Divergent Pathways and Convergent Roles. In *Mechanism of Plant Hormone Signaling under Stress*; Pandey, G.K., Ed.; Wiley Online Library: New York, NY, USA, 2017.
31. Jajic, I.; Sarna, T.; Strzalka, K. Senescence, Stress, and Reactive Oxygen Species. *Plants* **2015**, *4*, 393–411. [CrossRef]
32. Keegstra, K. Plant Cell Walls. *Plant Physiol.* **2010**, *154*, 483–486. [CrossRef]
33. Höfte, H. The Yin and Yang of Cell Wall Integrity Control: Brassinosteroid and FERONIA Signaling. *Plant Cell Physiol.* **2015**, *56*, 224–231. [CrossRef]
34. Qi, J.; Wang, J.; Gong, Z.; Zhou, J.M. Apoplastic ROS signaling in plant immunity. *Curr. Opin. Plant Biol.* **2017**, *38*, 92–100. [CrossRef]
35. Zipfel, C. Plant pattern-recognition receptors. *Trends Immunol.* **2014**, *35*, 345–351. [CrossRef] [PubMed]
36. Kimura, S.; Waszczak, C.; Hunter, K.; Wrzaczek, M. Bound by Fate: The Role of Reactive Oxygen Species in Receptor-Like Kinase Signaling. *Plant Cell* **2017**, *29*, 638–654. [CrossRef] [PubMed]
37. Schmidt, R.; Kunkowska, A.B.; Schippers, J.H. Role of Reactive Oxygen Species during Cell Expansion in Leaves. *Plant Physiol.* **2016**, *172*, 2098–2106. [CrossRef] [PubMed]
38. Novaković, L.; Guo, T.; Bacic, A.; Sampathkumar, A.; Johnson, K.L. Hitting the Wall—Sensing and Signaling Pathways Involved in Plant Cell Wall Remodeling in Response to Abiotic Stress. *Plants* **2018**, *7*, 89. [CrossRef]
39. Lazzarotto, F.; Turchetto-Zolet, A.C.; Margis-Pinheiro, M. Revisiting the Non-Animal Peroxidase Superfamily. *Trends Plant Sci.* **2015**, *20*, 807–813. [CrossRef] [PubMed]
40. Mathé, C.; Barre, A.; Jourda, C.; Dunand, C. Evolution and expression of class III peroxidases. *Arch. Biochem. Biophys.* **2010**, *500*, 58–65. [CrossRef]
41. Berglund, G.I.; Carlsson, G.H.; Smith, A.T.; Szöke, H.; Henriksen, A.; Hajdu, J. The catalytic pathway of horseradish peroxidase at high resolution. *Nature* **2002**, *417*, 463–468. [CrossRef] [PubMed]
42. Lüthje, S.; Martinez-Cortes, T. Membrane-Bound Class III Peroxidases: Unexpected Enzymes with Exciting Functions. *Int. J. Mol. Sci.* **2018**, *19*, 2876. [CrossRef]
43. O´Brien, J.A.; Daudi, A.; Finch, P.; Butt, V.S.; Whitelegge, J.P.; Souda, P.; Ausubel, F.M.; Bolwell, G.P. A peroxidase-dependent apoplastic oxidative burst in cultured Arabidopsis cells functions in MAMP-elicited defense. *Plant Physiol.* **2012**, *158*, 2013–2027. [CrossRef]
44. Francoz, E.; Ranocha, P.; Nguyen-Kim, H.; Jamet, E.; Burlat, V.; Dunand, C. Roles of cell wall peroxidases in plant development. *Phytochemistry* **2015**, *112*, 15–21. [CrossRef]
45. Veitch, N.C. Structural determinants of plant peroxidase function. *Phytochem. Rev.* **2004**, *3*, 3–18. [CrossRef]
46. Shigeto, J.; Tsutsumi, Y. Diverse functions and reactions of class III peroxidases. *New Phytol.* **2016**, *209*, 1395–1402. [CrossRef]
47. Dennes, L.; McKenna, J.F.; Segonzac, C.; Wormit, A.; Madhou, P.; Bennet, M.; Mansfield, J.; Zipfel, C.; Hamann, T. Cell wall damage-induced lignin biosynthesis is regulated by a reactive oxygen species- and jasmonic acid-dependent process in Arabidopsis. *Plant Physiol.* **2011**, *156*, 1364–1374. [CrossRef]
48. Survila, M.; Davidsson, P.R.; Pennanen, V.; Kariola, T.; Broberg, M.; Sipari, N.; Heino, P.; Palva, E.T. Peroxidase-Generated Apoplastic ROS Impair Cuticle Integrity and Contribute to DAMP-Elicited Defenses. *Front. Plant Sci.* **2016**, *23*, 1945. [CrossRef]
49. Kärkönen, A.; Kuchitsu, K. Reactive oxygen species in cell wall metabolism and development in plants. *Phytochemistry* **2015**, *112*, 22–32. [CrossRef] [PubMed]
50. Foyer, G.; Noctor, C.F. Intracellular redox compartmentation and ROS-related communication in regulation and signaling. *Plant Physiol.* **2016**, *171*, 1581–1592.
51. Das, K.; Roychoudhury, A. Reactive oxygen species (ROS) and response of antioxidants as ROS-scavengers during environmental stress in plants. *Front. Environ. Sci.* **2014**, *2*, 53. [CrossRef]

52. Ghuge, S.A.; Tisi, A.; Carucci, A.; Rodrigues-Pousada, R.A.; Franchi, S.; Tavladoraki, P.; Angelini, R.; Cona, A. Cell Wall Amine Oxidases: New Players in Root Xylem Differentiation under Stress Conditions. *Plants* **2015**, *4*, 489–504. [CrossRef] [PubMed]
53. Tavladoraki, P.; Cona, A.; Angelini, R. Copper-Containing Amine Oxidases and FAD-Dependent Polyamine Oxidases Are Key Players in Plant Tissue Differentiation and Organ Development. *Front. Plant Sci.* **2016**, *7*, 824. [CrossRef] [PubMed]
54. Chen, D.; Shao, Q.; Yin, L.; Younis, A.; Zheng, B. Polyamine Function in Plants: Metabolism, Regulation on Development, and Roles in Abiotic Stress Responses. *Front. Plant Sci.* **2019**, *9*, 1945. [CrossRef]
55. Cona, A.; Rea, G.; Angelini, R.; Federico, R.; Tavladoraki, P. Functions of amine oxidases in plant development and defence. *Trend Plant Sci.* **2006**, *11*, 80–88. [CrossRef] [PubMed]
56. Pei, Z.M.; Murata, Y.; Benning, G.; Thomine, S. Calcium channels activated by hydrogen peroxide mediate abscisic acid signalling in guard cells. *Nature* **2000**, *406*, 731–734. [CrossRef]
57. Kwak, J.M.; Mori, I.C.; Pei, Z.M.; Leonhardt, N.; Torres, M.A.; Dangl, J.L.; Schroeder, J.I. NADPH oxidase AtrbohD and AtrbohF genes function in ROS-dependent ABA signalling in Arabidopsis. *EMBO J.* **2003**, *22*, 2623–2633. [CrossRef]
58. Davidson, R.M.; Reeves, P.A.; Manosalva, P.M.; Leach, J.E. Germins: A diverse protein family important for crop improvement. *Plant Sci.* **2009**, *177*, 499–510. [CrossRef]
59. Zhang, X.Y.; Nie, Z.H.; Wang, W.J.; Leung, D.W.; Xu, D.G.; Chen, B.L.; Liu, E.E. Relationship between disease resistance and rice oxalate oxidases in transgenic rice. *PLoS ONE* **2013**, *8*, e78348. [CrossRef] [PubMed]
60. Berniers, F.; Berna, A. Germins and germin-like proteins: Plant do-all proteins. But what do they do exactly? *Plant Physiol. Biochem.* **2001**, *39*, 545–554. [CrossRef]
61. Suzuki, N.; Miller, G.; Morales, J.; Shulaev, V.; Torres, M.A.; Mittler, R. Respiratory burst oxidases: The engines of ROS signaling. *Curr. Opin. Plant Biol.* **2001**, *14*, 691–699. [CrossRef]
62. Sagi, M.; Fluhr, R. Production of reactive oxygen species by plant NADPH oxidases. *Plant Physiol.* **2006**, *141*, 336–340. [CrossRef]
63. Nühse, T.S.; Bottrill, A.R.; Jones, A.M.E.; Peck, S.C. Quantitive phosphoproteomic analysis of plasma membrane proteins reveals regulatory mechanism of pnat innate immune responses. *Plant J.* **2007**, *51*, 931–940. [CrossRef] [PubMed]
64. Kadota, Y.; Shirasu, K.; Zipfel, C. Regulation of the NADPH Oxidase RBOHD During Plant Immunity. *Plant Cell Physiol.* **2015**, *56*, 1472–1480. [CrossRef] [PubMed]
65. Hao, H.; Fan, L.; Chen, T.; Li, R.; Li, X.; He, Q.; Botella, M.A.; Lin, J. Clathrin and membrane microdomains cooperatively regulate RbohD dynamics and activity in Arabidopsis. *Plan Cell* **2014**, *26*, 1729–1745. [CrossRef] [PubMed]
66. Noirot, E.; Lherminier, J.; Robert, F.; Moricová, P.; Kieu, K.; Leborgne-Castel, N.; Simon-Plas, F.; Bouhidel, K. Dynamic changes in the subcellular distribution of the tobacco ROS-producing enzyme RBOHD in response to the oomycete elicitor cryptogein. *J. Exp. Bot.* **2014**, *65*, 5011–5022. [CrossRef] [PubMed]
67. Torres, M.A.; Onouchi, H.; Hamada, S.; Machida, C.; Hammond-Kosack, K.E.; Jones, J.D.G. Six Arabidopsis thaliana homologues of the human respiratory burst oxidase (gp91phox). *Plant J.* **1998**, *14*, 365–370. [CrossRef] [PubMed]
68. Ogasawara, Y.; Kaya, H.; Hiaroka, G.; Yumoto, F.; Kimura, S.; Kadota, Y.; Hishinuma, H.; Senzaki, E.; Yamagoe, S.; Nagata, K.; et al. Synergistic activation of the Arabidopsis NADPH oxidase AtrbohD by Ca2+ and phosphorylation. *J. Biol. Chem.* **2008**, *283*, 8885–8892. [CrossRef]
69. Kimura, S.; Kaya, H.; Kawarazaki, T.; Hiraoka, G.; Senzaki, G.; Michikawa, M.; Kuchitsu, K. Protein phosphorylation is a prerequisite for the Ca2+-dependent activation of Arabidopsis NADPH oxidase and may function as a trigger for the positive feedback regulation of Ca^{2+} and reactive oxygen species. *Biochim. Biophys. Acta* **2012**, *1823*, 398–405. [CrossRef]
70. Asai, S.; Ichikawa, T.; Nomura, H.; Kobayashi, M.; Kamiyoshihara, Y.; Mori, H.; Kadota, Y.; Zipfel, C.; Jones, J.D.G.; Yoshioka, H. The variable domain of plant calcium-dependent protein kinase (CDPK) confers subcellular localization and substrate recognition for NADPH oxidase. *J. Biol. Chem.* **2013**, *288*, 14332–14340. [CrossRef]
71. Ogawa, K.; Kanematsu, S.; Asada, K. Generation of superoxide anion and localization of CuZn superoxide dismutase in the vascular tissue of spinach hypocotyls: Their association with lignification. *Plant Cell Physiol.* **1997**, *38*, 1118–1126. [CrossRef]

72. Koffler, B.E.; Bloem, E.; Zellnig, G.; Zechmann, B. High resolution imaging of subcellular glutathione concentrations by quantitative immunoelectron microscopy in different leaf areas of Arabidopsis. *Micron* **2013**, *45*, 119–128. [CrossRef]
73. Booker, F.L.; Burkey, K.O.; Jones, A.M. Re-evaluating the role of ascorbic acid and phenolic glycosides in ozone scavenging in the leaf apoplast of *Arabidopsis thaliana* L. *Plant Cell Environ.* **2012**, *35*, 1456–1466. [CrossRef] [PubMed]
74. Tran, D.; Kadono, T.; Molas, M.L.; Errakhi, R.; Briand, J.; Biligui, B.; Kawano, T.; Bouteau, F. A role for oxalic acid generation in ozone-induced signallization in Arabidopis cells. *Plant Cell Environ* **2013**, *36*, 569–578. [CrossRef] [PubMed]
75. Zechmann, B. Compartment-Specific Importance of Ascorbate During Environmental Stress in Plants. *Antioxid. Redox Signal.* **2017**, *29*, 1488–1501. [CrossRef]
76. Tripathy, B.C.; Oelmüller, R. Reactive oxygen species generation and signaling in plants. *Plant Signal. Behav.* **2012**, *7*, 1621–1633. [CrossRef] [PubMed]
77. Yesbergenova, Z.; Yang, G.; Oron, E.; Soffer, D.; Fluhr, R.; Sagi, M. The plant Mo-hydroxylases aldehyde oxidase and xanthine dehydrogenase have distinct reactive oxygen species signatures and are induced by drought and abscisic acid. *Plant J.* **2005**, *42*, 862–876. [CrossRef] [PubMed]
78. Yergaliyev, T.M.; Nurbekova, Z.; Mukiyanova, G.; Akbassova, A.; Sutula, M.; Zhangazin, S.; Bari, A.; Tleukulova, Z.; Shamekova, M.; Masalimov, Z.K.; et al. The involvement of ROS producing aldehyde oxidase in plant response to Tombusvirus infection. *Plant Physiol. Biochem.* **2016**, *109*, 36–44. [CrossRef]
79. Srivastava, S.; Brychkova, G.; Yarmolinsky, D.; Soltabayeva, A.; Samani, T.; Sagi, M. Aldehyde Oxidase 4 Plays a Critical Role in Delaying Silique Senescence by Catalyzing Aldehyde Detoxification. *Plant Physiol.* **2017**, *173*, 1977–1997. [CrossRef] [PubMed]
80. Ma, X.; Wang, W.; Bittner, F.; Schmidt, N.; Berkey, R.; Zhang, L.; King, H.; Zhang, Y.; Feng, J.; Wen, Y.; et al. Dual and Opposing Roles of Xanthine Dehydrogenase in Defense-Associated Reactive Oxygen Species Metabolism in Arabidopsis. *Plant Cell* **2016**, *28*, 1108–1126. [CrossRef]
81. Van Breusegem, F.; Bailey-Serres, J.; Mittler, R. Unraveling the tapestry of networks involving reactive oxygen species in plants. *Plant Physiol.* **2008**, *147*, 978–984. [CrossRef]
82. Schmitt, F.J.; Renger, G.; Friedrich, T.; Kreslavski, V.D.; Zharmukhamedov, S.K.; Los, D.A.; Kuznetsov, V.V.; Allakhverdiev, S.I. Reactive oxygen species: Re-evaluation of generation, monitoring and role in stress-signaling in phototrophic organisms. *Biochim. Biophys. Acta* **2014**, *1837*, 835–848. [CrossRef] [PubMed]
83. Cezary, W.; Carmody, M.; Kangasjärvi, J. Reactive Oxygen Species in Plant Signaling. *Annu. Rev. Plant Biol.* **2018**, *69*, 209–236.
84. Navrot, N.; Rouhier, N.; Gelhaye, E.; Jacquot, J.P. Reactive oxygen species generation and antioxidant systems in plant mitochondria. *Physiol. Plant.* **2007**, *129*, 185–195. [CrossRef]
85. Morgan, M.J.; Lehmann, M.; Schwarzländer, M.; Baxter, C.J.; Sienkiewicz-Porzucek, A.; Williams, T.C.; Schauer, N.; Fernie, A.R.; Fricker, M.D.; Ratcliffe, R.G.; et al. Decrease in manganese superoxide dismutase leads to reduced root growth and affects tricarboxylic acid cycle flux and mitochondrial redox homeostasis. *Plant Physiol.* **2008**, *147*, 101–114. [CrossRef] [PubMed]
86. Murphy, M.P. How mitochondria produce reactive oxygen species. *Biochem. J.* **2009**, *47*, 1–13. [CrossRef] [PubMed]
87. Braidot, E.; Petrussa, E.; Vianello, A.; Macri, F. Hydrogen peroxide generation by higher plant mitochondria oxidizing complex I or complex II substrates. *FEBS Lett.* **1999**, *451*, 347–350. [CrossRef]
88. Umbach, A.L.; Fiorani, F.; Siedow, J.N. Characterization of transformed Arabidopsis with altered alternative oxidase levels and analysis of effects on reactive oxygen species in tissue. *Plant Physiol.* **2005**, *139*, 1806–1820. [CrossRef] [PubMed]
89. Gleason, C.; Huang, S.; Thatcher, L.F.; Foley, R.C.; Anderson, C.R.; Carroll, A.J.; Millar, H.A.; Singh, K.B. Mitochondrial complex II has a key role in mitochondrial-derived reactive oxygen species influence on plant stress gene regulation and defense. *Proc. Natl. Acad. Sci. USA* **2011**, *108*, 10768–10773. [CrossRef]
90. Huang, S.; Van Aken, O.; Schwarzländer, M.; Belt, K.; Millar, A.H. The Roles of Mitochondrial Reactive Oxygen Species in Cellular Signaling and Stress Response in Plants. *Plant Physiol.* **2016**, *171*, 1551–1559. [CrossRef]

91. Šírová, J.; Sedlářová, M.; Piterková, J.; Luhová, L.; Petřivalský, M. The role of nitric oxide in the germination of plant seeds and pollen. *Plant Sci.* **2011**, *181*, 560–572. [CrossRef]
92. Vanlerberghe, G.C. Alternative oxidase: A mitochondrial respiratory pathway to maintain metabolic and signaling homeostasis during abiotic and biotic stress in plants. *Int. J. Mol. Sci.* **2013**, *14*, 6805–6847. [CrossRef]
93. Barreto, P.; Okura, V.K.; Neshich, I.A.; Maia Ide, G.; Arruda, P. Overexpression of UCP1 in tobacco induces mitochondrial biogenesis and amplifies a broad stress response. *BMC Plant Biol.* **2014**, *14*, 144. [CrossRef]
94. Barreto, P.; Yassitepe, J.; Wilson, Z.A.; Arruda, P. Mitochondrial Uncoupling Protein 1 Overexpression Increases Yield in *Nicotiana tabacum* under Drought Stress by Improving Source and Sink Metabolism. *Front. Plant Sci.* **2017**, *8*, 1836. [CrossRef]
95. Triantaphylidès, C.; Havaux, M. Singlet oxygen in plants: Production, detoxification and signaling. *Trends Plant Sci.* **2009**, *14*, 219–228. [CrossRef]
96. Roach, T.; Krieger-Liszkay, A. Regulation of photosynthetic electron transport and photoinhibition. *Curr. Protein Pept. Sci.* **2014**, *15*, 351–362. [CrossRef]
97. Ramel, F.; Birtic, S.; Cuiné, S.; Triantaphylides, C.; Ravanat, J.L.; Havaux, M. Chemical quenching of singlet oxygen by carotenoids in plants. *Plant Physiol.* **2012**, *158*, 1268–1287. [CrossRef]
98. Asada, K. The water–water cycle in chloroplasts: Scavenging of active oxygens and dissipation of excess photons. *Annu. Rev. Plant Physiol. Plant Mol. Biol.* **1999**, *50*, 601–639. [CrossRef]
99. Pospíšil, P. Production of Reactive Oxygen Species by Photosystem II as a Response to Light and Temperature Stress. *Front. Plant Sci.* **2016**, *7*, 1950. [CrossRef]
100. Noctor, G.; Arisi, A.C.M.; Jouanin, L.; Foyer, C.H. Photorespiratory glycine enhances glutathione accumulation in both the chloroplastic and cytosolic compartments. *J. Exp. Bot.* **1999**, *50*, 1157–1167. [CrossRef]
101. Dat, J.; Vandenabeele, S.; Vranová, E.; Van Montagu, M.; Inzé, D.; Van Breusegem, F. Dual action of the active oxygen species during plant stress responses. *Cell. Mol. Life Sci.* **2000**, *57*, 779–795. [CrossRef]
102. Foyer, C.H.; Halliwell, B. The presence of glutathione and glutathione reductase in chloroplasts: A proposed role in ascorbic acid metabolism. *Planta* **1976**, *133*, 21–25. [CrossRef]
103. Foyer, C.H.; Noctor, G. Ascorbate and glutathione: The heart of the redox hub. *Plant Physiol.* **2011**, *155*, 2–18. [CrossRef]
104. Del Río, L.A.; Corpas, F.J.; Sandalio, L.M.; Palma, J.M.; Gómez, M.; Barroso, J.B. Reactive oxygen species, antioxidant systems and nitric oxide in peroxisomes. *J. Exp. Bot.* **2002**, *53*, 1255–1272. [CrossRef]
105. Noctor, G.; Foyer, C.H. Ascorbate and glutathione: Keeping active oxygen under control. *Annu. Rev. Plant Physiol. Plant Mol. Biol.* **1998**, *49*, 249–279. [CrossRef]
106. Romero-Puertas, M.C.; Corpas, F.J.; Sandalio, L.M.; Leterrier, M.; Rodriguez-Serrano, M.; del Río, L.A.; Palma, J.M. Glutathione reductase from pea leaves: Response to abiotic stress and characterization of the peroxisomal isozyme. *New Phytol.* **2006**, *170*, 43–52. [CrossRef]
107. Del Río, L.A.; López-Huertas, E. ROS Generation in Peroxisomes and its Role in Cell Signaling. *Plant Cell Physiol.* **2016**, *57*, 1364–1376. [CrossRef]
108. López-Huertas, E.; Corpas, F.J.; Sandalio, L.M.; del Río, L.A. Characterization of membrane polypeptides from pea leaf peroxisomes involved in superoxide radical generation. *Biochem. J.* **1999**, *337*, 531–536. [CrossRef]
109. Corpas, F.J.; Barroso, J.B.; del Río, L.A. Peroxisomes as a source of reactive oxygen species and nitric oxide signal molecules in plant cells. *Trends Plant Sci.* **2001**, *6*, 145–150. [CrossRef]
110. Mhamdi, A.; Noctor, G.; Baker, A. Plant catalases: Peroxisomal redox guardians. *Arch. Biochem. Biophys.* **2012**, *525*, 181–194. [CrossRef]
111. Michels, P.A.; Moyersoen, J.; Krazy, H.; Galland, N.; Herman, M.; Hannaert, V. Peroxisomes, glyoxysomes and glycosomes. *Mol. Membr. Biol.* **2005**, *22*, 133–145. [CrossRef]
112. Pracharoenwattana, I.; Smith, S.M. When is a peroxisome not a peroxisome? *Trends Plant Sci.* **2008**, *13*, 522–525. [CrossRef]
113. Corpas, F.J.; Barroso, J.B.; Palma, J.M.; Rodriguez-Ruiz, M. Plant peroxisomes: A nitro-oxidative cocktail. *Redox Biol.* **2017**, *11*, 535–542. [CrossRef]
114. Corpas, F.J.; del Río, L.A.; Palma, J.M. Impact of Nitric Oxide (NO) on the ROS Metabolism of Peroxisomes. *Plants* **2019**, *8*, 37. [CrossRef]
115. Titus, D.E.; Becker, W.M. Investigation of glyoxysome-peroxisome transition in germinating cucumber cotyledons using double-label immunoelectron microscopy. *J. Cell Biol.* **1985**, *101*, 1288–1299. [CrossRef]

116. Bailly, C.; El-Maarouf-Bouteau, H.; Corbineau, F. From intracellular signaling networks to cell death: The dual role of reactive oxygen species in seed physiology. *C. R. Biol.* **2008**, *331*, 806–814. [CrossRef]
117. Kumar, J.S.P.; Prasad, R.S.; Banerjee, R.; Thammineni, C. Seed birth to death: Dual functions of reactive oxygen species in seed physiology. *Ann. Bot.* **2015**, *116*, 663–668. [CrossRef]
118. Schwarz, D.S.; Blower, M.D. The endoplasmic reticulum: Structure, function and response to cellular signalling. *Cell. Mol. Life Sci.* **2016**, *73*, 79–94. [CrossRef]
119. Neve, E.P.; Ingelman-Sundberg, M. Cytochrome P450 proteins: Retention and distribution from the endoplasmic reticulum. *Curr. Opin. Drug Discov. Dev.* **2010**, *13*, 78 85.
120. Zeeshan, H.M.A.; Lee, G.H.; Kim, H.R.; Chae, H.J. Endoplasmic Reticulum Stress and Associated ROS. *Int. J. Mol. Sci.* **2016**, *17*, 327. [CrossRef]
121. Ozgur, R.; Turkan, I.; Uzilday, B.; Sekmen, A.H. Endoplasmic reticulum stress triggers ROS signalling, changes the redox state, and regulates the antioxidant defence of Arabidopsis thaliana. *J. Exp. Bot.* **2014**, *65*, 1377–1390. [CrossRef]
122. Howell, S.H. Endoplasmic reticulum stress responses in plants. *Annu. Rev. Plant Biol.* **2013**, *64*, 477–499. [CrossRef]
123. Duan, Y.; Zhang, W.; Li, B.; Wang, Y.; Li, K.; Han, C.; Zhang, Y.; Li, X. An endoplasmic reticulum response pathway mediates programmed cell death of root tip induced by water stress in Arabidopsis. *New Phytol.* **2010**, *186*, 681–695. [CrossRef]
124. Werck-Reichhart, D.; Feyereisen, R. Cytochromes P450: A success story. *Genome Biol.* **2000**, *1*, reviews3003.1. [CrossRef] [PubMed]
125. Chen, T.; Fluhr, R. Singlet Oxygen Plays an Essential Role in the Root's Response to Osmotic Stress. *Plant Physiol.* **2018**, *177*, 1717–1727. [CrossRef] [PubMed]
126. Noctor, G.; Foyer, C.H. Update on redox compartmentation intracellular redox compartmentation and ROS-related communication in regulation and signaling. *Plant Physiol.* **2017**, *171*, 1581–1592. [CrossRef] [PubMed]
127. Lim, S.D.; Kim, S.H.; Gilroy, S.; Cushman, J.C.; Choi, W.G. Quantitative ROS bioreporters: A robust toolkit for studying biological roles of ROS in response to abiotic and biotic stresses. *Physiol. Plant.* **2019**, *165*, 356–368. [CrossRef] [PubMed]

9

Short-Term Low Temperature Induces Nitro-Oxidative Stress that Deregulates the NADP-Malic Enzyme Function by Tyrosine Nitration in *Arabidopsis thaliana*

Juan C. Begara-Morales [1], Beatriz Sánchez-Calvo [1], María V. Gómez-Rodríguez [1], Mounira Chaki [1], Raquel Valderrama [1], Capilla Mata-Pérez [1], Javier López-Jaramillo [2], Francisco J. Corpas [3] and Juan B. Barroso [1,*]

1 Group of Biochemistry and Cell Signaling in Nitric Oxide, Department of Experimental Biology, Center for Advanced Studies in Olive Grove and Olive Oils, Faculty of Experimental Sciences, University of Jaén, Campus "Las Lagunillas", s/n, E-23071 Jaén, Spain; jbegara@ujaen.es (J.C.B.-M.); bscalvo@ujaen.es (B.S.-C.); mvgomez@ujaen.es (M.V.G.-R.); mounira@ujaen.es (M.C.); ravalde@ujaen.es (R.V.); mmata@ujaen.es (C.M.-P.)

2 Institute of Biotechnology, Department of Organic Chemistry, Faculty of Sciences, University of Granada, E-18071 Granada, Spain; fjljara@ugr.es

3 Group of Antioxidants, Free Radicals, and Nitric Oxide in Biotechnology, Food and Agriculture, Department of Biochemistry, Cell and Molecular Biology of Plants, Estación Experimental del Zaidín, CSIC, C/Profesor Albareda 1, E-18080 Granada, Spain; javier.corpas@eez.csic.es

* Correspondence: jbarroso@ujaen.es

Abstract: Low temperature (LT) negatively affects plant growth and development via the alteration of the metabolism of reactive oxygen and nitrogen species (ROS and RNS). Among RNS, tyrosine nitration, the addition of an NO_2 group to a tyrosine residue, can modulate reduced nicotinamide-dinucleotide phosphate (NADPH)-generating systems and, therefore, can alter the levels of NADPH, a key cofactor in cellular redox homeostasis. NADPH also acts as an indispensable electron donor within a wide range of enzymatic reactions, biosynthetic pathways, and detoxification processes, which could affect plant viability. To extend our knowledge about the regulation of this key cofactor by this nitric oxide (NO)-related post-translational modification, we analyzed the effect of tyrosine nitration on another NADPH-generating enzyme, the NADP-malic enzyme (NADP-ME), under LT stress. In *Arabidopsis thaliana* seedlings exposed to short-term LT (4 °C for 48 h), a 50% growth reduction accompanied by an increase in the content of superoxide, nitric oxide, and peroxynitrite, in addition to diminished cytosolic NADP-ME activity, were found. In vitro assays confirmed that peroxynitrite inhibits cytosolic NADP-ME2 activity due to tyrosine nitration. The mass spectrometric analysis of nitrated NADP-ME2 enabled us to determine that Tyr-73 was exclusively nitrated to 3-nitrotyrosine by peroxynitrite. The in silico analysis of the *Arabidopsis* NADP-ME2 protein sequence suggests that Tyr73 nitration could disrupt the interactions between the specific amino acids responsible for protein structure stability. In conclusion, the present data show that short-term LT stress affects the metabolism of ROS and RNS, which appears to negatively modulate the activity of cytosolic NADP-ME through the tyrosine nitration process.

Keywords: NADP malic enzyme; low temperature; nitric oxide; tyrosine nitration; peroxynitrite; reactive oxygen species; reactive nitrogen species; nitro-oxidative stress

---

1. Introduction

Low temperature (LT) is an environmental issue that affects plant physiology and biochemistry at different levels, including antioxidant enzymes, photosynthesis, gene expression, nutrients, and water uptake [1–9]. At the economic level, the impact of LT on plant crops, such as pepper, rice, tomato, maize, and olive, and tropical and subtropical fruits worldwide, is very significant as it influences both their production and quality [1,10,11]. However, model plants, such as *Arabidopsis thaliana,* act as a very useful tool to decipher the molecular mechanism of response to LT stress [12–16].

LT usually induces nitro-oxidative stress, mediated by the overproduction of reactive oxygen(ROS) and nitrogen (RNS) species [1]. Interestingly, an increasing number of reports suggest that certain reduced nicotinamide-dinucleotide phosphate(NADPH)-generating dehydrogenases might be involved in the protection mechanism against nitro-oxidative stresses induced by adverse environmental conditions [17–23]. In plants, several NADPH-generating systems come into play, such as ferredoxin-NADP reductase as a component of photosystem I, and a group of NADP-dehydrogenases that have been found in different subcellular locations. This group of enzymes includes NADP-isocitrate dehydrogenase (NADP-ICDH), glucose-6-phosphate dehydrogenase (G6PDH), 6-phosphogluconate dehydrogenase (6PGDH) and the NADP-malic enzyme (NADP-ME), also called NADP-malate dehydrogenase.

The NADP-malic enzyme, together with the other NADP-dehydrogenases, is a key component of the NADPH-production systems required to maintain the redox balance in cells. It has been identified from bacteria to humans as an enzyme that catalyzes the reversible oxidative decarboxylation of L-malate to pyruvate, CO_2, and NADPH [24–26]. In plants, different isoenzymes have been described in plastids and cytosol. In *Arabidopsis,* cytosolic NADP-ME2 is considered to be responsible for most NADP-ME activity in mature tissues [27–29] and has been linked to a wide range of processes [30], such as lignin biosynthesis, by providing NADPH [31], and to control cytosolic pH by balancing the synthesis and degradation of L-malate [32]. Other roles that have been suggested for NADP-ME include the control of stomatal closure through the degradation of L-malate during the daytime and seed germination [33]. The presence of a cytosolic NADP-ME isoform has been reported in the guard cell complexes of C_3 plant wheat. However, a more profound analysis of the NADP-ME isoforms in plants is still required. These studies will contribute to unraveling the biological role of plastidic and cytosolic isoenzymes in the same tissue, or even different NADP-MEs in the same subcellular location. Four NADP-ME isoforms have been identified in monocot rice (*Oryza sativa*) [34] that, unfortunately, have not yet been characterized at the molecular level. In addition, the transcripts of different NADP-ME isoforms, located in the cytosol and photosynthetic and non-photosynthetic organelles, have been identified in *Flaveria* sp., which exhibit different C_3 and C_4 photosynthetic pathways [35]. Interestingly, NADP-ME has also been proposed to be involved in plant responses to biotic and abiotic stress (reviewed by [30]).

One of the regulatory mechanisms of plant response to stress is protein function modulation via nitric oxide (NO)-related posttranslational modifications (PTMs) [36–38]. Interestingly, different NADPH-generating enzymes have been identified as being the target of these NO-PTMs [39–41], but information on the specific impact of these modifications to their function in the nitro-oxidative stress context is still scarce. Along this line, LT is one of the main abiotic stresses that modulates the metabolism of ROS and RNS, and also affects NADP-ME function [1], which suggests the regulation of this enzyme by NO-PTMs, such as tyrosine nitration, as reported for NADPH-generating systems [41,42]. S-nitrosylation, the attachment of NO to a specific cysteine residue, is an NO-PTMs that has been widely analyzed as a regulatory process during plant response to stress [43]. However, tyrosine nitration also appears to play an important role during plant response to the nitro-oxidative stress generated under environmental insults [44]. This NO-PTM is produced by the addition of a nitro group (-NO_2) to the tyrosine residue aromatic ring which gives rise to 3-nitrotyrosine. This results in significantly reducing local pKa, which can affect the tyrosine function [45]. Different factors have been proposed to regulate this PTM, including protein structure and environmental compartments. Although information on specific denitrase activity in plants that allow this PTM to be considered key in signaling processes is still lacking [46], these covalent changes may result in effects such as protein function loss and gain or no functional change [42,47–50] and, therefore, impact cellular function. Indeed, different NADPH-generating enzymes have been proposed to be modulated by tyrosine nitration [41,42], but the effect of NO on protein structure [42] has been analyzed only for NADP-ICDH, with NADP-ME2 being one of the least studied enzymes.

In this context, the main goals of this study were to determine whether LT stress could affect the cross-talk between ROS and RNS metabolisms in *Arabidopsis thaliana* seedlings and to study its potential correlation with the group of NADPH-generating dehydrogenases. As NADP-ME was the only NADP-dehydrogenase to be modulated under LT conditions, we focused on this enzyme to study the potential regulatory effect of tyrosine nitration by an in vitro approach.

2. Materials and Methods

2.1. Plant Material and Growth Conditions

Arabidopsis thaliana ecotype Columbia seeds were surface-sterilized and grown on Petri dishes containing Murashige and Skoog medium (Sigma-Aldrich-Fluka, St. Louis, MO, USA) according to [51]. *Arabidopsis* seeds were subsequently grown for 2 weeks in a 16 h light/dark regime at 22 °C/8 h and 18 °C (long-day conditions) with a light intensity of 50 $\mu E\ m^{-2}\ s^{-1}$. For cold treatment purposes, the Petri dishes containing *Arabidopsis* seedlings were placed inside different containers with ice and transferred to a cold room set at 4 ± 2 °C for 48 h.

2.2. Crude Extracts of Plant Tissues

Arabidopsis thaliana seedlings were frozen and ground in liquid N_2 using a pestle and mortar. Then an extraction buffer (1:3, *w/v*) containing 100 mM Tris-HCl, pH 7.8, 0.02 (*v/v*) Triton X-100,1mM EDTA and 10% (*v/v*) glycerol was added. The resulting homogenates were centrifuged at 17,000× *g* for 20 min, and supernatants were subjected to different analyses.

2.3. Detection of Nitric Oxide (NO), Peroxynitrite ($ONOO^-$) and Superoxide Radicals ($O_2^{\bullet-}$) by Confocal Laser Scanning Microscopy

The root tips of *Arabidopsis thaliana* were used to detect different reactive oxygen and nitrogen species by confocal laser scanning Microscopy (CLSM,) according to [52]. Briefly, samples were incubated in the dark for 1h with the following specific fluorescents freshly made in 10 mM Tris-HCl, pH 7.4: 10 μM 4-amino-5-methylamino-2′,7′-difluoro fluorescein diacetate (DAF-FM DA, Calbiochem, San Diego, CA, USA) to detect nitric oxide(NO); 10 μM 3-(*p*-aminophenyl) fluorescein (APF, Invitrogen, Carlsbad, CA, USA) to detect peroxynitrite ($ONOO^-$) and 10 μM dihydroethidium (DHE, Sigma-Aldrich-Fluka, St. Louis, MO, USA) to detect superoxide radicals ($O_2^{\bullet-}$). DAF-FM DA and APF were incubated at 25 °C and DHE at 37 °C. Then samples were washed twice in the same buffer for 15 min and mounted on a microscope slide for examination purposes with a CLSM (Leica TCS SP5 II). The following filters and collection modalities were used: 490 nm excitation and 520 nm emission for DHE, 495 nm excitation and 515 nm emission for DAF-FM DA, and 495 nm excitation and 515 nm emission for APF. The images obtained by CLSM from the control and LT-treated *Arabidopsis* roots were processed and analyzed by the Leica confocal software [52].

2.4. Enzyme Activity Assay

Glucose-6-phosphate dehydrogenase (G6PDH; EC 1.1.1.49), NADP-isocitrate dehydrogenase (NADP-ICDH, EC 1.1.1.42) and NADP-malic enzyme (NADP-ME; EC 1.1.1.40) activity were determined at 25 °C according to [17,23,53] by spectrophoto metrically recording the reduction in NADP at 340 nm. The reaction mixture (1 mL) contained 50 mM 4-(2-Hydroxyethyl)piperazine-1-ethanesulfonic acid (HEPES), pH 7.6, 2 mM $MgCl_2$, and 0.8 mM NADP. The reaction was initiated by adding 5 mM glucose-6-phosphate, 10 mM 2R,3S-isocitrate or 1 mM L-malate, respectively.

2.5. Expression and Purification of Cytosolic Arabidopsis Thaliana NADP-Dependent Malic Enzyme 2 (NADP-ME2)

c-DNA was obtained from the total *Arabidopsis* leaf RNA by using the first strand c-DNA synthesis kit (Roche, Basel, Switzerland) following the manufacturer's instruction. Then the amplification of the cytosolic NADP-ME2 gene (NM_121205.3) was performed by PCR using Fast Start High-Fidelity polymerase (Roche, Basel, Switzerland) and specific primer sets: 5′-AGAGATATGGAAGTACTCCGACTGATTTACC-3′ and 5′-ACAAAACTTTTTTAACGGTAGTTTCTGTACACAG-3′. The PCR product (1785 bp) was then cloned into the pALEXb vector (BiomedalSL, Sevilla, Spain) and the expression of the recombinant protein carrying an *N*-terminal choline-binding domain was performed according to [48]. Briefly, the recombinant protein was induced by adding 1 mM salicylate and 10 mM 3-methyl benzoate. Protein purification was then carried out using a 1-mL LYTRAP column (Biomedal SL, Sevilla, Spain), in which protein was eluted in 1-mL fractions using a discontinuous gradient of choline [48]. Finally, the purity grade of the recombinant protein expression was analyzed by 10% SDS-PAGE.

2.6. Treatment of NADP-ME with SIN-1 (Peroxynitrite Donor)

Molecule SIN-1 (3-morpholinosyl-nonimine) is considered a peroxynitrite donor that is a protein-nitrating compound [54]. Cytosolic recombinant NADP-ME2 was incubated at 37 °C for 1 h with increasing concentrations (0 to 5 mM) of SIN-1 (Calbiochem, San Diego, CA, USA) freshly made before use. A polyclonal antibody against 3-nitrotyrosine (dilution 1:2500) was employed to corroborate the nitration of NADP-ME2 by the immunoblot analysis.

2.7. Identification of Nitrated Tyrosine in Recombinant Cytosolic NADP-ME2 by Mass Spectrometric Techniques

To identify the potential targets of tyrosine nitration, the purified recombinant cytosolic NADP-ME2 treated with SIN-1 was processed by MS/MS as previously described [42]. The sample was subjected to a reduction with dithiothreitol (DTT), derivatization with iodoacetamide (IAA), and enzymatic digestion with trypsin (37 °C, 8 h). It was subsequently purified to eliminate the interferences from the digestion process. Then the peptide mixture was analyzed in a MALDI-TOF/TOF (matrix-assisted laser desorption ionization-time of flight/time of flight) mass spectrometer (4800, AB Sciex) to evaluate sample quality. To this end, peptide mass fingerprinting (PMF) and Uniprot databases were used to interpret the MALDI-TOF spectra and for identification purposes (release 2011_02), respectively. The resulting sample was subjected to a liquid chromatography–tandem mass spectrometry (LC-MS/MS) analysis, using a Velos-LTQ mass spectrometer equipped with a micro-ESI ion source (Thermo Fisher, San Jose, CA, USA). Briefly, the sample was evaporated and subsequently diluted in water containing methanol (5%) and acid formic (1%). Next a 10-cm long, 150-μm id Vydac C18 column (Vydac, IL, USA) was used for the chromatographic process, in which separation was performed at 1 $\mu L\ min^{-1}$ with a 3% to 40% acetonitrile gradient for 30 min (solvent A, 0.1% formic acid; solvent B, acetonitrile with 0.1% formic acid). The Velos-LTQ instrument was operated in the positive ion mode with a 2 kV spray voltage. The scan range of each full MS scan was *m/z* 400 to 2000. After each MS scan, the targeted MS/MS spectra were collected to identify both the unmodified and nitrated form of the predicted tyrosine-containing peptides. The parent mass list of the targeted scan was selected to ensure maximum coverage of the tyrosine-containing tryptic peptides for ME. To obtain a potential list of targeted *m/z* values, the protein was subjected to in silico digestion using nitrated tyrosine as a dynamic modification. The resulting list of predicted peptides (in both the nitrated and unmodified forms) was filtered to exclude any peptide not containing tyrosine residues. The MS/MS spectra were searched with the Proteome Discoverer software (Thermo Fisher, San Jose, CA, USA) under the following conditions: peptide mass tolerance 2 Da, fragment tolerance 0.8 Da, enzyme set as trypsin, no missed cleavages. The dynamic modifications were cysteine carbamidomethylation (+57 Da), methionine oxidation (+16 Da), and tyrosine nitration (+45). Searches were carried out using a database containing all the peptides listed in Table S1. Identifications were filtered with XCorr > 3, P(pep) < 15%. The MS spectra of the nitrated tyrosines were manually validated by comparing the spectra obtained for the unmodified peptide and the nitrated peptide.

2.8. Modeling and the Molecular Evolution Analysis

The tertiary structure of the NADP-malic enzyme from *Arabidopsis thaliana* was modeled at the Swiss Model Server [55], the CPHmodels 2.1 server [56]. HHpred [57], IntFold [58], I-Tasser [59], Phyre2 [60], and RaptorX [61]. Model quality was evaluated by three-dimensional profiles (Verify 3D) [62], the distribution of atom–atom contacts (Errat method) [63], Procheck [64], Qmean [65], and Qmean-Z score [66]. The coordinates of the quaternary structure were calculated by super positioning the best model on the Protein Data Bank (PDB)entry 1O0S with XtalView [67].

The Evolutionary Trace Server was employed to analyze molecular evolution [68] using the model of the tertiary structure as input. Both phylogenetic significance and evolutionary conservation were explored by a BLASTP [69] search the UniProtKB [70] release 2014_10. The prediction of pKa was carried out by PropKa 3.0 [71], and the phosphorylation score was computed on the NetPhos 2.0 Server [72].

2.9. Other Assays

Protein concentration was estimated by the Bio-Rad Protein Assay (Hercules, CA, USA) using bovine serum albumin as a standard. The statistical significance between means was analyzed by a Student's *t*-test.

3. Results

Figure 1A shows the phenotype of the 14-day-old *Arabidopsis thaliana* seedlings grown in MS medium subjected to low temperature (LT) for the last 48 h. These seedlings were smaller in size, and the leaves presented some chlorotic symptoms. The fresh weight of these seedlings also drastically reduced by 50% as a result of LT (Figure 1B), which indicates that the LT conditions caused significant stress to *Arabidopsis thaliana* seedlings.

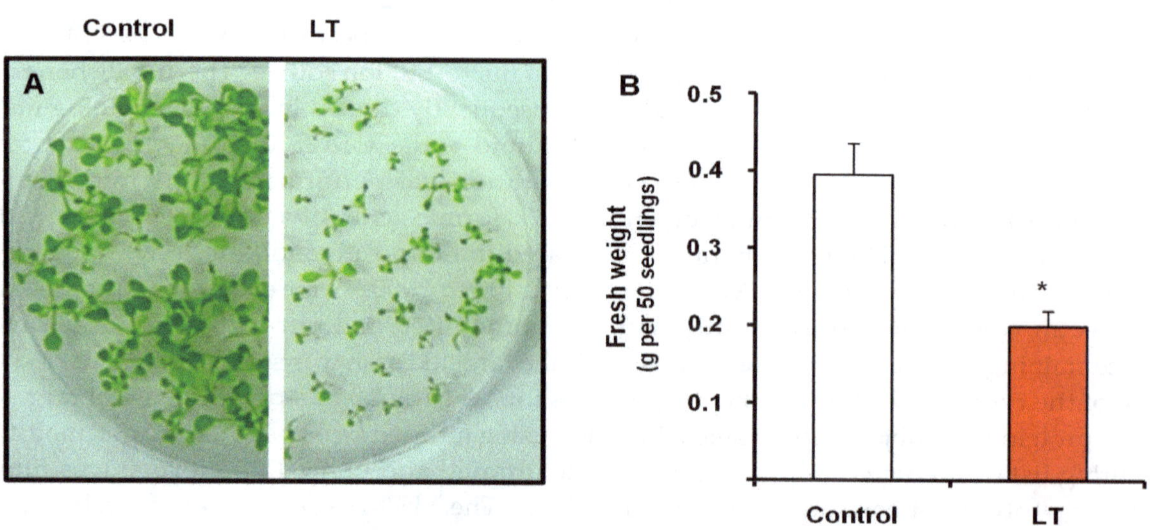

Figure 1. (**A**) Appearance and (**B**) fresh weight of the 14-day-old *Arabidopsis* seedlings grown in MS medium subjected to low temperature (LT) for the last 48 h. The results are the mean of at least three different experiments + SEM. For each experiment, 200 seedlings were used. * Differences in relation to the control values were significant at $p < 0.05$.

3.1. NADP-Dehydrogenase Activities under Short-Term LT Stress

It has been previously reported that, as part of the recycling system of NADPH, several NADP-dehydrogenases may be implicated in plant response to some abiotic conditions [23,30,73]. Hence, the activity of the main NADP-dehydrogenases at LT was evaluated. Figure 2A shows how NADP-ME lowered by 25% at LT, whereas a slight, but insignificant increase was recorded for NADP-ICDH (Figure 2B). G6PDH activity showed no change at LT (Figure 2C).

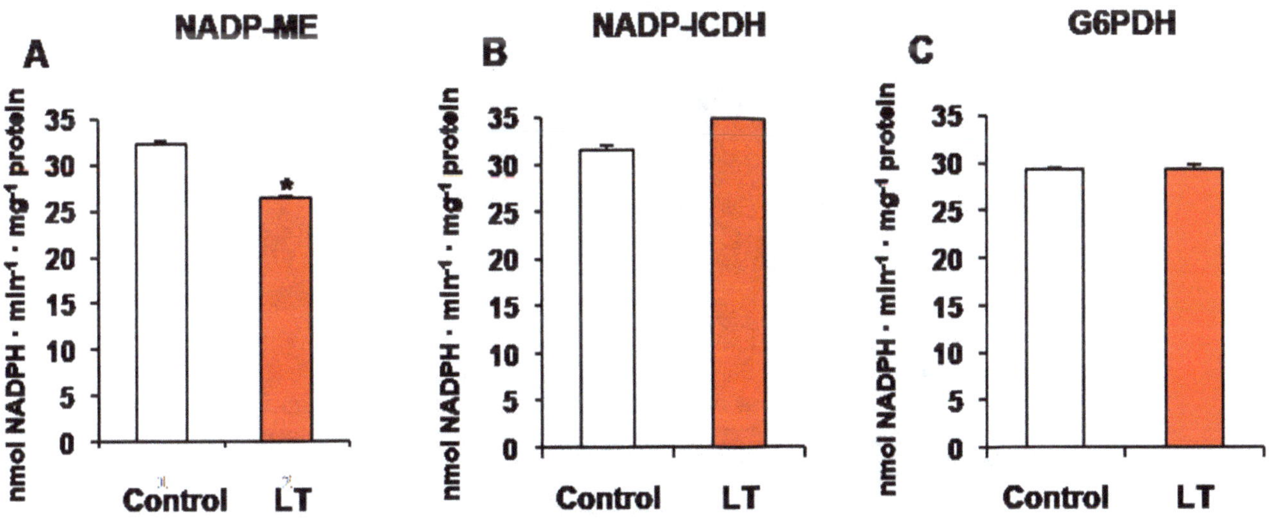

Figure 2. Activity of NADP-dehydrogenase enzymes of the 14-day-old *Arabidopsis* seedlings grown in MS medium subjected to low temperature (LT) for the last 48 h. (**A**) NADP-malic enzyme (NADP-ME). (**B**) NADP-isocitrate dehydrogenase (NADP-ICDH). (**C**) Glucose-6-phosphate dehydrogenase. The results are the mean of at least three different experiments + SEM. * Differences in relation to the control values were significant at $p < 0.05$. NADP: nicotinamide-dinucleotide phosphate.

3.2. Cellular Analysis of Superoxide Radical ($O_2^{\bullet-}$), Nitric Oxide (NO), and Peroxynitrite ($ONOO^-$) in the Roots of the Arabidopsis Seedlings Exposed to LT

Figure 3 (panels A–F) illustrates the examination by CLSM of the levels of $O_2^{\bullet-}$, NO, and $ONOO^-$ in the root tips of *Arabidopsis thaliana* seedlings after LT stress for 48 h. $O_2^{\bullet-}$ content was evaluated after incubating *Arabidopsis* seedlings with fluorescent probe DHE. In the roots of the control seedlings, slight green fluorescence related to $O_2^{\bullet-}$ was observed in the tips (Figure 3A). However, increased green fluorescence was detected after exposing seedlings to LT (Figure 3B). Regarding NO production, a significant increase was detected in the roots of the seedlings exposed to LT by using DAF-FM as the fluorescence probe compared to the control plants (Figure 3C,D). APF was used to analyze $ONOO^-$ content, which resulted from the reaction between $O_2^{\bullet-}$ and NO. In the control roots, $ONOO^-$ showed a very slight fluorescent signal (Figure 3E). Nevertheless, this RNS notably increased as a result of LT stress, with homogeneous distribution throughout roots (Figure 3F), similarly to the observed NO distribution.

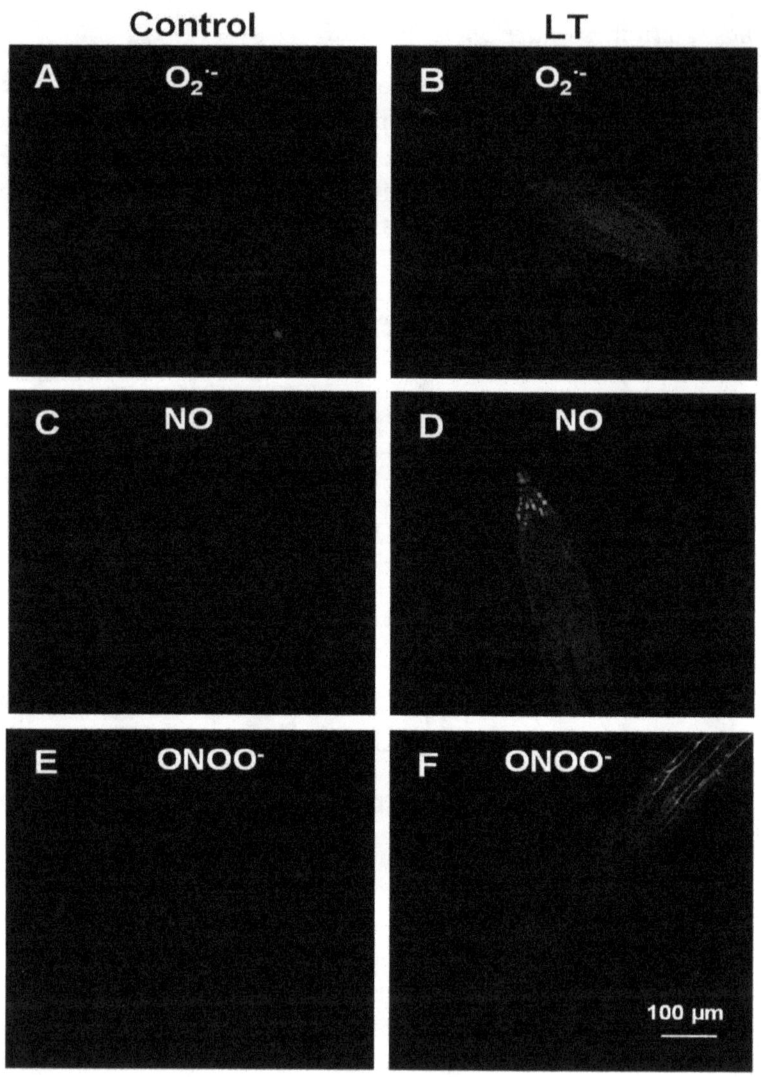

Figure 3. Fluorescence and confocal images showing the in vivo detection of superoxide radical ($O_2^{\bullet-}$) (panels **A** and **B**), nitric oxide (NO) (panels **C** and **D**), and peroxynitrite ($ONOO^-$) (panels **E** and **F**) in *Arabidopsis* root tips after exposing seedlings to low temperature (LT) for 48 h.

3.3. Expression and Purification of Cytosolic NADP-ME2. Effect of Peroxynitrite ($ONOO^-$)

As NADP-ME was the only NADP-dehydrogenase to be modulated under LT stress, and to gain further insight into cytosolic *Arabidopsis thaliana* NADP-ME2, which is the most abundant in this plant [74], the recombinant protein was obtained by cloning the cDNA encoding the enzyme and overexpression in *Escherichia coli*. Figure 4A shows the different fractions obtained after the expression and affinity purification of LYTAG recombinant NADP-ME. The molecular weight of this recombinant construction was 84.3 kDa and resulted from the cytosolic NADP-ME2 protein (63 kDa) containing Ly-tag (21.28 kDa). The fractions with an acceptable purity grade (E5 to E7) showed NADP-EM activity of 9850 µmol NADPH min^{-1} mg^{-1} protein and were used for the subsequent experiments. To analyze the potential action of peroxynitrite generated under LT stress, the recombinant protein was treated with SIN-1 (3-morpholinosydnonimine) as the peroxynitrite donor. Figure 4B depicts the inhibitory effect of $ONOO^-$ on NADP-ME activity in a dose-dependent manner, which ranged from 33% with 0.1 mM SIN-1 to 56% with 5 mM SIN-1. An immunoblot analysis, run with an antibody against 3-nitrotyrosine, was employed to corroborate the nitration of recombinant NADP-ME as a result of SIN-1 treatment (Figure 4C).

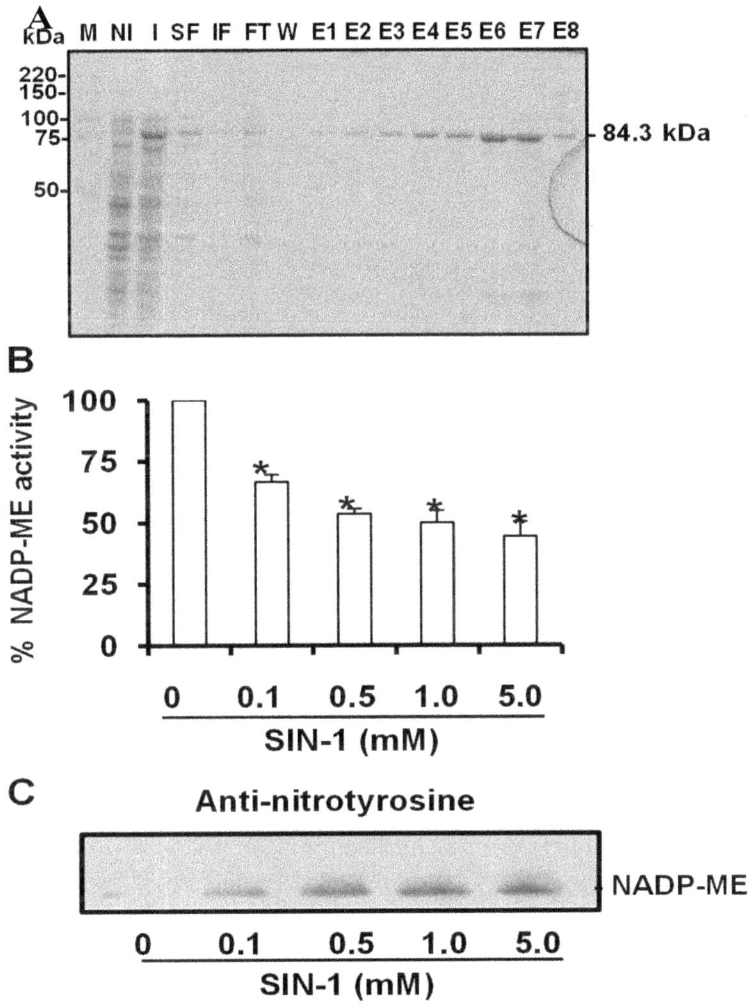

Figure 4. Purification of recombinant cytosolic NADP-ME2 and effect of SIN-1 (3-morpholinosydnonimine (peroxynitrite donor)) on its activity. (**A**) Purity of recombinant NADP-ME after expression and purification was evaluated by SDS-PAGE (10%) and Coomasie blue staining. M, molecular markers; NI, total protein in no-induced culture; I, total protein in induced culture; SF, soluble fraction; IF, insoluble fraction; FT, flow-through; W, wash; E1-E8, elution fractions. (**B**) Effect of SIN-1 on recombinant NADP-EM2 activity. (**C**) Representative immunoblot showing the degree of tyrosine nitration of NADP-EM2 treated with different concentrations of SIN-1 and detected with an antibody against 3-nitrotyrosine (dilution 1:2500). 3 μg of protein was used per line. The specific activity of recombinant NADP-ME was 9850 μmol NADPH min^{-1} mg^{-1} protein. Data are the means ± SEM of at least three replicates. * Differences from the control values were significant at $p < 0.05$.

3.4. Spectral Characterization of the Nitrated Recombinant Arabidopsis NADP-ME2

There are 25 tyrosine residues in the sequence of the cytosolic NADP-ME2 in *Arabidopsis* plants. Thus to be able to identify which of these residues is(are) (a) target(s) of this NO-related post-translational modification, the recombinant protein was treated with peroxynitrite, digested with trypsin, and the resulting peptides were analyzed by MALDI-TOF/TOF mass spectrometry. After the LC-MS/MS analysis, a list of scanned and identified peptides was obtained (Table S1), among which only one was identified to contain a nitrated tyrosine. Figure 5 shows the comparison of the nitrated (top) and unmodified (bottom) MS/MS spectra of these identified peptides from NADP-ME2. Nitrated peptide DAHYLTGLLPPVILSQDVQER (Z = 3) had 21 amino acids and a mass of 2409 Da (2364 Da, plus 45 Da), which is compatible with Tyr-73 nitration (Figure 5).

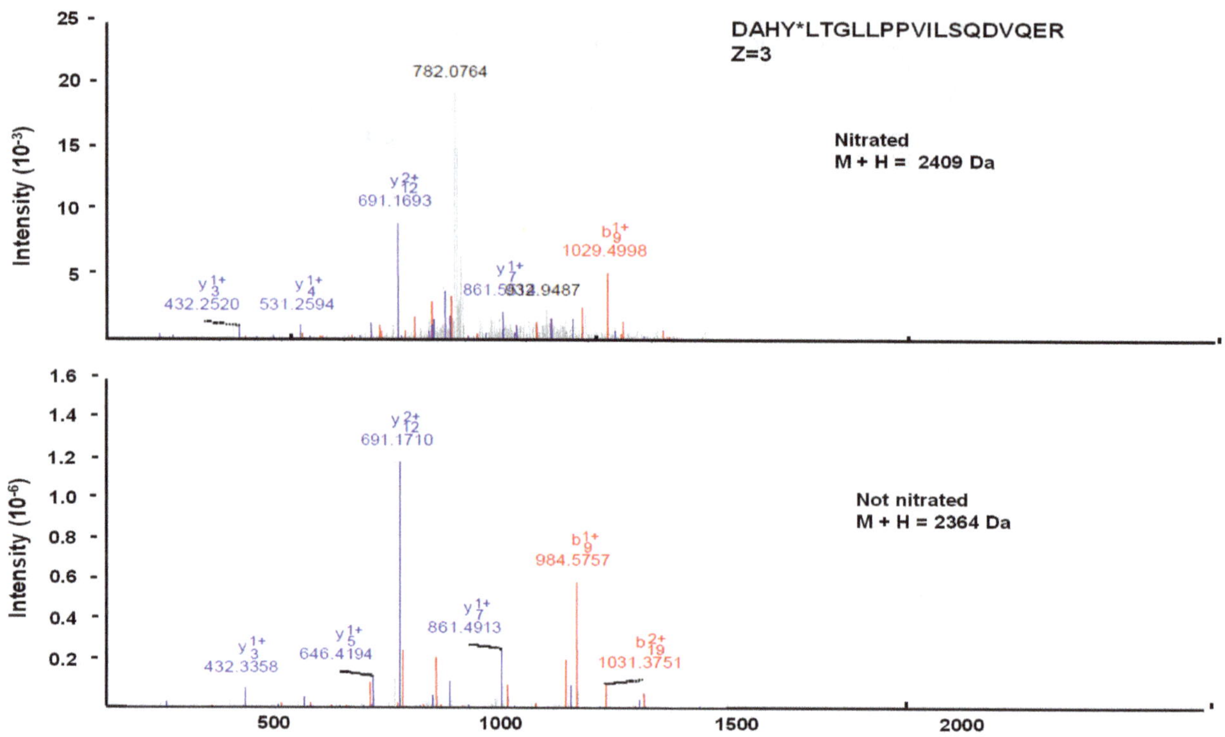

Figure 5. Comparison of the nitrated (top) and unmodified (bottom) MS/MS spectra of the identified peptide (DAHY*LTGLLPPVILSQDVQER) from *Arabidopsis thaliana* NADP-ME2. Peptide fragment ions are indicated by "b" if the charge is retained on the *N*-terminus (red) and by "y" if the charge is maintained on the *C*-terminus (blue). The subscript indicates the number of amino acid residues in the considered fragment from either *N*-terminus or *C*-terminus. The superscript indicates the charge (1+ or 2+) to the backbone fragmentation.

3.5. Modeling the NADP-ME2 from Arabidopsis Thaliana

The tertiary structure of NADP-ME2 from *A. thaliana* was modeled as described in the Materials and Methods section. The quality evaluation revealed that the models computed by servers RaptorX (95% coverage) and IntFold (100% coverage) gave scores, which led us to use IntFold for the final model and to carry out subsequent studies. This model was computed using Protein Data Bank (PDB) entries 2aw5 (human malic enzyme) and 1O0S (*Ascaris suum* malic enzyme complexed with NADH) as a template. These entries share 48% and 46% identity, respectively, with the sequence from *A. thaliana*; the RMSD (root-mean-square deviation) of the model with the templates being 0.62 Å with 2aw5 and 1.73 Å with 1o0s. The coordinates yielded a Qmean value [65] of 0.784, with a Qmean-Z score [66] of 0.2, and an Errat overall quality factor [63] of 78.26, and 89.3% of the residues had an averaged 3D-1D score above 0.2 (Verify3D) [62], while 93.7% of the residues were in the most favored regions in the Ramachandran plot and seven lay in outlier regions.

As NADP-malic enzymes are complex molecules with a double dimer structure that yields a dimer interface and a tetramer interface, the coordinates of the biological assembly of the enzyme from *Ascaris suum* (PDB code 1O0S) were used as a template to identify the interfaces in the *A. thaliana* model (Figure 6). At this point, it is noteworthy that no further refinements were carried out and that the overlapping regions were limited to face monomers from different dimers that did not share any interface.

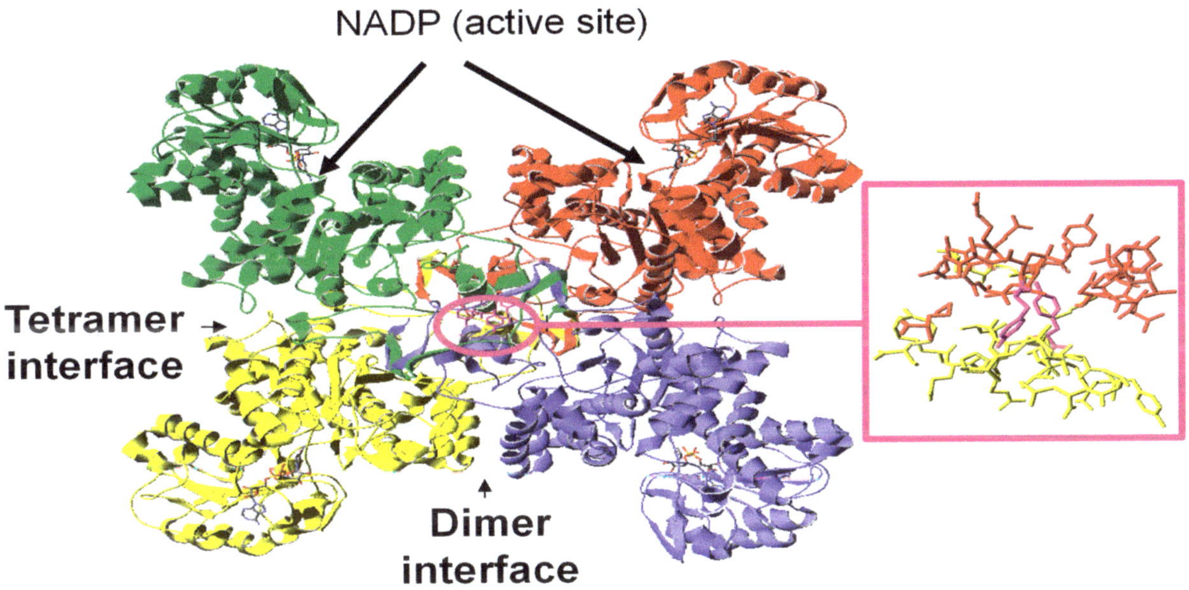

Figure 6. Schematic drawing of the model of the quaternary structure of the malic enzyme from *A. thaliana* showing Y73 (magenta) and details of the residues within 10 Å from Y73, which reveal its interfacial location. The four monomers are depicted in different colors.

3.6. Location of the Nitrated Tyrosine Site in NADP-ME2

The malic enzyme is a dimer of the dimer with close contacts at the dimer interface, whereas the association of the two dimers to yield the tetramer is weaker. In general, the tertiary structure of all malic enzymes is similar. However, small differences may influence catalytic and regulatory mechanisms. Each monomer is divided into four domains (A, B, C, and D) that behave as rigid bodies during the conformational transition from the open to the closed form upon the binding of the divalent cation and the substrate [75].

The evolutionary analysis of the 25 Tyr residues of the NADP-ME2 sequence from *A. thaliana* shows that 15 Tyr residues are preserved in plants, with Tyr-136 being absolutely preserved through evolution in plants, mammals, and bacteria (Table S2). Through these 15 preserved residues, Tyr 158, Tyr242, Tyr457, Tyr 491, and Tyr 592 are predicted as being highly susceptible to phosphorylation (score > 0.75). Among the Tyr residues preserved in plants and predicted as non-phosphorylated, Tyr73 was identified as being nitrated by mass spectrometry. Tyr73 was a particular residue showing the highest estimated pKa (Table S2) which, according to the model, was located on the surface by protruding into the opposite monomer (Figure 6).

4. Discussion

Low temperature (LT) is an important environmental factor that significantly affects plant development and ROS metabolism [76]. *Arabidopsis thaliana*, as a model plant, has been used to analyze the effects of LT on physiology, gene expression, and metabolism at different levels, including photosynthesis, polyamines, and ROS, carbon, and nitrogen metabolisms [77–83]. Previous data have shown that NADPH regeneration by a family of NADP-dehydrogenases can prove essential as a cofactor in cellular redox homeostasis as it is an indispensable electron donor in several enzymatic reactions, biosynthetic pathways, and detoxification processes [30]. The involvement of NADPH in the metabolism of ROS and RNS has also been established as being necessary for not only the functioning of the ascorbate–glutathione cycle, via its participation in the regeneration of reduced glutathione (GSH) by glutathione reductase [84] but also for the activity of the NADPH-dependent thioredoxin system [85]. Consequently, NADPH appears to be important in the functioning of these central cell antioxidants by playing a key role in plant response to oxidative damage.

Similarly, NADPH is also necessary for $O_2^{\bullet-}$ generation by NADPH oxidase (NOX) [86] and is a cofactor required for NO generation by L-arginine-dependent nitric oxide synthase activity [87,88]. In addition, NADPH-generating enzymes appear to be involved in the plant response to different abiotic stresses [17–23])

In this study, by using *Arabidopsis thaliana* as a model, we analyzed the mechanism of response to short-term LT, specifically in relation to the status of some ROS, RNS, and NADP-malic enzyme activity. Additionally, the potential regulatory effect of tyrosine nitration on NADP-ME2 activity was analyzed in particular by an in vitro approach.

4.1. Differential Behaviors of NADP-Dehydrogenases, ROS, and RNS Molecules during Short-Term LT

The ROS metabolism under LT stress has been analyzed in several studies using different plant species [76,89–95]. More recently, occurrence of NO in higher plants and its importance under different stress conditions have become a new area of intense research to understand the role of NO and its related molecules (RNS) in the signaling events leading to plant response to LT stress conditions and their interaction with the ROS metabolism [1,11,96–100].

Cold acclimation in LT situations is the result of different processes, such as cryoprotectant and phospholipid production, protein stabilization, ion homeostasis maintenance, and stress situations, which are usually related to the scavenging of reactive oxygen species (ROS) [11,83]. In fact, LT-induced oxidative stress in *Arabidopsis* has been previously described and is characterized by an increase in lipid oxidation and H_2O_2 content, and by variation in antioxidant enzymes and the reduced functionality of catalase activity, ascorbate peroxidase (APX) and also soluble antioxidants, such as ascorbate [101–103]. The combined effect of less antioxidant capacity and increased ROS is associated with the damage and degradation of photosystems I and II, which finally affect plant growth [81,104]. Our finding, which indicates increased superoxide radical, falls in line closely with all these previously reported data, while our experimental LT conditions (4 °C for 48 h) led to a highly significant reduction in growth (Figure 1) accompanied by oxidative stress.

Similarly, previous data have shown that LT also provokes increased NO content in different plant organs [98,105], particularly in *Arabidopsis thaliana* leaves [96,106]. The increase in NO content observed in the roots of *Arabidopsis* seedlings also falls closely in line with these studies. However, an increase in peroxynitrite ($ONOO^-$) content under LT stress conditions has not yet been reported. This increase would be expected to occur as $ONOO^-$ is quickly produced by a chemical reaction that takes place between superoxide and nitric oxide radicals [107], both of which increase at LT. $ONOO^-$ is the most powerful oxidant that can be produced in nitro-oxidative stress situations, such as salinity and cadmium stress [73,108], which can mediate the tyrosine nitration of proteins and affect the function of ferredoxin-NADP reductase, carbonic anhydrase, NADH-dependent hydroxypyruvate reductase, and different components of the ascorbate–glutathione cycle [37,41,47,48,109,110]. This NO-related posttranslational modification must, therefore, be a mechanism to extend the regulatory and signaling effects of NO under physiological and stress conditions.

Different NADP-dehydrogenases have been shown to be modulated by diverse types of abiotic stresses; e.g., heavy metal [17,111,112], ozone [113], salinity [18,23,114,115], drought [116], and wounding [117]. In pea plants left under natural senescence conditions in which ROS and RNS increased, we also recently identified Tyr392 to be a target of tyrosine nitration that inhibits cytosolic NADP-ICDH [42].

Under our experimental conditions, the analyzed NADP-dehydrogenase activities were differentially affected by LT in *Arabidopsis* seedlings due to the reduced activity caused by NADP-ME activity, while G6PDH and NADP-ICDH activities remained unaffected. These results suggest that NADP-ME could be the NADP-dehydrogenase enzyme that is more sensitive to the oxidation state generated by ROS and RNS production at LT stress. Interestingly, NADP-ME is also the most sensitive to H_2S treatment in sweet pepper fruits [118]. This suggests that NADP-ME could be affected at low oxidation levels during the course of this abiotic stress, whereas the function of other dehydrogenases

(G6PDH and ICDH) could remain in an attempt to maintain the cellular redox state by producing NADPH under nitro-oxidative stress. This could be crucial for plant survival under environmental insults as NADPH acts as a vital electron donor in several reductive and detoxification processes, and in antioxidant processes. Furthermore, NADPH is required in the metabolism of reactive oxygen and nitrogen species (ROS and RNS) [51] Similar to NADP-dehydrogenase enzymes, NO-PTMs differentially regulate Asc-GSH cycle components during plant response to the nitro-oxidative stress generated by salinity [47,48]. In this case, APX and monodehydroascorbate reductae (MDAR) are inactivated by tyrosine nitration, whereas glutathione reductase (GR) is not affected by this NO-PTM, which suggests a key role for GR to maintain the GSH level needed for the functioning of this antioxidant system. As the Asc-GSH cycle requires NADPH for functioning, this key antioxidant system could be regulated by tyrosine nitration at both levels: directly by regulating the function of its components; indirectly by regulating the enzymes that produce reduced power in the form of NADPH. However, these data differ from those on the behavior of NADP-dehydrogenase activities in other species exposed to LT. For example, pepper plants subjected to LT conditions (8 °C) showed an increase in the oxidative markers, such as lipid oxidation and protein nitration, together with an overall induction of the activity of major NADPH-generating systems (6PGDH, NADP-ICDH, G6PDH, and NADP-ME) after 24-hour treatments. These different results could be a result of the different exposure times to LT stress and, therefore, suggest that the oxidation level reached during this stress could be responsible for the specific modulation of NADPH-generating enzymes.

4.2. NADP-ME Activity Is Dysregulated by Nitration

As stated above, diverse types of environmental conditions have the capacity to modulate different NADP-dehydrogenases. Furthermore, nitro-oxidative stress situations can generate peroxynitrite ($ONOO^-$) by a quick chemical reaction between superoxide radical and nitric oxide [107], which can mediate tyrosine nitration and affect the function of several proteins [47,48,109,110]. As NADPH regeneration can be a key cofactor in cellular redox homeostasis, the regulation of NADPH-generating enzymes by NO could play a key role in plant response to these adverse situations as these systems may act as a second line of defense to maintain the functioning of the main antioxidant systems [51]. Indeed, the activity of different NADPH-generating enzymes is regulated by RNS. For instance, ferredoxin-NADP reductase activity is inhibited by tyrosine nitration under high-temperature stress [41], and NADP-dependent isocitrate dehydrogenase activity is also inhibited by Tyr392 nitration, accompanied by increased ROS and RNS under natural senescence conditions [42]. When taking all these data together, tyrosine nitration seems to inhibit NADPH-generating enzymes in these situations. With NADP-EM2, Tyr73 nitration also inhibits the activity of the recombinant protein in a concentration-dependent manner. The inactivation of different NADP-generating enzymes, such as FNR, NADP-ICDH, or NADP-ME2, suggests that this NO-PTM may impair NADPH levels and, therefore, jeopardize the cellular redox state. To determine the functional role of this Tyr73 as a target of tyrosine nitration, we performed an in silico analysis. Previous reports indicate that chemical tetranitromethane (TNM) is able to mediate both cysteine oxidation and tyrosine nitration [119]. In line with this, there are also reports that TNM induces the complete inhibition of NADP-ME activity [120]. Loss of enzyme activity has been observed at pH 8.0, but not at pH 6.3, while NADP-ME has almost been 90% deactivated by incubation with an 80-fold molar excess of TNM for 5 min at 30 °C. Substrate L-malate or Mg^{2+} alone offers no protection, while NADP provided considerable protection. As reported for other NADP-dehydrogenases, in the presence of L-malate and Mg^{2+}, NADP totally protects enzyme activity. This suggests that tyrosine residue may be located at or near the active site of maize NADP-ME. Spectral analysis of the modified enzyme has indicated that the modification of at least one tyrosine residue per subunit results in the complete loss of enzyme activity. The fluorescence study of unmodified and modified enzymes has postulated that the essential tyrosine residue of maize NADP-ME is possibly involved in L-malate binding. By using knockout *Arabidopsis* mutants of NADP-ME2, it has been recently suggested that this enzyme does not play a key role in the response mechanism to oxidative stress [121]. In this context, our data indicated that this enzyme was negatively modulated by several RNS produced under nitro-oxidative stress conditions caused by short-term LT. This could partly explain why NADP-ME does not appear to be involved in oxidative stress because it appears to be partially deactivated [121].

Tyrosine nitration is a NO-mediated posttranslational modification (NO-PTM) that, in some cases, modulates the function of a specified antioxidant enzyme [37,122]. For example, for pea cytosolic APX, a key enzyme for regulating H_2O_2, cellular levels under salinity and nitro-oxidative-associated stress conditions, the enzyme is jointly regulated by this NO-PTM. Thus peroxynitrite inhibits APX activity by the nitration of Tyr235, which is located at the bottom of the pocket where the haem group is enclosed [48]. Furthermore, MDAR activity, another component of the Asc-GSH cycle, is also inhibited by Try345 nitration, which could be involved in the cofactor binding site [47]. Interestingly, NADP-ICDH activity is also down-regulated by Tyr392 nitration as a result of the oxidative process generated during the natural senescence of pea plant roots. In this case, the nitration process also appears to be involved in the disruption of the enzyme-cofactor interaction [42]. In addition, peroxisomal hydroxypyruvate reductase, an enzyme involved in the photorespiration process, is down-regulated by tyrosine nitration. In this case, Tyr 198 nitration has been proposed to be responsible for this inhibition [110]. In this context, NADP-ME regulation by tyrosine nitration is another example of the key role played by tyrosine nitration in the regulation of the target protein function. It is noteworthy that this process becomes relevant when it regulates crucial processes in plants, such as antioxidant defense systems. However, the regulation mechanism is not always straightforward and, for NADP-ME, modulation is especially complex [75].

Tyr73 is a particular residue as determined by both location and estimated pKa. Residue Tyr73 is at a contact distance from Glu11 and Asp12 from the same monomer, and Glu78, Asp70, Tyr 587, and Arg588 from the neighbor monomer. The mechanism underlying the effect of Tyr73 nitration on functionality may be the disruption of these interactions as it has been reported for the pigeon malic enzyme that the residues located at the N-end play an important role in subunit interactions and Mn(II) malate binding [123].

5. Conclusions

The present data show that short-term LT stress affects the metabolism of both ROS and RNS in *Arabidopsis* seedlings, which appears to be accompanied by nitro-oxidative stress. Of all the different analyzed NADP-dehydrogenases, cytosolic NADP-ME was negatively modulated under LT conditions. In this context, the in vitro analysis that used recombinant *Arabidopsis* cytosolic NADP-ME2 treated with peroxynitrite enables us to demonstrate that this enzyme is negatively modulated by nitration, which falls in line with the reduction observed in NADP-ME activity under short-term LT stress and also limits NADPH supply. At the molecular level, Tyr73 was identified as the most likely residue to be involved in this negative NADP-ME regulation by this NO-PTM.

Author Contributions: Conceptualization, J.B.B.; Formal analysis, J.C.B.-M. and J.B.B.; Investigation, J.C.B.-M., B.S.-C., M.V.G.-R., M.C., R.V., C.M.-P. and J.L.-J.; Methodology, F.J.C. and J.B.B.; Writing—original draft, J.C.B.-M., F.J.C. and J.B.B.; Writing—review and editing, J.C.B.-M., F.J.C. and J.B.B.

Acknowledgments: Technical and human support provided by CICT of Universidad de Jaén (UJA, MINECO, Junta de Andalucía, and FEDER) is gratefully acknowledged.

References

1. Airaki, M.; Leterrier, M.; Mateos, R.M.; Valderrama, R.; Chaki, M.; Barroso, J.B.; Del Rio, L.A.; Palma, J.M.; Corpas, F.J. Metabolism of reactive oxygen species and reactive nitrogen species in pepper (*Capsicum annuum* L.) plants under low temperature stress. *Plant Cell Environ.* **2012**, *35*, 281–295. [CrossRef]
2. Chen, Y.; Jiang, J.; Chang, Q.; Gu, C.; Song, A.; Chen, S.; Dong, B.; Chen, F. Cold acclimation induces freezing tolerance via antioxidative enzymes, proline metabolism and gene expression changes in two chrysanthemum species. *Mol. Biol. Rep.* **2014**, *41*, 815–822. [CrossRef] [PubMed]
3. Fowler, S.; Thomashow, M.F. *Arabidopsis* transcriptome profiling indicates that multiple regulatory pathways are activated during cold acclimation in addition to the CBF cold response pathway. *Plant Cell* **2002**, *14*, 1675–1690. [CrossRef] [PubMed]
4. Guy, C.; Kaplan, F.; Kopka, J.; Selbig, J.; Hincha, D.K. Metabolomics of temperature stress. *Physiol. Plant.* **2008**, *132*, 220–235. [CrossRef] [PubMed]
5. Hughes, M.A.; Dunn, M.A. The molecular biology of plant acclimation to low temperature. *J. Exp. Bot.* **1996**, *47*, 291–305. [CrossRef]
6. Janská, A.; Marsík, P.; Zelenková, S.; Ovesná, J. Cold stress and acclimation -what is important for metabolic adjustment? *Plant Biol.* **2010**, *12*, 395–405. [CrossRef]
7. Knight, M.R.; Knight, H. Low-temperature perception leading to gene expression and cold tolerance in higher plants. *New Phytol.* **2012**, *195*, 737–751. [CrossRef]
8. Liu, Y.; Geng, J.; Sha, X.; Zhao, Y.; Yang, P.; Hu, T. Effect of rhizobium symbiosis on low-temperature tolerance and antioxidant response in alfalfa (*Medicago sativa* L.). *Front. Plant. Sci.* **2019**, *10*, 538. [CrossRef]
9. Xiong, L.; Schumaker, K.S.; Zhu, J.-K. Cell signaling during cold, drought, and salt stress. *Plant Cell* **2002**, *14*, S165–S183. [CrossRef]
10. Allen, D.J.; Ort, D.R. Impacts of chilling temperatures on photosynthesis in warm-climate plants. *Trends Plant Sci.* **2001**, *6*, 36–42. [CrossRef]
11. Leyva-Pérez, M.; Valverde-Corredor, A.; Valderrama, R.; Jiménez-Ruiz, J.; Muñoz-Merida, A.; Trelles, O.; Barroso, J.B.; Mercado-Blanco, J.; Luque, F. Early and delayed long-term transcriptional changes and short-term transient responses during cold acclimation in olive leaves. *DNA Res.* **2014**, *22*, 1–11. [CrossRef] [PubMed]
12. Doherty, C.J.; Van Buskirk, H.A.; Myers, S.J.; Thomashow, M.F. Roles for *Arabidopsis* CAMTA transcription factors in cold-regulated gene expression and freezing tolerance. *Plant Cell* **2009**, *21*, 972–984. [CrossRef] [PubMed]
13. Medina, J.; Catalá, R.; Salinas, J. The CBFs: Three *Arabidopsis* transcription factors to cold acclimate. *Plant Sci.* **2011**, *180*, 3–11. [CrossRef] [PubMed]
14. Moon, J.C.; Lee, S.; Shin, S.Y.; Chae, H.B.; Jung, Y.J.; Jung, H.S.; Lee, K.O.; Lee, J.R.; Lee, S.Y. Overexpression of *Arabidopsis* NADPH-dependent thioredoxin reductase C (AtNTRC) confers freezing and cold shock tolerance to plants. *Biochem. Biophys. Res. Commun.* **2015**, *463*, 1225–1229. [CrossRef] [PubMed]
15. Zhang, J.Z.; Creelman, R.A.; Zhu, J.-K. From laboratory to field. Using information from *Arabidopsis* to engineer salt, cold, and drought tolerance in crops. *Plant Physiol.* **2004**, *135*, 615–621. [CrossRef] [PubMed]
16. Zhu, J.; Dong, C.-H.; Zhu, J.-K. Interplay between cold-responsive gene regulation, metabolism and RNA processing during plant cold acclimation. *Curr. Opin. Plant Biol.* **2007**, *10*, 290–295. [CrossRef]
17. Leterrier, M.; del Río, L.A.; Corpas, F.J. Cytosolic NADP-isocitrate dehydrogenase of pea plants: Genomic clone characterization and functional analysis under abiotic stress conditions. *Free Radic. Res.* **2007**, *41*, 191–199. [CrossRef]
18. Manai, J.; Gouia, H.; Corpas, F.J. Redox and nitric oxide homeostasis are affected in tomato (*Solanum lycopersicum*) roots under salinity-induced oxidative stress. *J. Plant Physiol.* **2014**, *171*, 1028–1035. [CrossRef]
19. Marino, D.; González, E.M.; Frendo, P.; Puppo, A.; Arrese-Igor, C. NADPH recycling systems in oxidative stressed pea nodules: A key role for the NADP+-dependent isocitrate dehydrogenase. *Planta* **2007**, *225*, 413–421. [CrossRef]

20. Mhamdi, A.; Mauve, C.; Gouia, H.; Saindrenan, P.; Hodges, M.; Noctor, G. Cytosolic NADP-dependent isocitrate dehydrogenase contributes to redox homeostasis and the regulation of pathogen responses in *Arabidopsis* leaves. *Plant Cell Environ.* **2010**, *33*, 1112–1123.
21. Nemoto, Y.; Sasakuma, T. Specific expression of glucose-6-phosphate dehydrogenase (G6PDH) gene by salt stress in wheat (*Triticum aestivum* L.). *Plant Sci.* **2000**, *158*, 53–60. [CrossRef]
22. Noctor, G.; Queval, G.; Gakière, B. NAD (P) synthesis and pyridine nucleotide cycling in plants and their potential importance in stress conditions. *J. Exp. Bot.* **2006**, *57*, 1603–1620. [CrossRef] [PubMed]
23. Valderrama, R.; Corpas, F.J.; Carreras, A.; Gómez-Rodríguez, M.V.; Chaki, M.; Pedrajas, J.R.; Fernández-Ocaña, A.; del Río, L.A.; Barroso, J.B. The dehydrogenase-mediated recycling of NADPH is a key antioxidant system against salt-induced oxidative stress in olive plants. *Plant Cell Environ.* **2006**, *29*, 1449–1459. [CrossRef] [PubMed]
24. Barroso, J.B.; Peragon, J.; García-Salguero, L.; de la Higuera, M.; Lupiáñnez, J.A. Carbohydrate deprivation reduces NADPH-production in fish liver but not in adipose tissue. *Int. J. Biochem. Cell Biol.* **2001**, *33*, 785–796. [CrossRef]
25. Drincovich, M.F.; Casati, P.; Andreo, C.S. NADP-malic enzyme from plants: A ubiquitous enzyme involved in different metabolic pathways. *FEBS Lett.* **2001**, *490*, 1–6. [CrossRef]
26. Walker, D.A. Pyruvate carboxylation and plant metabolism. *Biol. Rev. Camb. Philos. Soc.* **1962**, *37*, 215–256. [CrossRef] [PubMed]
27. Casati, P.; Drincovich, M.F.; Edwards, G.E.; Andreo, C.S. Malate metabolism by NADP-malic enzyme in plant defense. *Photosynth. Res.* **1999**, *61*, 99–105. [CrossRef]
28. Liu, S.; Cheng, Y.; Zhang, X.; Guan, Q.; Nishiuchi, S.; Hase, K.; Takano, T. Expression of an NADP-malic enzyme gene in rice (*Oryza sativa* L) is induced by environmental stresses; over-expression of the gene in *Arabidopsis* confers salt and osmotic stress tolerance. *Plant Mol. Biol.* **2007**, *64*, 49–58. [CrossRef]
29. Voll, L.M.; Zell, M.B.; Engelsdorf, T.; Saur, A.; Wheeler, M.G.; Drincovich, M.F.; Weber, A.P.M.; Maurino, V.G. Loss of cytosolic NADP-malic enzyme 2 in *Arabidopsis* thaliana is associated with enhanced susceptibility to Colletotrichum higginsianum. *New Phytol.* **2012**, *195*, 189–202. [CrossRef]
30. Chen, Q.; Wang, B.; Ding, H.; Zhang, J.; Li, S. The role of NADP-malic enzyme in plants under stress. *Plant Sci.* **2019**, *281*, 206–212. [CrossRef]
31. Schaaf, J.; Walter, M.H.; Hess, D. Primary metabolism in plant defense (regulation of a bean malic enzyme gene promoter in transgenic tobacco by developmental and environmental cues). *Plant Physiol.* **1995**, *108*, 949–960. [CrossRef] [PubMed]
32. Martinoia, E.; Rentsch, D. Malate compartmentation-responses to a complex metabolism. *Annu. Rev. Plant Biol.* **1994**, *45*, 447–467. [CrossRef]
33. Yazdanpanah, F.; Maurino, V.G.; Mettler-Altmann, T.; Buijs, G.; Bailly, M.; Karimi Jashni, M.; Willems, L.; Sergeeva, L.I.; Rajjou, L.; Hilhorst, H.W.M. NADP-MALIC ENZYME 1 affects germination after seed storage in *Arabidopsis thaliana*. *Plant Cell Physiol.* **2018**, *60*, 318–328. [CrossRef] [PubMed]
34. Chi, W.; Yang, J.; Wu, N.; Zhang, F. Four rice genes encoding NADP malic enzyme exhibit distinct expression profiles. *Biosci. Biotechnol. Biochem.* **2004**, *68*, 1865–1874. [CrossRef] [PubMed]
35. Lai, L.B.; Tausta, S.L.; Nelson, T.M. Differential regulation of transcripts encoding cytosolic NADP-malic enzyme in C3 and C4 *Flaveria* species. *Plant Physiol.* **2002**, *128*, 140–149. [CrossRef] [PubMed]
36. Astier, J.; Lindermayr, C. Nitric oxide-dependent posttranslational modification in plants: An update. *Int. J. Mol. Sci.* **2012**, *13*, 15193–15208. [CrossRef]
37. Corpas, F.J.; Begara-Morales, J.C.; Sánchez-Calvo, B.; Chaki, M.; Barroso, J.B. Nitration and S-nitrosylation: Two post-translational modifications (PTMs) mediated by reactive nitrogen species (RNS) and their role in signalling processes of plant cells. In *Reactive Oxygen and Nitrogen Species Signaling and Communication in Plants*; Gupta, K.J., Igamberdiev, A.U., Eds.; Springer: Cham, Switzerland, 2015; pp. 267–281.
38. Lindermayr, C.; Durner, J. S-Nitrosylation in plants: Pattern and function. *J. Proteomics* **2009**, *73*, 1–9. [CrossRef]
39. Begara-Morales, J.C.; López-Jaramillo, F.J.; Sánchez-Calvo, B.; Carreras, A.; Ortega-Muñoz, M.; Santoyo-González, F.; Corpas, F.J.; Barroso, J.B. Vinyl sulfone silica: Application of an open preactivated support to the study of transnitrosylation of plant proteins by S-nitrosoglutathione. *BMC Plant Biol.* **2013**, *13*, 61. [CrossRef]

40. Begara-Morales, J.C.; Sánchez-Calvo, B.; Luque, F.; Leyva-Pérez, M.O.; Leterrier, M.; Corpas, F.J.; Barroso, J.B. Differential transcriptomic analysis by RNA-seq of GSNO-responsive genes between *Arabidopsis* roots and leaves. *Plant Cell Physiol.* **2014**, *55*, 1085–1095. [CrossRef]
41. Chaki, M.; Valderrama, R.; Fernández-Ocaña, A.; Carreras, A.; Gómez-Rodríguez, M.V.; López-Jaramillo, J.; Begara-Morales, J.C.; Sánchez-Calvo, B.; Luque, F.; Leterrier, M. High temperature triggers the metabolism of S-nitrosothiols in sunflower mediating a process of nitrosative stress which provokes the inhibition of ferredoxin-NADP reductase by tyrosine nitration. *Plant Cell Environ.* **2011**, *34*, 1803–1818. [CrossRef]
42. Begara-Morales, J.C.; Chaki, M.; Sánchez-Calvo, B.; Mata-Pérez, C.; Leterrier, M.; Palma, J.M.; Barroso, J.B.; Corpas, F.J. Protein tyrosine nitration in pea roots during development and senescence. *J. Exp. Bot.* **2013**, *64*, 1121–1134. [CrossRef] [PubMed]
43. Begara-Morales, J.C.; Chaki, M.; Valderrama, R.; Mata-Pérez, C.; Padilla, M.N.; Barroso, J.B. The function of S-nitrosothiols during abiotic stress in plants. *J. Exp. Bot.* **2019**, *70*, 4429–4439. [CrossRef] [PubMed]
44. Mata-Pérez, C.; Begara-Morales, J.C.; Chaki, M.; Sánchez-Calvo, B.; Valderrama, R.; Padilla, M.N.; Corpas, F.J.; Barroso, J.B. Protein Tyrosine Nitration during Development and Abiotic Stress Response in Plants. *Front. Plant. Sci.* **2016**, *7*. [CrossRef] [PubMed]
45. Turko, I.V.; Murad, F. Protein nitration in cardiovascular diseases. *Pharmacol. Rev.* **2002**, *54*, 619–634. [CrossRef] [PubMed]
46. Kolbert, Z.; Feigl, G.; Bordé, A.; Molnár, A.; Erdei, L. Protein tyrosine nitration in plants: Present knowledge, computational prediction and future perspectives. *Plant Physiol. Biochem.* **2017**, *113*, 56–63. [CrossRef] [PubMed]
47. Begara-Morales, J.C.; Sánchez-Calvo, B.; Chaki, M.; Mata-Pérez, C.; Valderrama, R.; Padilla, M.N.; López-Jaramillo, J.; Luque, F.; Corpas, F.J.; Barroso, J.B. Differential molecular response of monodehydroascorbate reductase and glutathione reductase by nitration and S-nitrosylation. *J. Exp. Bot.* **2015**, *66*, 5983–5996. [CrossRef] [PubMed]
48. Begara-Morales, J.C.; Sánchez-Calvo, B.; Chaki, M.; Valderrama, R.; Mata-Pérez, C.; López-Jaramillo, J.; Padilla, M.N.; Carreras, A.; Corpas, F.J.; Barroso, J.B. Dual regulation of cytosolic ascorbate peroxidase (APX) by tyrosine nitration and S-nitrosylation. *J. Exp. Bot.* **2014**, *65*, 527–538. [CrossRef] [PubMed]
49. Radi, R. Protein tyrosine nitration: Biochemical mechanisms and structural basis of functional effects. *Acc. Chem. Res.* **2013**, *46*, 550–559. [CrossRef] [PubMed]
50. Souza, J.M.; Peluffo, G.; Radi, R. Protein tyrosine nitration-functional alteration or just a biomarker? *Free Radic. Biol. Med.* **2008**, *45*, 357–366. [CrossRef]
51. Corpas, F.J.; Barroso, J.B. NADPH-generating dehydrogenases: Their role in the mechanism of protection against nitro-oxidative stress induced by adverse environmental conditions. *Front. Environ. Sci.* **2014**, *2*, 55. [CrossRef]
52. Corpas, F.J.; Barroso, J.B. Lead-induced stress, which triggers the production of nitric oxide (NO) and superoxide anion (O_2^-) in *Arabidopsis* peroxisomes, affects catalase activity. *Nitric Oxide* **2017**, *68*, 103–110. [CrossRef] [PubMed]
53. Barroso, J.B.; Peragón, J.; Contreras-Jurado, C.; García-Salguero, L.; Corpas, F.J.; Esteban, F.J.; Peinado, M.A.; De La Higuera, M.; Lupiáñez, J.A. Impact of starvation-refeeding on kinetics and protein expression of trout liver NADPH-production systems. *Am. J. Physiol. Regul. Integr. Comp. Physiol.* **1998**, *274*, R1578–R1587. [CrossRef] [PubMed]
54. Daiber, A.; Bachschmid, M.; Beckman, J.S.; Munzel, T.; Ullrich, V. The impact of metal catalysis on protein tyrosine nitration by peroxynitrite. *Biochem. Biophys. Res. Commun.* **2004**, *317*, 873–881. [CrossRef] [PubMed]
55. Arnold, K.; Bordoli, L.; Kopp, J.R.; Schwede, T. The SWISS-MODEL workspace: A web-based environment for protein structure homology modelling. *Bioinformatics* **2006**, *22*, 195–201. [CrossRef]
56. Nielsen, M.; Lundegaard, C.; Lund, O.; Petersen, T.N. CPHmodels-3.0—Remote homology modeling using structure-guided sequence profiles. *Nucleic Acids Res.* **2010**, *38*, W576–W581. [CrossRef]
57. Söding, J.; Biegert, A.; Lupas, A.N. The HHpred interactive server for protein homology detection and structure prediction. *Nucleic Acids Res.* **2005**, *33*, W244–W248. [CrossRef]
58. Roche, D.B.; Buenavista, M.T.; Tetchner, S.J.; McGuffin, L.J. The IntFOLD server: An integrated web resource for protein fold recognition, 3D model quality assessment, intrinsic disorder prediction, domain prediction and ligand binding site prediction. *Nucleic Acids Res.* **2011**, *39*, W171–W176. [CrossRef]
59. Zhang, Y. I-TASSER server for protein 3D structure prediction. *BMC Bioinform.* **2008**, *9*, 40. [CrossRef]

60. Kelley, L.A.; Sternberg, M.J.E. Protein structure prediction on the Web: A case study using the Phyre server. *Nat. Protoc.* **2009**, *4*, 363–371. [CrossRef]
61. Källberg, M.; Wang, H.; Wang, S.; Peng, J.; Wang, Z.; Lu, H.; Xu, J. Template-based protein structure modeling using the RaptorX web server. *Nat. Protoc.* **2012**, *7*, 1511–1522. [CrossRef]
62. Eisenberg, D.; Lüthy, R.; Bowie, J.U. VERIFY3D: Assessment of protein models with three-dimensional profiles. *Methods Enzymol.* **1997**, *277*, 396–404. [PubMed]
63. Colovos, C.; Yeates, T.O. Verification of protein structures: Patterns of nonbonded atomic interactions. *Protein Sci.* **1993**, *2*, 1511–1519. [CrossRef] [PubMed]
64. Laskowski, R.A.; MacArthur, M.W.; Moss, D.S.; Thornton, J.M. PROCHECK: A program to check the stereochemical quality of protein structures. *J. Appl. Crystallogr.* **1993**, *26*, 283–291. [CrossRef]
65. Benkert, P.; Tosatto, S.C.E.; Schomburg, D. QMEAN: A comprehensive scoring function for model quality assessment. *Proteins Struct. Funct. Bioinform.* **2008**, *71*, 261–277. [CrossRef] [PubMed]
66. Benkert, P.; Biasini, M.; Schwede, T. Toward the estimation of the absolute quality of individual protein structure models. *Bioinformatics* **2011**, *27*, 343–350. [CrossRef]
67. McRee, D.E. A visual protein crystallographic software system for X11/XView. *J. Mol. Graph.* **1992**, *10*, 44–47. [CrossRef]
68. Mihalek, I.; Res, I.; Lichtarge, O. A family of evolution-entropy hybrid methods for ranking protein residues by importance. *J. Mol. Biol.* **2004**, *336*, 1265–1282. [CrossRef]
69. Altschul, S.F.; Madden, T.L.; Schäffer, A.A.; Zhang, J.; Zhang, Z.; Miller, W.; Lipman, D.J. Gapped BLAST and PSI-BLAST: A new generation of protein database search programs. *Nucleic Acids Res.* **1997**, *25*, 3389–3402. [CrossRef]
70. UniProt, C. The universal protein resource (UniProt). *Nucleic Acids Res.* **2008**, *36*, D190–D195.
71. Olsson, M.H.M.; Sondergaard, C.R.; Rostkowski, M.; Jensen, J.H. PROPKA3: Consistent treatment of internal and surface residues in empirical pKa predictions. *J. Chem. Theory Comput.* **2011**, *7*, 525–537. [CrossRef]
72. Blom, N.; Gammeltoft, S.; Brunak, S. Sequence and structure-based prediction of eukaryotic protein phosphorylation sites. *J. Mol. Biol.* **1999**, *294*, 1351–1362. [CrossRef] [PubMed]
73. Corpas, F.J.; Barroso, J.B. Peroxynitrite ($ONOO^-$) is endogenously produced in *Arabidopsis peroxisomes* and is overproduced under cadmium stress. *Ann. Bot.* **2014**, *113*, 87–96. [CrossRef] [PubMed]
74. Badia, M.B.; Arias, C.L.; Tronconi, M.A.; Maurino, V.G.; Andreo, C.S.; Drincovich, M.F.; Wheeler, M.C.G. Enhanced cytosolic NADP-ME2 activity in *A. thaliana* affects plant development, stress tolerance and specific diurnal and nocturnal cellular processes. *Plant Sci.* **2015**, *240*, 193–203. [CrossRef] [PubMed]
75. Chang, G.-G.; Tong, L. Structure and function of malic enzymes, a new class of oxidative decarboxylases. *Biochemistry* **2003**, *42*, 12721–12733. [CrossRef] [PubMed]
76. Dreyer, A.; Dietz, K.-J. Reactive oxygen species and the redox-regulatory network in cold stress acclimation. *Antioxidants* **2018**, *7*, 169. [CrossRef] [PubMed]
77. Alcázar, R.; Cuevas, J.C.; Planas, J.; Zarza, X.; Bortolotti, C.; Carrasco, P.; Salinas, J.; Tiburcio, A.F.; Altabella, T. Integration of polyamines in the cold acclimation response. *Plant Sci.* **2011**, *180*, 31–38. [CrossRef]
78. Goulas, E.; Schubert, M.; Kieselbach, T.; Kleczkowski, L.A.; Gardeström, P.; Schröder, W.; Hurry, V. The chloroplast lumen and stromal proteomes of *Arabidopsis thaliana* show differential sensitivity to short- and long-term exposure to low temperature. *Plant J.* **2006**, *47*, 720–734. [CrossRef]
79. Le, M.Q.; Pagter, M.; Hincha, D.K. Global changes in gene expression, assayed by microarray hybridization and quantitative RT-PCR, during acclimation of three *Arabidopsis thaliana* accessions to sub-zero temperatures after cold acclimation. *Plant Mol. Biol.* **2015**, *87*, 1–15. [CrossRef]
80. Nakaminami, K.; Matsui, A.; Nakagami, H.; Minami, A.; Nomura, Y.; Tanaka, M.; Morosawa, T.; Ishida, J.; Takahashi, S.; Uemura, M. Analysis of Differential Expression Patterns of mRNA and Protein During Cold-acclimation and De-acclimation in *Arabidopsis*. *Mol. Cell Proteomics* **2014**, *13*, 3602–3611. [CrossRef]
81. Savitch, L.V.; Barker-Astrom, J.; Ivanov, A.G.; Hurry, V.; Oquist, G.; Huner, N.P.; Gardeström, P. Cold acclimation of *Arabidopsis thaliana* results in incomplete recovery of photosynthetic capacity, associated with an increased reduction of the chloroplast stroma. *Planta* **2001**, *214*, 295–303. [CrossRef]
82. Talts, P.; Pärnik, T.; Gardeström, P.; Keerberg, O. Respiratory acclimation in *Arabidopsis thaliana* leaves at low temperature. *J. Plant Physiol.* **2004**, *161*, 573–579. [CrossRef] [PubMed]

83. Zuther, E.; Lee, Y.P.; Erban, A.; Kopka, J.; Hincha, D.K. Natural variation in freezing tolerance and cold acclimation response in *Arabidopsis thaliana* and related species. *Adv. Exp. Med. Biol.* **2018**, *1081*, 81–98. [PubMed]
84. Halliwell, B.; Foyer, C.H. Properties and physiological function of a glutathione reductase purified from spinach leaves by affinity chromatography. *Planta* **1978**, *139*, 9–17. [CrossRef] [PubMed]
85. Cha, J.-Y.; Kim, J.Y.; Jung, I.J.; Kim, M.R.; Melencion, A.; Alam, S.S.; Yun, D.-J.; Lee, S.Y.; Kim, M.G.; Kim, W.-Y. NADPH-dependent thioredoxin reductase A (NTRA) confers elevated tolerance to oxidative stress and drought. *Plant Physiol. Biochem.* **2014**, *80*, 184–191. [CrossRef] [PubMed]
86. Sagi, M.; Fluhr, R. Production of reactive oxygen species by plant NADPH oxidases. *Plant Physiol.* **2006**, *141*, 336–340. [CrossRef]
87. Barroso, J.B.; Corpas, F.J.; Carreras, A.; Sandalio, L.M.; Valderrama, R.; Palma, J.M.; Lupiáñez, J.A.; del Río, L.A. Localization of nitric-oxide synthase in plant peroxisomes. *J. Biol. Chem.* **1999**, *274*, 36729–36733. [CrossRef] [PubMed]
88. Corpas, F.J.; Palma, J.M.; Del Río, L.A.; Barroso, J.B. Evidence supporting the existence of L-arginine-dependent nitric oxide synthase activity in plants. *New Phytol.* **2009**, *184*, 9–14. [CrossRef]
89. Burdon, R.H.; O'Kane, D.; Fadzillah, N.; Gill, V.; Boyd, P.A.; Finch, R.R. Oxidative stress and responses in *Arabidopsis thaliana* and *Oryza sativa* subjected to chilling and salinity stress. *Biochem. Soc. Trans.* **1996**, *24*, 469–472. [CrossRef]
90. Holá, D.; Kocová, M.; Rothová, O.; Wilhelmová, N.; Benesová, M. Recovery of maize (*Zea mays* L.) inbreds and hybrids from chilling stress of various duration: Photosynthesis and antioxidant enzymes. *J. Plant Physiol.* **2007**, *164*, 868–877. [CrossRef]
91. Li, X.; Cai, J.; Liu, F.; Dai, T.; Cao, W.; Jiang, D. Cold priming drives the sub-cellular antioxidant systems to protect photosynthetic electron transport against subsequent low temperature stress in winter wheat. *Plant Physiol. Biochem.* **2014**, *82*, 34–43. [CrossRef]
92. McKersie, B.D.; Chen, Y.; de Beus, M.; Bowley, S.R.; Bowler, C.; Inzé, D.; D'Halluin, K.; Botterman, J. Superoxide dismutase enhances tolerance of freezing stress in transgenic alfalfa (*Medicago sativa* L.). *Plant Physiol.* **1993**, *103*, 1155–1163. [CrossRef] [PubMed]
93. Pastori, G.; Foyer, C.H.; Mullineaux, P. Low temperature-induced changes in the distribution of H_2O_2 and antioxidants between the bundle sheath and mesophyll cells of maize leaves. *J. Exp. Bot.* **2000**, *51*, 107–113. [CrossRef] [PubMed]
94. Payton, P.; Webb, R.; Kornyeyev, D.; Allen, R.; Holaday, A.S. Protecting cotton photosynthesis during moderate chilling at high light intensity by increasing chloroplastic antioxidant enzyme activity. *J. Exp. Bot.* **2001**, *52*, 2345–2354. [CrossRef] [PubMed]
95. Tsang, E.W.; Bowler, C.; Hérouart, D.; Van Camp, W.; Villarroel, R.; Genetello, C.; Van Montagu, M.; Inzé, D. Differential regulation of superoxide dismutases in plants exposed to environmental stress. *Plant Cell* **1991**, *3*, 783–792. [PubMed]
96. Cantrel, C.; Vazquez, T.; Puyaubert, J.; Rezé, N.; Lesch, M.; Kaiser, W.M.; Dutilleul, C.; Guillas, I.; Zachowski, A.; Baudouin, E. Nitric oxide participates in cold-responsive phosphosphingolipid formation and gene expression in Arabidopsis thaliana. *New Phytol.* **2011**, *189*, 415–427. [CrossRef] [PubMed]
97. Corpas, F.J.; Chaki, M.; Fernández-Ocaña, A.; Valderrama, R.; Palma, J.M.; Carreras, A.; Begara-Morales, J.C.; Airaki, M.; del Río, L.A.; Barroso, J.B. Metabolism of reactive nitrogen species in pea plants under abiotic stress conditions. *Plant Cell Physiol.* **2008**, *49*, 1711–1722. [CrossRef] [PubMed]
98. Puyaubert, J.; Baudouin, E. New clues for a cold case: Nitric oxide response to low temperature. *Plant Cell Environ.* **2014**, *37*, 2623–2630. [CrossRef] [PubMed]
99. Puyaubert, J.; Fares, A.; Rézé, N.; Peltier, J.B.; Baudouin, E. Identification of endogenously S-nitrosylated proteins in *Arabidopsis* plantlets: Effect of cold stress on cysteine nitrosylation level. *Plant Sci.* **2014**, *215*, 150–156. [CrossRef] [PubMed]
100. Tan, J.; Zhuo, C.; Guo, Z. Nitric oxide mediates cold- and dehydration-induced expression of a novel MfHyPRP that confers tolerance to abiotic stress. *Physiol. Plant.* **2013**, *149*, 310–320.
101. Todorova, D.; Sergiev, I.; Moskova, I.; Alexieva, V.; Hall, M. Oxidative stress provoked by low and high temperatures in wild type and ethylene-insensitive mutant eti5 of *Arabidopsis thaliana* (L.) Heynh. *Oxid. Commun.* **2012**, *35*, 651–661.

102. Wang, L.Y.; Zhang, Q.Y.; Wang, F.; Meng, X.; Meng, Q.W. Ascorbate plays a key role in alleviating low temperature-induced oxidative stress in *Arabidopsis*. *Photosynthetica* **2012**, *50*, 602–612. [CrossRef]
103. Zhu, W.; Zhao, D.-X.; Miao, Q.; Xue, T.-T.; Li, X.-Z.; Zheng, C.-C. Arabidopsis thaliana metallothionein, AtMT2a, mediates ROS balance during oxidative stress. *J. Plant Biol.* **2009**, *52*, 585–592. [CrossRef]
104. Distelbarth, H.; Nägele, T.; Heyer, A.G. Responses of antioxidant enzymes to cold and high light are not correlated to freezing tolerance in natural accessions of Arabidopsis thaliana. *Plant Biol.* **2013**, *15*, 982–990. [CrossRef] [PubMed]
105. Ziogas, V.; Tanou, G.; Filippou, P.; Diamantidis, G.; Vasilakakis, M.; Fotopoulos, V.; Molassiotis, A. Nitrosative responses in citrus plants exposed to six abiotic stress conditions. *Plant Physiol. Biochem.* **2013**, *68*, 118–126. [CrossRef] [PubMed]
106. Zhao, M.-G.; Chen, L.; Zhang, L.-L.; Zhang, W.-H. Nitric reductase-dependent nitric oxide production is involved in cold acclimation and freezing tolerance in *Arabidopsis*. *Plant Physiol.* **2009**, *151*, 755–767. [CrossRef]
107. Kissner, R.; Nauser, T.; Bugnon, P.; Lye, P.G.; Koppenol, W.H. Formation and properties of peroxynitrite as studied by laser flash photolysis, high-pressure stopped-flow technique, and pulse radiolysis. *Chem. Res. Toxicol.* **1997**, *10*, 1285–1292. [CrossRef]
108. Valderrama, R.; Corpas, F.J.; Carreras, A.; Fernández-Ocaña, A.M.; Chaki, M.; Luque, F.; Gómez-Rodríguez, M.V.; Colmenero-Varea, P.; Luis, A.; Barroso, J.B. Nitrosative stress in plants. *FEBS Lett.* **2007**, *581*, 453–461. [CrossRef]
109. Chaki, M.; Carreras, A.; López-Jaramillo, J.; Begara-Morales, J.C.; Sánchez-Calvo, B.; Valderrama, R.; Corpas, F.J.; Barroso, J.B. Tyrosine nitration provokes inhibition of sunflower carbonic anhydrase (β-CA) activity under high temperature stress. *Nitric Oxide* **2013**, *29*, 30–33. [CrossRef]
110. Corpas, F.J.; Leterrier, M.; Begara-Morales, J.C.; Valderrama, R.; Chaki, M.; López-Jaramillo, J.; Luque, F.; Palma, J.M.; Padilla, M.N.; Sánchez-Calvo, B.; et al. Inhibition of peroxisomal hydroxypyruvate reductase (HPR1) by tyrosine nitration. *Biochim. Biophys. Acta Gen. Subj.* **2013**, *1830*, 4981–4989. [CrossRef]
111. Leterrier, M.; Airaki, M.; Palma, J.M.; Chaki, M.; Barroso, J.B.; Corpas, F.J. Arsenic triggers the nitric oxide (NO) and S-nitrosoglutathione (GSNO) metabolism in Arabidopsis. *Environ. Pollut.* **2012**, *166*, 136–143. [CrossRef]
112. Ruíz-Torres, C.; Feriche-Linares, R.; Rodríguez-Ruíz, M.; Palma, J.M.; Corpas, F.J. Arsenic-induced stress activates sulfur metabolism in different organs of garlic (*Allium sativum* L.) plants accompanied by a general decline of the NADPH-generating systems in roots. *J. Plant Physiol.* **2017**, *211*, 27–35. [CrossRef] [PubMed]
113. Dghim, A.A.; Dumont, J.; Hasenfratz-Sauder, M.-P.; Dizengremel, P.; Le Thiec, D.; Jolivet, Y. Capacity for NADPH regeneration in the leaves of two poplar genotypes differing in ozone sensitivity. *Physiol. Plant.* **2013**, *148*, 36–50. [CrossRef] [PubMed]
114. Bouthour, D.; Kalai, T.; Chaffei, H.C.; Gouia, H.; Corpas, F.J. Differential response of NADP-dehydrogenases and carbon metabolism in leaves and roots of two durum wheat (*Triticum durum* Desf.) cultivars (Karim and Azizi) with different sensitivities to salt stress. *J. Plant Physiol.* **2015**, *179*, 56–63. [CrossRef] [PubMed]
115. Hýsková, V.; Plisková, V.; Cervený, V.; Ryslavá, H. NADP-dependent enzymes are involved in response to salt and hypoosmotic stress in cucumber plants. *Gen. Physiol. Biophys.* **2017**, *36*, 247–258. [CrossRef] [PubMed]
116. Hýsková, V.; Miedzinska, L.; Dobra, J.; Vankova, R.; Rýslavá, H. Phosphoenolpyruvate carboxylase, NADP-malic enzyme, and pyruvate, phosphate dikinase are involved in the acclimation of *Nicotiana tabacum* L. to drought stress. *J. Plant Physiol.* **2014**, *171*, 19–25. [CrossRef] [PubMed]
117. Houmani, H.; Rodriguez-Ruiz, M.; Palma, J.M.; Corpas, F.J. Mechanical wounding promotes local and long distance response in the halophyte Cakile maritima through the involvement of the ROS and RNS metabolism. *Nitric Oxide* **2018**, *74*, 93–101. [CrossRef]
118. Muñoz-Vargas, M.A.; González-Gordo, S.; Palma, J.M.; Corpas, F.J. Inhibition of NADP-malic enzyme activity by H_2S and NO in sweet pepper (*Capsicum annuum* L.) fruits. *Physiol. Plant.* **2019**. [CrossRef]
119. Sokolovsky, M.; Riordan, J.F.; Vallee, B.L. Tetranitromethane. A Reagent for the Nitration of Tyrosyl Residues in Proteins. *Biochemistry* **1966**, *5*, 3582–3589. [CrossRef] [PubMed]
120. Rao, S.R.; Kamath, B.G.; Bhagwat, A.S. Tyrosyl residue at or near the active site of maize NADP-malic enzyme. *Photosynthetica* **1999**, *36*, 225–231. [CrossRef]

121. Li, S.; Mhamdi, A.; Clement, C.; Jolivet, Y.; Noctor, G. Analysis of knockout mutants suggests that Arabidopsis NADP-MALIC ENZYME2 does not play an essential role in responses to oxidative stress of intracellular or extracellular origin. *J. Exp. Bot.* **2013**, *64*, 3605–3614. [CrossRef]
122. Corpas, F.J.; Palma, J.M.; del Río, L.A.; Barroso, J.B. Protein tyrosine nitration in higher plants grown under natural and stress conditions. *Front. Plant. Sci.* **2013**, *4*, 29. [CrossRef] [PubMed]
123. Chou, W.-Y.; Huang, S.-M.; Chang, G.-G. Functional roles of the N-terminal amino acid residues in the Mn (II)-L-malate binding and subunit interactions of pigeon liver malic enzyme. *Protein Eng.* **1997**, *10*, 1205–1211. [CrossRef] [PubMed]

10

Bifunctional Chloroplastic DJ-1B from *Arabidopsis thaliana* is an Oxidation-Robust Holdase and a Glyoxalase Sensitive to H_2O_2

Aleksandra Lewandowska [1,2,3,4,5,†], Trung Nghia Vo [1,2,3,†], Thuy-Dung Ho Nguyen [1,2,3], Khadija Wahni [1,2,3], Didier Vertommen [6], Frank Van Breusegem [2,4,5,*], David Young [1,2,3,*] and Joris Messens [1,2,3,*]

1 VIB-VUB Center for Structural Biology, VIB, 1050 Brussels, Belgium; allew@psb.ugent.be (A.L.); Trung.Nghia.Vo@vub.be (T.N.V.); Ho.Thuy.Dung.Nguyen@vub.be (T.-D.H.N.); khadija.wahni@vib-vub.be (K.W.)
2 Brussels Center for Redox Biology, 1050 Brussels, Belgium
3 Structural Biology Brussels, Vrije Universiteit Brussel, 1050 Brussels, Belgium
4 Department of Plant Biotechnology and Bioinformatics, Ghent University, 9052 Ghent, Belgium
5 VIB Center for Plant Systems Biology, VIB, 9052 Ghent, Belgium
6 de Duve Institute, Université Catholique de Louvain, 1200 Brussels, Belgium; didier.vertommen@uclouvain.be
* Correspondence: frbre@psb.vib-ugent.be (F.V.B.); David.Young@vub.be (D.Y.); joris.messens@vib-vub.be (J.M.)
† These authors contributed equally to this work.

Abstract: Members of the DJ-1 protein family are multifunctional enzymes whose loss increases the susceptibility of the cell to oxidative stress. However, little is known about the function of the plant DJ-1 homologs. Therefore, we analyzed the effect of oxidation on the structure and function of chloroplastic AtDJ-1B and studied the phenotype of T-DNA lines lacking the protein. In vitro oxidation of AtDJ-1B with H_2O_2 lowers its glyoxalase activity, but has no effect on its holdase chaperone function. Remarkably, upon oxidation, the thermostability of AtDJ-1B increases with no significant alteration of the overall secondary structure. Moreover, we found that *AtDJ-1B* transcript levels are invariable, and loss of AtDJ-1B does not affect plant viability, growth and stress response. All in all, two discrete functions of AtDJ-1B respond differently to H_2O_2, and AtDJ-1B is not essential for plant development under stress.

Keywords: chaperone; glyoxalase; holdase; redox

---

1. Introduction

α-dicarbonyls, such as glyoxal (GO) and methylglyoxal (MG), are toxic compounds produced during glycolysis, metal-catalyzed glucose auto-oxidation, and lipid peroxidation. When they react with proteins, they form advanced glycation end-products (AGEs), which have been implicated in the progression of diseases such as diabetes, atherosclerosis, and neurological disorders [1–3]. Plants also accumulate sugar-derived MG as a byproduct of the Calvin cycle [4]. GO and MG are being removed by a glutathione (GSH)-dependent two-enzyme system, consisting of glyoxalase I (GLYI) and glyoxalase II (GLYII) [5]. A third glyoxalase, initially termed GLYIII [6], is a GSH-independent enzyme. This GLYIII protein is a member of the DJ-1/ThiJ/PfpI superfamily, and it is conserved across all kingdoms. In humans and plants, this enzyme is termed DJ-1, and it is known to exhibit a variety of cellular functions. Human DJ-1 protects rat neurons against oxidative stress [7], has chaperone [8], glyoxalase [9], and deglycase activity [10], and forms a scaffold for RNA-associated proteins [11].

DJ-1 has a conserved cysteine (Cys106 in human DJ-1), which is essential for its glyoxalase activity [12], and is easily oxidized to sulfinic and sulfonic acid by H_2O_2 [13,14].

Humans have only one isoform of homodimeric DJ-1, while plants have six isoforms encoded by tandem repeats and resulting in DJ-1 pseudodimers with varying subcellular localization [15]. In plants, the role of these DJ-1 isoforms in protecting against oxidative stress is poorly understood. Among Arabidopsis DJ-1 proteins, the experimental evidence is limited to cytosolic AtDJ-1A and chloroplastic AtDJ1-C. AtDJ-1A protects the plant from high light by acting as an antioxidant enzyme and a copper chaperone for superoxide dismutase (SOD) [16]. However, the SOD chaperone activity of DJ-1 proteins has been disputed [17]. Another DJ-1 homolog, AtDJ-1C, was found to be essential for plant viability and chloroplast development [18]. Previously, we identified AtDJ-1A to be sulfenylated in vivo [19], and more recently we also identified AtDJ-1B as a redox-sensitive sulfenylated protein in the choroplasts [20].

In this study, we focus on AtDJ-1B, and evaluate the effect of H_2O_2 on the activity of this enzyme. We found that oxidation increases the thermal stability of AtDJ-1B, but inactivates its glyoxalase activity. We further demonstrated that AtDJ-1B is a holdase and that oxidation of AtDJ-1B has no effect on its chaperone activity. Finally, the analysis of two independent SALK T-DNA lines showed for the first time that the loss of AtDJ-1B does not affect the viability, growth, or stress response of Arabidopsis plants.

2. Materials and Methods

A summarized table of buffers referred to in this section can be found in Supplementary Information.

2.1. *Cloning and Purification of Recombinant AtDJ-1B*

An N-terminal Tobacco Etch Virus (TEV)-cleavage site was introduced into a codon-optimized Arabidopsis AtDJ-1B open reading frame by polymerase chain reaction (PCR), before subcloning into a pDEST15 expression vector with an N-terminal glutathione-S-transferase (GST)-tag (Gateway technology, Thermo Fisher Scientific, Waltham, MA, USA)). This expression vector was transformed into SHuffle® T7 competent *E. coli* (New England Biolabs, Ipswitch, MA, USA) and plated on agar to obtain a single colony, which was grown overnight at 30 °C in Luria-Bertani Broth (LB) supplemented with 100 μg/mL ampicillin. 1 L Terrific Broth (TB, with 100 μg/mL ampicillin) cultures were inoculated with this overnight pre-culture and grown at 30 °C until they reached the exponential growth phase, then cooled to 20 °C, supplemented with 0.4 mM isopropyl-β-D-1-thiogalactopyranoside and further grown overnight at 20 °C.

Cells were pelleted and resuspended in Lysis Buffer, lysed in a cell cracker at 20 kilopounds per square inch, and centrifuged at 40,000 × *g* for 30 min, 4 °C. The supernatant was passed through a 0.45-μm filter and loaded onto Glutathione (GSH) Sepharose High Performance resin (GE Healthcare Europe, Diegem, Belgium) equilibrated with Binding Buffer A. After 1 h of incubation, unbound material was removed by washing the resin with 10 column volumes Binding Buffer, and the GST-tag was cleaved by incubating the resin with an 1 mg of TEV protease per estimated 10 mg of total bound protein overnight at 4 °C. Non-bound material (containing cleaved AtDJ-1B) was collected and incubated for 1 h at 4 °C with Ni^{2+}-Sepharose 6 Fast Flow resin (GE Healthcare Europe, Diegem, Belgium) equilibrated with Binding Buffer B, for the purpose of capturing the His-tagged TEV protease. Non-binding protein was collected and concentrated by Vivaspin 20 (Sartorius, Göttingen, Germany) centrifugal filtration and injected onto a size-exclusion (SE) Superdex75 16/60 column (GE Healthcare Europe, Diegem, Belgium) equilibrated with SE Buffer. Protein eluates containing AtDJ-1B were collected and concentrated by centrifugal filtration, and protein concentration determined by Bradford assay [21]. Protein was then flash-frozen in liquid nitrogen and stored at −80 °C.

2.2. Circular Dichroism (CD)

Purified AtDJ-1B was buffer-exchanged into 50 mM 4-(2-hydroxyethyl)-1-piperazineethanesulfonic acid (HEPES) pH 7.3, 500 mM NaCl using Bio-Spin® 6 columns (Bio-Rad Laboratories N.V., Temse, Belgium), the protein concentration measured by Bradford assay, and treated with either 5 mM tris(2-carboxyethyl)phosphine (TCEP), or with several H_2O_2 molar ratios (2:1, 10:1; 100:1). Treatments with TCEP or H_2O_2 were carried out at 25 °C for 1 h, after which samples were buffer-exchanged into degassed, filtered 20 mM sodium phosphate pH 7.4, 100 mM KF on Bio-Spin® 6 (Bio-Rad Laboratories N.V., Temse, Belgium). These samples were centrifuged for 10 min at 20,000 × *g*, and the protein concentration was determined with the Bradford assay. Samples were vacuum-degassed for 10 min and CD spectra were collected at 20 °C using a J-715 spectropolarimeter (JASCO Inc., Easton, MD, USA) with a 0.1 mm cuvette from 250–185 nm using 1.0 nm sampling and averaged over 6 technical replicates. The spectrum of the buffer alone was subtracted from the resulting data, and the data converted to mean residue ellipticity. Spectra were deconvoluted by CONTIN ridge-regression analysis [22] within the DICHROWEB server [23,24], using a reference data set optimized for 190–240 nm.

2.3. Thermal Unfolding

The thermal stability of AtDJ-1B was assessed by monitoring the change in the absorbance of circularly polarized light at 222 nm as a function of temperature. Immediately following spectra acquisition by CD as described above, the same samples were subjected to a temperature gradient from 15–85 °C at a ramp of 1 °C min^{-1}, and the absorption at 222 nm measured at 0.2 °C intervals. The linear slopes of the initial and final baselines of the sigmoidal unfolding curves were fitted with equation 1. For equation 1, A is ellipticity in mdeg (θ), m and k are the pre-unfolding intercept and slope respectively, T is temperature, n and l are the post-unfolding intercept and slope respectively.

$$(A - (m + k * T))/((n + l * T) - (m + k * T)) \tag{1}$$

The resulting data of [U]/[F] was then plotted against T (Kelvin), and fitted with a Boltzmann Sigmoidal equation (Equation 2) in GraphPad Prism (version 7.0), where y is the data for [U]/[F], *slope* is the steepness of the curve, x is temperature in Kelvin, and Tm is the parameter for melting temperature (T_M).

$$y = ymin + (ymax - ymin)/(1 + \exp\left(\frac{Tm - x}{\text{slope}}\right)) \tag{2}$$

2.4. Glyoxalase Assay

The glyoxalase activity of AtDJ-1B was estimated by monitoring the degradation of its substrate, glyoxal, in function of time. The amount of glyoxal in the sample was measured by derivatizing it with 1,2-diaminobenzene, as described for methylglyoxal [25].

A reaction mix containing 7.62 mM glyoxal and 571 nM AtDJ-1B in Assay Buffer A was incubated for 2 h at 30 °C on a thermoblock with mixing. Every 20 min an aliquot of the reaction mix was taken, 2-fold diluted in Assay Buffer A, and 1,2-diaminobenzene (Sigma-Aldrich, Overijse, Belgium) and $HClO_4$ were added to give a final concentration of 186 nM AtDJ-1B, 575 μM 1,2-diaminobenzene (Sigma-Aldrich, Overijse, Belgium) and 0.5 M $HClO_4$. After 5 min mixing, the absorbance of the derivatized sample at 340 nm was measured. For each time point measurements from control standard solutions of 0, 2, 4, 6, 8, 10 mM of glyoxal derivatized in the same manner as the samples containing AtDJ-1B were taken for generation of a standard curve. The assays were performed for three experimental replicates with averaging of two technical replicates.

Purified AtDJ-1B was either reduced with 5 mM TCEP or oxidized by treatment with varying molar excess (2-fold, 4-fold, 6-fold, 8-fold, 10-fold, or 100-fold) of H_2O_2, or with 5 mM diamide (Sigma-Aldrich, Overijse, Belgium). After 1 h of incubation at 25 °C the oxidizing or reducing agents were removed by buffer exchange to Assay Buffer A using Bio-Spin® columns (Bio-Rad Laboratories N.V., Temse, Belgium).

2.5. Chaperone Assay

AtDJ-1B protein sample was buffer exchanged to Assay Buffer B and either reduced with 5 mM TCEP or oxidized by treatment with either 2-fold or 10-fold molar excess H_2O_2 to AtDJ-1B of 2:1 and 10:1 for 1 h at 25 °C. The samples were again buffer-exchanged into Assay Buffer B.

The thermal unfolding of citrate synthase (CS; from porcine heart; Sigma-Aldrich, Overijse, Belgium) was induced at 44 °C in Assay Buffer B with 0.24 μM citrate synthase (20 μg mL^{-1}) and either 3-fold molar excess of Hsp90 (Sigma-Aldrich, Overijse, Belgium) as positive control, 5-fold molar excess of lysozyme (from chicken egg white, (Sigma-Aldrich, Overijse, Belgium)) as negative control, or 5-fold or 20-fold molar excess of AtDJ-1B protein. Aliquots (17 μL) were taken every 20 min and added to a cuvette containing the CS Assay Buffer to a final volume of 500 μL. Citrate synthase (CS) activity was followed by monitoring the rate of decrease of acetyl-CoA at 233 nm.

2.6. Oxidation of AtDJ-1B for Mass Spectrometric Analysis

AtDJ-1B was reduced with 50 mM TCEP for 30 min at room temperature and excess of TCEP was removed by Bio-Spin® (Bio-Rad Laboratories N.V., Temse, Belgium) to a buffer containing 50 mM HEPES pH 7.3, 0.5 M NaCl. 73 μg of protein (25 μM) was mixed with 125-fold molar excess of dimedone prior to the addition of 10-fold molar excess of H_2O_2, and incubated at 37 °C for 1 h. H_2O_2 and dimedone were removed by Bio-Spin™ and the protein concentration determined. Iodoacetamide (IAM) was added at 400-fold molar excess (i.e., final concentration 8 mM) and the sample incubated for 1 h at room temperature in the dark.

2.7. Determination of Cysteine Oxidation States by Mass Spectrometry

For the identification of modified residues in AtDJ-1B, 10 μg of desalted proteins were denatured by methanol/chloroform precipitation and digested O/N with trypsin or chymotrypsin at 30 °C in 50 mM NH_4HCO_3 (pH 8.0).

After dissolving in eluent C (0.1% (*v*/*v*) trifluoroacetic acid in 2% (*v*/*v*) acetonitrile (ACN)), peptides were directly loaded onto reversed-phase pre-column (Acclaim PepMap 100, Thermo Fisher Scientific, Waltham, MA, USA) and eluted in backflush mode. Peptides were separated using a reversed-phase analytical column (Acclaim PepMap RSLC, 0.075 × 250 mm, Thermo Scientific), equilibrated in eluent A (0.1% (*v*/*v*) hydrofluoric acid (FA) with 4% eluent B (0.1% (*v*/*v*) FA in 80% (*v*/*v*) ACN, with a linear gradient of 4%–27.5% eluent B (0.1% hydrofluoric acid in 98% acetonitrile) for 100 min, 27.5%–40% eluent B for 10 min, 40%–95% eluent B for 1 min and holding at 95% for the last 10 min at a constant flow rate of 300 nl/min on an EASY-nLC 1000 ultra performance liquid chromatography (UPLC) system (Thermo Fisher Scientific, Waltham, MA, USA). Orbitrap Fusion Lumos tribrid mass spectrometer (Thermo Fisher Scientific, Waltham, MA, USA) was used for the analysis of the resulting peptides, which were then subjected to NanoSpray Ionization (NSI) source followed by tandem mass spectrometry (MS/MS) in Fusion Lumos coupled online to the UPLC. Orbitrap at a resolution of 120,000 was used for intact peptide detection. Peptides were selected for MS/MS using HCD (higher energy collisional dissociation) setting as 30; Orbitrap at a resolution of 30,000 was used for ion fragment detection. The the top 20 precursor ions above a threshold ion count of 5.0^3 in the MS survey scan with 20.0 s dynamic exclusion were subjected to a data-dependent procedure that alternated between one MS scan followed by 20 MS/MS scans. The electrospray voltage applied was 2.1 kV. MS1 and MS2 spectra were obtained with an AGC target of 4E5 ions and a maximum injection time of 50ms and an AGC target of 5E4 ions and a maximum injection time of 100 ms, respectively. The *m/z* scan range for MS scans was 350 to 1500. MS/MS data processing was performed using Sequest HT search engine within Proteome Discoverer 2.2 against a homemade protein database containing the recombinant AtDJ-1A and AtDJ-1B sequences. Trypsin or chymotrypsin was specified as cleavage enzyme allowing up to 2 missed cleavages, 5 modifications per peptide, and up to 7 charges. Mass error was set to 10 ppm for precursor ions and 0.2 Da for fragment ions. Oxidation on Met, sulfenic-dimedone, sulfinic or sulfonic on Cys were considered as variable modifications. False discovery rate (FDR) was assessed using a fixed value PSM validator and thresholds for protein, peptide and modification site were specified at 1%. Site of covalent modification were manually validated. The mixed disulfides were identified by the use of the DBond software (Hanyang University, South Korea) [26].

2.8. Gene Expression Levels

The transcript level changes upon various treatments were obtained from several RNA Sequencing (RNA-Seq) experimental datasets, including: 3-h high light stress on *cat2-2* plants [27], treatment of Col-0 plants with 50 μM Antimycin A [28], Restricted Gas and Continuous Light (RGCL) treatment [29] of *cat2-2* or Col-0 plants for 24 h [30], 24 h of methyl viologen treatment of Col-0 plants (He et al., submitted) and *Pseudomonas syringae* infection of Col-0 plants (Stael et al., personal communication). Data visualization was performed using Heatmapper [31], genes were hierarchically clustered using Euclidean distance with average linkage.

2.9. Plant Material

All mutants used in this study are SALK T-DNA insertion lines in Col-0 background: *dj1a* (SALK_049637), *dj1b-4* (SALK_046449), *dj1b-9* (SALK_093414). All lines were genotyped prior to further analysis and homozygous plants were selected for phenotyping and confirmation of gene expression levels by RT-qPCR.

2.10. RT-qPCR

To confirm that T-DNA lines are true knockout lines, their RNA was extracted from three-week-old plants rosettes using TRIzol solubilization and extraction [32] followed by RNeasy Plant Mini Kit (Qiagen, Venlo, The Netherlands). First strand cDNA synthesis was performed using iScript cDNA Synthesis Kit (Bio-Rad Laboratories N.V., Temse, Belgium) using 1 μg of total RNA used as input material. The 5-fold diluted cDNA and gene-specific primers (Supplementary Table) were used for RT-qPCR performed by iCycler iQ (Bio-Rad Laboratories N.V., Temse, Belgium), with SYBR Green I Master Kit according to manufacturer's instructions. Data was analyzed by qBASE+ (Biogazelle, Zwijnaarde, Belgium), using *ELONGATION FACTOR 1α (EF-1α)* and *POLYUBIQUITIN 5 (UBIQ5)* as reference genes. For each data point three biological and three technical replicates were used.

2.11. Growth Conditions and Plant Stress Assays

To assess the effect of high light, plants were grown in soil for 21 days in a controlled chamber (100 $\mu mol \cdot m^{-2} \cdot s^{-1}$ light intensity,16 h/8 h light/dark, 21 °C, 50% relative humidity). Three-week-old plants were transferred o high light (600 $\mu mol \cdot m^{-2} \cdot s^{-1}$) for 72 h. The bright-light pictures of the plants, as well as measurements of photosystem II (PSII) maximum efficiency (Fv'/Fm') using an Imaging-PAM-Series chlorophyll fluorescence system (HeinzWalz, Effeltrich, Germany) were taken every 24 h.

To assess mutants' response to various stress conditions in vitro, plants were surface-sterilized by fumigation, vernalized for 3–4 days at 4 °C and grown under controlled conditions (16 h/8 h light/dark, 100 $\mu mol \cdot m^{-2} \cdot s^{-1}$ light intensity, 21 °C, 70% relative humidity) for 21 days. The medium was half-strength Murashige-Skoog medium (1/2 MS, 1% *w*/*v* sucrose, 0.8% *w*/*v* agar) containing one of the following additives: 25 mM mannitol, 50 mM NaCl, 25 nM methyl viologen or 1 μM 3-aminotriazole (3-AT). Plants were also grown on control plates, i.e., 1/2 MS with no additives. Plant rosette measurements were performed using ImageJ software [33] from 9 until 19 days after germination.

The Restricted Gas exchange and Continuous Light (RGCL) treatment was used to trigger photorespiratory stress in plants growing on 1/2 MS agar plates [34]. 21 days after vernalization, plates (either 1/2 MS or 1/2 MS + 3-AT) were sealed with multiple layers of Parafilm® M (Bemis

Company Inc., Oshkosh, WI, USA) in order to restrict gas exchange and transferred to continuous light (100 μmol·m^{-2}·s^{-1} light intensity, 21 °C, 70% relative humidity) for 10 days. The bright-light pictures of the plants, as well as measurements of photosystem II (PSII) maximum efficiency (Fv′/Fm′) using an Imaging- PAM-Series chlorophyll fluorescence system (HeinzWalz, Effeltrich, Germany) were taken every 2–3 days.

3. Results

3.1. AtDJ-1B Contains Multiple Oxidant-Sensitive Cysteines

The conserved catalytic cysteine of human DJ-1 is recognized as being highly prone to oxidation, with the most commonly observed oxidation state of this cysteine being sulfinic acid [35]. Arabidopsis AtDJ-1B is sulfenylated in planta [20], but nothing was known about its sulfenylation sites. AtDJ-1B is a pseudodimer containing two conserved DJ-1/PfpI domains in which there are eight cysteines, six within the N-terminal subunit, and two in the C-terminal subunit (Figure 1A). Of these eight cysteines, two correspond to the conserved catalytic cysteines of homodimeric human DJ-1 (Cys109 and Cys314), and seven are predicted to be solvent-exposed based on a homology model of AtDJ-1B (Cys129 being the sole buried residue). Hence, we sought to characterize the extent of cysteine oxidation in recombinantly expressed AtDJ-1B after exposure to H_2O_2. Mass spectrometry (MS) analysis of AtDJ-1B treated for 1 h with a 10-molar excess of H_2O_2 showed, as expected, sulfinylation of the conserved active-site cysteines (Cys109 and Cys314) (Figure 1B). Sulfinylation also occurred at Cys110, and to a lesser extent at Cys129 and Cys339 (Figure 1B).

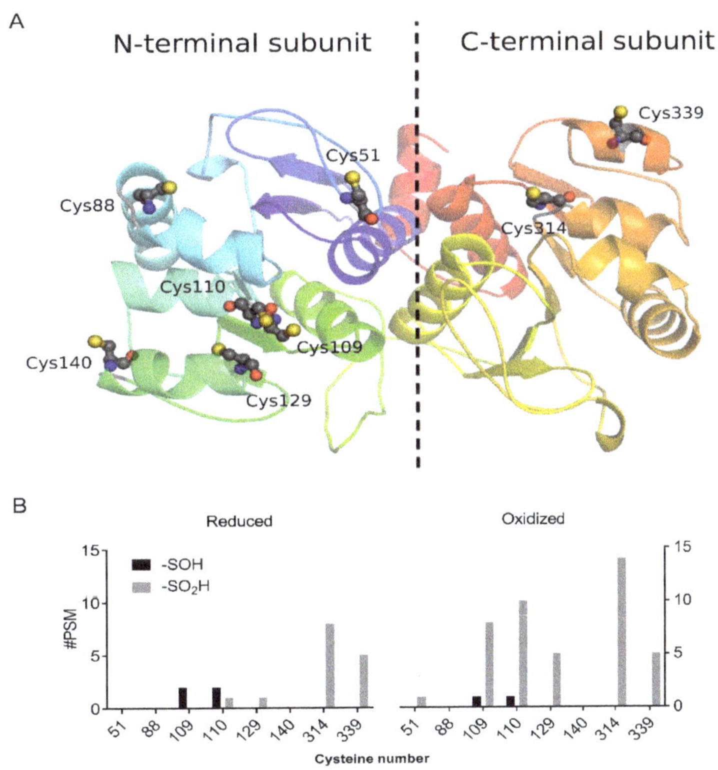

Figure 1. AtDJ-1B cysteines are differentially oxidized upon H_2O_2 treatment. (**A**) Homology model of AtDJ-1B with annotated cysteines in ball-and-stick representation. A dotted line approximately marks the 2-fold pseudosymmetry axis arising from the pseudo-dimeric AtDJ-1B monomer. Model generated using the I-TASSER structural prediction server, wherein the crystal structure of AtDJ-1D (PDB ID, 3UK7) was applied as a threading template. (**B**) Number of peptides (#PSM) detected by mass spectrometry that contained sulfenylated (-SOH) and sulfinylated (-SO$_2$H) cysteines, either in a sample reduced by 5 mM TCEP or oxidized by 10-fold molar excess of H_2O_2 for 1 h.

3.2. Oxidized AtDJ-1B Becomes More Thermostable

To assess the effect of the redox state of AtDJ-1B on its structural stability, we used circular dichroism (CD). We compared the relative secondary structure content of reduced and oxidized AtDJ-1B using increasing molar ratios of H_2O_2 (Figure 2A). The secondary structure of AtDJ-1B was not significantly altered upon oxidation, and both reduced and oxidized AtDJ-1B are characterized by 36–37% α-helical and 15–17% β-strand content, in agreement with previous findings, and comparable to what was observed for human DJ-1 [36].

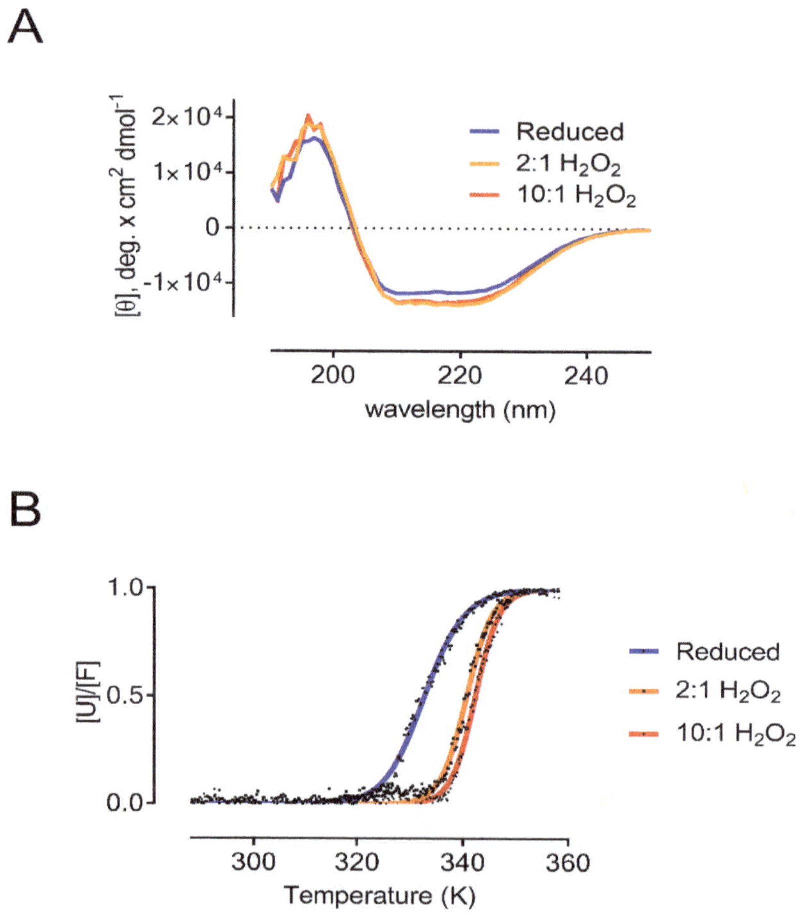

Figure 2. Oxidation does not influence the secondary structure of AtDJ-1B but increases its thermostability. (**A**) Comparative CD spectra of AtDJ-1B reduced with TCEP (blue trace), and DJ-1B oxidized by treatment with either a 2-fold (orange) or 10-fold (red) molar excess of H_2O_2. A slight spectral shift can be observed for the H_2O_2-treated protein relative to the reduced protein, though this spectral difference does not ultimately relate to any significant change in secondary structure. (**B**) Thermal unfolding of AtDJ-1B fitted to the Gibbs-Helmholtz equation. The change in ellipticity at 222 nm was followed by CD as a function of temperature for reduced AtDJ-1B (5 mM TCEP), and oxidized AtDJ-1B (either 2-fold, or 10-fold molar excess of H_2O_2). Data were converted to fraction of unfolded protein ([U]/[F]) and baseline-subtracted as described in Materials & Methods.

To evaluate the effect of oxidation on the thermostability of AtDJ-1B, we determined the melting temperature (T_M) of AtDJ-1B by following the change in ellipticity at 222 nm in function of temperature (Figure 2B). After fitting the data with a Boltzmann Sigmoidal equation, we obtained a T_M of 59.7 °C for reduced AtDJ-1B. For oxidized AtDJ-1B, the T_M increased to 67.6 °C and 69.3 °C following treatment with 2-fold and 10-fold molar excess of H_2O_2, respectively. This demonstrates that the oxidized form of AtDJ-1B is significantly more thermostable than the reduced form.

3.3. Oxidation Inactivates the Glyoxalase Activity of AtDJ-1B

One of the demonstrated functions of DJ-1 proteins is as a GSH-independent glyoxalase enzyme [9,37,38]. To measure the glyoxalase activity of AtDJ-1B, we monitored the consumption of glyoxal in function of time through derivatization with 1,2-diaminobenzene (Figure 3). Then, under near-steady-state conditions (8 mM glyoxal) the effect of increasing molar ratios of H_2O_2 on the glyoxalase activity of AtDJ-1B was evaluated. Pre-treatment of AtDJ-1B with a 2-fold molar excess H_2O_2 resulted in a decrease of the relative activity to ~64% of the activity of the reduced enzyme, and a further decrease to ~16% after treatment with 10-fold molar excess H_2O_2. This dose-dependent oxidative inactivation of AtDJ-1B was further confirmed by complete inactivation of glyoxalase activity following treatment with a 100-fold molar excess of H_2O_2. Oxidation of AtDJ-1B by treatment with 5 mM diamide (disulfide bond formation) also elicited a decrease to ~17% of full activity.

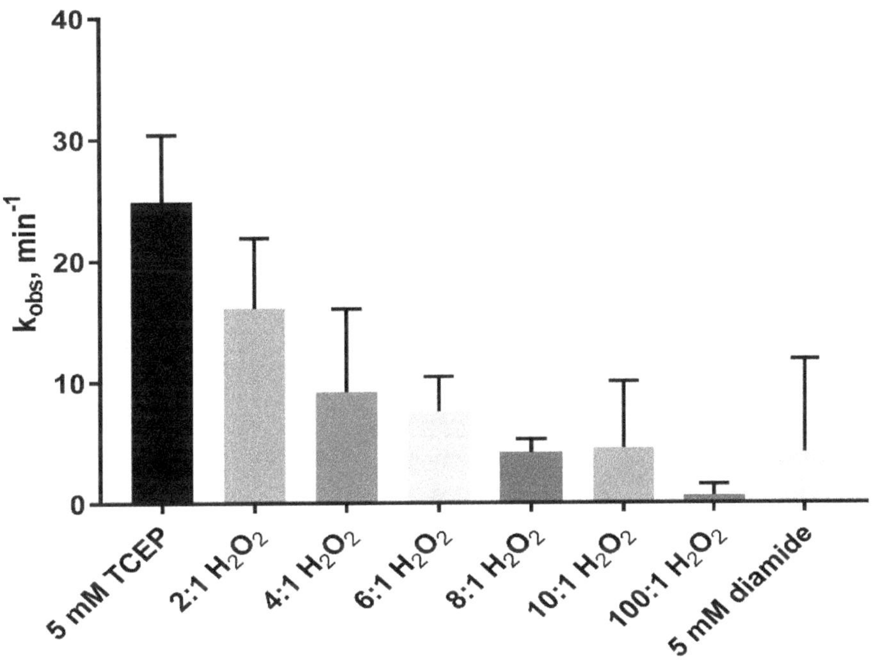

Figure 3. Glyoxalase activity of AtDJ-1B is lost after oxidation. Reduced and oxidized AtDJ-1B were prepared by pretreatment with either TCEP, varying molar ratios of H_2O_2: protein, or diamide. The reductant/oxidant was then removed prior to the activity assay. Displayed are observed rate constant (k_{obs}) values averaged from experimental triplicates with ± standard deviation (SD) indicated.

3.4. Oxidation does not Affect the Chaperone Activity of AtDJ-1B

Hsp31, a close homolog of AtDJ-1B in *Saccharomyces cerevisiae*, has previously been demonstrated to function as a holdase and confer stress resistance [39–41]. To explore whether AtDJ-1B is capable of similar functionality, we assessed the ability of AtDJ-1B to act as a holdase towards citrate synthase (CS). Here we found that AtDJ-1B protects CS against thermal inactivation at 44 °C, with 5-fold molar excess of AtDJ-1B preserving 35% of CS activity after 40 min heat treatment compared to only 10% preservation when using a 5-fold molar excess lysozyme as a negative control (Figure 4A). Although AtDJ-1B offered less protection than the positive control of Hsp90 (preserving 68% of activity at a 3-fold molar excess), the extent to which AtDJ-1B protected CS against inactivation was proportional to the relative concentration of AtDJ-1B, with 20-fold molar excess resulting in greater protection than 5-fold excess (Figure 4A). Considering a putative role of AtDJ-1B in the oxidative stress response, we evaluated the impact of oxidation on the ability of AtDJ-1B to act as a holdase. Remarkably, no significant difference was observed between the chaperone effectiveness of reduced (TCEP-treated) and oxidized (H_2O_2-treated) AtDJ-1B, with all samples showing a similar level of protection against thermal inactivation of CS (Figure 4B).

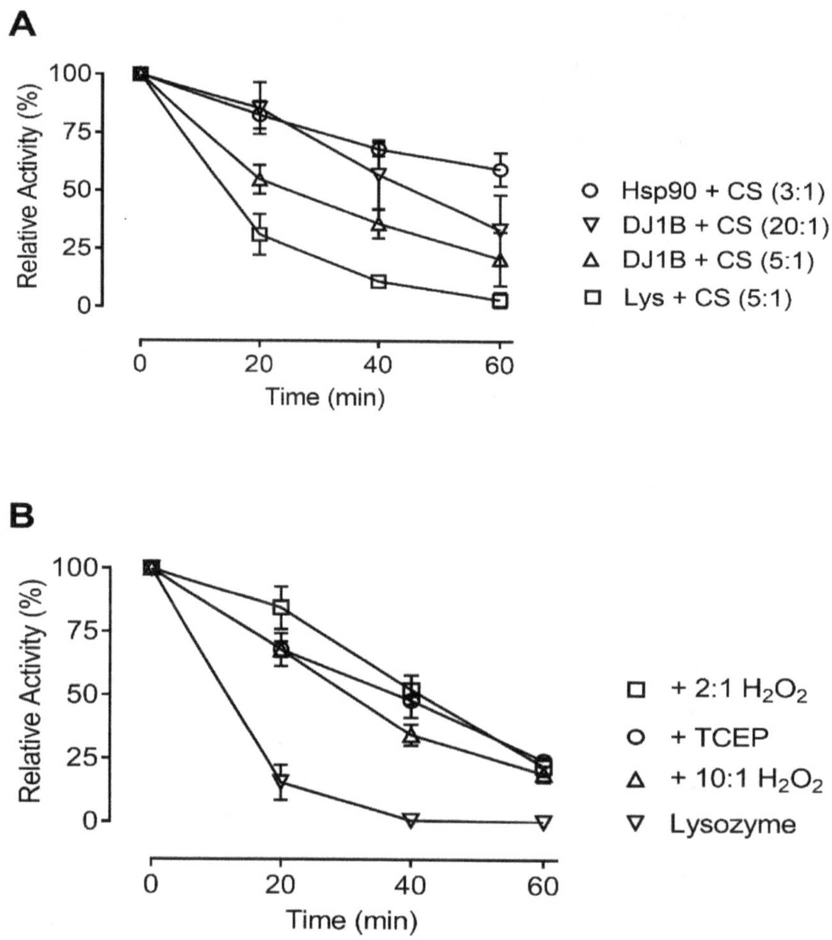

Figure 4. Suppression of thermal inactivation of citrate synthase (CS) by AtDJ-1B is independent of the oxidation state of AtDJ-1B. (**A**) At 44 °C, CS was incubated with 5-fold molar excess of lysozyme, 3-fold molar excess of Hsp90, 20- or 5-fold molar excess of DJ-1B at 44 °C, and activity measurements taken every 20 min for 1 h. (**B**) CS was incubated with 5-fold molar excess of lysozyme, or 20-fold molar excess of DJ-1B, either reduced with 5 mM TCEP, or oxidized with 10:1 or 2:1 molar excess of H_2O_2 at 44 °C, and activity measurements taken every 20 min for 1 h. Given is relative activity as a percentage of the full enzymatic activity of CS, just prior to heat-block incubation. Each data point represents average ± SD for triplicates.

3.5. *DJ-1B-Deficient Plants are Phenotypically Identical to Wildtype*

Having established the possible activities of DJ-1B and their redox regulation, we sought to determine the phenotype of Arabidopsis T-DNA *dj-1b* insertion lines. Two SALK T-DNA *dj1b* lines (further referred to as *dj1b-4* and *dj1b-9*, see Materials and Methods) and one SALK T-DNA *dj-1a* line were genotyped and confirmed to be homozygous. The *dj1b-4* and *dj1b-9* mutants contain the T-DNA insert in the 4th exon and the 4th intron, respectively, while in the *dj1a* line the insert is located in the promoter region (Supplementary Figure S1). The lack of gene transcripts was validated by reverse transcription quantitative PCR (RT-qPCR) (see Supplementary data).

When grown in soil in a growth chamber (16 h of light, 8 h of dark, 100 $\mu mol \cdot m^{-2} \cdot s^{-1}$ light intensity), the *dj1b* and *dj1a* mutants were phenotypically identical to wildtype Col-0 plants (Figure 5). Since *dj1a* mutants were previously reported to show increased susceptibility to high light stress [16], we sought to confirm these findings and checked whether plants lacking AtDJ-1B show the same phenotype. After 3 days of high light stress treatment, both wildtype and mutant plants had no leaf lesions and their photosystem II efficiency, inferred from Fv′/Fm′ ratios, remained unchanged (Figure 6).

Figure 5. 3-week old plants lacking AtDJ-1A or AtDJ-1B are phenotypically identical to wildtype Col-0 plants. Col-0, *dj1a* and *dj1b-9* plants were grown for 3 weeks in a long-day light regimen (16 h/8 h light/dark, 21 °C for 3 weeks. Representative bright-light images were taken 21 days after sowing.

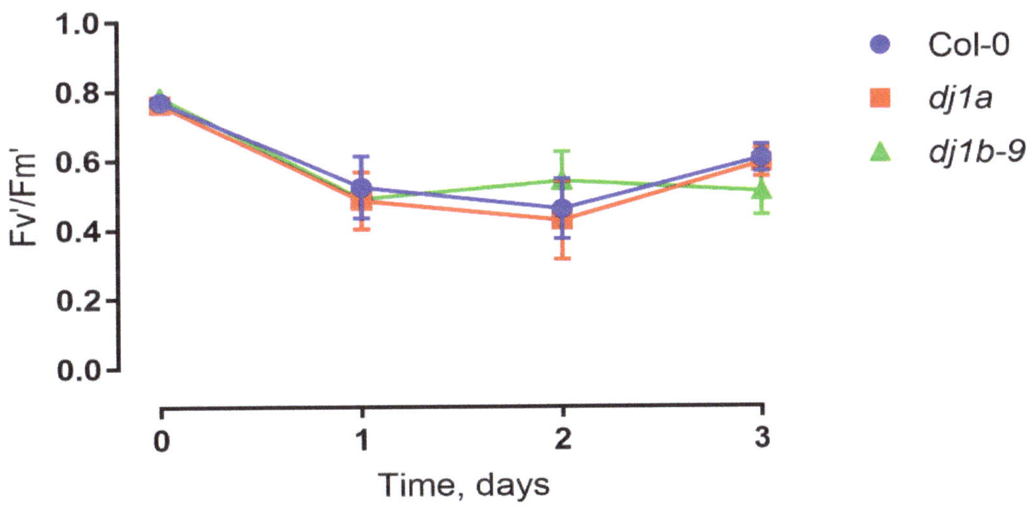

Figure 6. Photosystem II efficiency of mutants subjected to high light treatment was the same as that of wild type (WT). Three-week-old plants were exposed to high light intensities (600 $\mu mol \cdot m^{-2} \cdot s^{-1}$) for 72 h and the Fv′/Fm′ ratios were measured. Each data point represents average from biological triplicates ± SD.

To further characterize the phenotypes of *dj1a* and *dj1b* mutants, we grew them on 1/2 Murashige-Skoog (MS) agar medium for 3 weeks and assessed their rosette sizes when plants were subjected to different stress agents in the medium: NaCl, mannitol, methyl viologen, 3-amino-1,2, 4-triazole. In all cases, mutants grown on control 1/2 MS medium or subjected to stress were characterized by the same rosette size as wildtype (Figure 7). Moreover, when plants were subjected to photorespiratory stress by a Restricted Gas exchange and Continuous Light (RGCL) treatment [34] for up to 10 days, the Fv′/Fm′ ratio of the mutants was the same as for wildtype (Figure 8). Taken together, these results indicate that the loss of DJ-1A or DJ-1B does not result in altered growth rates and photosynthetic capacities when exposed to the described abiotic stress conditions.

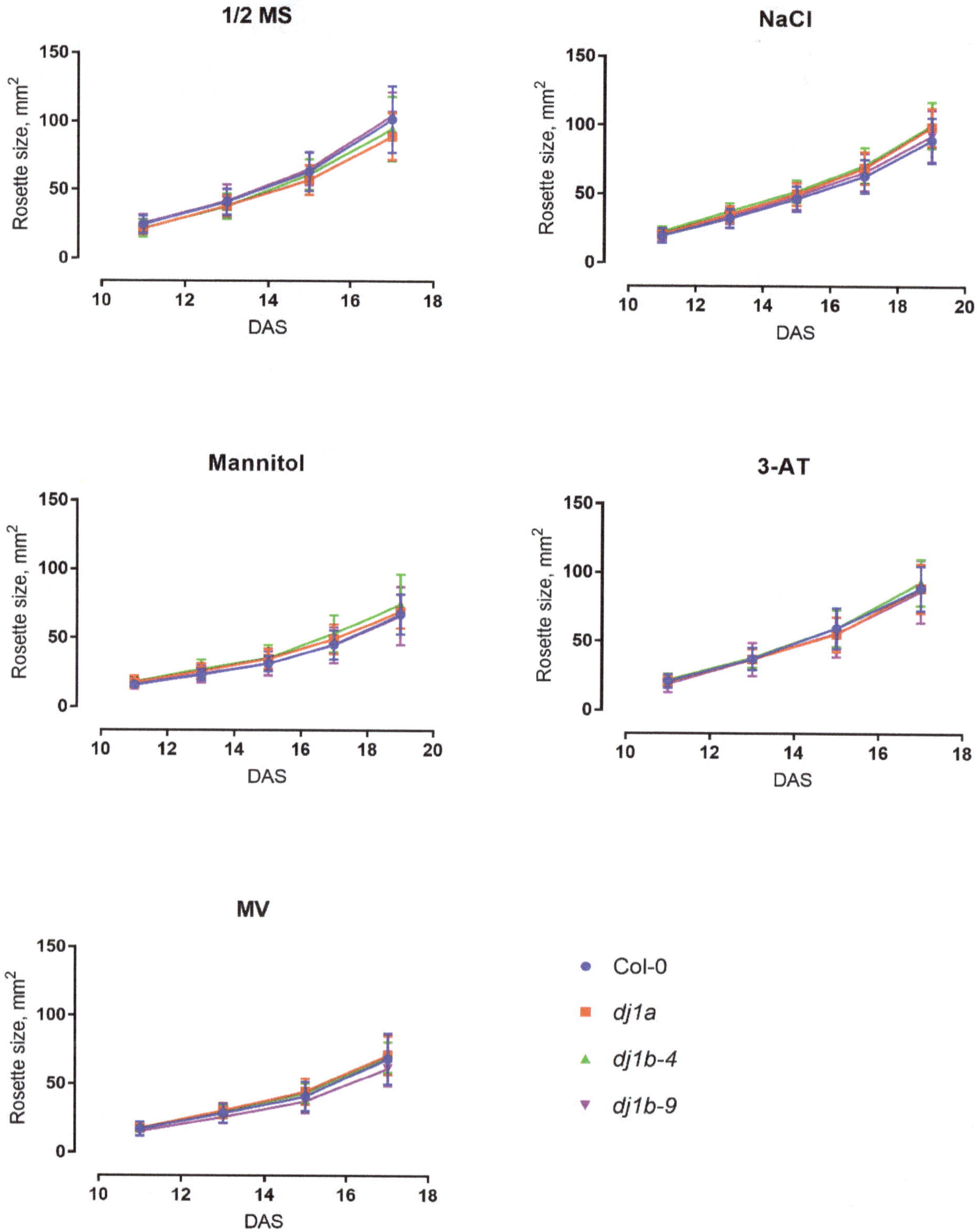

Figure 7. T-DNA insertion lines lacking AtDJ-1A or AtDJ-1B have no growth defects when subjected to various stresses. The plants were grown on 1/2 MS medium with stress-triggering additives: NaCl, mannitol, 3-amino-1,2,4-triazole (3-AT) or methyl viologen (MV) and their rosette sizes were analysed from bright-light images using ImageJ Software. Data points represent biological replicates ± SD.

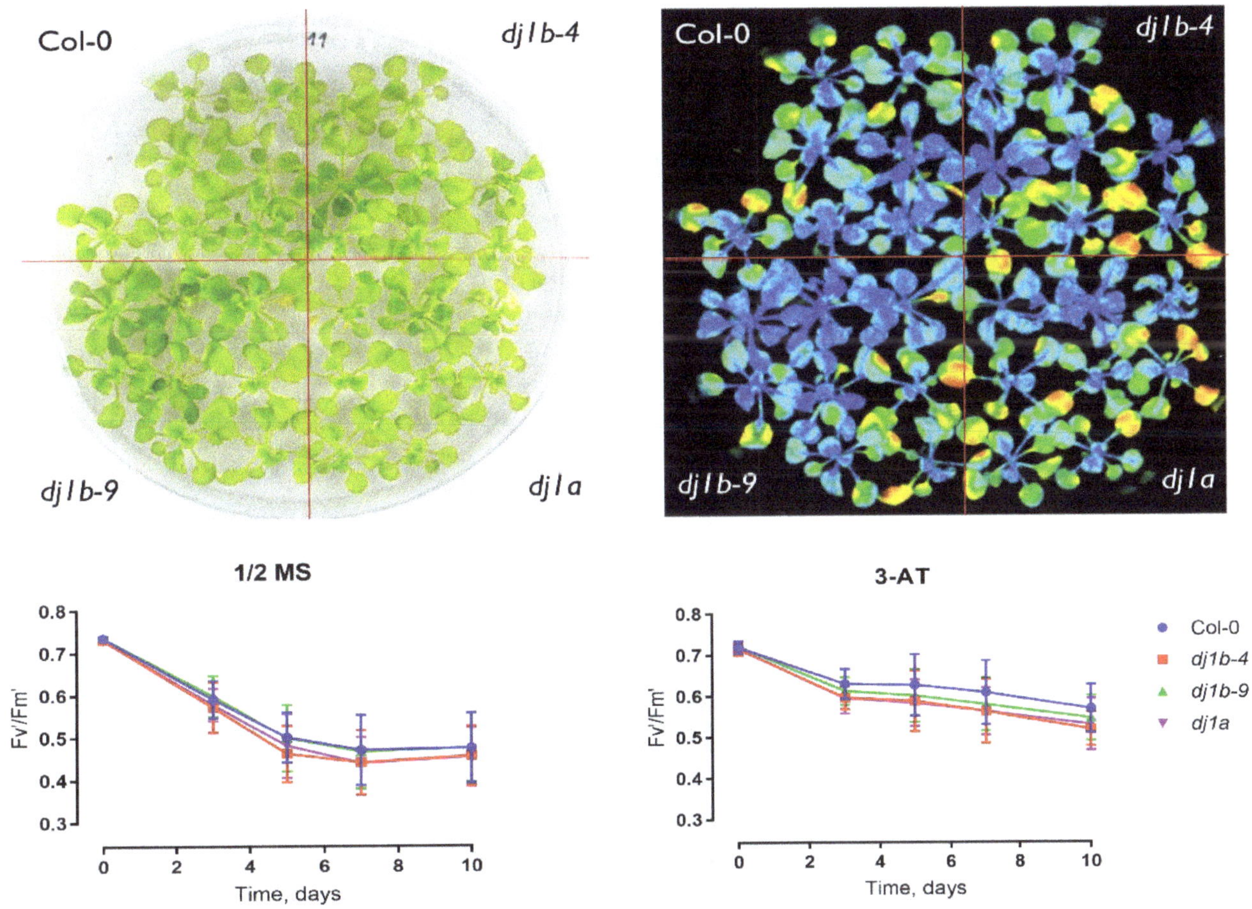

Figure 8. Plants deficient in AtDJ-1A orAt DJ-1B show no photorespiratory phenotype. (Upper panel) T-DNA *dj1a* and *dj1b* insertion lines do not differ from WT, as shown on a representative bright-light image (left) and color-coded Fv'/Fm' image (right) after 7 days of Restricted Gas exchange and Continuous Light (RGCL) treatment. (Lower panel) Fv'/Fm' decrease during the RGCL treatment. Data points represent averages from biological replicates ± SD.

3.6. Transcript Levels of DJ-1B are Stress-Independent

Since various oxidative stress treatments trigger transcriptome-wide changes in Arabidopsis mRNA levels [42], we analyzed several RNA-Seq datasets, aiming to determine whether mRNA levels of all Arabidopsis DJ-1 homologs are also susceptible to such treatments (Figure 9). We concentrated on treatments triggering oxidative stress, such as RGCL [34], high light [27], methyl viologen treatment, and *Pseudomonas syringae* infection. In addition to wildtype Col-0, we also analyzed the transcriptome of *cat2-2* mutants, which are commonly used as stress-inducible systems to study oxidative stress in vivo [43].

The RNA-Seq dataset analysis confirmed the earlier finding that AtDJ-1A is upregulated under stress [16]. The same trend was observed for AtDJ-1E, whose mRNA levels were also upregulated in the majority of the stress conditions. Remarkably, for AtDJ-1F, which shares 76% protein sequence identity with AtDJ-1E, the mRNA levels are significantly decreased upon stress. The mRNA levels of the other three AtDJ-1 homologs are less susceptible to stress treatments; in particular AtDJ-1B transcription levels are not affected. This result is in line with previous studies of plant DJ-1 homologs showing different expression patterns and transcriptional responses to stress [44].

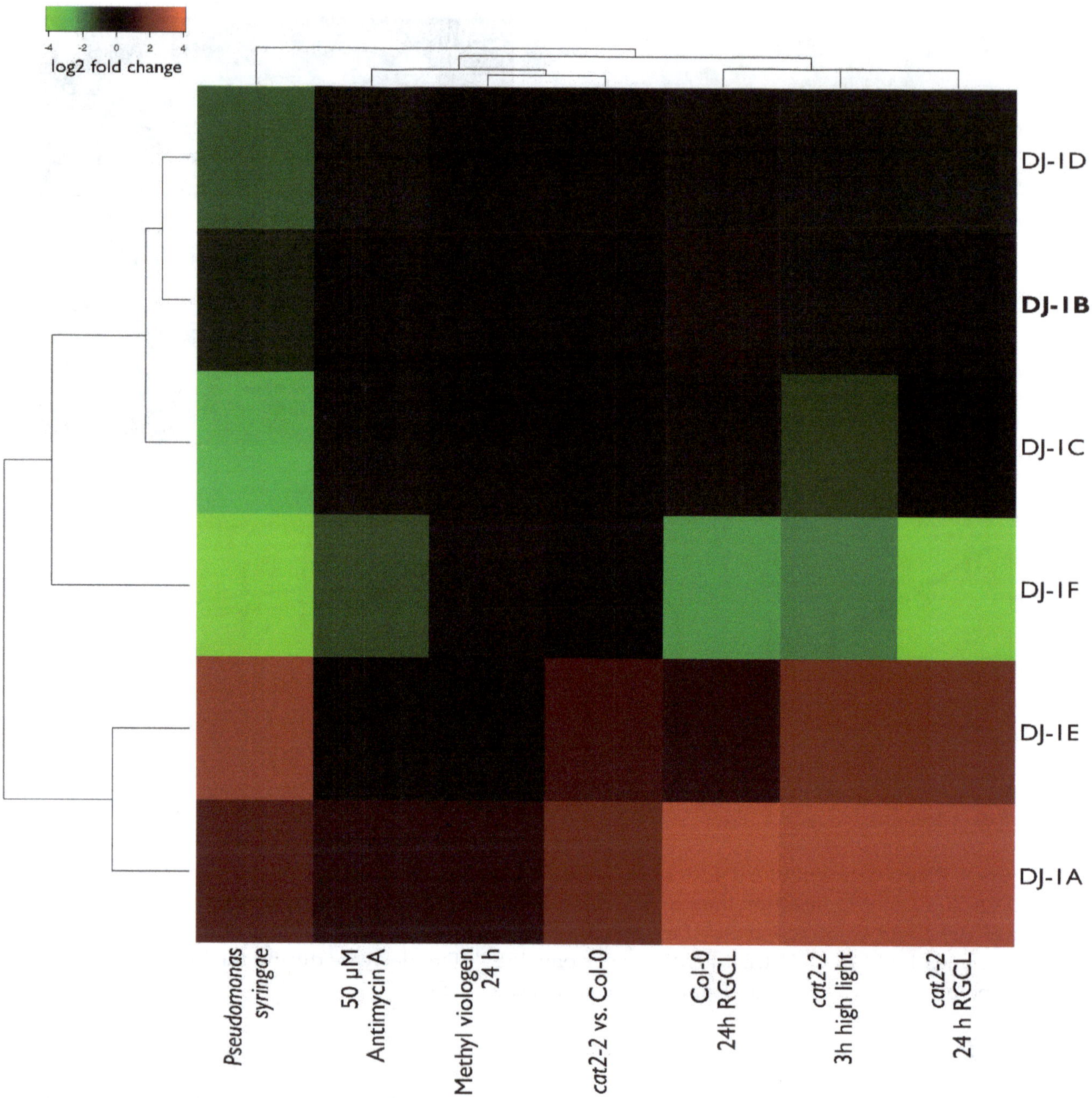

Figure 9. Arabidopsis DJ-1 gene expression levels respond differently after various stress treatments. Arabidopsis plants were either treated by a stress agent (pathogen *Pseudomonas syringage*, methyl viologen, high light, RGCL—Restricted Gas Continuous Light—treatment [29], or two lines (Col-0 and *cat2-2*) were compared to each other. The RNASeq experiments from which the data was obtained, as well as the detailed descriptions of the treatments are referred to in Materials and Methods. Colors represent log2 fold change.

4. Discussion

We report reduced AtDJ-1B as a glyoxalase with specific activity of 600 nmol·min^{-1}·mg protein^{-1} (Supplementary Table). For non-reduced AtDJ-1B, a specific activity of 310 nmol·min^{-1}·mg protein^{-1} was reported [12]. The glyoxalase activity of AtDJ-1B is ~35 times lower than for AtDJ-1D when glyoxal is used as substrate (Supplementary Table) [12]. This lower glyoxalase activity could be due to the absence of a conserved histidine, which has been suggested to facilitate proton transfer and glyoxal stereospecificity [45]. In plants under biotic and abiotic stress, and after H_2O_2 treatment, the intracellular concentration of glyoxals has been shown to increase from 100 to 2000 μmol/g tissue [46,47]. When methylglyoxal is photoreduced by photosystem I during photosynthesis, it donates electrons to O_2, producing superoxide ($O_2^{\bullet -}$) [48]. Thus, by removing superoxide-generating glyoxals, enhanced glyoxalase activity contributes to cellular protection during oxidative stress. However, we clearly showed inactivation of the glyoxalase activity of AtDJ-1B by H_2O_2, the predominant oxidant under oxidative stress conditions (Figure 4). To our knowledge, this is the first report of a DJ-1 protein losing its glyoxalase activity upon oxidation. Moreover, as the GLYI/II enzyme system catalyzes glyoxal/methylglyoxal detoxification with far greater catalytic efficiency than even AtDJ-1D [49], we propose that in plants the majority of glyoxals is detoxified by GLYI/II enzymes, rather than by H_2O_2-sensitive DJ-1 proteins.

AtDJ-1C has been reported with no glyoxalase activity [12], and its knockout results in non-viable plants [18]. Therefore, we sought to explore possible alternative functions for AtDJ-1B, which shares its chloroplastic localization with AtDJ-1C. We found that AtDJ-1B acts as an oxidation-robust holdase, though a molar excess of AtDJ-1B is needed to protect CS from thermal inactivation. Since porcine CS used for the holdase activity assay is not its natural client protein, it still remains unclear whether the holdase function of AtDJ-1B would be its real physiological function. Furthermore, it is known that dedicated chaperones suppress aggregation of client proteins at stoichiometric concentrations [50], whereas AtDJ-1B showed holdase activity at 5 to 20-fold molar excess. Similarly to human DJ-1 [8], AtDJ-1B could have specific client proteins in Arabidopsis that still have to be identified.

Furthermore, human DJ-1 has been demonstrated to be a redox-dependent holdase, with the oxidized form of DJ-1 able to suppress α-synuclein aggregation significantly better than the reduced form, offering 20 to 60% more protection [8]. On the other hand, the yeast homolog Hsp31 is a redox-independent holdase and this chaperone activity is required for protecting yeast from cytotoxic stress [39]. We also found the holdase activity of AtDJ-1B to be redox-independent. Remarkably, we observed that the thermostability of AtDJ-1B increases upon oxidation, which indicates that the oxidized form of AtDJ-1B is more likely than reduced AtDJ-1B to retain its chaperone activity under sustained heat stress. Other DJ-1 family members have also been shown to be highly thermostable, with a T_M values ranging from 64 to 77 °C [36,51,52]. Here, we determined a comparatively lower T_M for AtDJ-1B of 59.7 °C, and observe an increase in T_M of approximately 8–10 °C for oxidized AtDJ-1B. Similar increase in thermostability has also been observed for human DJ-1 and drosophila DJ-1β, with respective T_M increases of 13.3 and 11.5 °C upon oxidation in the presence of a 7-fold molar excess of H_2O_2 [51].

Finally, it is worthy of mention that both human and prokaryotic DJ-1 homologs also have deglycase activity [53–55], and it was proposed to be one of the core physiological functional roles of DJ-1 proteins [56]. As glyoxalase function is required for deglycase activity, when glyoxalase activity is lost upon oxidation, the deglycase function will also be abolished. In both its role as a deglycase and as a holdase, it is reasonable to assume that AtDJ-1B has preferential target proteins. Further understanding of the in vivo role of AtDJ-1B will require the identification of its interaction partners in Arabidopsis.

In contrast to AtDJ-1C, whose loss results in infertile plants with numerous developmental defects [18], AtDJ-1B loss does not influence plant viability. We also found that *dj1b* plants show wild-type resistance to stress, at least under the conditions tested. Moreover, while it was previously reported that AtDJ-1A deletion leads to increased susceptibility to high light [16], the phenotype of the *dj1a* SALK T-DNA line described here is identical to wildtype, which directly contradicts earlier findings [16].

The Arabidopsis DJ-1 family originates from one ancestor gene and diverged into six genes through both whole-genome and tandem duplications [57]. The same mechanism is responsible for higher duplication rates in plants in general, when compared to other eukaryotes [58,59]. After duplication, proteins evolve new functions; one of the copies might acquire a novel function, or if the ancestral protein is multifunctional, the duplicates divide the original function, for instance by differential expression patterns [59]. The analysis of RNASeq data presented in this paper shows that Arabidopsis DJ-1 proteins differ greatly in their expression patterns. These differences will be better understood after the analysis of their cis-regulatory elements in promoter sequences, like described for other plant species [44,60,61]. Moreover, since the mRNA levels are not representative for the protein levels, especially during dynamic phases such as cellular differentiation and stress response [62], it would be very informative to quantify all DJ-1 protein levels, both in stress and control conditions. Generating multiple mutant lines and biochemically characterizing all six Arabidopsis DJ-1 proteins will explain to what degree their functions overlap and whether the findings about the role of AtDJ-1B presented here are applicable to the other homologs as well.

5. Conclusions

In summary, we showed for the first time that Arabidopsis DJ-1B is a bifunctional protein, having both glyoxalase and holdase activity. Importantly, these two functions are differently regulated by H_2O_2; while the glyoxalase activity is lost upon oxidation, the holdase activity is not affected by H_2O_2. We also, for the first time, report the *dj1b* plant phenotype and prove that AtDJ-1B is not necessary for viability, development, or stress resistance of Arabidopsis plants, which might be due to redundant functions of all DJ-1 homologs. To explain why Arabidopsis has retained six DJ-1 homologs, the physiological relevance of the bifunctionality of AtDJ-1B still need to be confirmed in vivo and compared to those of the other AtDJ-1 proteins.

Supplementary Materials
Figure S1: *AtDJ-1A* and *AtDJ-1B* gene models, Figure S2: *AtDJ-1B* and *DJ-1A* transcript levels (left and right, respectively) in WT and KO T-DNA lines, Table S1: Component summary of buffering solutions used, Table S2: AtDJ-1B specific glyoxalase activities and corresponding observed reaction rates determined during glyoxalase assay, Table S3: Arabidopsis thaliana DJ-1 specific activities [12], Table S4: Primers used for the study, Table S5: Log2 fold change values of DJ-1 mRNA expression levels, as visualised on Figure 9.

Author Contributions: Conceptualization, A.L., D.Y., F.V.B. and J.M.; methodology, A.L., D.V. and D.Y.; formal analysis, A.L., T.N.V., D.V and D.Y.; investigation, A.L., T.N.V., T.-D.H.N., K.W., D.Y; resources, F.V.B. and J.M.; writing—original draft preparation, A.L., T.N.V and D.Y.; writing—review and editing, A.L., D.Y., F.V.B. and J.M.; supervision, F.V.B. and J.M.; project administration, F.V.B. and J.M.; funding acquisition, F.V.B. and J.M.

References

1. Vistoli, G.; De Maddis, D.; Cipak, A.; Zarkovic, N.; Carini, M.; Aldini, G. Advanced glycoxidation and lipoxidation end products (AGEs and ALEs): An overview of their mechanisms of formation. *Free Radic. Res.* **2013**, *47* (Suppl. 1), 3–27. [CrossRef] [PubMed]
2. Wetzels, S.; Wouters, K.; Schalkwijk, C.G.; Vanmierlo, T.; Hendriks, J.J. Methylglyoxal-Derived Advanced Glycation Endproducts in Multiple Sclerosis. *Int. J. Mol. Sci.* **2017**, *18*, 421. [CrossRef] [PubMed]
3. Chetyrkin, S.; Mathis, M.; Pedchenko, V.; Sanchez, O.A.; McDonald, W.H.; Hachey, D.L.; Madu, H.; Stec, D.; Hudson, B.; Voziyan, P. Glucose autoxidation induces functional damage to proteins via modification of critical arginine residues. *Biochemistry* **2011**, *50*, 6102–6112. [CrossRef] [PubMed]
4. Takagi, D.; Inoue, H.; Odawara, M.; Shimakawa, G.; Miyake, C. The Calvin cycle inevitably produces sugar-derived reactive carbonyl methylglyoxal during photosynthesis: A potential cause of plant diabetes. *Plant Cell Physiol.* **2014**, *55*, 333–340. [CrossRef] [PubMed]
5. Schmitz, J.; Dittmar, I.C.; Brockmann, J.D.; Schmidt, M.; Hudig, M.; Rossoni, A.W.; Maurino, V.G. Defense against Reactive Carbonyl Species Involves at Least Three Subcellular Compartments Where Individual Components of the System Respond to Cellular Sugar Status. *Plant Cell* **2017**, *29*, 3234–3254. [CrossRef] [PubMed]

6. Misra, K.; Banerjee, A.B.; Ray, S.; Ray, M. Glyoxalase III from *Escherichia coli*: A single novel enzyme for the conversion of methylglyoxal into D-lactate without reduced glutathione. *Biochem. J.* **1995**, *305 Pt 3*, 999–1003. [CrossRef]
7. Kahle, P.J.; Waak, J.; Gasser, T. DJ-1 and prevention of oxidative stress in Parkinson's disease and other age-related disorders. *Free Radic. Biol. Med.* **2009**, *47*, 1354–1361. [CrossRef]
8. Shendelman, S.; Jonason, A.; Martinat, C.; Leete, T.; Abeliovich, A. DJ-1 is a redox-dependent molecular chaperone that inhibits alpha-synuclein aggregate formation. *PLoS Biol.* **2004**, *2*, e362. [CrossRef]
9. Lee, J.-Y.; Song, J.; Kwon, K.; Jang, S.; Kim, C.; Baek, K.; Kim, J.; Park, C. Human DJ-1 and its homologs are novel glyoxalases. *Hum. Mol. Genet.* **2012**, *21*, 3215–3225. [CrossRef]
10. Richarme, G.; Liu, C.; Mihoub, M.; Abdallah, J.; Leger, T.; Joly, N.; Liebart, J.-C.; Jurkunas, U.V.; Nadal, M.; Bouloc, P.; et al. Guanine glycation repair by DJ-1/Park7 and its bacterial homologs. *Science* **2017**, *357*, 208–211. [CrossRef]
11. Hod, Y.; Pentyala, S.N.; Whyard, T.C.; El-Maghrabi, M.R. Identification and characterization of a novel protein that regulates RNA-protein interaction. *J. Cell Biochem.* **1999**, *72*, 435–444. [CrossRef]
12. Kwon, K.; Choi, D.; Hyun, J.K.; Jung, H.S.; Baek, K.; Park, C. Novel glyoxalases from *Arabidopsis thaliana*. *FEBS J.* **2013**, *280*, 3328–3339. [CrossRef] [PubMed]
13. Kinumi, T.; Kimata, J.; Taira, T.; Ariga, H.; Niki, E. Cysteine-106 of DJ-1 is the most sensitive cysteine residue to hydrogen peroxide-mediated oxidation in vivo in human umbilical vein endothelial cells. *Biochem. Biophys. Res. Commun.* **2004**, *317*, 722–728. [CrossRef] [PubMed]
14. Canet-Aviles, R.M.; Wilson, M.A.; Miller, D.W.; Ahmad, R.; McLendon, C.; Bandyopadhyay, S.; Baptista, M.J.; Ringe, D.; Petsko, G.A.; Cookson, M.R. The Parkinson's disease protein DJ-1 is neuroprotective due to cysteine-sulfinic acid-driven mitochondrial localization. *Proc. Natl. Acad. Sci. USA* **2004**, *101*, 9103–9108. [CrossRef] [PubMed]
15. Punta, M.; Coggill, P.C.; Eberhardt, R.Y.; Mistry, J.; Tate, J.; Boursnell, C.; Pang, N.; Forslund, K.; Ceric, G.; Clements, J.; et al. The Pfam protein families database. *Nucleic Acids Res.* **2012**, *40*, D290–D301. [CrossRef] [PubMed]
16. Xu, X.M.; Lin, H.; Maple, J.; Björkblom, B.; Alves, G.; Larsen, J.P.; Møller, S.G. The Arabidopsis DJ-1a protein confers stress protection through cytosolic SOD activation. *J Cell Sci* **2010**, *123*, 1644–1651. [CrossRef] [PubMed]
17. Barbieri, L.; Luchinat, E.; Banci, L. Intracellular metal binding and redox behavior of human DJ-1. *J. Biol. Inorg. Chem.* **2018**, *23*, 61–69. [CrossRef]
18. Lin, J.; Nazarenus, T.J.; Frey, J.L.; Liang, X.; Wilson, M.A.; Stone, J.M. A plant DJ-1 homolog is essential for *Arabidopsis thaliana* chloroplast development. *PLoS ONE* **2011**, *6*, e23731. [CrossRef]
19. Akter, S.; Huang, J.; Bodra, N.; De Smet, B.; Wahni, K.; Rombaut, D.; Pauwels, J.; Gevaert, K.; Carroll, K.; Van Breusegem, F.; et al. DYn-2 based identification of Arabidopsis sulfenomes. *Mol. Cell. Proteom.* **2015**, *14*, 1183–1200. [CrossRef]
20. De Smet, B.; Willems, P.; Fernandez-Fernandez, A.D.; Alseekh, S.; Fernie, A.R.; Messens, J.; Van Breusegem, F. In vivo detection of protein cysteine sulfenylation in plastids. *Plant J.* **2018**. [CrossRef]
21. Bradford, M.M. A rapid and sensitive method for the quantitation of microgram quantities of protein utilizing the principle of protein-dye binding. *Anal. Biochem.* **1976**, *72*, 248–254. [CrossRef]
22. Provencher, S.W.; Glöckner, J. Estimation of globular protein secondary structure from circular dichroism. *Biochemistry* **1981**, *20*, 33–37. [CrossRef] [PubMed]
23. Lobley, A.; Whitmore, L.; Wallace, B.A. DICHROWEB: An interactive website for the analysis of protein secondary structure from circular dichroism spectra. *Bioinformatics* **2002**, *18*, 211–212. [CrossRef] [PubMed]
24. Whitmore, L.; Wallace, B.A. DICHROWEB, an online server for protein secondary structure analyses from circular dichroism spectroscopic data. *Nucleic Acids Res.* **2004**, *32*, W668–W673. [CrossRef] [PubMed]
25. Cordeiro, C.; Ponces Freire, A. Methylglyoxal assay in cells as 2-methylquinoxaline using 1,2-diaminobenzene as derivatizing reagent. *Anal. Biochem.* **1996**, *234*, 221–224. [CrossRef]
26. Choi, S.; Jeong, J.; Na, S.; Lee, H.S.; Kim, H.Y.; Lee, K.J.; Paek, E. New algorithm for the identification of intact disulfide linkages based on fragmentation characteristics in tandem mass spectra. *J. Proteome Res.* **2010**, *9*, 626–635. [CrossRef]

27. Kerchev, P.; Waszczak, C.; Lewandowska, A.; Willems, P.; Shapiguzov, A.; Li, Z.; Alseekh, S.; Mühlenbock, P.; Hoeberichts, F.A.; Huang, J.; et al. Lack of GLYCOLATE OXIDASE1, but not GLYCOLATE OXIDASE2, attenuates the photorespiratory phenotype of catalase2-deficient arabidopsis. *Plant Physiol.* **2016**, *171*, 1704–1719. [CrossRef]
28. Zhang, X.; Ivanova, A.; Vandepoele, K.; Radomiljac, J.; Van de Velde, J.; Berkowitz, O.; Willems, P.; Xu, Y.; Ng, S.; Van Aken, O.; et al. The Transcription Factor MYB29 Is a Regulator of ALTERNATIVE OXIDASE1a. *Plant Physiol.* **2017**, *173*, 1824–1843. [CrossRef]
29. Kerchev, P.I.; De Clercq, I.; Denecker, J.; Muhlenbock, P.; Kumpf, R.; Nguyen, L.; Audenaert, D.; Dejonghe, W.; Van Breusegem, F. Mitochondrial perturbation negatively affects auxin signaling. *Mol. Plant* **2014**, *7*, 1138–1150. [CrossRef]
30. Waszczak, C.; Kerchev, P.I.; Muhlenbock, P.; Hoeberichts, F.A.; Van Der Kelen, K.; Mhamdi, A.; Willems, P.; Denecker, J.; Kumpf, R.P.; Noctor, G.; et al. SHORT-ROOT Deficiency Alleviates the Cell Death Phenotype of the Arabidopsis catalase2 Mutant under Photorespiration-Promoting Conditions. *Plant Cell* **2016**, *28*, 1844–1859. [CrossRef]
31. Babicki, S.; Arndt, D.; Marcu, A.; Liang, Y.; Grant, J.R.; Maciejewski, A.; Wishart, D.S. Heatmapper: Web-enabled heat mapping for all. *Nucleic Acids Res.* **2016**, *44*, W147–W153. [CrossRef] [PubMed]
32. Rio, D.C.; Ares, M., Jr.; Hannon, G.J.; Nilsen, T.W. Purification of RNA using TRIzol (TRI reagent). *Cold Spring Harb. Protoc.* **2010**, *2010*. [CrossRef] [PubMed]
33. Rueden, C.T.; Schindelin, J.; Hiner, M.C.; DeZonia, B.E.; Walter, A.E.; Arena, E.T.; Eliceiri, K.W. ImageJ2: ImageJ for the next generation of scientific image data. *BMC Bioinform.* **2017**, *18*, 529. [CrossRef] [PubMed]
34. Kerchev, P.; Muhlenbock, P.; Denecker, J.; Morreel, K.; Hoeberichts, F.A.; Van Der Kelen, K.; Vandorpe, M.; Nguyen, L.; Audenaert, D.; Van Breusegem, F. Activation of auxin signalling counteracts photorespiratory H_2O_2-dependent cell death. *Plant Cell Environ.* **2015**, *38*, 253–265. [CrossRef] [PubMed]
35. Wilson, M.A. The role of cysteine oxidation in DJ-1 function and dysfunction. *Antioxid. Redox Signal.* **2011**, *15*, 111–122. [CrossRef] [PubMed]
36. Lakshminarasimhan, M.; Maldonado, M.T.; Zhou, W.; Fink, A.L.; Wilson, M.A. Structural impact of three Parkinsonism-associated missense mutations on human DJ-1. *Biochemistry* **2008**, *47*, 1381–1392. [CrossRef]
37. Hasim, S.; Hussin, N.A.; Alomar, F.; Bidasee, K.R.; Nickerson, K.W.; Wilson, M.A. A glutathione-independent glyoxalase of the DJ-1 superfamily plays an important role in managing metabolically generated methylglyoxal in *Candida albicans*. *J. Biol. Chem.* **2014**, *289*, 1662–1674. [CrossRef]
38. Bankapalli, K.; Saladi, S.; Awadia, S.S.; Goswami, A.V.; Samaddar, M.; D'Silva, P. Robust glyoxalase activity of Hsp31, a ThiJ/DJ-1/PfpI family member protein, is critical for oxidative stress resistance in *Saccharomyces cerevisiae*. *J. Biol. Chem.* **2015**, *290*, 26491–26507. [CrossRef]
39. Tsai, C.-J.; Aslam, K.; Drendel, H.M.; Asiago, J.M.; Goode, K.M.; Paul, L.N.; Rochet, J.-C.; Hazbun, T.R. Hsp31 is a stress response chaperone that intervenes in the protein misfolding process. *J. Biol. Chem.* **2015**, *290*, 24816–24834. [CrossRef]
40. Amm, I.; Norell, D.; Wolf, D.H. Absence of the Yeast Hsp31 chaperones of the DJ-1 superfamily perturbs cytoplasmic protein quality control in late growth phase. *PLoS ONE* **2015**, *10*, e0140363. [CrossRef]
41. Aslam, K.; Hazbun, T.R. Hsp31, a member of the DJ-1 superfamily, is a multitasking stress responder with chaperone activity. *Prion* **2016**, *10*, 103–111. [CrossRef] [PubMed]
42. Willems, P.; Mhamdi, A.; Stael, S.; Storme, V.; Kerchev, P.; Noctor, G.; Gevaert, K.; Van Breusegem, F. The ROS wheel: Refining ROS transcriptional footprints. *Plant Physiol.* **2016**, *171*, 1720–1733. [CrossRef] [PubMed]
43. Mhamdi, A.; Queval, G.; Chaouch, S.; Vanderauwera, S.; Van Breusegem, F.; Noctor, G. Catalase function in plants: A focus on Arabidopsis mutants as stress-mimic models. *J. Exp. Bot.* **2010**, *61*, 4197–4220. [CrossRef] [PubMed]
44. Ghosh, A. Genome-Wide Identification of Glyoxalase Genes in Medicago truncatula and Their Expression Profiling in Response to Various Developmental and Environmental Stimuli. *Front. Plant Sci.* **2017**, *8*, 836. [CrossRef] [PubMed]
45. Choi, D.; Kim, J.; Ha, S.; Kwon, K.; Kim, E.-H.; Lee, H.-Y.; Ryu, K.-S.; Park, C. Stereospecific mechanism of DJ-1 glyoxalases inferred from their hemithioacetal-containing crystal structures. *FEBS J.* **2014**, *281*, 5447–5462. [CrossRef] [PubMed]

46. Yadav, S.K.; Singla-Pareek, S.L.; Ray, M.; Reddy, M.K.; Sopory, S.K. Methylglyoxal levels in plants under salinity stress are dependent on glyoxalase I and glutathione. *Biochem. Biophys. Res. Commun.* **2005**, *337*, 61–67. [CrossRef] [PubMed]
47. Melvin, P.; Bankapalli, K.; D'Silva, P.; Shivaprasad, P.V. Methylglyoxal detoxification by a DJ-1 family protein provides dual abiotic and biotic stress tolerance in transgenic plants. *Plant Mol. Biol.* **2017**, *94*, 381–397. [CrossRef]
48. Saito, R.; Yamamoto, H.; Makino, A.; Sugimoto, T.; Miyake, C. Methylglyoxal functions as Hill oxidant and stimulates the photoreduction of O(2) at photosystem I: A symptom of plant diabetes. *Plant Cell Environ.* **2011**, *34*, 1454–1464. [CrossRef]
49. Ghosh, A.; Pareek, A.; Sopory, S.K.; Singla-Pareek, S.L. A glutathione responsive rice glyoxalase II, OsGLYII-2, functions in salinity adaptation by maintaining better photosynthesis efficiency and anti-oxidant pool. *Plant J.* **2014**, *80*, 93–105. [CrossRef]
50. Haslbeck, M.; Buchner, J. Assays to characterize molecular chaperone function in vitro. *Methods Mol. Biol.* **2015**, *1292*, 39–51. [CrossRef]
51. Lin, J.; Prahlad, J.; Wilson, M.A. Conservation of oxidative protein stabilization in an insect homologue of parkinsonism-associated protein DJ-1. *Biochemistry* **2012**, *51*, 3799–3807. [CrossRef] [PubMed]
52. Culleton, B.A.; Lall, P.; Kinsella, G.K.; Doyle, S.; McCaffrey, J.; Fitzpatrick, D.A.; Burnell, A.M. A role for the Parkinson's disease protein DJ-1 as a chaperone and antioxidant in the anhydrobiotic nematode *Panagrolaimus superbus*. *Cell Stress Chaperones* **2015**, *20*, 121–137. [CrossRef] [PubMed]
53. Richarme, G.; Marguet, E.; Forterre, P.; Ishino, S.; Ishino, Y. DJ-1 family Maillard deglycases prevent acrylamide formation. *Biochem. Biophys. Res. Commun.* **2016**, *478*, 1111–1116. [CrossRef] [PubMed]
54. Mihoub, M.; Abdallah, J.; Gontero, B.; Dairou, J.; Richarme, G. The DJ-1 superfamily member Hsp31 repairs proteins from glycation by methylglyoxal and glyoxal. *Biochem. Biophys. Res. Commun.* **2015**, *463*, 1305–1310. [CrossRef] [PubMed]
55. Abdallah, J.; Mihoub, M.; Gautier, V.; Richarme, G. The DJ-1 superfamily members YhbO and YajL from *Escherichia coli* repair proteins from glycation by methylglyoxal and glyoxal. *Biochem. Biophys. Res. Commun.* **2016**, *470*, 282–286. [CrossRef] [PubMed]
56. Richarme, G.; Dairou, J. Parkinsonism-associated protein DJ-1 is a bona fide deglycase. *Biochem. Biophys. Res. Commun.* **2017**, *483*, 387–391. [CrossRef] [PubMed]
57. Arsovski, A.A.; Pradinuk, J.; Guo, X.Q.; Wang, S.; Adams, K.L. Evolution of Cis-Regulatory Elements and Regulatory Networks in Duplicated Genes of Arabidopsis. *Plant Physiol.* **2015**, *169*, 2982–2991. [CrossRef] [PubMed]
58. Hanada, K.; Zou, C.; Lehti-Shiu, M.D.; Shinozaki, K.; Shiu, S.H. Importance of lineage-specific expansion of plant tandem duplicates in the adaptive response to environmental stimuli. *Plant Physiol.* **2008**, *148*, 993–1003. [CrossRef]
59. Rensing, S.A. Gene duplication as a driver of plant morphogenetic evolution. *Curr. Opin. Plant Biol.* **2014**, *17*, 43–48. [CrossRef]
60. Islam, T.; Ghosh, A. Genome-wide dissection and expression profiling of unique glyoxalase III genes in soybean reveal the differential pattern of transcriptional regulation. *Sci. Rep.* **2018**, *8*, 4848. [CrossRef]
61. Ghosh, A.; Kushwaha, H.R.; Hasan, M.R.; Pareek, A.; Sopory, S.K.; Singla-Pareek, S.L. Presence of unique glyoxalase III proteins in plants indicates the existence of shorter route for methylglyoxal detoxification. *Sci. Rep.* **2016**, *6*, 18358. [CrossRef] [PubMed]
62. Liu, Y.; Beyer, A.; Aebersold, R. On the Dependency of Cellular Protein Levels on mRNA Abundance. *Cell* **2016**, *165*, 535–550. [CrossRef] [PubMed]

11

Regulation of Ascorbate-Glutathione Pathway in Mitigating Oxidative Damage in Plants under Abiotic Stress

Mirza Hasanuzzaman [1,*], M. H. M. Borhannuddin Bhuyan [2,3], Taufika Islam Anee [1], Khursheda Parvin [2,4], Kamrun Nahar [5], Jubayer Al Mahmud [6] and Masayuki Fujita [2,*]

1. Department of Agronomy, Faculty of Agriculture, Sher-e-Bangla Agricultural University, Dhaka 1207, Bangladesh
2. Laboratory of Plant Stress Responses, Department of Applied Biological Science, Faculty of Agriculture, Kagawa University, Miki-cho, Kita-gun, Kagawa 761-0795, Japan
3. Citrus Research Station, Bangladesh Agricultural Research Institute, Jaintapur, Sylhet 3156, Bangladesh
4. Department of Horticulture, Faculty of Agriculture, Sher-e-Bangla Agricultural University, Dhaka 1207, Bangladesh
5. Department of Agricultural Botany, Faculty of Agriculture, Sher-e-Bangla Agricultural University, Dhaka 1207, Bangladesh
6. Department of Agroforestry and Environmental Science, Faculty of Agriculture, Sher-e-Bangla Agricultural University, Dhaka 1207, Bangladesh

* Correspondence: mhzsauag@yahoo.com (M.H.); fujita@ag.kagawa-u.ac.jp (M.F.)

Abstract: Reactive oxygen species (ROS) generation is a usual phenomenon in a plant both under a normal and stressed condition. However, under unfavorable or adverse conditions, ROS production exceeds the capacity of the antioxidant defense system. Both non-enzymatic and enzymatic components of the antioxidant defense system either detoxify or scavenge ROS and mitigate their deleterious effects. The Ascorbate-Glutathione (AsA-GSH) pathway, also known as Asada–Halliwell pathway comprises of AsA, GSH, and four enzymes viz. ascorbate peroxidase, monodehydroascorbate reductase, dehydroascorbate reductase, and glutathione reductase, play a vital role in detoxifying ROS. Apart from ROS detoxification, they also interact with other defense systems in plants and protect the plants from various abiotic stress-induced damages. Several plant studies revealed that the upregulation or overexpression of AsA-GSH pathway enzymes and the enhancement of the AsA and GSH levels conferred plants better tolerance to abiotic stresses by reducing the ROS. In this review, we summarize the recent progress of the research on AsA-GSH pathway in terms of oxidative stress tolerance in plants. We also focus on the defense mechanisms as well as molecular interactions.

Keywords: antioxidant defense; free radicals; glyoxalase system; hydrogen peroxide; plant abiotic stress; reactive oxygen species; redox biology; stress signaling

---

1. Introduction

With the advancement of lifestyle the natural resources are being exploited and the interruption of natural environment is increasing the extremity of various kinds of abiotic stress, including salt stress, drought stress, waterlogging, temperature extremes, including high and low, excess and low light intensity, radiation stress, ozone, metal and metalloid toxicity, and other organic or inorganic pollutants. Environmental extremity narrows ways to increase plant productivity. Ever-increasing population demands newly cultivable areas, including the adverse land areas, even the higher crop production in per unit area.

Any abiotic stress impaired stomatal function, photosystem activity, Calvin cycle, or photosynthetic enzyme activities, as well as altered electron transport chain reactions. Moreover, unfavorable peroxisomal or cytosolic atmosphere led to overwhelm of electron absorption and generate ROS as a common outcome and subsequently causes oxidative damage [1,2]. If the challenges of plant scientists are increasing productivity against the abiotic stresses, their concentrations are moving to the depth for breaking the obstacles at the cellular or organelles levels, where abiotic stresses impose common types of barrier to hinder their function. Reactive oxygen species is an inescapable outcome of aerobic reactions, which are partly reduced or activated by the appearance of oxygen. Reactive oxygen species is a combined name that indicates different, highly active components. Superoxide (O_2^-), hydroxyl ($OH^•$), and peroxyl ($ROO^•$) are some examples of oxygen radicals. Hydrogen peroxide (H_2O_2), singlet oxygen ($^1O_2$), and ozone (O_3) are the non-radical types of ROS [3]. Reactive oxygen species are important for plants. They have dual role in plants: a small amount of those acts as a signal for inducing abiotic stress responses towards adaptation process, while the excess generation of those causes oxidative damage. However, in severe cases, oxidative damages to membranes (lipid peroxidation), proteins, nucleic acid, including RNA and DNA, and even directs to the oxidative obliteration of the cell [4]. Chloroplast, mitochondrion, membranes of the cell or its ultrastructural organelles, apoplast, and nucleolus are the locations of ROS production. Nonetheless, peroxisome is also considered as a powerful source of ROS since the electron transport chain (ETC) and photochemical reactions are the majority of the processes generating ROS [5–7].

Plants have an antioxidant defense system having non-enzymatic and enzymatic antioxidants in cellular organelles, which scavenges different ROS up to a certain level. If the ROS generation is higher than the scavenging ability of the antioxidant system, then oxidative damage occurs. Antioxidant defense system comprises ascorbate (AsA), glutathione (GSH), carotenoids, tocopherols, flavonoids, etc., which are some commonly known non-enzymatic antioxidants [5]. Ascorbate peroxidase (APX), monodehydroascorbate reductase (MDHAR), dehydroascorbate reductase (DHAR), glutathione reductase (GR), superoxide dismutase (SOD), catalase (CAT), glutathione peroxidase (GPX), glutathione *S*-transferase (GST), and peroxiredoxin (PRX) are well known enzymatic antioxidant components [8,9]. Among all of these, AsA, GSH, APX, MDHAR, DHAR, and GR comprise the AsA-GSH cycle.

Ascorbate is one of the most powerful substrates for scavenging H_2O_2. Ascorbate maintains the reduced state of α-tocopherol. Ascorbate is supposed to be concerned in zeaxanthin biosynthesis dissipating excess light energy in the thylakoid membranes of chloroplast and prevents oxidative stress. Ascorbate sustains reduce the state of prosthetic metal ions and maintain the activity of antioxidant enzymes [6]. Glutathione regulates various metabolic functions; it acts as an antioxidant. Glutathione peroxidase and GST utilize GSH as substrate; GPX is responsible for ROS detoxification, whereas GST is liable for xenobiotic detoxification [1]. The glyoxalase system consisting of glyoxalase I (Gly I) and glyoxalase II (Gly II) enzymes detoxifies cytotoxic and oxidative stress creator methylglyoxal (MG), where Gly I uses GSH and after finishing MG detoxification, GSH is recycled [2]. The positive role of AsA-GSH cycle components has been documented in many plants that are affected by abiotic stresses [1,2]. Participation of the GSH/glutathione disulfide (GSSG, the oxidized form of GSH) redox in maintaining a favorable cellular environment and in stress signal and adaptation were discussed in some previous findings. Glutathione participates in signal transduction, the proper pathway, of which remains unrevealed. The presence of AsA and GSH has been reported to improve osmoregulation, plant water status and nutrient status, water use efficiency, photosynthetic performance, and the overall productivity of plants. Exogenous AsA and GSH applications have been reported to enhance the antioxidant defense as well as the overall tolerance of plants against abiotic stresses. Accordingly, the enzymatic antioxidants of AsA-GSH cycle participates in scavenging ROS, whereas AsA and GSH not only directly scavenge a range of ROS but also perform many other functions to maintain a favorable state in cytosol and other cellular organelles to enhance antioxidant capacity and to reduce oxidative stress, which is induced by different abiotic stresses; AsA and GSH also improve the physiological performance of plants. Since the discovery of the AsA-GSH cycle, its most discussed topics are related to antioxidative protection. However, in this aspect, various other factors should be revealed like physiological factors/processes involved in generating oxidative stress, role of AsA-GSH cycle components in regulating those physiological processes and ultimately the oxidative stress. Considering the multiple vital roles of AsA-GSH cycle in mitigating oxidative stress, this review accommodates presently available and updates of research findings and perspectives.

2. Ascorbate-Glutathione Pathway—An Overview

Ascorbate-Glutathione pathway (also called as Asada–Halliwell pathway) is the major pathway of antioxidant defense, which mainly detoxify the H_2O_2 in a plant cell. Apart from AsA and GSH, its enzymes—APX, MDHAR, DHAR, and GR [6]—have significant roles. Both AsA and GSH are found in the cytosol, nucleus, chloroplast, mitochondria, and peroxisome, where they operate the functions assisted by four enzymes and, therefore, each enzyme has several isoforms that are based on the cellular localization [10]. Both AsA and GSH are present in cellular organelles in a millimolar range, for instance, in *Arabidopsis thaliana*, AsA concentration is the highest (22.8 mM) in the peroxisome, where GSH is highest (14.9 mM) in mitochondria [11,12]. AsA and GSH both have high redox potentials and, therefore, interact with many components and pathways towards the maintenance of a generally reduced state. There are few steps, by which AsA and GSH work coordinately to detoxify H_2O_2, and at the same time, both AsA and GSH are regenerated. First, the enzyme APX converts H_2O_2 into the water with the help of AsA as an electron donor, which is also converted into monodehydroascorbate (MDHA). This MDHA again regenerates AsA by the activity of MDHAR and a part of this is spontaneously converted into dehydroascorbate (DHA). Later, DHA is reduced to AsA again by using GSH, which results in its oxidation to produce GSSG. Finally, this GSSG regenerates GSH by the activity of GR using NADPH as the electron donor (Figure 1). Both AsA and GSH are strong antioxidants, but the maintenance of their redox state is important in conferring stress tolerance in plants, which largely depends on the activities of the four enzymes that are associated with the AsA-GSH cycle [6,13]. In the next sections, we have described all of the components of the AsA-GSH pathway.

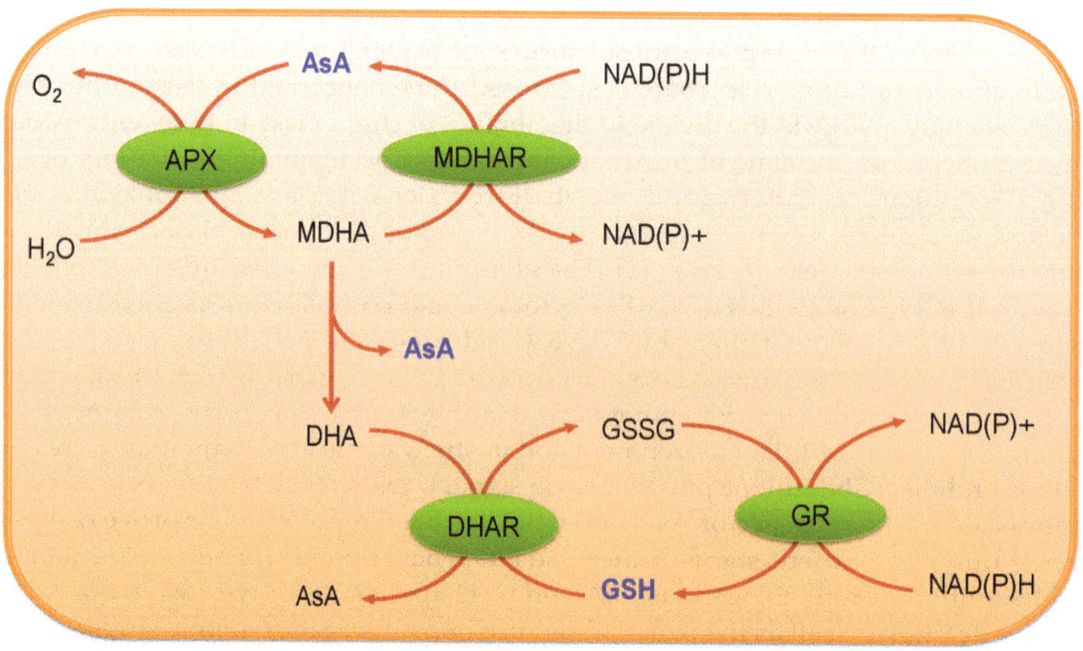

Figure 1. Ascorbate-Glutathione (AsA-GSH) (Ascorbate-Glutathione) pathway [ascorbate, AsA; ascorbate peroxidase, APX; monodehydroascorbate, MDHA; monodehydroascorbate reductase, MDHAR; dehydroascorbate, DHA; dehydroascorbate reductase, DHAR; glutathione, GSH; oxidized glutathione, GSSG; glutathione reductase, GR; Nicotinamide adenine dinucleotide phosphate (reduced form), NAD(P)H; Nicotinamide adenine dinucleotide phosphate (oxidized form), $NAD(P)^+$].

3. Components of AsA-GSH Pathway

3.1. Ascorbate

All living organisms either make AsA (also known as Vitamin C) or get it in their foodstuffs. Naturally abundant L-AsA is of a simplest chemical structure and is related to C6 sugars. It is a hexonic acid aldono-1,4-lactone (either L-galacturonic or L-gulonic acid), having an enediol group at C2 and C3 [14]. The enediol group enables L-AsA for donating one or two electrons to form an initial oxidized intermediate (MDHA) and further to an oxidized (DHA) form. The C5 and C6–OH group serves to provide alcoholic nature. They can react with produced acetals, ketals by reacting with aldehydes and ketones, respectively. Having two asymmetric C, L-AsA illustratesa positive optical rotation, which is unaltered by the acidicpH of solution but greatly affected by alkaline pH, which increases over +160° in 2N NaOH solution [14].

In solid-state, L-AsA is stable but oxidizes readily in solution, in particular in the presence of Cu, Fe, or alkali to form DHA. Afterward, two MDHA can undergo spontaneous reaction to rejuvenate one molecule of L-AsA and one molecule of DHA [15].

Ascorbate biosynthesis system is one of the ancient pathways and formed in very primitive life process on this planet. In plant tissue, AsA can be synthesized from several biochemical pathways. D-glucose is the primary substrate for producing AsA, and in this pathway, a set of ten reactions occurred (Figure 2). Ascorbate can be formed via four pathways viz. L-galactose, L-gulose, D-galacturonic acid, and myo-inositol pathway [16–18]. The biosynthesis of AsA is lineal with the cell wall formation. After the initial reactions, the D-galacturonic acid and L-galactose pathways both yielded L-galactono-1,4-lactone. Besides, in L-gulose and myo-inositol pathway, L-gulonic acid is produced, which is further hydrolyzes to form L-gulono-1,4-lactone, are catalyzed for the synthesizing of AsA in the mitochondria (Figure 2) [19].

In an organism, AsA metabolism comprises of biosynthesis (catabolism) and degradation (anabolism), and the balance between catabolism and anabolism determines the intracellular concentration of AsA. In the previous paragraph, we briefly discussed the biosynthesis of AsA. Hence, we will discuss the degradation and turnover of AsA in this paragraph. In some plants, the AsA turnover rate is relatively very high [20]. On the other hand, AsA is neither stable nor is restricted to oxido–reduction, which changes the equilibrium of AsA and DHA in plant tissue. The AsA pool undergoes turnover in plants. As AsA has prominent responsibility in the redox function metabolism, therefore the recycling of AsA from MDHA (catalyzed by MDHAR using NADPH) and DHA (DHAR using GSH) is the necessity to keep the redox balance as well as the higher total AsA pool [21]; and the functioning of the water-water cycle to optimize photosynthesis [22]. If the oxidized forms are not recovered, they will undergo further degradation to form oxalic or L-tartaric acids (Figure 2) [23]. In a plant cell, AsA act as a multifunctional biosynthetic precursor. While using radioactive ^{14}C AsA, some studies tried to understand the degradation of AsA and DHA, but the mechanism is still not fully understood (Figure 2) [24]. However, it is well understood that the cleavage between C2 and C3 results in oxalate formation, whereas the cleavage between C3 to C6 produces L-threonate, via L-idonate [25,26]. Furthermore, DHA can be hydrolyzed into 2,3-L-diketogulonate, being further oxidized to unknown intermediate (Figure 2) and catalyzed by ascorbate oxidase (AO). Sometimes, this intermediate produces toxic H_2O_2 non-enzymatically and it may inhibit peroxidase [24,27].

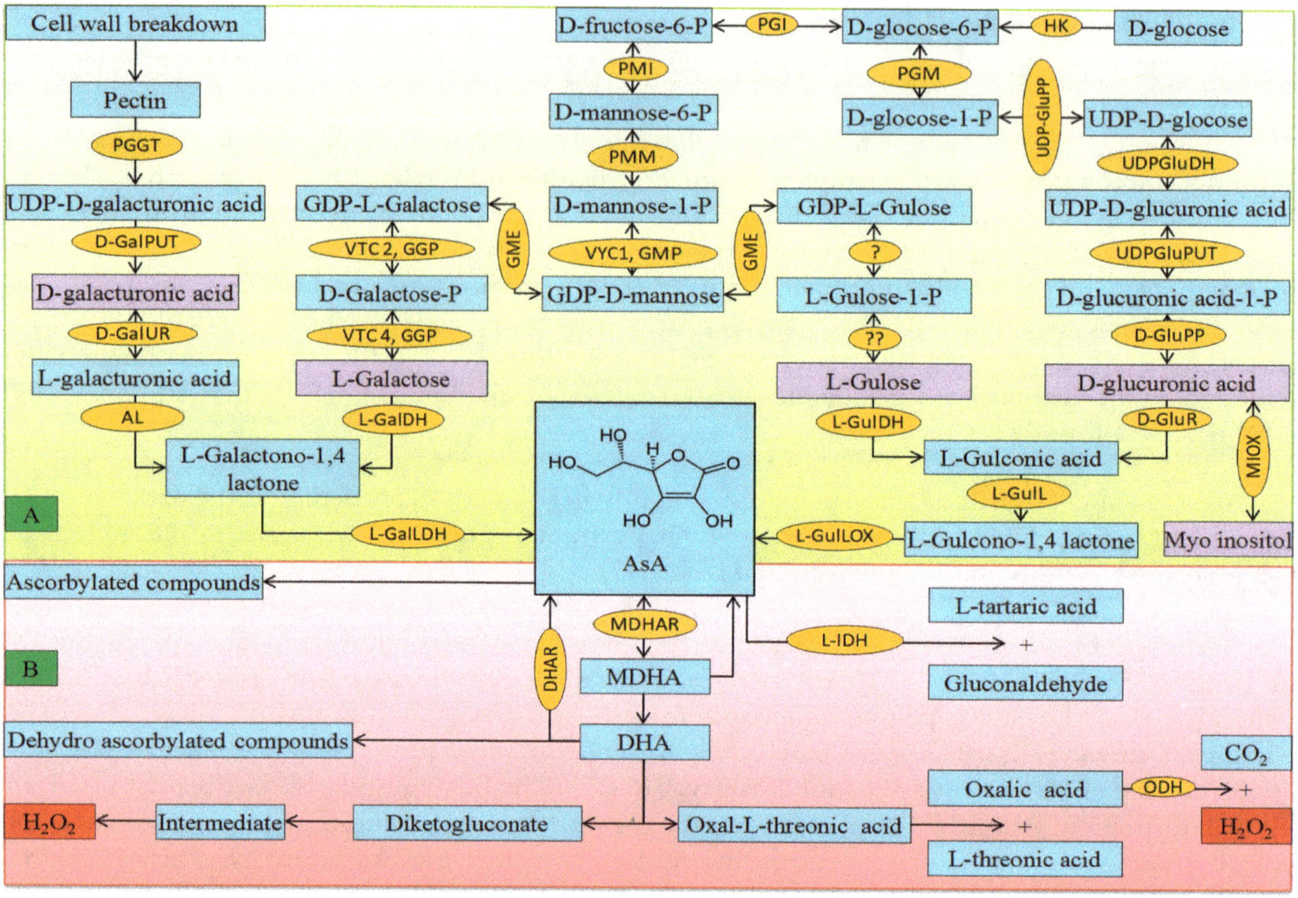

Figure 2. Ascorbate biosynthesis and metabolism is a complex set of reactions, some involving unidentified enzymes; some of the products are reactive and potentially damaging carbonyl compounds, (**A**) biosynthetic pathway; and, (**B**) regeneration and degradation pathways in plants. The metabolites in the violate box represent the name of each biosynthetic pathway. The elaborated name of enzymes are as follows (HK: Hexokinase; PGI: glucose-6-phosphate isomerase; PMI: mannose-6-phosphate isomeras; PMM: phosphomannomutase; TC1 or GMP: GDP-D-mannose pyrophosphorylase/mannose-1-phosphate guanylyltransferase; VTC2 or GGP: GDP-D-mannose 3′,5′-epimerase, GME: GDP-L-galactose phosphorylase;VTC4 or GPP:L-galactose-1-phosphate phosphatase; GalDH: L-galactose dehydrogenase; L-GalLDH: L-galactono-1,4-lactone dehydrogenase; ?: nucleotide pyrophosphatase or sugar-1-phosphate guanyltransferase; ??: sugar phosphatase; L-GulDH: L-gulose dehydrogenase; L-GulL: L-gulonolactonase; L-GulLOX: L-gulono-1,4-lactone oxidase; PPGT: polygalacturonate 4-alpha-galacturonosyltransferase; D-GalPUT: D-galacturonate-1-phosphate uridyltransferase; D-GalUR: D-galacturonate reductase; AL: aldonolactonase; PGM: phosphoglucomutase; UDPGluPP: UDP-glucose-pyrophosphorylase; UDP-GluDH: UDP-glucose dehydrogenase; UDP-GluPUT: glucuronate-1-phosphate uridylyltransferase; D-GluPP: D-glucurono-1-phosphate phosphatase; MIOX: myo-inositol oxygenase;D-GluR: D-glucuronate reductase; MDHAR: monodehydroascorbate reductase; DHAR: dehydroascorbate reductase;L-IDH: L-Idodonate dehydrogenase).

3.2. Glutathione

Glutathione is an omnipresent low molecular weight tripeptide (γ_{-L}-glutamyl-$_{L}$-cysteinylglycine; γ-Glu-Cys-Gly), which is a strong antioxidant and an essential metabolite with a multifarious role in plants [28,29]. It was first discovered from yeast cells subsequently in many plants and animal tissue. Later on, in 1936, it was found as the reducing agent present in the plant tissue [30].

Although GSH is composed of glutamine (Glu), cysteine (Cys), and glycine (Gly), three essential amino acids; but some plant may contain homologues of GSH, where Gly is replaced by other amino acids [31]. In plants, reduced GSH accounts for >98% of total GSH [10]. Generally, cells possess three major reservoirs of GSH cytosol (80–85%), mitochondria (10–15%), and endoplasmic reticulum [32]. The thiol group is specific to γ-glutamyltranspeptidase (GGT) and it allows GSH a higher degree of stability [32,33]. Nevertheless, GSH plays a vital role, including antioxidant defense, xenobiotics detoxification, cell cycle regulation, and apoptosis, reserving cysteine, maintaining redox balance as well as immunity modulation and fibrogenesis [10,29].

In plants, GSH biosynthesis involves two enzymatic steps, which require ATP and the constituent amino acids (Figure 3). In the earliest stratum, γ-glutamylcysteine (γ-EC) is produced by γ-glutamylcyteine synthetase (γ-ECS, EC 6.3.2.2) catalysis and participation from Glu and Cys. In the next stratum, GSH is synthesized from γ-EC and Gly via bonding from the Cys residue of γ-EC with α-amino group of Gly, catalyzed by GSH synthetase (GSH-S, EC 6.3.2.3, also known as GSH synthase). After synthesis in the cytoplasm, GSH is transported to other cellular organelles [34].

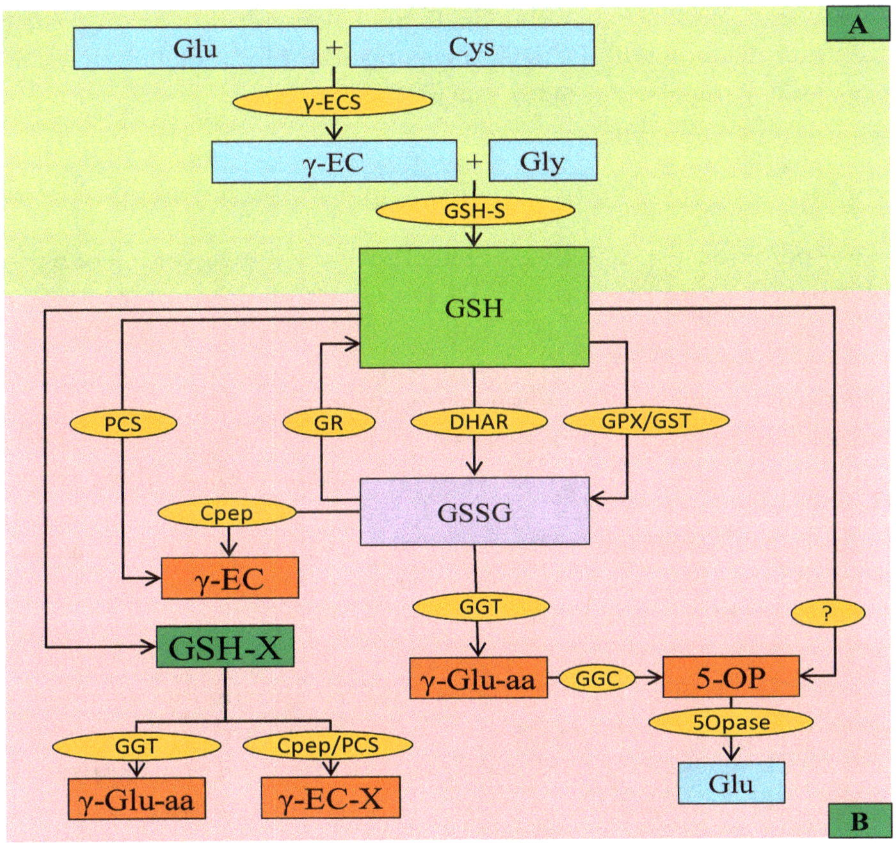

Figure 3. Glutathione biosynthesis, metabolism, and degradation in plants. (**A**) Biosynthesis the first step occurred in plastid: Glu and Cys form γ-glutamylcysteine (γ-EC) catalyzed by γ-EC synthetase (γ-ECS). The second step occurred in the cytosol or in plastid: γ-EC and Gly bond together to form GSH catalyzed by GSH-S (glutathione synthase). Further, GSH participates in ROS scavenging and is converted into GSH/glutathione disulfide (GSSG) by the enzyme glutathione peroxidase (GPX), glutathione *S*-transferase (GST), and DHAR. Further GSSG can be recycled to GSH by the activity of glutathione reductase (GR). (**B**) In the degradation pathway, GSH and S-conjugated compound (GS-X) can be degraded to γ-EC and γ-EC-X by phytochelatin synthase (PCS). While, carboxypeptidase (Cpep) and γ-glutamyl transpeptidase (GGT) both could degrade GS-X to form γ-Glu-aa (aa, amino acid) and γ-EC-X, respectively. Similarly, GSSG is degraded by GGT and Cpep to form γ-Glu-aa and γ-EC, respectively. Further, the produced γ-Glu-aa is converted to 5-oxoproline (5-OP) by γ-glutamyl cyclotransferase (GGC). Besides, GSH is also converted to 5-OP. Although it is thought that this reaction is catalyzed by GGC, still it is unclear. 5-OP is converted to Glu in the next step by the action of 5-oxoprolinase (OPase).

Glutathione is very important for various physiological processes, especially during abiotic stress; it coordinates with AsA turnover and is oxidized to GSSG [35]. Moreover, some other thiol-dependent enzymes, GPX and GST use GSH as co-factor, hence converted to GSSG, which is further reduced back to GSH, with GR catalysis. In higher plants, there are two genes that are reported to encode GRs (*GR1* and *GR2*), where *GR2* is essential for plant development [31,34].

The degradation of GSH is another important phenomenon of GSH metabolism (Figure 3). Up to now, as many as four types of GSH degrading enzymes have been described, which either use GSH or act on GSSG or other GSH-conjugates. Among them, carboxypeptidase activity could degrade GSH itself or GSH-conjugates. Cytosolic PCS is another enzyme that is responsible for the breakdown of GSH-conjugates that are mainly activated during metal/metalloid(s) stress. Another enzyme γ-glutamyl transpeptidase (GGT) acts in GSH transpeptidation or hydrolysis, which is further converted to free Glu by the action of GGC (γ-glutamyl cyclotransferases) and 5-oxoprolinase (5-OPase) (Figure 3) [31,36]. Moreover, another vacuolar GGTs have been reported in *Arabidopsis*, which breaks GSH-conjugates [37]; hence, along with PCS, GGTs is important for metabolizing GSH-conjugates that is formed during secondary metabolites synthesis [37,38].

3.3. Ascorbate Peroxidase

The class I heme-peroxidases; APX (EC 1.11.1.11) occurred in several isoforms in plant cell, viz. cytosolic APX isoform (cAPX), mitochondrial APX isoform (mitAPX), peroxisomal and glyoxysomal APX isoform (mAPX), and chloroplastic APX isoform (chAPX) differed in their substrate specificity, molecular weight, optimal pH ranges between 7 and 8 for maximum activity and stability [39–41]. More importantly, isoforms activity is not stable when AsA is absent. For example, AsA concentrations those are lower than 20 μM greatlyreduced chAPX activity. All of the APX isoforms are heme peroxidase they are inhibited by cyanide and azide. Iron plays a vital role in the APX catalytic site; hence, despite the presence of high AsA concentration, Fe deficiency reduced the activity of cAPX [42]. If the single Cys32 residue near arginine (Arg) 172 residue is altered, APX loses the ability to oxidize AsA to DHA, but it can oxidize other small aromatic molecules. Therefore, the APX properties differ with the guaiacol peroxidases, but they are 33% identical with cytochrome c peroxidase (CCP) [43].

During ROS (H_2O_2) detoxification, APX binds H_2O_2 producing intermediate (I), and heme iron [Fe(V)] is oxidized forming oxyferryl species ($Fe^{4+} = O$). Afterward, APX is regenerated from I, in a two-step reaction withAsA, where the AsA donate an electron and become oxidized. Detailed reactions are shown below (HS = Substrate, S = One electron oxidized form of the substrate).

$$\text{APX} + H_2O_2 \rightarrow \text{Intermediate (I)} + H_2O$$

$$\text{Intermediate I} + \text{HS} \rightarrow \text{Intermediate II} + \text{S}$$

$$\text{Intermediate II} + \text{HS} \rightarrow \text{APX} + \text{S} + H_2O$$

In plant cell, APX scavenges H_2O_2, which mainly participates in AsA-GSH cycle catalyzes the reactions produces MDHA (Figure 4) and, subsequently, MDHA yields DHA [6,10,44].

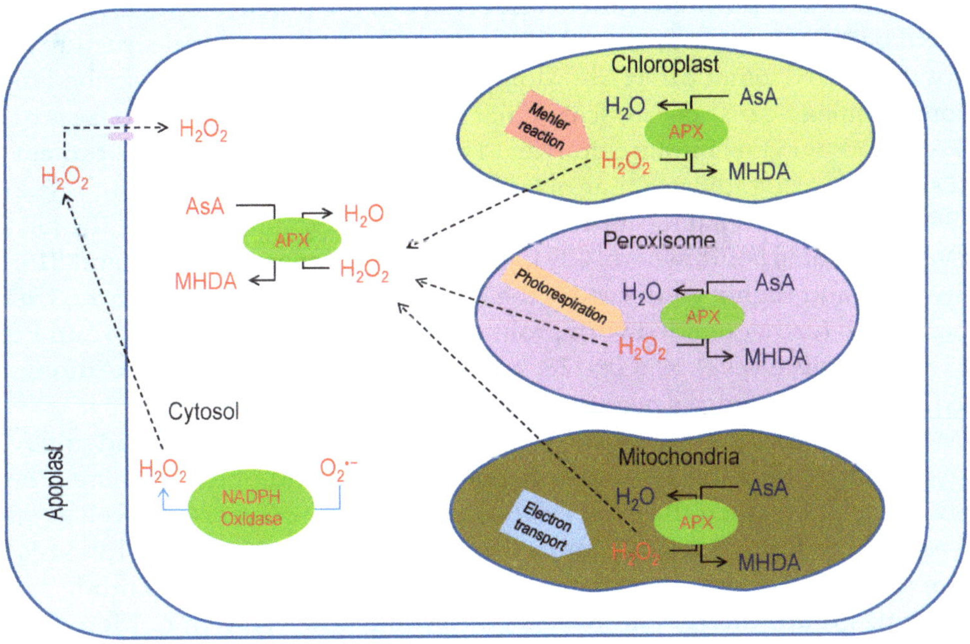

Figure 4. The function of Ascorbate peroxidase (APX) for the abolition of excess reactive oxygen species (ROS) generation in various cellular compartments. Additional details are in the text.

3.4. Monodehydroascorbate Reductase

The MDHAR (EC 1.6.5.4) helps in the revival of AsA [35], having several isoforms that are found in different organelles. Reports suggested that there are three MDHAR genes in tomato, five genes and six isoforms in *Arabidopsis* and rice, and as many as nineteen genes in wheat. In the plant cell, MDHAR activity was detected in different cell organelles, for instance, cytosol, mitochondria, chloroplasts, peroxisomes, and glyoxysomes (Figure 5) [45].

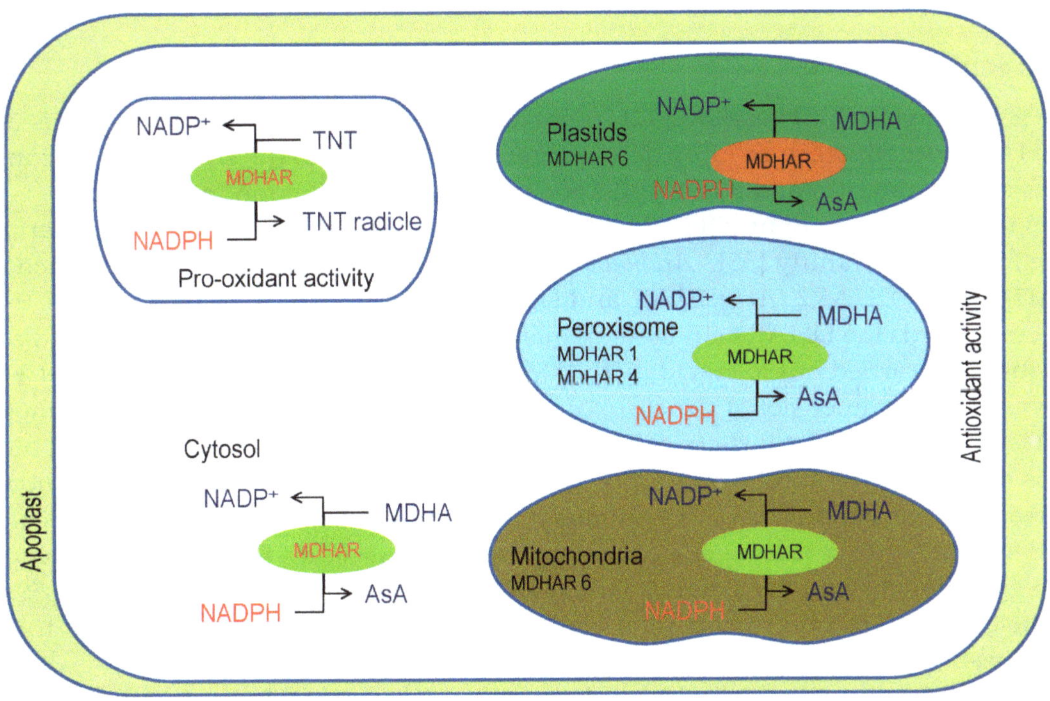

Figure 5. The antioxidant of MDHAR in regenerating AsA to support the removal of reactive oxygen species (ROS) (lower left) contrasts the pro-oxidant role of MDHAR creating 2,4,6-trinitrotoluene (TNT) toxicity.

Although the enzyme is purified from several sources, the detailed structure of this enzyme was published in recent past. Begara-morales et al. [46] coined three-dimensional structure after conducting silico analysis of pea peroxisomal MDHAR. More recently, Park et al. [47] described details MDHAR composition from japonica rice. Those indicated that the structure of MDHAR consists of flavin adenine dinucleotide (FAD) and pyridine binding domain. The structure resembles Fe–S protein reductase [47]. The elucidated structure also indicates that rice MDHAR contains a typical α/β fold, and Arg320 and tyrosine (Tyr) 349 residues are vital for its activity. On the fad-binding domain, the fad is bounded by hydrogen and van der waals bond, where, Gly13, 15 and 297, alanine (Ala) 122 and 319, and threonine (Thr) 123 are involved and highly conservedin the bottom of the crevice. Moreover, Lys53 and Glu178 bridged together, which further bonded with proline (Pro) 49, where both Glu178 and Pro49, are highly conserved, as well as lysine (Lys) 53, Glu178, and FAD bonded each other. Among others, Arg48 compensate FAD phosphate group's negative charge.

In rice, MDHAR α-helices surround β-sheets in the nicotinamide adenine dineucleotide (NAD)-binding domain, where the Tyr174, histidine (His) 315, and phenylalanine (Phe) 348 residues are shifted. Sandwiching between the FAD isoalloxazine ring and Tyr174, as well as steric hindrance moved Tyr174 away. The Phe348 residue shifts outward while His315 comes towards NAD binding site. In addition, the hydrogen bond is formed between Glu178 and nicotinamide ring; Arg202 and ribose and phosphate group, Glu314 and ribose; Glu196 and adenosine ring. Moreover, Glu196 offers rice MDHAR selectivity to NAD; preferring NADH over NADPH [47].

Interestingly, MDHAR can bind substrates other than MDHA, such as isoascorbic acid Evidence showed that phenoxyl radicals, like ferulic acid, quercetin, chlorogenic acid, and coniferyl alcohol, might be reduced by MDHAR [48]. In *Arabidopsis*, MDHAR activity reduces 2,4,6-trinitrotoluene (TNT) and creates its toxicity (Figure 5), but the MDHAR6 mutants are more tolerant, as TNT could not reduce and thus autooxidizes to creates O_2^- [49].

Reports imply that MDHAR response to abiotic stress conditions by reducing MDHA that produces by the excess ROS scavenging, which was observed in many test species (Figure 5) [50,51]. The chlMDHAR is involved in photosynthetic activity during lack of peroxiredoxin [52]. In addition, chlMDAHR activity increased by three- to six-fold during pepper fruit ripening [53].

3.5. Dehydroascorbate Reductase

A major enzyme for GSH assisted DHA recycling is DHAR (EC 1.8.5.1), which is also known by GSH:DHA oxidoreductase or GSH dehydrogenase (AsA) [51,54]. This regeneration process is accomplished at alkaline pH and it is a well known biochemical reaction in plants.

The plant GSH-dependent DHAR is a monomeric enzyme, which is a member of the GSHS-transferase superfamily [55]. *Arabidopsis* possesses three functional DHAR encoding genes, DHAR1 (*At1g19570*), DHAR2 (*At1g75270*), and DHAR3 (*At5g16710*). In recent decades, the attention of researchers towards the DHAR activity in plants for regenerating DHA increased, and a number of investigations were carried out to elucidate the structure and molecular mechanism of DHAR.

The overall three-dimensional structure of DHAR from different plant origin is almost identical, except with some additional short-chain before the α1-helix. The enzyme has several binding sites. The G site is responsible for binding the GSH in the enzyme. The GSH cysteinyl sulfur bonded Cys20 and occupied disulfide bond. The GSH γ-glutamyl is stabilized, via H-bonds H_2O molecule, and then forms the backbone with serine (Ser) 73 and aspartic acid (Asp) 72. The Phe22 is engaged with the γ-glutamyl group by the van der Waals bond, in addition to hydrogen bonds with Lys59. The Val60 stabilizes the central cysteinyl region. The glycinyl group of GSH is loosely bound, forming a salt bridge with Lys47 [54].

The substrate-binding site or DHA binding site or H-site of DHAR enzyme typically exhibits more structural plasticity, but not simultaneously. From the structure of *Pennisetum glaucum* DHAR gene (*PgDHAR1)*, it was observed that Lys8 and Asp19 are responsible for DHA binding [56].

The DHAR catalyzing accomplished by the following three reactions (Figure 6):

$$\text{DHAR-S} + \text{DHA} \rightarrow \text{DHAR-SOH} + \text{AsA (Reaction 1)}$$

$$\text{DHAR-SOH} + \text{GSH} \rightarrow \text{DHAR-S-SG (Reaction 2)}$$

$$\text{DHAR-S-SG} + \text{GSH} \rightarrow \text{DHAR-S} + \text{GSSG (Reaction 3)}$$

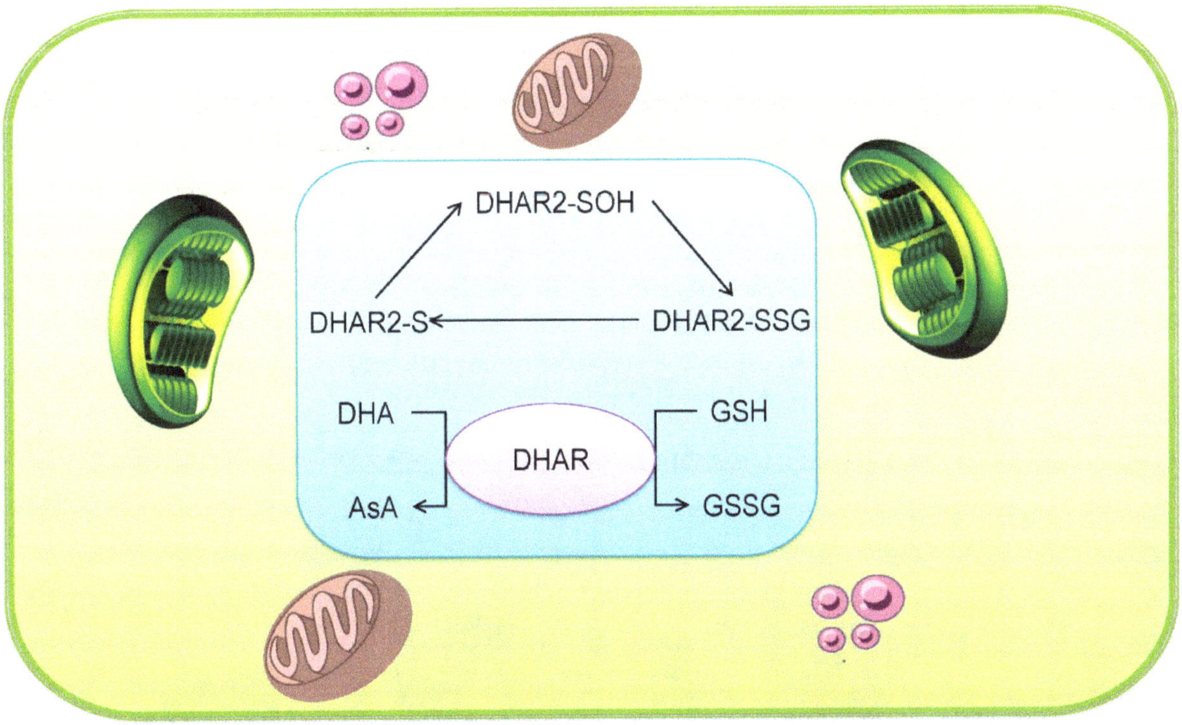

Figure 6. The mechanistic scheme, the ping-pong mechanism for the enzymatic reduction of dehydroascorbate (DHA).

Therefore, the process can be summarized by the following reaction:

$$\text{2GSH} + \text{DHA} \rightarrow \text{GSSG} + \text{AsA (Reaction 4)}$$

As stated earlier, the integral function of DHAR is to reduce DHA to regenerate AsA. During this process, the active site of Cys is oxidized by DHA and further converted to the sulfenic acid. The reaction requires one molecule of H_2O. Knockout mutants of *Arabidopsis DHAR1*, *DHAR2*, and *DHAR3*, did not show any significant differences in total AsA content until facing the abiotic stress condition, which confirmed the necessity of DHAR in reducing the DHA during stress [55].

3.6. Glutathione Reductase

The flavoprotein oxidoreductase GR (glutathione reductase, EC 1.8.1.7) and reduced GSSG to GSH, also known by the term GSR or $NADP^+$ oxidoreductase, as it employs NADPH for its cellular activity. Although GR is stated as a dimer, the monomeric, heterodimeric, and heterotetrameric forms have also been illustrated [29].

No less than two genes that encoded GR were reported, viz. GR1 and GR2 from higher plants. Where GR1 is cytosolic or peroxisomal and shorter, contrary, GR2 comprises a long N-terminal sequence and mitochondrial or chloroplastic [57]. Up to date, several researchers reported GR isoforms in many plants, for instance, tobacco, spinach, etc. [10]. Although being found in above-stated organelles of the cell, the chloroplastic isoforms are accounted for 80% of GR [58]. The enzyme possesses a different quaternary structure that is based on the source from which it was purified [29].

Resembles with flavin-containing enzymes, GR exhibits the Rossmann folds, which is very much conserved and serves as the FAD and NADPH binding domains [59]. There is a controversy regarding the number of domain present in GR. Some reports suggested three, while some suggested four. Some researchers suggested an interface domain in GR protein; therefore, the enzyme has four domains, viz FAD-binding domain, NADPH-binding domain, GSSG-binding domain, and an interface domain [60]. Two Arg residues Arg287 and Arg293 are exclusively necessary for the enzymatic activity of GR [61]. Two Cys residue formed a disulfide bridge, which is redox-active and highly conserved. Serl64 replaces Cys residue in higher plants [62].

The enzyme shows high specificity to substrate binding, although the enzyme reduces GSH conjugates as well as mixed GSSG. Although Plant GR can employ $NADP^{+-}$, its affinity towards NADPH is high [29]. The catalytic mechanism of GR accomplishes in two steps. The first step involves NADPH dependent reduction of the flavin moiety, which is further oxidized, meanwhile the disulfide bridge in active site reduced to form an anion–thiolate and release Cys. In the next step, GSSG molecule binds in the active site forming a disulfide bond together with a Cys and histidine (His) separately of the active site. Afterward, one GSH leaves the His, while another followed it and then leaves the Cys residue leaving the disulfide the bridge in the enzyme active site [63].

The brief reaction catalyzing by GR is as follows (Figure 7):

$$GSSG + NADPH + H^+ = GSH + NADP^+$$

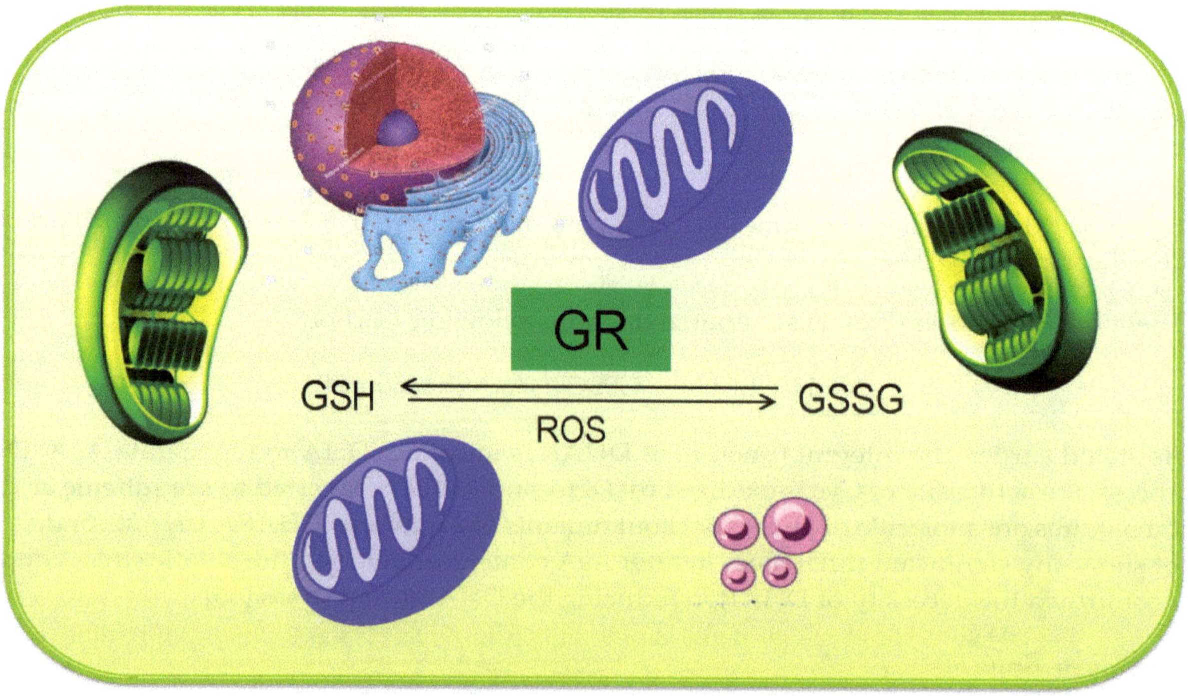

Figure 7. Mechanistic scheme for the enzymatic reduction of glutathione/oxidized glutathione (GSSG) in a plant cell.

During catalysis, pH, and NADPH, and GSSG concentration modulate GR activity. It was reported that low NADPH concentration reduces the GR activity, while below pH 5.5 and over 7.0 is not suitable for GR actions. On the other hand, NADPH-induced GR inhibition was prevented by GSSG [64].

4. Ascorbate and glutathione Redox and its Role in Plant Metabolism

Balanced metabolism is the prerequisite for better productivity in plants, which is always disturbed due to biotic and abiotic stresses. Thus, redox balance is one of the key features of life, by which oxidized products are reduced for further oxidization and energy supply. Moreover, plant cells should counter the oxidation of vital cellular component that occurs continuously due to the presence of 21% atmospheric molecular O_2, which is further complicated due to light-induced overproduction of ROS during photosynthesis. In addition, to keep the electron transport cascades active, simultaneous conversion of electron carriers between reduced and oxidized forms are required. Furthermore, photosynthetic and respiration needs regular electron flux to the electron transport chains from a different site. Therefore, the primary consequence is the generation of $O_2^-$and, subsequently, other ROS, from different enzyme catalysis [10,65]. Although playing a signaling role, over generation of ROS is harmful to cells; thus should be regulated to govern the redox homeostasis [66,67]. For example, AsA, GSH, tocopherols, thioredoxin, glutaredoxin, and peroxiredoxin, and energy metabolism mediators and electron carriers, for example, AsA/DHA, GSH/GSSG, FADH/FAD^+, NADPH/$NADP^+$, and NAD^+/NADH play vital roles in plant cell for maintaining the redox balance and are termed as redox managers [68]. Among the redox managers, there are significant contributions drawn by AsA and GSH, hence in this section; we will discuss their role to keep redox balance as well as maintaining smooth cellular metabolism.

Reports suggested that, under control condition, the AsA/DHA ratio remains >9. Ascorbate becomes oxidized during ROS scavenging, electron donation to photosystem II (PSII), violaxanthin de-epoxidation, and α-tocopherol reductive quenching [69,70]. While the direct reduction of MDHA by ferredoxin at photosystem I (PSI) and by MDHAR, as well as DHA reduction by GSH dependent DHAR activity, maintains a highly reduced state of AsA pool [2,71]. The biosynthesis and metabolism of AsA are discussed earlier in this article (Section 3.1). In the apoplast and vacuoles, AsA pool is an important redox buffer for ROS detoxification, where AsA recycling is mainly accomplished in the cytosol, and AsA/DHA acts as an oxidative stress sensor [72].

The GSH redox potential depends on the GSH concentration and the ratio of GSH/GSSG. In the GSH pool [GSH/GSSG], if the GSSG remains constant, but total GSH decreases, the equilibrium position dropped and redox balance is disrupted. Thus, proper judgment of the GSH/GSSG could give the idea of the redox ratio [73]. Glutathione serves in a multiplicity of metabolic functions; for instance, it participates in the regeneration of AsA from DHA in the chloroplast by DHAR [74]. Moreover, GSH plays a role in reducing glutaredoxins, functions as a precursor of phytochelatin (PC) synthesis for chelating heavy metal, signal transduction, sulfur metabolism, xenobiotics detoxification, and protects protein thiols against irreversible oxidation with disulfide formation or glutathionylation, which inhibited enzymes, like enolase and 6-phosphogluconolactonase [75]. Plant cells have distinct compartmentation of GSH. Although all other cellular compartments, except vacuole, contain GSH/GSSG redox buffer, the only vacuole is the storehouse of GSH where the GSH-conjugates are degraded [76].

As discussed in earlier (Section 2), both AsA and GSH are connected to the reactions network, the AsA-GSH pathway, and the cellular redox homeostasis depends on the pathway components [76]. In this cycle, AsA performs electron donation for APX that works for H_2O_2 detoxification. Due to its high attraction for H_2O_2, APX is capable of efficient ROS scavenging, even in a low concentration, which gives rise to DHA. The produced DHA is further recycled back, which maintains a high ratio of AsA/DHA. If DHA cannot reduce, it might further be irreversibly hydrolyzed, which decreases the ability of AsA redox pool [77]. In the catalysis process GSSG produced, which is further recycled back, which maintains not only a high GSH/GSSG ratio, but also the balance between GSH and AsA pools [10].

In addition, the redox couples of AsA/DHA and GSH/GSSG can also function in another way for accomplishing redox signaling [78]. As discussed earlier, the AsA/DHA couples create redox balance inside cells. Moreover, AO converts AsA to DHA in the apoplast [79], and it creates a redox gradient to connect intra- and extra-cellular atmosphere transverse the plasma membrane. Hence, AsA/DHA redox pair functions in apoplastic and cytoplasmic signals [80]. In contrast, the GSH/GSSG couple plays their functions in balancing intracellular redox potential, which in the intracellular redox signaling [81]. In this regard, the GSH distribution in different cellular organelle is very important for understanding cellular redox situation, for which signaling, as well as cellular metabolism, are smoothly going on [12].

5. Overview of Oxidative Stress and Antioxidant Defense in Plants

The production of ROS in living organisms is a usual cellular metabolism, and it is found in a large number in the internal constituents of the cell-like chloroplast, mitochondria, cytosol, peroxisomes, etc. [82–84].

Each plant cell maintains a dynamic balance between ROS and ROS-scavenging antioxidants. Abiotic stress destroys such cellular balance in favor of oxidative reactions by producing a huge amount of ROS [85]. Insufficient energy indulgence in the photosynthetic process during abiotic stresses reduces molecular oxygen and then produces a large amount of ROS, including H_2O_2, O_2^-, $^1O_2$, $OH^•$, and so on (Figure 8) [10,86]. Reactive oxygen species are extremely reactive molecules and they can damage a large variety of cellular biomolecules, including carbohydrates, nucleic acids, lipids, proteins, etc., and alter their functions [85,87]. In addition, MG, a cytotoxic compound and reactive oxidizer, spontaneously produced in a cell in little amount but under abiotic stresses, its production increased and participated in developing oxidative stress (Figure 8). Similar to ROS, MG production is increased under abiotic stress, which can damage the ultra-structural constituents of cell and cause mutation, and ultimately provokes programmed cell death (PCD) [10].

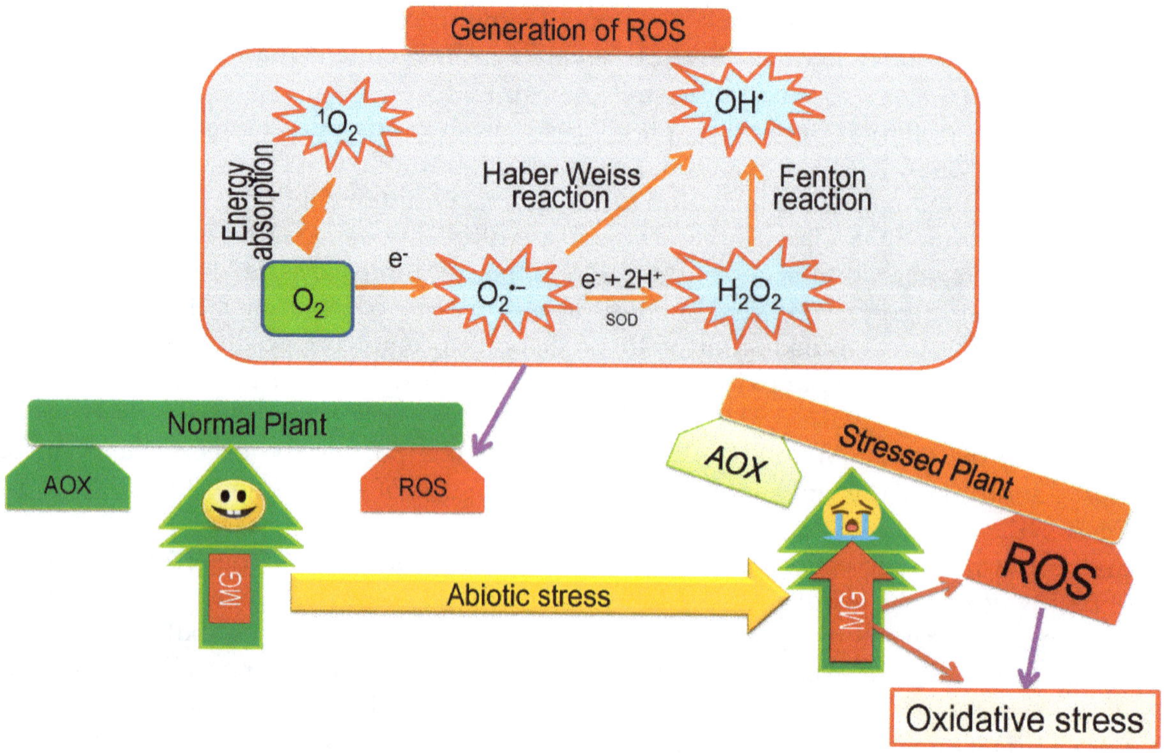

Figure 8. Abiotic stress-induced oxidative stress through the generation of ROS. Additional details are in the text.

Besides causing oxidative stress, ROS and MG play signaling roles for stress tolerance, which controls acclimation and defense responses by modulating some antioxidants and their respective genes [10,86]. The excess generation of ROS and MG is also able to activate interruption in redox homeostasis, which can give the signal for cellular death or shortening plant life cycle [10,82].

Plant cells have well established antioxidant defense and glyoxalase system for scavenging toxic ROS and MG, respectively. The antioxidant defense system consists of some non-enzymatic components (AsA, GSH, alkaloids, α-tocopherol, non-protein amino acids, and phenolic compounds) and enzymatic components (SOD, CAT, APX, MDHAR, DHAR, GR, GST, and GPX [5,28]). Within the antioxidant defense system, the AsA-GSH pool performs the direct and significant role for minimizing stress effect through scavenging of ROS by using key four enzymes, e.g. APX, MDHAR, DHAR, and GR [28,88,89]. In our previous section (Section 3.3, 3.4, 3.5, and 3.6), we elaborately discussed the function of these four enzymes in ROS detoxification. Usually, in the antioxidant defense system, SOD gives frontline protection against ROS by converting O_2^- to H_2O_2. Subsequently, CAT and APX scavenge H_2O_2 to H_2O. Glutathione peroxidase and GST also scavenge H_2O_2 to H_2O with the help of GSH (Figure 9) [28].

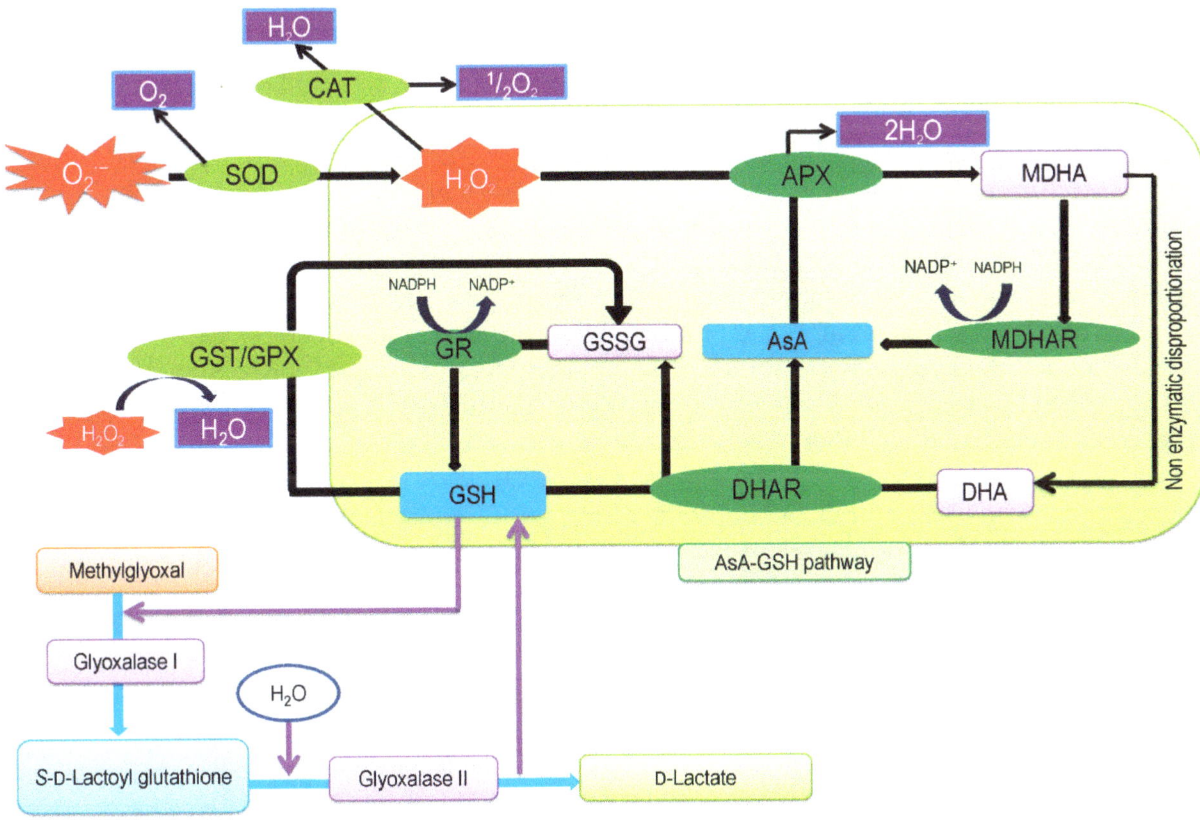

Figure 9. AsA-GSH pathway of the antioxidant defense system and its relation with the glyoxalase system. Additional details are in the text.

Toxic MG is detoxified in the cell by glyoxalase system. Glutathione is not only the major element of AsA-GSH cycle, but it also plays a significant function in the MG detoxification system. Glyoxalase system is composed of two vital enzymes, Gly I and Gly II. In glyoxalase system, MG is detoxified to non-toxic compound in two steps reactions; in the initial step, MG is transformed to *S*-D-lactoyl-glutathione through the utilization of GSH and in the final step *S*-D-lactoyl-glutathione transformed in to D-Lactate, where GSH is recycled back [11]. Moreover, GSH contributes to metal chelation. It enhances the amount of PC under stress condition, which makes a complex with metal and drives into the cell vacuole as inert form [90].

6. Role of AsA-GSH in Regulating Oxidative Stress under Abiotic Stresses

Abiotic stress-induced excess ROS causes oxidative stress in plants followed by cellular damage, even death. Hence, the plant itself defends against this higher ROS accumulation by their defense mechanism. Plant significantly activates the AsA-GSH pathway for ROS detoxification. In this section, we will discuss the involvement of AsA-GSH cycle for alleviating oxidative stress upon various abiotic stresses reviewing recently published articles (Tables 1–3).

6.1. Salinity

One of the most devastating abiotic stress factors—salinity by which cultivable land is becoming barren thus reduces total crop production day by day. Oxidative stress is the most dangerous event under salt inundation is imposed by salinity-induced ionic and osmotic stress [10]. Hence, these ionic and osmotic stress both disturb the photosystem, and thus cause excess ROS, such as $^1O_2$, O_2^-, H_2O_2, and OH. Salinity-persuaded acute ROS accumulations, then bother cellular redox followed by cellular damage counting membrane dysfunction, DNA damage, collapse the enzymatic action, along with distraction of the antioxidant defense system [91,92]. At this point, the plant synthesizes cellular AsA and GSH, which act as non-enzymatic antioxidants by involving their enzymatic components to detoxify ROS up to tolerable levels (Table 1).

However, the enzymes of AsA-GSH pathway showed their differential responses intolerant and sensitive varieties due to saline toxicity. Among salt-tolerant (Pokkali) and sensitive (BRRI dhan29) rice cultivars. Pokkali responded by enhancing the enzymatic activities of the AsA-GSH cycle, where, lowered APX and higher DHAR activity along with unchanged MDHAR and GR activities were found from BRRI dhan29. Rahman et al. [91,93] reported about the well involvement of AsA-GSH cycle in salt-stressed *O. sativa* where ROS generation was extreme. Here, salt exposed rice enhanced the reduced and oxidized GSH content with a lesser amount of AsA by the higher APX, MDHAR, DHAR, and GR activities against overproduced ROS. *Vigna radiata* was grown under the saline condition [94] and where salt-induced oxidative stress was marked with extreme O_2^- and H_2O_2 overgeneration. Salt-stressed *V. radiata* augmented GSH and GSSG contents along with lowered AsA, whereas up-regulated the activity of all enzymatic antioxidants of AsA-GSH cycle and thus responded with elevated ROS [95]. Salt exposed *Lens culinaris* up-stimulated both MDHAR and DHAR activities, which resulted in a lesser amount of AsA and indicated the overproduced H_2O_2 detoxification [96]. Recently, Singh et al. [97] disclosed the incremental activity of enzymatic antioxidants, including APX, DHAR, and GR, with lower AsA, GSH, and GSSG contents, because of salt-induced higher ROS accumulation in *Solanum lycopersicum*. Similarly, 150 mM salt-treated *S. lycopersicum* also decreased AsA content, which might be used in H_2O_2 detoxification, while better GSH showed its role in lowering H_2O_2. Ahmad et al. [98] also observed higher APX, and GR activities, while MDHAR and DHAR activities again reduced as well as supported AsA-GSH mediated ROS regulation. Ahanger et al. [99] reported the same response of *S. lycopersicum* upon saline toxicity. Both activities of APX and GR were enhanced in salt-treated *Triticum aestivum* besides elevated H_2O_2 generation and resulted in higher GSH accumulation [100]. The activity of APX, MDHAR, DHAR, and GR enhanced in salt-stressed *S. lycopersicum* to check the excessive H_2O_2 generation, which resulted in lowered AsA and GSH contents [92].

The changes in AsA-GSH pathway were investigated in salt-stressed *Nitraria tangutorum* by applying a varied level of NaCl (100, 200, 300, and 400 mM) [101]. They noticed a gradual enhancement of AsA, DHA, GSH, and GSSG contents by keeping pace with sequential increment of salt-induced H_2O_2. Here, increased MDHAR and DHAR activities in stressed seedlings also contributed to increasing AsA, and higher DHAR and GR were responsible for better GSH and GSSG contents [92,102]. Talaat et al. corroborated these results with salt-exposed *Phaseolus vulgaris* [103]. Thus, as a part of plant antioxidant defense under salinity, AsA-GSH pathway is very efficient to regulate extra ROS for being tolerant.

6.2. Drought

Drought is another most important abiotic stress, which generates excess ROS accumulation and thus causes variation in the enzymatic activities of AsA-GSH pathway for ROS detoxification. The enzymatic responses of AsA-GSH pathways varied, depending upon plant species, plant age, drought intensity, and duration [10]. Commonly, drought up-regulated the enzymatic antioxidant activities of AsA-GSH pool [10,104]. Plant tolerance to drought stress is categorized based on stress-induced endogenous antioxidants contents along with enzymatic activities (Table 2). *Dendranthema grandiflorum* responded differentially according to their tolerant and sensitive varieties, where tolerant one comparatively displayed better enzyme activity of antioxidants than sensitive [105]. Lou et al. [106] demonstrated how *T. aestivum* responded upon drought exposure. Hence, they noticed that the AsA-GSH cycle responded considerably with excess ROS generation by significant variation of GSH/GSSG and AsA/DHA redox along with the steady increment of H_2O_2. Their team also observed the enzymatic up-stimulation of AsA-GSH pathway to alleviate stress by scavenging excess ROS in *T. aestivum* spike. Thus, *T. aestivum* showed higher participation of AsA with higher APX activity in drought exposure for scavenging extra H_2O_2, as well as higher enzymatic activity to run the AsA-GSH pathway systematically [107].

Table 1. Role of AsA-GSH in regulating oxidative stress under salinity and drought.

Plant Species	Stress Levels	Status of AsA-GSH Component(s)	ROS Regulation	References
Triticum aestivum L.	100 mM NaCl	GSH content increased by 15%; Stimulated APX and GR activities by 78% and 56%, respectively	Increased H_2O_2 content about 79%	[100]
T. aestivum L. cv. BARI Gom-21	12% PEG for 48 and 72 h	Decreased AsA content at 48 h, but after 72 h, AsA content again enhanced; Increased GSH and GSSG content where GSH/GSSG ratio decreased time-dependently; Enhanced the activities of APX, MDHAR, and GR	Enhanced the H_2O_2 content by 62% and increased O_2^- accumulation	[113]
T. aestivum L.	10% PEG	Reduced AsA/DHA and GSH/GSSG redox; Increased enzymatic antioxidants actions of AsA-GSH cycle	Increased H_2O_2 production	[107]
T. aestivum L.	35–40% field capacity (FC) water	Increased GSH/GSSG by 64% while decreased AsA/DHA by 52% respective with a duration of stress; Enhanced APX, MDHAR, DHAR and GR activities	Increased H_2O_2 along with stress duration	[106]
T. aestivum cv. Pradip	150 and 300 mM NaCl	Reduced AsA content upto 52%; Increased reduced and oxidized GSH accumulation by 55% and 18%, respectively with 32% higher GSH/GSSG ratio; Increased APX activity with 29% reduction of GR activity; Slightly increased MDHAR and DHAR activity	Enhanced H_2O_2 generation by 60%	[28]
Oryza sativa L. cv. BRRI dhan47	150 mM NaCl	Increased GSH accumulation while reduced AsA content by 49% Increased GSH content and lowered the redox status of both AsA/DHA and GSH/GSSG; Upregulated the activity of APX, MDHAR, DHAR, and GR	Increased the production of O_2^- with 82% higher H_2O_2 accumulation	[93]

Table 1. *Cont.*

Plant Species	Stress Levels	Status of AsA-GSH Component(s)	ROS Regulation	References
O. sativa L. cv. BRRI dhan49	300 mM NaCl	Reduced AsA and GSH accumulation by 51% and 57%, respectively; Decrease GSH/GSSG redox by 87%; Showed lowered APX (27%), MDHAR (24%), DHAR and GR (25%) activities	Increased H_2O_2 content upto 69%	[114]
O. sativa L. cv. BRRI dhan54	300 mM NaCl	Improved AsA content by 51% with higher GSH content; Decreased GSH/GSSG ratio by 53%; Showed higher APX (27%) and DHAR activities while decreased both GR (23%) and MDHAR activities	Accumulated 63% higher H_2O_2 content	[114]
Brassica napus L. cv. BINA sharisha 3	100 and mM NaCl	Reduced the AsA content by 22%; Increased GSH content by 72% and GSSG content by 88%; Unaltered the GSH/GSSG ratio; Amplified APX activity by 32%, decreased DHAR activity by 17%; Slightly increased GR activity	Accumulated higher H_2O_2 content by 76%	[115]
B. napus L. cv. BINA sharisha 3	200 mM NaCl	Reduced the AsA content (40%) along with increased GSH (43%) and GSSG (136%) contents; Decreased the GSH/GSSG ratio (40%); Amplified the APX activity (39%) and reduced the MDHAR (29%) and DHAR (35%) activities; Improved GR activity (18%)	Showed 90% more H_2O_2 content	[115]
B. napus L.	15% PEG	The AsA accumulation remained unaltered and reduced the AsA/DHA ratio; Enhanced GSH content by 19% and GSSG by 67% and decreased GSH/GSSG ratio; Increased APX, MDHAR, DHAR and GR activities	Higher accumulation of H_2O_2 by 55%	[116]
B. campestris L.	15% PEG	Decreased AsA content by 27% with a decrease of AsA/DHA ratio; Increased GSH content by 33% with higher GSSG content by 79% and lowered GSH/GSSG ratio; Decreased DHAR activity	Higher accumulation of H_2O_2 about 109%	[116]
B. juncea L.	15% PEG	Increased the AsA content and did not affect the AsA/DHA ratio; Increased GSH content by 48% and GSSG by 83% and decreased GSH/GSSG ratio; Increased APX, MDHAR, DHAR and GR activities	Accumulation of 37% higher H_2O_2	[116]
B. juncea L. cv. BARI Sharisha 11	10% PEG	Reduced AsA content (14%) while increased both GSH (32%) and GSSG (48%) contents; Enhanced APX activity (24%); Decreased MDHAR and DHAR (33%) activities along with 31% increased GR activity	Acute generation of H_2O_2 (41%)	[117]
B. juncea L. cv. BARI Sharisha 11	20% PEG	Decreased AsA content by 34% while increased the content of GSH by 25% and GSSG by 101%; Up-regulated APX activity by 33%; Decreased activity of MDHAR and DHAR (30%)	Extreme generation of H_2O_2 by 95%	[117]

Table 1. *Cont.*

Plant Species	Stress Levels	Status of AsA-GSH Component(s)	ROS Regulation	References
B. napus L. cv. BINA Sarisha 3	10% PEG	Increased AsA (21%), GSH (55%) and GSSG contents while decreased GSH/GSSG ratio Unaltered the activities of APX, and increased the activity of MDHAR, DHAR, and GR (26%)	Elevated the H_2O_2 production	[11]
B. napus L. cv. BINA Sarisha 3	20% PEG	Unaltered AsA content along with higher content of GSH (46%) and GSSG and reduced GSH/GSSG ratio; Reduced the APX and MDHAR activities along with the higher activity of DHAR and GR (23%)	Showed higher H_2O_2 production	[11]
B. napus L. cv. BINA sharisha 3	10% PEG	Increased AsA, GSH (31%) and GSSG (83%) accumulation with lowered GSH/GSSG ratio; Increased APX activity while reduced MDHAR and DHAR activities, but GR activity remained unaltered	Increased H_2O_2 content by 53%	[52]
B. napus L. cv. BINA Sharisha 3	20% PEG	Slightly increased AsA content with 26% and 225% increase of GSH and GSSG content, respectively; Reduced GSH/GSSG ratio; Increased APX activity while decreased the activity of MDHAR, DHAR, and GR (30%)	Increased about 93% H_2O_2 content	[52]
B. rapa L. cv. BARI Sharisha-15	20% PEG	Slightly increased AsA content with 72% and 178% increase of GSH and GSSG content, respectively; Reduced GSH/GSSG ratio by 38%; Increased APX, MDHAR, DHAR, and GR activity	Increased about 131% H_2O_2 content	[104]
Cucumis melo L. cv. Yipintianxia No. 208	50 mM of $NaCl{:}Na_2SO_4{:}NaHCO_3{:}Na_2CO_3$ (1:9:9:1 M)	Improved AsA, GSSG and DHA contents; Lowered GSH content; Reduced the ratio of AsA/DHA and GSH/GSSG; Stimulated the activity of APX by 96% and DHAR by 38% while reducing the activity of MDHAR and GR by 48% and 34%, respectively	Increased H_2O_2 accumulation	[118]
Solanum lycopersicum L., var. Lakshmi	0.3 and 0.5 g NaCl kg^{-1} soil	Reduced AsA and AsA/DHA ratio; Lowered GSH and GSSG accumulation with decreased GSH/GSSG redox; Increased APX activity by 28%, DHAR activity by 28% and GR activity by 14%	Enhanced H_2O_2 and O_2^- accumulation	[97]
S. lycopersicum L.cv. Boludo	60 mM NaCl, 30 days	Reduced the activities of APX, DHAR, and GR; Increased MDHAR activity	Higher H_2O_2 generation	[119]
S. lycopersicum L. *var.* Pusa Ruby	150 mM NaCl	Decreased AsA and GSH content with a higher content of DHA and GSSG; Increased APX, MDHAR, DHAr and GR activities	Higher generation of H_2O_2 and O_2^-	[92]
S. lycopersicum L. var. Pusa Rohini	150 mM NaCl	Reduced AsA content by 42%; Increased both GSH and GSSG accumulation; Enhanced the activity of APX and GR by 86% and 29%, respectively with reduction of the activity of MDHAR and DHAR by 38% and 32%, respectively	Accumulated about 3 fold higher H_2O_2 content	[99]

Table 1. *Cont.*

Plant Species	Stress Levels	Status of AsA-GSH Component(s)	ROS Regulation	References
S. lycopersicon L. cv.K-21	150 mM NaCl	Reduced AsA content by 40% with 50% higher GSH content; Lowered GSSG content by 23% while increased GSH/GSSG ratio by 112%; Increased APX (86%) and GR (92%) activity along with the lowered activity of MDHAR (32%) and DHAR (30%)	Elevated H_2O_2 content about 175%	[98]
Nitraria Tangutorum Bobr.	100,200, 300 and 400 mM NaCl	Increased AsA, DHA, GSH and GSSG accumulation decreased their redox status; Enhanced the activity of APX and GR; Unvaried the activity of DHAR and MDHAR but increased DHAR activity only at 300 mM NaCl	Increased O_2^- and H_2O_2 content by 38–98 and 49–102% respectively	[101]
Camellia sinensis (L.) O.Kuntze	300 mM NaCl	Enhanced the AsA and GSH content; Increased APX activity	Elevated H_2O_2 and O_2^- content	[120]
Phaseolus vulgaris L. cv. Nebraska	2.5 and 5.0 dS m^{-1} prepared from a mixture of NaCl, $CaCl_2$, and $MgSO_4$	Increased AsA, GSH, DHA and GSSG accumulations; Enhanced AsA/DHA and GSH/GSSG status; Stimulated the enzymatic activity of APX, MDHAR, DHAR and GR activities	Accumulated higher H_2O_2 content	[103]
Vigna radiate L. cv. BINA moog-1	25% PEG	Reduced AsA content along with higher GSH content of 92%; Increased GSSG content by 236% and reduced GSH/GSSG ratio; Amplified the activity of APX (21%) and GR while reduced MDHAR and DHAR activities	Elevated H_2O_2 content by 114% with higher O_2^- generation	[111]
V. radiata L.	200 mM NaCl	Reduced AsA content; Increased GSSG and GSH accumulation and lowered GSH/GSSG ratio; Amplified the activity of APX, MDHAR, DHAR, and GR	Increased H_2O_2 content by 80% and O_2^- generation by 86%	[95]
V. radiata L. cv. BARI Mung-2	5% PEG	Reduced AsA content where decreased AsA/DHA ratio by 54%; Increased GSSG content; Upregulated the activity of APX and GR (42%) while downregulated the MDHAR (26%) and DHAR activities	Elevated H_2O_2 and O_2^- accumulation	[50]
Lens culinaris Medik cv. BARI Lentil-7	20% PEG	Lowered AsA content with higher total GSH content; Unaltered the APX and GR activities while the increased activity of MDHAR and DHAR (64%)	Accumulated higher H_2O_2 content	[96]
L. culinaris Medik cv. BARI Lentil-7	100 mM NaCl	Reduced AsA content by 87% while increased total GSH content by 260%; Improved the activity of APX, MDHAR, DHAR (286%) and GR (162%)	Increased H_2O_2 content by 15%	[96]
Anacardium occidentale L.	21-day water withdrawal	Enhanced total AsA and GSH content; Increased APX activity	Reduced H_2O_2 generation	[112]
Arabidopsis	12-day water withhold	Showed higher GSH and GSSG accumulation; Reduced GSH/GSSG ratio; Increased GR activity	Increased H_2O_2 accumulation rate	[108]
Cajanus cajan L.	Complete water withholding for 3, 6 and 9 days	Decreased GSH/GSSG ratio; Increased the activity of APX, DHAR, and GR	Higher H_2O_2 content	[109]

Table 1. *Cont.*

Plant Species	Stress Levels	Status of AsA-GSH Component(s)	ROS Regulation	References
Amaranthus tricolor L.cv. VA13	30% FC	Increased AsA and GSH contents by 286% and 98%, respectively; Improved APX, MDHAR, DHAR, and GR activity by 371%, 379%, 375%, and 375%, respectively	No increment of H_2O_2 content	[110]
A. tricolor L.cv. VA15	30% FC	Increased AsA and GSH contents along with higher redox status of AsA/total AsA and GSH/total GSH; Enhanced the activity of APX, MDHAR, DHAR, and GR by 37%, 45%, 40%, and 2%, respectively	Accumulated higher H_2O_2 content by 137%	[110]
C. sinensis (L.) O. Kuntze	20% PEG	Higher contents of both AsA and GSH; Enhanced the APX activity	Higher accumulation of H_2O_2 and O_2^-	[120]

Drought-stressed *A. thaliana* enhanced GSH and GSSG content along with the higher GR activity [108]. Hence, *Arabidopsis* showed the GSH dependent H_2O_2 detoxification to attain tolerance. Higher total AsA was accumulated in *Cajanus cajan* upon complete water restriction conditions for up to nine days to defend against excess H_2O_2 toxicity [109]. Hence, drought enhanced the enzymatic activity of APX, DHAR, and GR for decreasing GSH/GSSG, as well as controlling ROS level.

Similarly, tolerant genotype VA13 of *Amaranthus tricolor* showed comparatively better tolerance under drought stress than sensitive one (VA15) by expressing differential responses of the enzymatic and non-enzymatic ROS detoxification pathways [110]. Hence, VA13 expressed remarkable increment in AsA-GSH redox by accelerating the enzymatic antioxidative actions by which increased non-enzymatic antioxidants (AsA and GSH) accumulation, which are vital for ROS detoxification.

Vigna radiata responded differently regarding different drought intensity [111] to control diverse levels of ROS. Moderate drought imposed by 10% polyethylene glycol (PEG) induced comparatively lowered ROS than severe drought (by 20% PEG). Therefore, severe drought-stressed *Brassica* showed a larger use of AsA-GSH pathways against higher H_2O_2 generation than moderate stress. Here, higher stress caused a higher increase of APX activity along with lowest MDHAR and DHAR activity, while GR activity reduced differently than lower stress exposure to rapeseeds seedlings. Additionally, Hasanuzzaman et al. [52] also observed AsA and GSH both antioxidants contents reduced under severe drought condition, but increased under moderate stress. Bhuiyan et al. [104] found increased AsA content in *B. rapa* under drought (20% PEG). They also observed increased APX activity in drought-stressed seedlings, which assisted in efficiently scavenging the H_2O_2. Another two enzymes related to AsA regeneration MDHAR and DHAR also upregulated, as a result the AsA level was increased and strongly maintained its redox balance during oxidative stress situation. Nahar et al. [111] narrated the function of AsA as ROS detoxifier under drought stress where AsA content reduced in *V. radiata* with the increasing of ROS generation. Here, drought-induced higher APX activity enhanced the oxidation of AsA by scavenging H_2O_2 and improved GR activity increased the supply of GSH for involving ROS detoxification. *Anacardium occidentale* also showed the active participation of AsA-GSH cycle by integrative responses of both non-enzymatic and enzymatic antioxidants for drought-induced excess ROS regulation, where the higher accumulation of AsA and GSH, along with APX activity, coordinately reduced the overproduced H_2O_2 [112]. Thus, the AsA-GSH pathways involve in ROS detoxification as well as ROS homeostasis by eliminating excess ROS for keeping them up to the requirement of functioning cell signals.

6.3. Toxic Metals/Metalloids

Due to fast industrialization of the modern world and unrestrained anthropogenic activities, toxic metals/metalloids stresses have become a gargantuan problem for the plant growth and development [121]. Plants experience toxic metals/metalloids stress try to survive to some extent by using their well established antioxidant defense system. But, the activity and performance of defense system differ with stress concentration, stress duration, plant type, and age of the plant.

The enzymes of AsA-GSH pathway confirmed their differential responses to different toxic metals/metalloids stress (Table 2). Mahmud et al. [122] confirmed that due to Cr stress, the few components of AsA-GSH pathway increased their amount or activity in *B. juncea* L. cv. BARI Sharisha-11. They found five days duration of 0.15 mM and 0.3 mM K_2CrO_4 treatment decreased the content of AsA, but did not change the GSH content. Moreover, activities of APX and GR were enhanced; however, the activities of MDHAR and DHAR were diminished. The higher APX and GR activity might play a function in scavenging excess ROS. A similar upregulation of APX and GR was also recorded in *B. napus* L. cv. BINA sharisha 3 due to Cd treatment [123]. From two separate experiments, they also found Cd stress (0.5 mM and 1.0 mM $CdCl_2$) for 48 h decreased the AsA content, but increased GSH content only under 0.5 mM $CdCl_2$ treatment. Exposure of *Gossypium* to 50 and 100 μM $Pb(NO_3)_2$ for six weeks increased the H_2O_2 content and APX activity [124]. The addition of 150 μM $NiCl_2{\cdot}6H_2O$ in growing media of *B. juncea* L. for one week increased the H_2O_2 content. Moreover, Ni stress decreased the AsA level but augmented the content of GSH and GSSG. Nickel also diminished the function of DHAR and MDHAR, however enhanced APX and GR activity [125]. Similar differential responses of AsA-GSH pathway components were also observed under As [126] and Al [50] toxicity. It can be stated that overproduced ROS plays the signaling role to some extent and inaugurate the higher activity of AsA-GSH enzymes under metals/metalloids toxicity. The upregulation of enzymes plays a significant role in maintaining the redox balance of AsA-GSH pathway under stress condition.

Table 2. Status of AsA-GSH in regulating oxidative stress under metal/metalloid stress.

Plant Species	Stress Levels	Status of AsA-GSH Component(s)	ROS Regulation	References
Brassica napus L. cv. BINA sharisha 3	Cd (0.5 mM and 1.0 mM $CdCl_2$), 48 h	Reduced AsA content by 20% under 0.5 mM and 32% under 1.0 mM $CdCl_2$ treatment; Increased GSH content only under 0.5 mM $CdCl_2$ stress but enhanced level of GSSG by 34% under 0.5 mM and 65% under 1.0 mM $CdCl_2$ treatment; Increased function of APX by 39% and 43% under 0.5 mM and 1.0 mM $CdCl_2$ treatment but MDHAR and DHAR activity were diminished in dose dependant fashion; GR activity increased by 66% due to 0.5 mM $CdCl_2$ treatment but reduced by 24% due to 1.0 mM $CdCl_2$ treatment	Enhanced H_2O_2 content by 37% under 0.5 mM and 60% under 1.0 mM $CdCl_2$ treatment	[88]
Gossypium spp. (genotype MNH 886)	Pb [50 and 100 μM $Pb(NO_3)_2$], 6 weeks	Increased APX activity	Increased H_2O_2 content	[124]
T. aestivum L. cv. Pradip	As (0.25 and 0.5 mM $Na_2HAsO_47H_2O$), 72 h	Reduced AsA content by 14% under 0.25 and 34% underd 0.5 mM $Na_2HAsO_4{\cdot}7H_2O$ treatment; Increased GSH content by 46% and 34%, GSSG content by 50 and 101% under 0.25 and 0.5 mM $Na_2HAsO_4{\cdot}7H_2O$ stress; Enhanced APX function by 39% and 43% but decreased DHAR function by 33% and 30% under 0.25 and 0.5 mM $Na_2HAsO_4{\cdot}7H_2O$ treatment; Increased GR function by 31% under 0.25 mM	Increased H_2O_2 content by 41% under 0.25 and 95% under 0.5 mM $Na_2HAsO_4{\cdot}7H_2O$ treatment	[127]

Table 2. *Cont.*

Plant Species	Stress Levels	Status of AsA-GSH Component(s)	ROS Regulation	References
B. napus L. viz. ZS 758, Zheda619, ZY 50 and Zheda 622	Cr (400 μM), 15 days	Increased GSH and GSSG content; Increased APX activity	Increased H_2O_2 content	[128]
Oryza sativa L. cv. BRRI dhan29	As (0.5 mM and 1 mM Na_2HAsO_4), 5 days	Decreased AsA content by 33 and 51% and increased DHA content by 27% and 40% under 0.5mM and 1mM Na_2HAsO_4 treatment, respectively; Decreased ratio of AsA/DHA; Enhanced GSH content by 48 and 82% under 0.5mM and 1mM Na_2HAsO_4 treatment, respectively; Enhanced GSSG content whereas lessened GSH/GSSG ratio by 25% under 0.5mM and 41% under 1mM Na_2HAsO_4 treatment; Augmented the function of APX, MDHAR, and GR, however, reduced the activity of DHAR	Increased H_2O_2 content by 65% and 89% under 0.5mM and 1mM Na_2HAsO_4 treatment, respectively	[126]
O. sativa L. cv. Disang (tolerant)	100 μM $AlCl_3$, 48 h	Increased AsA content in both roots and shoots; Enhanced the GSH content in shoots; Higher activities of APX, MDHAR, DHAR, and GR,	Elevated the generation of H_2O_2 and O_2^-	[129]
O. sativa L. cv. Joymati (sensitive)	100 μM $AlCl_3$, 48 h	Higher accumulation of AsA in both roots and shoots; Reduced the GSH content in roots while shoots content was unaltered; Increased APX, MDHAR, DHAR activities; Slightly increased GR activities	Higher accumulation of H_2O_2 and O_2^-	[129]
V. radiata L. cv. BARI Mung-2	Cd (mild: 1.0 mM $CdCl_2$, severer: 1.5 mM $CdCl_2$), 48 h	Declined AsA content by 31% due to mild and 41% due to severe stress; Enhanced DHA level and reduced AsA/DHA ratio; GSH content did not change due to mild stress but enhanced owing to stress severity; GSSG level enhanced, and GSH/GSSG ratio decreased in dose-dependent manner; Increased function of APX but lessened MDHAR and DHAR function due to both level of stress; GR activity increased only due to severe stress	H_2O_2 level and O_2^- generation rate was augmented by 73% and 127% due to mild and 69% and 120% due to severe Cd stresses	[130]
V. radiata L. cv. BARI Mung-2	Cd (1.5 mM $CdCl_2$), 48 h	AsA content decreased by 27%, and the ratio of AsA/DHA reduced by 80% whereas DHA content increased considerably; Augmented the function of APX and GR however lessened function of MDHAR and DHAR	Increased H_2O_2 level and O_2^- generation rate	[131]
O. sativa L. cv. BRRI dhan29	Cd (0.25 mM and 0.5 mM $CdCl_2$), 3 days	AsA content and AsA/DHA ratio reduced by 37% and 57% due to 0.25 mM $CdCl_2$ and reduced by 51% and 68% due to 0.5 mM $CdCl_2$, respectively; DHA content increased significantly; GSH content enhanced due to 0.25 mM $CdCl_2$ stress, but reduced due to 0.5 mM $CdCl_2$ stress; GSSG content enhanced by 76% under 0.25 mM and 108% under 0.5 mM $CdCl_2$ stress; Reduced ratio of GSH/GSSG in dose dependant manner; Enhanced APX, MDHAR and GR activity	Enhenced H_2O_2 by 46% under 0.25 mM $CdCl_2$ and 84% under 0.5 mM $CdCl_2$ treatmen whereas O_2^- generation rate increased in dose dependant manner	[132]

Table 2. *Cont.*

O. sativa L. cv. BRRI dhan29	Cd (0.3 mM $CdCl_2$), 3 days	Lessened level of AsA and AsA/DHA ratio but enhanced DHA level; Enhanced the level of GSH and GSSG however lessened GSH/GSSG ratio; Enhanced the action of APX, MDHAR, and GR whereas declined DHAR function	Overproduced ROS (H_2O_2 and O_2^-)	[133]
O. sativa L. Zhunliangyou 608	Cd (5 µM $Cd(NO_3)_2{\cdot}4H_2O$), 6 days	Reduced AsA content; Increased GSH content; Slightly reduced the APX activity	H_2O_2 content increased by 22.73%	[134]
Abelmoschus esculentus L. Moench	Pb (100 mg L^{-1}), 21 days	Increased AsA content	Enhanced H_2O_2 content	[135]
B. juncea L. cv. BARI Sharisha-11	Cr (mild: 0.15 mM K_2CrO_4, severe: 0.3 mM K_2CrO_4), 5 days	AsA content lessened by 19% due to mild and 32% due to severe stress whereas DHA level enhanced by 83% due to mild and 133% due to severe stress as well as AsA/DHA ratio lessened by 47% due to mild and 82% due to severe stress; GSH content did not change considerably but GSSG content enhanced by 42% due to mild and 67% due to severe stress as well as GSH/GSSG ratio lessened by 26% due to mild and 41% due to severe stress; The function of APX enhanced by 21% due to mild and 28% due to severe stress; The activity of MDHAR and DHAR reduced by 25 and 32% under mild and 31 and 50%, under severe stress, respectively; Mild stress increased the activity of GR by 19% while severe stress increased by 16%	H_2O_2 level enhanced by 24% and 46% due to mild and severe stress. Similarly, O_2^- generation rate also raised in a dose-dependent manner	[122]
B. campestris L. cv. BARISharisha 9, *B. napus* L. cv. BARI Sharisha 13 and *B. juncea* L. cv. BARI Sharisha 16	Cd (mild: 0.25 mM $CdCl_2$, severer: 0.5 mM $CdCl_2$), 3 days	Decreased level ofAsA, augmented level of DHA as well as decreased AsA/DHA ratio in all studied cultivars; GSH and GSSG level enhanced, but GSH/GSSG ratio lessened in all studied cultivars; APX and GR activities of all species increased significantly under both levels of Cd toxicity	Enhanced H_2O_2 level and O_2^- production rate in all tested cultivars in a concentration-dependent fashion	[136]
B. juncea L. BARI Sharisha-11	Cd (mild: 0.5 mM $CdCl_2$, severer: 1.0 mM $CdCl_2$), 3 days	Reduced AsA content with higher DHA content and thus decreased AsA/DHA ratio; Increased GSH and GSSG levels as well as declined GSH/GSSG ratio; APX activity increased where GR increased at mild stress but remained unaltered at severe stress; Decreased MDHAR and DHAR activities	Enhanced the H_2O_2 and O_2^-level	[137]
V. radiata L. cv. BARI Mung-2	Al ($AlCl_3$, 0.5 mM), 48 and 72 h	Enhanced DHA content but reduced AsA level and AsA/DHA ratio; Increased level of GSH and GSSG but the diminished ratio of GSH/GSSG; Augmented APX activity but decreased MDHAR and DHAR activity	Enhanced H_2O_2 level by 83% and O_2^- generation rate by 110%	[50]

Table 2. *Cont.*

T. aestivum L. cv. Pradip	Pb [mild: 0.5 mM $Pb(NO_3)_2$, severer: 1.0 mM $Pb(NO_3)_2$], 2 days	AsA decreased in a dose-dependent manner; Mild stress improved the GSH level, but severe stress reduced it; Increased GSSG content; Increased APX activity; Diminished activity of MDHAR and DHAR in a concentration-dependent fashion; Mild stress improved GR activity but severe stress reduced it	Mild stress increased H_2O_2 levels by 41%, but severe stress enhanced it by 95% while O_2^- generation rate also increased in a dose-dependent manner	[35]
B. juncea L. cv. BARI Sharisha-11	Cd (mild: 0.5 mM $CdCl_2$, severer: 1.0 mM $CdCl_2$), 3 days	AsA content decreased by 24% due to mild and 42% due to severe stress whereas DHA level enhanced by 79% due mild and 200% due to severe stress; Decreased AsA/DHA ratio in dose-dependent manner; GSH and GSSG content enhanced by 19% and 44%, respectively, due to mild stress, while only GSSG content enhanced due to severe stress by 72%; The ratio of GSH/GSSG declined by 17% due to mild and 43% due to severe stress; Enhanced APX by 15% due to mild and 24% due to severe stress; The activity of MDHAR and DHAR reduced by 12% and 14% due to mild stress whereas 17% and 24%, due to severe stress, respectively; The activity of GR enhanced under mild stress by 16% and lessened under severe stress by 9%	Level of H_2O_2 enhanced by 43% due to mild and 54% due to severe stress. Augmented O_2^- generation rate in a dose-dependent manner	[138]
B. juncea L. cv. varuna	Ni, (150 µM $NiCl_2.6H_2O$), 1 week	AsA content decreased by 61% whereas GSH and GSSG content increased by 75% and 151%, respectively; Enhanced function of APX by 60% and GR by 72%; DHAR and MDHAR activities were decreased by 62% and 65%, respectively	Increased H_2O_2 by 3.23-fold	[125]
Pisum sativum L. cv. Corne de Bélier	Pb (500 mg $PbCl_2$ kg^{-1}), 28 days	Increased APX and GR activity	Increased H_2O_2 content	[139]
O. sativa L. cv. BRRI dhan54	Ni (0.25 mMand 0.5 mM $NiSO_4{\cdot}7H_2O$)	Diminished content of AsA and enhanced content of DHA as well as the lessened ratio of AsA/DHA by 73% and 92% under 0.25 mM and 0.5 mM $NiSO_4{\cdot}7H_2O$ stress; GSH and GSSG level enhanced in a dose-dependent manner. However, the GSH/GSSG ratio reduced only under 0.5 mM $NiSO_4{\cdot}7H_2O$ treatment; Increased APX, MDHAR, DHAR and GR activity by 70%, 61%, 19% and 37% under 0.25 mM $NiSO_4{\cdot}7H_2O$ and 114%, 115%, 31% and 104% under 0.5 mM $NiSO_4{\cdot}7H_2O$ treatment, respectively	Increased H_2O_2 content by 28% and 35% due to 0.25 mM and 0.5 mM $NiSO_4{\cdot}7H_2O$ treatment	[68]
Capsicum annuum L.cv. Semerkand	Cd (0.1 mM $CdCl_2$), 3 weeks	Enhanced AsA and GSH content	Increased H_2O_2 content	[140]
C. annuum L. cv. Semerkand	Pb (0.1 mM $PbCl_2$), 3 weeks	Enhanced AsA and GSH content	Increased H_2O_2 content	[140]
Zea mays L. cv. Run Nong 35 and Wan Dan 13	Cd (50 mg $3CdSO_4{\cdot}8H_2O$ kg^{-1} soil), 6 months	Decreased GSH content	Increased accumulation of H_2O_2	[141]

6.4. Extreme Temperature

Along with the rise in average global temperature, HT stress has been turned into a topic to be concerned about among environmentalists and researchers worldwide. In general, a 5 °C temperature rise above the optimum temperature of growth is considered to be extreme temperature stress or HT stress or heat shock to any plant species [142,143]. Heat stress causes denaturation of protein and membrane lipids, enzyme inactivation, inhibited protein synthesis, and loss of membrane integrity [144], which results from the disruption of cellular homeostasis through the ROS formed in a mass amount under heat stress [143,145]. Focusing on the role of AsA-GSH pathway to scavenge these ROS, different crop species under different levels of extreme or HT stress have been studied (Table 3).

Khanna-Chopra and Chauhan [146] selected a warmer season to induce HT stress to two different cultivars of wheat (*T. aestivum*), which are Hindi62 (heat-tolerant) and PBW343 (heat-sensitive). They sowed the wheat seeds in mid-January and considered it as heat stress environment, while the control plants were sown in mid-November and considered as the non-stress environment. Data were collected at seven days interval up to 35 days after anthesis (DAA), and the results showed a sharp increase in H_2O_2 content up to 14 days, but then declined. Whereas, MDHAR and DHAR enzymes' activity only increased in Hindi62, but APX and GR activities showed a fluctuating pattern of alteration in both cultivars [146]. Another cereal *Z. mays* when experimented similarly with two different cultivars; LM-11 (heat-sensitive) and CML-32 (heat-tolerant), exposed to 40 °C for 72 h, resulted in higher APX and GR activities in CML-32 roots, while a reduction occurred in the shoot. In LM-11, none of the enzyme activity or AsA content was affected [147]. Higher levels of O_2^- production rate and H_2O_2 content were observed in *Ficus concinna* seedlings under 48 h of HT (35 °C and 40 °C) stress condition, where AsA and GSH contents were unaffected at 35 °C, while declining AsA at 40 °C temperature [148]. The activity of APX, MDHAR, DHAR, and GR enzymes increased at 35 °C, but then again reduced at 40 °C to the level of control plants [148]. Under similar heat stress condition (40 °C, 48 h), *V. radiata* seedlings resulted in decreased GSH content and MDHAR-DHAR activities, but higher APX-GR activities [50]. Kiwi fruit (*Actinidia deliciosa*) seedlings, when exposed to 45 °C in an incubator for 8 h, resulted in higher AsA content and enhanced activity of all the AsA-GSH cycle enzymes [143]. Tomato seedlings were studied in two different aspects: short-term heat shock (40 °C, 9 h) [149] and long-term heat stress (38/28 °C day/night, seven days) [150]. In both experiments, the enhancement of O_2^- generation rate and H_2O_2 content were recorded, but enzyme (APX and GR) activity was only increased at short-term stress condition [149], while the long-term heat exposure reduced all four enzymes activities and GSH content [150]. Similar enzymatic activity was observed in *Nicotiana tabacum* seedlings after seven days of heat (35 °C) stress [151]. From the above discussion, it can be stated that heat stress prevailing for longer duration is less likely to have the capability to modulate AsA-GSH pathway as compared to short-term heat stress.

Table 3. Role of AsA-GSH in regulating oxidative stress under extreme temperature, flooding, and atmospheric pollutant.

Plant Species	Stress Levels	Status of AsA-GSH Component(s)	ROS Mitigation	References
Actinidia deliciosa	45 °C, 8 h	Increased content of AsA; Higher activity of APX, MDHAR, DHAR, and GR	Increased H_2O_2 content	[143]
Zea mays L. cv. Ludan No. 8	46 °C, 16 h	Decreased GSH, and GSSG content, but interestingly GSH/(GSH + GSSG) ratio increased; Reduced GR activity	-	[154]
Cinnamonum camphora	40 °C, 2 days	Reduced AsA content with higher DHA content; Increased GSH and GSSG content; Enhanced the activities of APX, MDHAr, DHAR, and GR	Higher content of H_2O_2 and O_2^-	[166]
S. lycopersicum L. cv. Ailsa Craig	40 °C, 9 h	Higher APX and GR activities by 74% and 45%, respectively	H_2O_2 content increased by 49%	[149]

Table 3. *Cont.*

Plant Species	Stress Levels	Status of AsA-GSH Component(s)	ROS Mitigation	References
S. lycopersicum L.cv. Boludo	35 °C, 30 days	Increased the APX, DHAR and GR activities; Reduced the MDHAR activity	Increased H_2O_2 content	[119]
Vicia faba L. cv. *C5*	42 °C, 48 h	Enhanced the AsA, GSH ans GSSG content significantly; The enzymatic activity of APX and GR also enhanced	Extreme accumulation of O_2^- and H_2O_2	[167]
V.radiata L. cv. BARI Mung-2	40 °C, 48 h	Decreased 64% in AsA/DHA ratio; GSSG pool increased; Higher APX (42%) and GR (50%) activities but declined activities of MDHAR (17%) and DHAR	Higher H_2O_2 content and O_2^-production rate	[50]
Z. mays cv. CML-32 and LM-11	40 °C, 72 h	Increased AsA content in both shoot and root of tolerant (CML-32) one, but unaffected in the susceptible (LM-11) one; Both APX and GR activity increased in roots of CML-32 but reduced in the shoot	Higher H_2O_2 accumulation, especially in shoots	[147]
L. esculentum Mill. cv. Puhong 968	38/28 °C day/night, 7 days	AsA+DHA and DHA increased by 220% and 99% respectively; AsA/DHA ratio decreased by 33%.; Higher GSSG (25%), but reduced GSH content (23.4%) and GSH/GSSG ratio (39%); APX, MDHAR, DHAR and GR activities declined	Enhanced O_2^- generation rate and H_2O_2 content by 129% and 33% respectively	[150]
Nicotiana tabacum cv. BY-2	35 °C, 7 days	Total GSH and AsA contents rose after 7 days heat stress; Increased MDHAR. DHAR and GR activities up to 72 h	The increasing trend of H_2O_2 generation was observed up to 72 h, and then a sharp decline occurred	[151]
Ficus concinna var. subsessilis	35 °C and 40 °C, 48 h	AsA content reduced at 40 °C but GSH content similar to control at both 35 and 40 °C; DHA content enhanced by 49% at 35 °C and by 70% at 40 °C; APX activity increased by 51% and 30% at 35 °C and 40 °C; Activities of MDHAR, DHAR, and GR increased at 35 °C, but GR activity decreased by 34% at 40 °C	At 35 °C, 103% higher H_2O_2 content and 58% higher O_2^-production rate and at 40 °C those were 3.3- and 2.2-fold respectively	[148]
T. aestivum cv. Hindi62 and PBW343	Heat stress environment, Late sown (Mid-January)	Higher activities of MDHAR and DHAR was observed in heat-tolerant (Hindi62) one whereas other enzyme activities seemed mostly to decline with time	The content of H_2O_2 was higher up to 14 DAA compared to non-stressed seedlings	[147]
G. hirsutum cv. Siza	Waterlogged pot for 3 days and 6 days	Increased content of AsA by 20% at 3 days and 30% at 6 days of waterlogging; Lower APX, MDHAR and GR activities	Enhanced O_2^- generation rate by 22 and 53% and H_2O_2 content by 10 and 39% at 3 and 6 days of waterlogging, respectively	[154]
Sesamum indicum L. cv. BARI Til-4	Waterlogged pot by 2 cm standing water on the soil surface for 2, 4, 6 and 8 days	Reduced AsA content upto 38%; Enhanced GSH and GSSG content significantly; Increased APX and MDHAR activities; Reduced DHAR activity upto 59%; GR activity decreased upto 23%	Increased H_2O_2content sharply	[168]
Z. mays cv. Huzum-265 and Huzum-55	Root portions waterlogged for 21 h	Reduced AsA content in both cultivars; Increased APX activity in both cultivars	-	[159]

Table 3. *Cont.*

Plant Species	Stress Levels	Status of AsA-GSH Component(s)	ROS Mitigation	References
Glycine max L.	Waterlogged pot for 14 days	GSH activity declined sharply in roots but shoots unaffected; Reduced GR activity in shoots but roots unaffected	-	[153]
Trifolium repens L. cv. Rivendel and *T. pratense* L. cv. Raya	2 cm standing water on the soil surface for 14 days and 21 days	Increased contents of both oxidized and reduced AsA observed in both genotypes	Higher H_2O_2 generation in both genotypes	[158]
V. radiata L. cvs. T-44 and Pusa Baisakhi; and *V. luteola*	Pot filled with water to 1–2 cm height below the soil level, 8 days	Increased activities of both APX and GR in tolerant genotypes but in susceptible one, activities reduced	Reduced contents of $O_2^-$and H_2O_2 in susceptible (Pusa Baisakhi) cultivar	[156]
O. sativa L. MR219-4, MR219-9 and FR13A	Complete submergence for 4, 8 and 12 days	APX activity declined by 88% in FR13A under 4 days of submergence but decreased about 64 and 83% under 8 and 12 days of submergence; GR activity increased in FR13A and MR219-4 cultivars by 10- and 13-fold respectively after 8 days	-	[152]
Allium fistulosum L. cv. Erhan	Waterlogging (5 cm) at substrate surface for 10 days	Lower APX and GR activities	Increased rate of O_2^- generation by 240.4% and 289.8% higher H_2O_2 content	[157]
C. cajan L. genotypes ICPL 84,023 and ICP 7035	Soil surface waterlogged (1–2 cm) for 6 days	Reduced APX and GR activities in susceptible genotype, which was higher in tolerant one	Lower accumulation of H_2O_2 and rate of O_2^- generation	[155]
S. melongena L. cv. EG117 and EG203	Flooding with a water level of 5 cm, 72 h	Increased AsA content in susceptible EG117 genotype GSH content in both genotypes; Increased APX activity but decreased GR activity	-	[169]
S. lycopersicum cv. ASVEG and L4422	Flooding with a water level of 5 cm, 72 h	Increase in both AsA and GSH contents; Non-significant changes in APX and GR activities	-	[169]
Lolium perenne	Grown in an area with high air pollution	APX and DHAR activities decreased while MDHAR and GR activities increased	A higher concentration of H_2O_2 in pollens	[162]
Populus deltoides × *Populus nigra* cvs. Carpaccio and Robusta	O_3 treatment (120 nmol mol^{-1} for 13 h), 17 days	No impact on AsA and GSH contents; DHAR activity decreased while GR and MDHAR activity increased	-	[164]
Fragaria x *anansa*	High dose of carbon monoxide (CO) nitroxide (NO_x) and sulfur dioxide (SO_2)	The activity of both APX and GR increased upto medium dose but reduced under high dose	H_2O_2 content as well as O_2^- generation rate increased	[165]
O. sativa L. cvs. SY63 and WXJ14	Continuous O_3 exposure for up to 79 days	Both AsA and GSH contents are more likely to decrease; APX, MDHAR, DHAR, and GR activity increased up to 70 days of O_3 exposure	Both O_2^- generation rate and H_2O_2 contents increased	[163]
Prosopis juliflora	Grown in the polluted industrial region	The content of AsA and APX activity increased under polluted environment	-	[161]
Erythrina orientalis	Grown in a polluted industrial area	Increased activities of both APX and GR enzymes recorded	-	[160]
T. aestivum L. cv. BARI Gom-26	Acidic pH (4.5) of growing media	Increased AsA and GSH content; Improved redox balance of GSH/GSSG; Increased activity of APX, MDHAR, DHR, and GR	H_2O_2 contents increased by 209%	[170]

6.5. Flooding

Changes in global climate result in the frequent or unexpected occurrence of heavy rainfall in different regions of the globe, which causes a sudden flood and disrupts the normal ecosystem [2]. Such changes in the ecosystem may cause the extinction of plants species and imbalance in the natural environment [2]. Flooding induced production of ROS and subsequent cellular damage has been authenticated in many studies so far [152–154]. Following are the discussion regarding crop species facing flooding stresses and modulation of their AsA-GSH pathway by flooding stress (Table 3).

Pigeon pea (*C. cajan*) seedlings that are exposed to waterlogged condition for six days revealed that tolerant cultivar could increase APX and GR activities, but a susceptible one cannot [155]. They also observed that, unlike other cases, waterlogging caused a lower accumulation of H_2O_2 and O_2^- [155]. In another experiment with *V. radiata*, Sairam et al. [156] showed that waterlogging similarly reduced the H_2O_2 and O_2^-production rate in susceptible cultivar, while the tolerant ones remained unaffected. However, both APX and GR enzymes' activity increased in tolerant genotypes, while the susceptible one got reduced [156]. The enhanced production rate of $O_2^-$and H_2O_2 content under flooding stress have been reported in cotton [154], Welsh onion [157], and clover [158] plants. Cotton (*G. hirsutum* cv. Siza) plants after three and six days of flood exposure raised the AsA content, but reduced the activity of APX, MDHAR, and GR [154]. A similar reduction in APX and GR enzymes activities was also recorded in Welsh onion (*Allium fistulosum* L.) after 10 days of waterlogging stress [157]. When *Z. mays* seedlings were waterlogged for 21 h at their root portions, they resulted in reduced AsA content and increased APX activity [159]. On the other hand, under long duration (14 days) flooding stress, *Glycine max* L. plants showed a reduction of GSH activity in roots and GR activity in the shoot, but the GSH in shoot and GR in root were not affected [153]. In case of complete submergence of *O. sativa* L. plants for two, four, or eight days, elevated levels GR enzyme activity was recorded, while APX enzyme activity increased only in tolerant cultivar [152]. Accordingly, the discussion reveals that the impact of flooding stress on AsA-GSH pathway varies depending upon the plant species and duration.

6.6. Atmospheric Pollutants

Atmospheric pollutants are the substances that are assembled in the air to a level or magnitude that is dangerous for living beings. Plants that are grown under different levels of atmospheric pollution have shown their oxidative stress responses and AsA-GSH pathway regulation in different manners (Table 3).

Erythrina orientalis plants were grown in three different locations of Philippines: La Mesa (a non-polluted area); and, Makati and Quezon (highly air-polluted cities). The results revealed that plants grown in the non-polluted area had lower activities of APX and GR as compared to the ones grown in highly polluted areas [160]. A similar increase in APX activity along with higher AsA content was recorded in *Prosopis juliflora* plants grown under polluted industrial region [161]. In a recent experiment, Lucas et al. [162] studied *Lolium perenne* plants that were grown under two different areas of Spain, Madrid, and Ciudad Real, where Madrid was considered to be more polluted than Ciudad Real. The findings indicated that the pollens of *L. perenne* accumulated higher concentration of H_2O_2 and in shoots APX and DHAR activity declined, but the activity of MDHAR and GR increased in the shoot of *L. perenne* plants that were grown in Madrid [162]. When rice seedlings were exposed to continuous O_3 treatment, the results showed a remarkable increase in both O_2^- generation rate and H_2O_2 content. In addition, contents of AsA and GSH reduced, while APX, MDHAR, DHAR, and GR activity increased upto 70 days of O_3 exposure in SY63 cultivar and upto 79 days of O_3 exposure in WXJ14 cultivar [163]. Ascorbate and GSH contents were not affected by O_3 exposure in the *Populus* seedlings, but DHAR activity was lower, while the activity of GR and MDHAR was higher after 17 days of O_3 treatment [164]. Young strawberry (*Fragaria* x *anansa*) seedlings were exposed to three different levels of CO, NO_x, and SO_2, which are as follows: CO @ 133, 267, and 533 ppm, NO_x and SO_2 @ 25, 50, and 199 ppm corresponding to low, medium, and high dose, respectively. As a result of exposure to these atmospheric pollutants, H_2O_2 content as well as O_2^- generation rate increased.

However, at low and medium doses of their exposure APX and GR activity increased, while at a high dose that decreased [165]. All sorts of atmospheric pollutants have a remarkable effect on AsA-GSH pathway, but further studies are required to demonstrate that those pollutants completely induced the modification of the AsA-GSH pathway.

6.7. Other Stress

Conklin et al. confirmed the positive role of AsA in protecting plants from ultraviolet (UV) radiation [159], where they found that Vit-C deficient mutant of *A. thaliana* was suffered by stress-induced damages than that of wild type. AsA-deficient mutants also showed sensitivity to O_3 stress due to a lower biosynthesis of AsA [171]. Gao and Zhang [172] reported that vitc1 mutants of *A. thaliana* showed physiological disorders and greater oxidative damages than the wild type, which was due to lower activities of antioxidant enzymes. Mutant plants also showed lower GSH/GSSG and higher DHA/(AsA+DHA) ratio than the wild type. Singh et al. [173] observed a decrease in AsA-GSH cycle enzymes in UV-exposed plants, which in turn affected the plants with oxidative stress. Similar to higher plants, marine macroalga *Ulva fasciata* also showed a positive correlation between enhanced the functions of AsA-GSH cycle and better tolerance of plants to UV radiation [174]. In their study, scavenging of H_2O_2 was regulated by AsA-GSH cycle components, especially APX and GR. Noshi et al. [175] reported that AsA-GSH redox pool provided better protection of *Arabidopsis* from high-light mediated oxidative stress, which was mainly attained due to the higher activities of DHAR. However, both AsA and GSH were found to be responsible for conferring high light (HL) stress [175]. Later, Zheng et al. [176] that susceptibility of *Arabidopsis* mutant was to HL stress were related to the deficiency of AsA and GSH. When AsA deficient *A. thaliana* mutant (vtc2-1) exposed to HL, they generated a high level of H_2O_2 (an oxidative stress marker) than the wild type, which was highly and negatively correlated with the total AsA content. The lack of AsA also resulted in lower chlorophyll (chl) content, chl fluorescence parameters, and PSII photochemistry [176]. Recently, Choudhury et al. [177] studied the metabolomics of *A. thaliana* grown under HL and found that the increased biosynthesis of GSH supports the photochemistry that supports *Arabidopsis* better survival under HL stress.

The pivotal role of the AsA-GSH cycle was also observed in low pH stress also. Bhuyan et al. [170] tested five spring wheat cultivars at different levels of low pH stress. Their observation exhibited that low-pH stress resulted in elevated $O_2^-$and H_2O_2 generation. A decrease in AsA content with increased DHA content was observed, although the APX activity decreased. Increased MDHAR activity was observed, but the ratio of AsA/DHA was not increased. Decreased GSH content and increased GSSG content were found where DHAR and GR activity decreased, resulting in a drop of the GSH/GSSG ratio.

7. Exogenous Use of AsA and GSH in Conferring Abiotic Stress Tolerance

While considering the vital role of both AsA and GSH and their redox researches have been trying to explore the possibilities of using exogenous AsA and GSH in protective plants from abiotic stress. However, the effects are not straightforward due to their species and dose dependency. In the next sections, we provided a summary of the recent results on plant abiotic stress tolerance while applying exogenous AsA and GSH.

7.1. Exogenous AsA

As a non-enzymatic antioxidant, AsA is vital for plant defense mechanism by involving in stress perception and subsequent signaling, and therefore plant responses [178]. Besides its regenerative nature, AsA is also able to donate electrons with and/or without the help of enzymes, and thus significantly detoxifies ROS [179]. Thus, exogenous AsA application is the most prominent for enhancing plant tolerance due to its efficient protection against lipids and proteins oxidation under abiotic stresses [180].

Ascorbate can be exogenously applied as a foliage application, seed treatment, and co-treatment for the alleviation of stress-induced damages [181]. Many researchers reported about the supplemental

AsA-mediated antioxidant defense regulation in various plant species under different stressors, such as salt stress [182], drought [183], extreme temperature [184], ozone [185], and heavy metal stress [186].

Supplemental AsA application effectively lowered the oxidative stress in salt-stressed *Phaseolus vulgaris*, as indicated by the reduction of malondialdehyde (MDA) and ROS accumulation through activating their immune systems related with up-regulation of SOD, CAT and GR activities [187]. The AsA recovered salinity-induced oxidative damage in *Caralluma tuberculata* by lowering the activity of APX, POD, CAT, and GR, which were increased upon saline toxicity [188]. Exogenous AsA-induced plant tolerance, especially on the AsA-GSH pathway, is cultivar dependent [189]. Hence, they used both salt-tolerant *O. sativa* cv. Pokkali and salt-sensitive *O. sativa* cv. Peta to exogenously apply AsA as co-treatment with salinity and found a reduction of H_2O_2 generation in both cultivars. Here, AsA enhanced endogenous AsA and GSH, along with higher SOD, APX, and GR activities in salt-stressed both cultivars in line with lowered ROS and MDA production. However, Pokkali showed more prominent responses of salt tolerance than Peta. Finally, Wang et al. [189] suggested that exogenous AsA differentially increased the salt tolerance mechanism, and thus lessened salt-induced ROS in two rice cultivars. Exogenous AsA enhanced the salinity tolerance of *Z. mays* through protecting oxidative stress with stimulation of plant antioxidant defense [190]. In this study, AsA was used as seed priming against 100 mM NaCl, and AsA restored the salt-induced membrane damage. Hence, external AsA improved the non-enzymatic antioxidants, including Pro, AsA, and GSH accumulation, where SOD and GPX activities increased. Rady and Hemida [190] found lowered CAT activity in AsA treated seedlings under salt stress, which pointed out the AsA-induced decline of H_2O_2 generation.

Plants get relief from drought stress by exogenous AsA application, which was reported by previous researchers [114,191]. Alam et al. [114] studied the AsA induced attenuation of oxidative stress in *B. napus*, *B. campestris*, and *B. juncea* under 15% PEG, indicated by decreasing lipoxygenase (LOX) activity, H_2O_2, and MDA contents. This AsA mediated oxidative stress mitigation was described by AsA caused the strengthening of plant antioxidant defense mechanisms. Hence, exogenous AsA not only responsible for modulating AsA-GSH cycle, but also increased other enzymatic antioxidants activities, such as CAT, GPX, Gly I, and Gly II in all plant species, except GST, which was only increased in *B. napus*. Exogenous AsA mitigated the PEG-induced oxidative stress in *Z. mays* where AsA used as co-treatment, later endogenous AsA content increased, followed by scavenging surplus H_2O_2 generation and a reduction of lipid peroxidation [191].The higher transcript levels of SOD, CAT, APX, GR, MDHAR, and DHAR were induced in tall fescue by AsA application under PEG-induced water crisis, in respect with the only stressed condition [192]. Subsequently, Xu et al. [192] recommended AsA as a phytoprotectant to improved plants tolerances upon drought stress.

Exogenously applied 50 μM AsA decreased high temperature (HT, 45/35°C)-induced elevated H_2O_2 and MDA contents with lowered electrolyte leakage (EL) in *V. radiata* [184]. Hence, supplemental AsA altered the heat-induced lowered SOD, CAT, APX, and GR activities with increasing endogenous AsA and GSH contents.The AsA also enhanced the antioxidant capacity of tomato to cope with low-temperature stress [193]. The foliar application of AsA decreased the EL and MDA content in *T. aestivum* seedlings when exposed to the combined stress of herbicide and low temperature (−2 °C) [194]. This AsA-induced lowered oxidative damage might be because of ROS scavenging under stress indicated by AsA mediated lowered O_2^- and H_2O_2, which were attributed by increasing POD, APX, and GR activities.

Seed priming with AsA also increased plant tolerance to metal stress. Hence seed priming with AsA of *A. esculentus* showed the alleviation of Pb-induced oxidative stress that was confirmed by lowered H_2O_2 and MDA contents [135]. This AsA-induced alleviation of oxidative stress supported by exogenous AsA mediated increment of endogenous AsA contents, as well as upregulation of SOD, POD, and CAT activities in Pb-stressed *A. esculentus*. The AsA priming also increased the anthocyanins content in Pb-exposed seedlings, which again enhanced the metal tolerance by checking ROS production. Previously, exogenous foliar application of AsA on rice seedlings increased AsA, and

GSH contents, while enhanced both AsA/DHA and GSH/GSSG redox status, along with higher APX and GR activities under Cd stress [195].

Alamri et al. [196] investigated the potentiality of exogenous AsA to remove the metal-induced oxidative stress in *T. aestivum*. They observed that AsA suppressed the higher content of MDA and H_2O_2 in Pb exposed seedlings by improving the antioxidant enzymes activities, including SOD, CAT, and GR. Thus, AsA-mediated higher activity of enzymatic antioxidants could be responsible for the lowered membrane damage indicated by EL as well as Pb tolerance.

The AsA supplementation also showed its effective role in the mitigation of Cd-induced oxidative stress. Zhang et al. [141] reported that AsA application, as a foliar spray, could become a potent tool to alleviate Cd toxicity in *Z. mays*. They used 0.1, 0.3, and 0.5 mM of AsA in against 3.36 mM Cd contamination, while they observed a remarkable gradual reduction of both H_2O_2 and MDA contents under stressed conditions with increasing AsA levels. Foliar AsA application improved endogenous GSH along with the augmentation of SOD, POD, CAT, and GR activities, which are in line with AsA-induced lessening of oxidative stress in Cd-exposed *Z. mays*.

Thus, exogenous AsA application scavenges ROS in the plant under abiotic stresses, and then protects cell membrane stability. Therefore, reduced MDA content and EL were reported with AsA application as a sign of AsA-induced alleviation of oxidative damages. Accordingly, such exogenous AsA-induced strengthening of plant antioxidant defense, along with lessening oxidative stress, explained the potential of AsA for conferring abiotic stresses.

7.2. Exogenous GSH

At endogenous level, being an active participant of AsA-GSH cycle, GSH scavenges H_2O_2 in enzyme-dependent pathways; GSH is a substrate for GPX; GSH detoxifies lipid hydroperoxides together with GSTs; and, GSH/GSSG induces signals for abiotic stress adaptation [6]. Moreover, several research studies reported that the exogenous application of GSH proved to confirm the additional beneficial effects for enhanced the antioxidant defense system and abiotic stress tolerance.

After exogenous GSH pretreatment, mung bean plants were imposed with HT (42 °C), and beneficial effects were noticed. It enhanced chl and leaf RWC; increased cellular GSH content and GSH/GSSG lowering GSSG content; amplified APX, MDHAR, DHAR, GR, GPX, GST, and CAT activities; exogenous GSH pretreatment upheld the activity of Gly I and Gly II of MG detoxification system. The upregulation of both antioxidant and glyoxalase system ensured the HT tolerance. Meanwhile, GSH supplementation with HT decreased H_2O_2, O_2^-, MDA level, the activity of LOX, and MG content [95]. Increased temperature (35 °C) in root-zone variably affected physiological processes, growth, and Calvin cycle, which mediated inconsistency in antioxidant components; HT also affected antioxidant enzymes' gene expression of *Cucumis sativus* L. seedlings. HT-induced reduction of GSH content, the ratio of GSH/GSSG, photosynthetic pigments level, photosynthesis, and changes of linked gene expression were evident. HT also augmented soluble protein, proline (Pro), O_2^- generation and MDA level, expression of genes, and antioxidant enzymes functioning. The application of supplemental GSH with HT upheld soluble protein, Pro, antioxidant enzymes activity, and its linked gene expression, as well as inhibited O_2^- generation and lipid peroxidation, than to HT treatment without GSH [197].

Exogenous GSH improved AsA, and GSH contents, GSH/GSSG, APX, MDHAR, DHAR, and GPX activity of antioxidant system in drought exaggerated mung bean seedlings, which helped to relieve the adverse effect reducing the ROS including H_2O_2 and O_2^-, both in content and visually in the leaf spots of which were visualized through histochemical detection. Exogenous GSH also decreased the LOX activity, which caused the oxidation of lipid. Exogenous GSH also up-regulated the activity of Gly I and Gly II, therefore, reduce the toxic consequence of MG, and MG-induced oxidative damage [111].

Exogenous GSH (1.0 mM) positively regulated an antioxidant system in wheat plants facing lead (Pb) stress. The imposition of Pb diminished growth, the relative water content of leaf, and chl *a* and *b* content; amplified Pro level, H_2O_2, and O_2^- generation, and lipid peroxidation. Glutathione supplementation with Pb stress improved the AsA and GSH contents, GSH/GSSG, activities of MDHAR,

DHAR, GR, SOD, CAT, and GPX, and decreased oxidative damage. The decline of H_2O_2 and O_2^- generation and membrane lipid peroxidation was clear evidence, together with an increased level of Pro and chl, which contributed overall tolerance to Pb toxicity [35]. Pretreatment with 100 μM GSH with 50 μM Cd reversed growth reduction and concealed Cd-provoked MDA buildup. In contrast to Cd compelled plants, GSH pretreatment reversed photosynthetic pigment destruction, downregulated Cd accumulation in root and shoot. Exogenous GSH considerably increased the functioning of POD and SOD. In contrast to the Cd affected plants, exogenous GSH pretreated plants extensively reassured decrease in Cu or augmented in Fe levels, which were due to Cd [198]. Exogenous GSH and Cys were applied on lead (100 and 500 mg L^{-1}) affected *Iris lactea* var. chinensis, growth, accumulation of Pb, and nonprotein thiol (NPT) accumulation pattern were observed. The addition of GSH improved GSH biosynthesis in root and shoot. Endogenous shoot level Cys was recorded for exogenous Cys addition. Exogenous GSH application, together with buthionine sulfoximine (BSO) addition, regulated enzymes involved in GSH biosynthesis. This GSH played an imperative function in Pb accumulation and adaptation to this stress [199]. Exogenous GSH application improved the germination and growth of *Arabidopsis*, tobacco, and pepper under mercury (Hg) stress. Exogenous GSH also conferred Cd, Cu, and Zn stress tolerance. Exogenous GSH downregulated H_2O_2 and O_2^- generation and MDA content, whereas upregulated chl level under Hg. Outstandingly, exogenous GSH reduced Hg accumulation in *Arabidopsis*. GSH showed high binding empathy to Hg, as compared to Cd, Cu, or Zn [200].

Salt-tolerant Pokkali and sensitive cultivar Peta of rice were scrutinized for the role of exogenous GSH on them. Exogenous GSH increased the activity of SOD, APX, and GR, the amount of AsA and GSH, and reversed chloroplasts' H_2O_2 and MDA accumulation in either cultivar affected by salinity (200 mM NaCl). However, tolerance was prominent in cv. Pokkali [163]. The supplementation of GSH inverted the pessimistic properties of salinity stressed (NaCl 100 mM) tomato plants improving the transcript levels and activities of enzymes that are linked to GSH biosynthesis and metabolism. The biosynthesis-related enzymes were gamma-glutamylcysteine synthetase (γ-ECS), glutathione synthetase (GS), whereas, others were GST, GPX, and GR. Exogenous supplemental GSH helped to upregulate the activity of SOD, peroxidase (POD), CAT, APX, MDHAR, DHAR and GR, GSH level, and GSH/GSSG in salt-stressed plants [201]. Externally applied GSH lessened the oxidative damage in different soybean genotypes via reducing H_2O_2 and MDA level, which were produced due to salinity. Glutathione supplementation minimizing the oxidative damage further contributed in yield attributes, and yield performance, which was seeds plant^{-1} and pods plant^{-1}, 100-seed weight and yield [202]. Defensive function of supplemental GSH (1 mM GSH) was examined for salt (200 mM NaCl) stressed mung bean. Mung bean plants when imposed with exogenous GSH and NaCl elevated AsA and GSH levels, GSH/GSSG, enhanced APX, MDHAR, DHAR, GR, SOD, CAT, GPX, and GST activities were recoded. Exogenous supplemental GSH also augmented the activity of Gly I and Gly II under salinity. Enhanced antioxidant and glyoxalase system components that resulted from the effect of exogenous GSH application had several beneficial effects. MDA, H_2O_2, and MG, O_2^-production turned down, and leaf RWC and chl level raise; all of which made mung bean seedlings capable to perform better under saline growing media [95]. Glutathione was exogenously applied on tomato plants affected by salinity. Additional of GSH decreased oxidative stress. The reason behind this was revealed as redistribution of light energy in PSII, higher cellular GSH, GSH/GSSH ratio and activities of SOD, CAT and APX, MDHAR, DHAR, GR, and GRx. Glutathione supplementation revolutionized growth inhibition, $Na^+$and Cl^-ions balance, and Na^+/K^+. Choloplast, as well as stomatal function related to photosynthetic performance, were documented to improve after the application of GSH with salinity [203].

8. Interaction of Other Pathways with AsA-GSH Pathways in Regulating ROS Metabolism

Beside oxidative stress mitigation, the AsA-GSH cycle also interacts other pathways to reduce ROS and oxidative stress. Therefore, in this section, we will discuss the potentiality of AsA-GSH pathway components and their interaction with other pathways to modulate the ROS metabolism in plants.

8.1. Interaction of AsA-GSH Cycle with NO Metabolic Pathway

Although AsA-GSH cycle protects cellular components from oxidative damage, its components, especially the proteins (APX, MDHAR, DHAR, and GR), are also vulnerable to the oxidative damage, which can modify their activity, hence breaking down the antioxidant defense. In plants, nitric oxide can be produced from several biochemical pathways, both oxidative and reductive [35]. In the GSH pool, the reduced form of GSH can interact with NO and produce GSNO, which is further catalyzed by the action of GSNO reductase (GSNOR) and release NO and GSSG, and maintains the equilibrium of NO and nitrosothiols, as well as balance the redox state in the cell [204]. Moreover, *S*-nitrosylation could modify protein interactions, thus tinkering the antioxidant response [205]. Reports suggest that all of the proteins of the AsA-GSH cycle are influenced through *S*-nitrosylation and/or nitration, which are accomplished from the interaction with NO. Among the AsA-GSH cycle, enzymes APX is the most studied, which is directly influenced by NO metabolism. For example, the inactivation of APX1 is caused due to the oxidation of Cys32 [205]. Contrary, nitrosylation of APX1 active-site Cys32 increases its activity and this post translation modification (PTM) is performed during salinity stress, which increases oxidative stress as well as *S*-nitrosothiols [206]. Among the other enzymes, MDHAR is negatively modulated by nitration, which cuses enzymatic inactivity by altering the position of the cofactor binding site [206], and hence disturb the AsA recycling process. Although information is available on the nitration and activity modulation of DHAR proteins [207,208], but the involvement of Try in this process, as well as the structural alteration impact of the enzyme, is still unclear. Moreover, GR is also targeted for nitration, which is reported to inhibit its activity in a mammalian cell, but the chloroplasic and cytosolic GR of pea is not affected by nitration [206].

8.2. Signaling Role of AsA-GSH Cycle Components and Interaction with Other Pathways

Ascorbate serves as the co-factor for redox enzymes, as well as a precursor for several biosynthetic pathways. In addition, AsA is an important reducing agent for Fe, Cu, and Mn, thus act as a pro-oxidant controlling toxic $OH^{\bullet}$ production from the Haber–Weiss and Fenton reaction [209]. Besides, its role as an antioxidant is the most important part of detoxifying ROS. As a pro-oxidant AsA regenerate α-tocopherol. Moreover, AsA also works in the photo-protection that is mediated by the xanthophylls cycle, where violaxanthin de-epoxidase use AsA as a co-factor [210]. Moreover, AsA is employed as the substrate for organic acid (oxalate and tartrate) biosynthesis (Figure 2). Rapid cell expansion is correlated with AO activity, which oxidized AsA [211]. During cell expansion, Pro residues present in the glycoproteins of cell wall undergo hydroxylation where prolyl hydroxylase use AsA as a cofactor [210]. Furthermore, AsA can potentially upregulate cytosolic free Ca^{2+} via anion channels and play a signaling role [212]. More than this breakdown of AsA to DHA by APX or AO creates an electrochemical gradient over the plasma membrane, which also has a signaling role.

Under abiotic stress conditions, GSH triggers adaptation or PCD by intercellular signaling [213]. Glutathionylation of protein Cys residues suggests its redox signaling role, which alters the transcription of proteins [214]. In *Arabidopsis*, stomatal movement induced from methyl jasmonate (MeJA) is regulated by intracellular GSH [215]. In tobacco, both GSH and GSSG application induce Ca^{2+} signaling as well as the expression of a specific gene, which supports the involvement of GSH with signal pathways that connect the Ca^{2+}-dependent protein kinase [216]. The protein family peroxiredoxins (Prxs) are also GSH dependent and catalyze the reduction of H_2O_2 [217]. The GSSG can be exchanged with sulfhydryl groups of proteins and produce protein–GSH disulfide conjugates, which has a long half-life and plays a vital role in cellular signaling [218]. Moreover, GSH influences translation, and PTM of proteins, modulation of metabolism, and gene expression [219]. Hence, the mechanistic process of GSH signaling role should be focused on in future studies.

8.3. AsA-GSH Cycle Interaction with Phytohormone Biosynthesis Pathways

Ascorbate regulates phytohormone biosynthesis; hence, modulating plant development [220]. Ascorbate shows activity, where cell developments are affected by hormonal signaling and modulate effective signaling processes [221]. The abscisic acid (ABA) involvement in stagnating growth and metabolism suggests the crucial role of AsA sensing for plant survival [222]. In addition, a number of dioxygenases that are directly related to hormonal biosynthesis require AsA is a cofactor [223]. Moreover, a low AsA induces PR proteins, but do not alter antioxidative enzymes. Thus, AsA acts as a "crosstalking" signal, where ABA acts as an important intermediary signal induces PR1 proteins in many plants. Hence, phytohormone signaling arises the AsA-dependent PR genes regulation [224]. On the other hand, 1-aminocyclopropane-1-carboxylate (ACC) synthase (*ACS*) and ACC oxidase (*ACO*) genesencoding ethylene biosynthetic enzymes is induced by GSH. Further, GSH increases serine acetyl transferase (SAT) level and confers Ni toxicity tolerance [225]. In rice, the overexpression of SA metabolism genes gave raise to both SA and GSH content under oxidative stress [226]. Therefore, GSH triggers phytohormones, and vice versa, along with other signaling genes [227].

8.4. Interaction of AsA-GSH Pathway with Glyoxalase Pathway

There is an intimate relationship between AsA-GSH cycle and the glyoxalase pathway through GSH, where it plays a vital role in the detoxification of MG. Methylglyoxal is a respiratory byproduct and produced usually in plants and detoxifies by the glyoxalase system. However, MG is overproduced under stress, which causes toxicity [51]. Moreover, MG can disfunctionate antioxidant enzymes [1]. In the MG detoxification process, Gly I (EC 4.4.1.5) and Gly II (EC 3.1.2.6) work simultaneously to detoxify MG (Discussed in Section 6). In this pathway, Gly I uses GSH as a cofactor and conjugates MG with GSH to form *S*-D-lactoylglutathione (SLG), Gly II, and then produce D-lactate breaking SLG, and regenerate GSH (Figure 9) [1], thus playing important interaction with glyoxalase system.

8.5. Interaction AsA-GSH Pathway with Xenobiotics Detoxification Pathways

Xenobiotic detoxification involves the conjugation of toxic xenobiotics with GSH, which are further transferred to the vacuole by using ATP driven tonoplast transporter. This detoxification enables secondary metabolites biosynthesis as well as storing in the vacuole, such as anthocyanin. Plants are having GSH-dependent enzyme, GST, which detoxify herbicides by conjugating it with GSH. Therefore, the glutathionylated metabolites are imported to vacuolar by ABC (ATP-binding cassette) transporters. However, the GST mainly functions in catalyzing natural products that were observed with xenobiotics and, similar to those, catalyzes alternative GSH-dependent biotransformation reactions and binds and carries phytochemicals between cellular compartments [228].

8.6. AsA-GSH Cycle Interaction with Metal Chelation Process

Maintaining lower metal/metalloid(s) level inside the cell involves metal sequestration by low molecular weight thiols, for instance, metallothioneins (MTs) and phytochelatins (PCs). The two important enzymes involved in this process, glutaredoxin (GRx) and thioredoxin (TRx), are GSH dependent and neutralize H_2O_2 or controls of protein thiols [229]. On the other hand, PCs are another important chelating agent containing that thiol group that are upregulated by different metal/metalloid(s) [32]. The basic component for this PC is GSH. The biosynthesis of PCs is accomplished by the enzyme PC synthase (PCS), which requires GSH as a substrate. The enzyme is crucial for metal detoxification, metal homeostasis, and stress tolerance [137].

9. Genetic Manipulation of AsA-GSH Pathway and Its Role in Abiotic Stress Tolerance

The regulation of AsA and GSH pool plays an important role in mitigating oxidative stress in plants. To attain this, the regulation of the enzymes that are related to the AsA-GSH pathway is vital. There are many plant studies that considered the genetic manipulation of AsA-GSH pathway. These studied revealed that the overexpression of AsA-GSH pathway enzymes provided the plants better protection against oxidative stress under various environmental adversities (Table 4). Transgenic tobacco plants overexpressing *PcAPX* showed enhanced tolerance to salt and drought [230]. Transgenic plants exhibited a 347% increase in APX activities under drought stress, as compared to control, which resulted in a remarkable decrease in H_2O_2 content (136%) than that of wild type (309%). The ascorbate content was also higher 63%) when compared to wild type (42%). Similar results were also observed in the case of salt stress [230]. Chin et al. [231] found that transgenic *Arabidopsis* overexpressing *OgCytAPX1* scavenged ROS effectively and showed enhanced tolerance to salt and heat. The overexpression of *Malpighia glabraMDHAR* gene resulted in a higher biosynthesis of AsA, which provided tobacco plants tolerance to salt stress [232]. Shin et al. [233] observed that the coexpression of *B. rapaBrMDHAR* and *BrDHAR* genes provided a remarkable improvement of oxidative stress in *A. thaliana*. The overexpression of *BrMDHAR* and *BrDHAR* showed enhanced MDHAR and DHAR activities and higher AsA/DHA ration. These plants also provided better radical scavenging capacity, which resulted in lower H_2O_2 content. Yin et al. [234] found that *Arabidopsis* plants overexpressing the gene *AtGR1* conferred Al stress tolerance by reducing reactive carnoyl species, which was mainly due to higher GSH level and GR activity. The plant that overexpressed *AtGR1* also maintained the activity of H_2O_2-scavenging enzymes. For instance, GPX and APX activities in Al-treated plants were decreased by 21 and 46%, respectively, but the wild-type plants only showed 8 and 30% decreases in such activities [234].

Modulating several *NAC* genes [*NAC* domain consists of three different genes; *NAM* (no apical meristem)-*ATAF* (*Arabidopsis* transcription activation factor)-*CUC* (cup-shaped cotyledon)] are also an efficient way to transform the AsA-GSH cycle, consequently enhancing stress tolerance. The overexpression of wheat *TaNAC2* in *Arabidopsis* lines showed tolerance against freezing, salt, and drought stress by modulating the AsA-GSH cycle [235]. Moreover, ectopic expression of *SlNAC2* conferred both salt (200 mM) and drought (20% PEG) tolerance up to 10 days in transgenic *Arabidopsis* lines, which is correlated with the lower accumulation of ROS. In addition, the transcriptomic abundance GSH metabolizing genes was also observed in transgenic lines, leading to increased GSH synthesis and lesser oxidative damage [236].

Table 4. Overexpression of genes related to AsA-GSH pathway and their role in ROS scavenging.

Enzymes	Gene	Gene Source	Target Plants	Regulatory Effects	References
APX	*OsAPX2*	Rice	Alfalfa	Decreased H_2O_2 and MDA contents; Enhanced salt tolerance	[237]
APX	*PcAPX*	*Populus tomentosa*	Tobacco	Increased AsA content; $NADP^+$/NADPH ratio; Decreased lipid peroxidation and H_2O_2 contents; Increased salt and drought stress tolerance	[230]
APX	*CaAPX*	*Camellia azalea*	Tobacco	Enhanced MDHAR and DHAR activity; Regulated ROS generation; Enhanced cold and HT tolerances	[238]
APX	*ScAPX6*	Sugarcane	*Nicotiana benthamiana*	Hormonal regulation; Lower ROS generation; Enhanced tolerance to Cu stress	[239]
APX	*OgCytAPX1*	*Oncidium*	*Arabidopsis*	Efficient ROS scavenging capacity; Maintained redox homeostasis and increased GPX activities which resulted in lower H_2O_2	[240]
MDHAR	*Am-MDAR*	*Avicennia marina*	*A. marina*	Increase in MDHAR and GR activity; Increased AsA content; Decreased lipid peroxidation; Improved salt-induced oxidative stress	[241]

Table 4. *Cont.*

Enzymes	Gene	Gene Source	Target Plants	Regulatory Effects	References
MDHAR	*MgMDHAR*	*Malpighia glabra*	Tobacco	DHAR activity increased by 1.8–2.1 fold; AsA/DHA increased by 81–84%; Lipid peroxidation decreased by 41–62%	[232]
MDHAR	*BrMDHAR*	Rapeseed	*Arabidopsis*	AsA/DHA ratio increased by 7%; Decrease of H_2O_2 content by 55%; Radical scavenging was 16% higher	[233]
DHAR	*DHAR*	Human	Tobacco	No changes in AsA content; Enhanced tolerance to chilling and salt stress	[242]
DHAR	*DHAR1*	Rice	*Arabidopsis*	AsA content increased by 1.4 fold; Enhanced tolerance to salt	[243]
DHAR	*DHAR*	Human	Tobacco	Increase in AsA content; Increased the activities of SOD and APX; Enhanced salt tolerance	[244]
DHAR	*DHAR*	*Arabidopsis*	*Arabidopsis*	AsA content increased by 2.0–4.25 fold; Enhanced tolerance to HT stress	[245]
DHAR	*DHAR*	Rice	Tobacco	AsA content increased by 1.6-fold; Improved tolerance to salinity and chilling	[246]
DHAR	*BrDHAR*	Rapeseed	*Arabidopsis*	AsA/DHA ratio increased by 11%; H_2O_2 content decreased by 56%; Radical scavenging was 16% higher	[233]
GR	*AtGR1*	*Arabidopsis*	*Arabidopsis*	Enhanced GSH content and GR activity (0.4- to 1.0-fold higher); H_2O_2 content reduced by 26%	[234]

10. Conclusions and Outlook

Ascorbate and GSH have roles in decreasing oxidative stress, and it has been reported in numbers of research findings. Most of the research findings reported about their roles in antioxidant defense system for scavenging ROS. However, exogenous GSH related research on antioxidant defense system needs further confirmation at the genetic and molecular level. Moreover, without the commonly known ROS, like H_2O_2, OH, O_2^-, etc. some other oxidative stress-inducing agents, like reactive nitrogen species, MG, etc., should be brought under consideration for research. How GSH can affect the generation of other kinds of oxidative stress-inducing damage. Research that is related to exogenous AsA or GSH-induced GST activity concerning xenobiotic detoxification is rare. The regulation of tocopherol by AsA or GSH can be an interesting area of research. For the reduction of metal-induced oxidative stress protection, GSH plays a vital role by producing PCs and inducing vacuolar sequestration. The credible function of AsA-GSH cycle in this area is so far to explicate. The GSH/GSSG redox is a well-reported term when discussing stress-induced oxidative damage and signal transduction process towards adaptation though the process is not well revealed. Interaction between and among the AsA-GSH cycle components and the hormones, other signaling molecules or any other molecules in oxidative stress, redox regulation, or plant adaptation process is not well understood. It is well known that chloroplast, its photosystem, and Calvin cycle activity or photosynthesis process is the maximum contributor of most of the ROS and oxidative stress under any abiotic stress condition. Several research findings reported about the role of AsA and GSH in improving the chl or carotenoid levels. However, very few of them reported regarding the roles of AsA and GSH in regulating stomatal conductance, Calvin cycle, RuBP activity/regeneration, or photosystem efficiency, which directly generates ROS and results in oxidative stress [203]. Some of the research findings show the positive roles of GSH improving/regulating Pro, which is cellular ROS scavenger or cytosol stabilizer to reduce ROS generation. These are the promising area of future research, which not only will alleviate the oxidative stress, but also improve the photosynthetic efficiency of plants for increasing plant production for the constantly growing population of the planet.

Although, in this review article, we focused on abiotic stress-induced oxidative damage and the role of AsA-GSH cycle to mitigate such adversities, biotic stress (fungi, bacteria, virus, nematodes, and parasitic organisms, etc.) might also alter the essential plant processes as well as cellular metabolism.For example, the production of ROS and oxidative stress, disruption of membranes, hampering photosynthesis, changing enzyme activities, cell death, and yield loss might also be attributed to biotic stress, which is in line with abiotic stress.Biotic stress also disrupts signal transduction, as well as transfigures signal pathways that are associated with stress acclimation. Over the past decade, AsA-GSH cycle has also emerged as an important component for the plant biotic stress response. Similar to abiotic stresses, biotic stresses also alters the metabolism and changes in antioxidant activity. Therefore, AsA-GSH cycle also directly impacts the important metabolomic processes, thus providing an important link between metabolism, signal transduction, and acclimation to plants during biotic stress.

Author Contributions: M.H. constructed the main conceptual ideas and proof outline. All authors participated in the drafting of this paper as individual subject matter experts in their fields. M.H., M.H.M.B.B., and J.A.M. prepared the figures. M.H., T.I.A., K.P. and K.N. prepared the tables. M.H. and M.F. has contributed critically to the improvement and editing of the manuscript. All the authors contributed to collecting the literature, improving the paper, and approved the final manuscript.

Acknowledgments: We thank Khadeja Sultana Sathi, Tonusree Saha, Mira Rahman and Naznin Ahmed for the formatting and proof checking of the manuscript.

References

1. Hasanuzzaman, M.; Nahar, K.; Anee, T.I.; Fujita, M. Glutathione in plants: Biosynthesis and physiological role in environmental stress tolerance. *PMBP* **2017**, *23*, 249–268. [CrossRef] [PubMed]
2. Hasanuzzaman, M.; Mahmud, J.A.; Nahar, K.; Anee, T.I.; Inafuku, M.; Oku, H.; Fujita, M. Responses, adaptation, and ROS metabolism in plants exposed to waterlogging stress. In *Reactive Oxygen Species and Antioxidant Systems in Plants: Role and Regulation under Abiotic Stress*; Khan, M.I.R., Khan, N.A., Eds.; Springer: New York, NY, USA, 2017.
3. Halliwell, B.; Gutteridge, J.M.C. Antioxidant defences: Endogenous and diet derived. *Free Radic. Biol. Med.* **2007**, *4*, 79–186.
4. Mittler, R. Oxidative stress, antioxidants and stress tolerance. *Trends Plant Sci.* **2002**, *7*, 405–410. [CrossRef]
5. Gill, S.S.; Tuteja, N. Reactive oxygen species and antioxidant machinery in abiotic stress tolerance in crop plants. *Plant Physiol. Biochem.* **2010**, *48*, 909–930. [CrossRef] [PubMed]
6. Hasanuzzaman, M.; Hossain, M.A.; Teixeira da Silva, J.A.; Fujita, M. Plant responses and tolerance to abiotic oxidative stress: Antioxidant defense is a key factor. In *Crop Stress and its Management: Perspectives and Strategies*; Bandi, V., Shanker, A.K., Shanker, C., Mandapaka, M., Eds.; Springer: Dordrecht, The Netherlands, 2012.
7. Sandalio, L.M.; Rodríguez-Serrano, M.; Romero-Puertas, M.C.; Luis, A. Role of peroxisomes as a source of reactive oxygen species (ROS) signaling molecules. In *Peroxisomes and Their Key Role in Cellular Signaling and Metabolism*; del Río, L., Ed.; Springer: Dordrecht, The Netherlands, 2013; pp. 231–255.
8. Hasanuzzaman, M.; Nahar, K.; Anee, T.I.; Fujita, M. Exogenous silicon attenuates cadmium-induced oxidative stress in *Brassica napus* L. by modulating AsA-GSH pathway and glyoxalase system. *Front. Plant Sci.* **2017**, *8*, 1061. [CrossRef] [PubMed]
9. Noctor, G.; Mhamdi, A.; Foyer, C.H. The roles of reactive oxygen metabolism in drought: Not so cut and dried. *Plant Physiol.* **2014**, *164*, 1636–1648. [CrossRef] [PubMed]
10. Hasanuzzaman, M.; Nahar, K.; Hossain, M.S.; Mahmud, J.A.; Rahman, A.; Inafuku, M.; Oku, H.; Fujita, M. Coordinated actions of glyoxalase and antioxidant defense systems in conferring abiotic stress tolerance in plants. *Int. J. Mol. Sci.* **2017**, *18*, 200. [CrossRef]

11. Hasanuzzaman, M.; Nahar, K.; Hossain, M.S.; Anee, T.I.; Parvin, K.; Fujita, M. Nitric oxide pretreatment enhances antioxidant defense and glyoxalase systems to confer PEG-induced oxidative stress in rapeseed. *J. Plant Interact.* **2017**, *12*, 323–331. [CrossRef]
12. Zechmann, B. Compartment-specific importance of glutathione during abiotic and biotic stress. *Front. Plant Sci.* **2014**, *5*, 566. [CrossRef]
13. Szarka, A.; Tomasskovics, B.; Bánhegyi, G. The ascorbate-glutathione-α-tocopherol triad in abiotic stress response. *Int. J. Mol. Sci.* **2012**, *13*, 4458–4483. [CrossRef]
14. Tripathi, R.P.; Singh, B.; Bisht, S.S.; Pandey, J. L-Ascorbic acid in organic synthesis: An overview. *Curr. Org. Chem.* **2009**, *13*, 99–122. [CrossRef]
15. Davey, M.W.; Montagu, M.V.; Inze, D.; Sanmartin, M.; Kanellis, A.; Smirnoff, N.; Benzie, I.J.J.; Strain, J.J.; Favell, D.; Fletcher, J. Plant L-ascorbic acid: Chemistry, function, metabolism, bioavailability and effects of processing. *J. Sci. Food Agric.* **2000**, *80*, 825–860. [CrossRef]
16. Mellidou, I.; Keulemans, J.; Kanellis, A.K.; Davey, M.W. Regulation of fruit ascorbic acid concentrations during ripening in high and low vitamin C tomato cultivars. *BMC Plant Biol.* **2012**, *12*, 239. [CrossRef] [PubMed]
17. Cronje, C.; George, G.M.; Fernie, A.R.; Bekker, J.; Kossmann, J.; Bauer, R. Manipulation of L-ascorbic acid biosynthesis pathways in *Solanum lycopersicum*: Elevated GDP-mannose pyrophosphorylase activity enhances L-ascorbate levels in red fruit. *Planta* **2012**, *23*, 553–564. [CrossRef] [PubMed]
18. Ishikawa, T.; Maruta, T.; Yoshimura, K.; Smirnoff, N. Biosynthesis and regulation of ascorbic acid in plants. In *Antioxidants and Antioxidant Enzymes in Higher Plants*; Gupta, D., Palma, J., Corpas, F., Eds.; Springer: Cham, Switzerland, 2018; pp. 163–179.
19. Wheeler, G.; Ishikawa, T.; Pornsaksit, V.; Smirnoff, N. Evolution of alternative biosynthetic pathways for vitamin C following plastid acquisition in photosynthetic eukaryotes. *Elife* **2015**, *4*, e06369. [CrossRef] [PubMed]
20. Pallanca, J.E.; Smirnoff, N. The control of ascorbic acid synthesis and turnover in pea seedlings. *J. Exp. Bot.* **2000**, *345*, 669–674. [CrossRef]
21. Asada, K. Production and scavenging of reactive oxygen species in chloroplasts and their functions. *Plant Physiol.* **2006**, *141*, 391–396. [CrossRef]
22. Miyake, C. Alternative electron flows (water-water cycle and cyclic electron flow around PSI) in photosynthesis: Molecular mechanisms and physiological functions. *Plant Cell Physiol.* **2010**, *51*, 1951–1963. [CrossRef]
23. Loewus, F.A. Biosynthesis and metabolism of ascorbic acid and of analogs of ascorbic acid in fungi. *Phytochemistry* **1999**, *52*, 193–210. [CrossRef]
24. Melino, V.J.; Soole, K.L.; Ford, C.M. Ascorbate metabolism and the developmental demand for tartaric and oxalic acids in ripening grape berries. *BMC Plant Biol.* **2009**, *9*, 145. [CrossRef]
25. Truffault, V.; Fry, S.C.; Stevens, R.G.; Gautier, H. Ascorbate degradation in tomato leads to accumulation of oxalate, threonate and oxalyl threonate. *Plant J.* **2017**, *89*, 996–1008. [CrossRef] [PubMed]
26. Dewhirst, R.A.; Clarkson, G.J.J.; Rothwell, S.D.; Fry, S.C. Novel insights into ascorbate retention and degradation during the washing and post-harvest storage of spinach and other salad leaves. *Food Chem.* **2017**, *233*, 237–246. [CrossRef] [PubMed]
27. Kärkönen, A.; Dewhirst, R.A.; Mackay, C.L.; Fry, S.C. Metabolites of 2,3-diketogulonate delay peroxidase action and induce non-enzymic H_2O_2 generation: Potential roles in the plant cell wall. *Arch. Biochem. Biophys.* **2017**, *620*, 12–22. [CrossRef] [PubMed]
28. Hasanuzzaman, M.; Hossain, M.A.; Fujita, M. Nitric oxide modulates antioxidant defense and the methylglyoxal detoxification system and reduces salinity-induced damage of wheat seedlings. *Plant Biotechnol. Rep.* **2011**, *5*, 353–365. [CrossRef]
29. Gill, S.S.; Anjum, N.A.; Hasanuzzaman, M.; Gill, R.; Trived, D.K.; Ahmad, I.; Pereira, E.; Tuteja, N. Glutathione reductase and glutathione: A boon in disguise for plant abiotic stress defense operations. *Plant Physiol. Biochem.* **2013**, *70*, 204–212. [CrossRef] [PubMed]
30. Noctor, G.; Queval, G.; Mhamdi, A.; Chaouch, S.; Foyer, C.H. Glutathione. The *Arabidopsis* Book. *Am. Soc. Plant Biol.* **2011**, *9*, e0142.
31. Noctor, G.; Mhamdi, A.; Chaouch, S.; Han, Y.I.; Neukermans, J.; Marquez-Garcia, B.E.L.E.N.; Queval, G.; Foyer, C.H. Glutathione in plants: An integrated overview. *Plant Cell Environ.* **2012**, *35*, 454–484. [CrossRef]

32. Nahar, K.; Hasanuzzaman, M.; Fujita, M. Physiological roles of glutathione in conferring abiotic stress tolerance to plants. In *Abiotic Stress Response in Plants*; Gill, S.S., Tuteja, N., Eds.; Wiley: Weinheim, Germany, 2016; pp. 151–179.
33. Zagorchev, L.; Seal, C.; Kranner, I.; Odjakova, M. A central role for thiols in plant tolerance to abiotic stress. *Int. J. Mol. Sci.* **2013**, *14*, 7405–7432. [CrossRef]
34. Mahmood, Q.; Ahmad, R.; Kwak, S.S.; Rashid, A.; Anjum, N.A. Ascorbate and glutathione: Protectors of plants in oxidative stress. In *Ascorbate–Glutathione Pathway and Stress Tolerance in Plants*; Mahmood, Q., Ahmad, R., Kwak, S.S., Rashid, A., Anjum, N.A., Eds.; Springer: Berlin, Germany, 2010; pp. 209–229.
35. Hasanuzzaman, M.; Nahar, K.; Rahman, A.; Mahmud, J.A.; Alharby, H.F.; Fujita, M. Exogenous glutathione attenuates lead-induced oxidative stress in wheat by improving antioxidant defense and physiological mechanisms. *J. Plant Interact.* **2018**, *13*, 203–212. [CrossRef]
36. Ferretti, M.; Destro, T.; Tosatto, S.C.E.; La Rocca, N.; Rascio, N.; Masi, A. Gamma-glutamyl transferase in the cell wall participates in extracellular glutathione salvage from the root apoplast. *New Phytol.* **2009**, *181*, 115–126. [CrossRef]
37. Ohkama-Ohtsu, N.; Sasaki-Sekimoto, Y.; Oikawam, A.; Jikumaru, Y.; Shinoda, S.; Inoue, E.; Kamide, Y.; Yokoyama, T.; Hirai, M.Y.; Shirasu, K.; et al. 12-oxo-phytodienoic acid-glutathione conjugate is transported into the vacuole in *Arabidopsis*. *Plant Cell Physiol.* **2011**, *52*, 205–209. [CrossRef] [PubMed]
38. Su, T.; Xu, J.; Li, Y.; Lei, L.; Zhao, L.; Yang, H.; Feng, J.; Liu, G.; Ren, D. Glutathione-indole-3-acetonitrile is required for camalexin biosynthesis in *Arabidopsis thaliana*. *Plant Cell* **2011**, *23*, 364–380. [CrossRef] [PubMed]
39. Yadav, P.; Yadav, T.; Kumar, S.; Rani, B.; Jain, V.; Malhotra, S.P. Partial purification and characterization of ascorbate peroxidase from ripening ber (*Ziziphus mauritiana* L) fruits. *Afr. J. Biotechnol.* **2014**, *13*, 3323–3331.
40. del Río, L.A.; López-Huertas, E. ROS generation in peroxisomes and its role in cell signaling. *Plant Cell Physiol.* **2016**, *57*, 1364–1376. [CrossRef] [PubMed]
41. Anjum, N.A.; Sharma, P.; Gill, S.S.; Hasanuzzaman, M.; Khan, E.A.; Kachhap, K.; Mohamed, A.A.; Thangavel, P.; Devi, G.D.; Vasudhevan, P.; et al. Catalase and ascorbate peroxidase—Representative H_2O_2-detoxifying heme enzymes in plants. *Environ. Sci. Pollut. Res.* **2016**, *23*, 19002–19029. [CrossRef]
42. Guo, K.; Tu, L.; Wang, P.; Du, X.; Ye, S.; Luo, M.; Zhang, X. Ascorbate alleviates Fe deficiency-induced stress in cotton (*Gossypium hirsutum*) by modulating ABA levels. *Front. Plant Sci.* **2017**, *7*, 1997. [CrossRef] [PubMed]
43. Lagoa, R.; Samhan-Arias, A.K.; Gutierrez-Merino, C. Correlation between the potency of flavonoids for cytochrome c reduction and inhibition of cardiolipin-induced peroxidase activity. *Biofactors* **2017**, *43*, 451–468. [CrossRef]
44. Ullah, S.; Kolo, Z.; Egbichi, I.; Keyster, M.; Ludidi, N. Nitric oxide influences glycine betaine content and ascorbate peroxidase activity in maize. *S. Afr. J. Bot.* **2016**, *105*, 218–225. [CrossRef]
45. Leterrier, M.; Cagnac, O. Function of the various MDAR isoforms in higher plants. In *Antioxidants and Antioxidant Enzymes in Higher Plants*; Gupta, D., Palma, J., Corpas, F., Eds.; Springer: Cham, Switzerland, 2018; pp. 83–94.
46. Begara-Morales, J.C.; Chaki, M.; Valderrama, R.; Mata-Pérez, C.; Padilla, M.; Barroso, J.B. Transcriptional regulation of gene expression related to hydrogen peroxide (H_2O_2) and nitric oxide (NO). In *Nitric Oxide and Hydrogen Peroxide Signaling in Higher Plants*; Gupta, D., Palma, J., Corpas, F.J., Eds.; Springer: Cham, Switzerland, 2019; pp. 69–90.
47. Park, A.K.; Kim, I.S.; Do, H.; Jeon, B.W.; Lee, C.W.; Roh, S.J.; Shin, S.C.; Park, H.; Kim, Y.S.; Kim, Y.H.; et al. Structure and catalytic mechanism of monodehydroascorbate reductase, MDHAR, from *Oryza sativa* L. *japonica*. *Sci. Rep.* **2016**, *6*, 33903. [CrossRef]
48. Sano, S. Molecular and functional characterization of monodehydro-ascorbate and dehydroascorbate reductases. In *Ascorbic Acid in Plant Growth, Development and Stress Tolerance*; Hossain, M., Munné-Bosch, S., Burritt, D., Diaz-Vivancos, P., Fujita, M., Lorence, A., Eds.; Springer: Cham, Switzerland, 2017; pp. 129–156.
49. Johnston, E.J.; Rylott, E.L.; Beynon, E.; Lorenz, A.; Chechik, V.; Bruce, N.C. Monodehydroascorbate reductase mediates TNT toxicity in plants. *Science* **2015**, *349*, 1072–1075. [CrossRef]
50. Nahar, K.; Hasanuzzaman, M.; Alam, M.; Rahman, A.; Mahmud, J.-A.; Suzuki, T.; Fujita, M. Insights into spermine-induced combined high temperature and drought tolerance in mung bean: Osmoregulation and roles of antioxidant and glyoxalase system. *Protoplasma* **2017**, *254*, 445–460. [CrossRef] [PubMed]

51. Bhuyan, M.H.M.B.; Hasanuzzaman, M.; Mahmud, J.A.; Hossain, M.S.; Bhuiyan, T.F.; Fijuta, F. Unraveling morphophysiological and biochemical responses of *Triticum aestivum* L. to extreme pH: Coordinated actions of antioxidant defense and glyoxalase systems. *Plants* **2019**, *8*, 24. [CrossRef] [PubMed]
52. Hasanuzzaman, M.; Nahar, K.; Anee, T.I.; Khan, M.I.R.; Fujita, M. Silicon-mediated regulation of antioxidant defense and glyoxalase systems confers drought stress tolerance in *Brassica napus* L. *S. Afr. J. Bot.* **2018**, *115*, 50–57. [CrossRef]
53. Foyer, C.H.; Shigeoka, S. Understanding oxidative stress and antioxidant functions to enhance photosynthesis. *Plant Physiol.* **2011**, *155*, 93–100. [CrossRef] [PubMed]
54. Palma, F.; Carvajal, F.; Jamilena, M.; Garrido, D. Putrescine treatment increases the antioxidant response and carbohydrate content in zucchini fruit stored at low temperature. *Postharvest Biol. Technol.* **2016**, *118*, 68–70. [CrossRef]
55. Krishna Das, B.; Kumar, A.; Maindola, P.; Mahanty, S.; Jain, S.K.; Reddy, M.K.; Arockiasamy, A. Non-native ligands define the active site of *Pennisetum glaucum* (L.) R. Br dehydroascorbate reductase. *Biochem. Biophys. Res. Commun.* **2016**, *473*, 1152–1157. [CrossRef] [PubMed]
56. Noshi, M.; Yamada, H.; Hatanaka, R.; Tanabe, N.; Tamoi, M.; Shigeoka, S. *Arabidopsis* dehydroascorbate reductase 1 and 2 modulate redox states of ascorbate-glutathione cycle in the cytosol in response to photooxidative stress. *Biosci. Biotechnol. Biochem.* **2017**, *81*, 523–533. [CrossRef]
57. Do, H.; Kim, I.S.; Jeon, B.W.; Lee, C.W.; Park, A.K.; Wi, A.R.; Shin, S.C.; Park, H.; Kim, Y.S.; Yoon, H.S.; et al. Structural understanding of the recycling of oxidized ascorbate by dehydroascorbate reductase (*OsDHAR*) from *Oryza sativa* L. *japonica*. *Sci. Rep.* **2016**, *6*, 19498. [CrossRef]
58. Kataya, A.R.; Reumann, S. *Arabidopsis* glutathione reductase 1 is dually targeted to peroxisomes and the cytosol. *Plant Signal. Behav.* **2010**, *5*, 171–175. [CrossRef]
59. Yousuf, P.Y.; Hakeem, K.U.R.; Chandna, R.; Ahmad, P. Role of glutathione reductase in plant abiotic stress. In *Abiotic Stress Responses in Plants*; Ahmad, P., Prasad, M.N.V., Eds.; Springer: New York, NY, USA, 2012; pp. 149–158.
60. Hanukoglu, I. Conservation of the enzyme–coenzyme interfaces in FAD and NADP binding adrenodoxin reductase—A ubiquitous enzyme. *J. Mol. Evol.* **2017**, *85*, 205–218. [CrossRef]
61. Tang, X.; Webb, M.A. Soybean root nodule cDNA encoding glutathione reductase. *Plant Physiol.* **1994**, *104*, 1081–1082. [CrossRef] [PubMed]
62. Pang, C.H.; Wang, B.S. Role of ascorbate peroxidase and glutathione reductase in ascorbate–glutathione cycle and stress tolerance in plants. In *Ascorbate-Glutathione Pathway and Stress Tolerance in Plants*; Anjum, N., Chan, M.T., Umar, S., Eds.; Springer: Dordrecht, The Netherlands, 2010; pp. 91–113.
63. Couto, N.; Wood, J.; Barber, J. The role of glutathione reductase and related enzymes on cellular redox homoeostasis network. *Free Radic. Biol. Med.* **2016**, *95*, 27–42. [CrossRef] [PubMed]
64. Foyer, C.H.; Pellny, T.K.; Locato, V.; Hull, J.; De Gara, L. Analysis of redox relationships in the plant cell cycle: Determination of ascorbate, glutathione, and poly (ADPribose) polymerase (PARP) in plant cell cultures. In *Redox-Mediated Signal Transduction, Methods in Molecular Biology*; Hancock, J., Conway, M., Eds.; Springer: New York, NY, USA, 2019; pp. 165–181.
65. Zhu, S.; He, L.; Zhang, F.; Li, M.; Jiao, S.; Li, Y.; Chen, M.; Zhao, X.E.; Wang, H. Fluorimetric evaluation of glutathione reductase activity and its inhibitors using carbon quantum dots. *Talanta* **2016**, *161*, 769–774. [CrossRef] [PubMed]
66. Hasanuzzaman, M.; Oku, H.; Nahar, K.; Bhuyan, M.B.; Mahmud, J.A.; Baluska, F.; Fujita, M. Nitric oxide-induced salt stress tolerance in plants: ROS metabolism, signaling, and molecular interactions. *Plant Biotechnol. Rep.* **2018**, *12*, 77–92. [CrossRef]
67. Nahar, K.; Hasanuzzaman, M.; Suzuki, T.; Fujita, M. Polyamines-induced aluminum tolerance in mung bean: A study on antioxidant defense and methylglyoxal detoxification systems. *Ecotoxicology* **2017**, *26*, 58–73. [CrossRef] [PubMed]
68. Hasanuzzaman, M.; Alam, M.M.; Nahar, K.; Mohsin, S.M.; Bhuyan, M.H.M.B.; Parvin, K.; Fujita, M. Silicon-induced antioxidant defense and methylglyoxal detoxification works coordinately in alleviating nickel toxicity in *Oryza sativa* L. *Ecotoxicology* **2019**, *28*, 261–276. [CrossRef] [PubMed]
69. Anjum, N.A.; Khan, N.A.; Sofo, A.; Baier, M.; Kizek, R. Redox homeostasis managers in plants under environmental stresses. *Front. Environ. Sci.* **2016**, *4*, 35. [CrossRef]

70. Smirnoff, N. Ascorbic acid: Metabolism and functions of a multi-facetted molecule. *Curr. Opin. Plant Biol.* **2000**, *3*, 229–235. [CrossRef]
71. Foyer, C.H. Redox homeostasis: Opening up ascorbate transport. *Nat. Plants* **2015**, *1*, 114012. [CrossRef]
72. Debolt, S.; Melino, V.; Ford, C.M. Ascorbate as a biosynthetic precursor in plants. *Ann. Bot.* **2006**, *99*, 3–8. [CrossRef]
73. Zechmann, B. Diurnal changes of subcellular glutathione content in *Arabidopsis thaliana*. *Biol. Plant.* **2017**, *61*, 791–796. [CrossRef]
74. Hernández, L.E.; Sobrino-Plata, J.; Montero-Palmero, M.B.; Carrasco-Gil, S.; Flores-Cáceres, M.L.; Ortega-Villasante, C.; Escobar, C. Contribution of glutathione to the control of cellular redox homeostasis under toxic metal and metalloid stress. *J. Exp. Bot.* **2015**, *66*, 2901–2911. [CrossRef] [PubMed]
75. Lallement, P.A.; Roret, T.; Tsan, P.; Gualberto, J.M.; Girardet, J.M.; Didierjean, C.; Rouhier, N.; Hecker, A. Insights into ascorbate regeneration in plants: Investigating the redox and structural properties of dehydroascorbate reductases from *Populus trichocarpa*. *Biochem. J.* **2016**, *473*, 717–731. [CrossRef] [PubMed]
76. Foyer, C.H.; Noctor, G. Ascorbate and glutathione: The heart of the redox hub. *Plant Physiol* **2011**, *155*, 2–18. [CrossRef] [PubMed]
77. Begara-Morales, J.C.; Sánchez-Calvo, B.; Chaki, M.; Valderrama, R.; Mata-Pérez, C.; Corpas, F.J.; Barroso, J.B. Protein *S*-nitrosylation and *S*-glutathionylation as regulators of redox homeostasis during abiotic stress response. In *Redox State as a Central Regulator of Plant-Cell Stress Responses*; Gupta, D., Palma, J., Corpas, F.J., Eds.; Springer: Cham, Switzerland, 2016; pp. 365–386.
78. Miret, J.A.; Müller, M. AsA/DHA redox pair influencing plant growth and stress tolerance. In *Ascorbic Acid in Plant Growth, Development and Stress Tolerance*; Hossain, M., Munné-Bosch, S., Burritt, D., Diaz-Vivancos, P., Fujita, M., Lorence, A., Eds.; Springer: Cham, Switzerland, 2017; pp. 297–319.
79. Munné-Bosch, S.; Queval, G.; Foyer, C.H. The impact of global change factors on redox signaling underpinning stress tolerance. *Plant Physiol.* **2013**, *161*, 5–19. [CrossRef] [PubMed]
80. Parsons, H.T.; Fry, S.C. Oxidation of dehydroascorbic acid and 2,3-diketogulonate under plant apoplastic conditions. *Phytochemistry* **2012**, *75*, 41–49. [CrossRef] [PubMed]
81. Li, R.; Xin, S.; Tao, C.; Jin, X.; Li, H. Cotton ascorbate oxidase promotes cell growth in cultured tobacco bright yellow-2 cells through generation of apoplast oxidation. *Int. J. Mol. Sci.* **2017**, *18*, 1346. [CrossRef] [PubMed]
82. Han, Y.; Chaouch, S.; Mhamdi, A.; Queval, G.; Zechmann, B.; Noctor, G. Functional analysis of Arabidopsis mutants points to novel roles for glutathione in coupling H_2O_2 to activation of salicylic acid accumulation and signaling. *Antioxid. Redox Signal.* **2013**, *18*, 2106–2121. [CrossRef] [PubMed]
83. Czarnocka, W.; Karpiński, S. Friend or foe? Reactive oxygen species production, scavenging and signaling in plant response to environmental stresses. *Free Radic. Biol. Medic.* **2018**, *122*, 4–20. [CrossRef] [PubMed]
84. Nafees, M.; Fahad, S.; Shah, F.; Sha, A.M.; Bukhari, M.A.; Maryam; Ahmed, I.; Ahmad, S.; Hussain, S. Reactive Oxygen Species Signaling in Plants. In *Reactive Oxygen Species Signaling in Plants*; Hasanuzzaman, M., Hakeem, K., Nahar, K., Alharby, H., Eds.; Springer: Cham, Switzerland, 2019; pp. 259–272.
85. Noctor, G.; Reichheld, J.-P.; Foyer, C.H. ROS-related redox regulation and signaling in plants. *Sem. Cell Develop. Biol.* **2019**, *80*, 3–12. [CrossRef]
86. Singh, A.; Kumar, A.; Yadav, S.; Singh, I.K. Reactive oxygen species-mediated signaling during abiotic stress. *Plant Gene* **2019**, *18*, 100173. [CrossRef]
87. Soares, C.; Carvalho, M.E.A.; Azevedo, R.A.; Fidalgo, F. Plants facing oxidative challenges—A little help from the antioxidant networks. *Environ. Exp. Bot.* **2019**, *161*, 4–25. [CrossRef]
88. Hasanuzzaman, M.; Hossain, M.A.; Fujita, M. Exogenous selenium pretreatment protects rapeseed seedlings from cadmium-induced oxidative stress by upregulating antioxidant defense and methylglyoxal detoxification systems. *Biol. Trace Elem. Res.* **2012**, *149*, 248–261. [CrossRef] [PubMed]
89. Latowski, D.; Surówka, E.; Strzałka, K. Regulatory role of components of ascorbate–glutathione pathway in plant stress tolerance. In *Ascorbate-Glutathione Pathway and Stress Tolerance in Plants*; Anjum, N., Chan, M.T., Umar, S., Eds.; Springer: Dordrecht, The Netherlands, 2010; pp. 1–53.
90. Hernández, J.A.; Barba-Espín, G.; Diaz-Vivancos, P. Glutathione-mediated biotic stress tolerance in plants. In *Glutathione in Plant Growth, Development, and Stress Tolerance*; Hossain, M., Mostofa, M., Diaz-Vivancos, P., Burritt, D., Fujita, M., Tran, L.S., Eds.; Springer: Cham, Switzerland, 2017; pp. 309–319.

91. Rahman, A.K.; Al Mahmud, J.; Hasanuzzaman, M.; Hossain, M.S.; Fujita, M. Salt stress tolerance in rice: Emerging role of exogenous phytoprotectants. In *Advances in International Rice Research*; Li, Q., Ed.; InTech: Rijeka, Croatia, 2017; pp. 139–174.
92. Parvin, K.; Hasanuzzaman, M.; Bhuyan, M.H.M.; Mohsin, S.M.; Fujita, M. Quercetin mediated salt tolerance in tomato through the enhancement of plant antioxidant defense and glyoxalase systems. *Plants* **2019**, *8*, 247. [CrossRef] [PubMed]
93. Rahman, A.; Nahar, K.; Hasanuzzaman, M.; Fujita, M. Calcium supplementation improves Na^+/K^+ ratio, antioxidant defense and glyoxalase systems in salt-stressed rice seedlings. *Front. Plant Sci.* **2016**, *7*, 609. [CrossRef] [PubMed]
94. Rahman, A.; Nahar, K.; Hasanuzzaman, M.; Fujita, M. Manganese-induced cadmium stress tolerance in rice seedlings: Coordinated action of antioxidant defense, glyoxalase system and nutrient homeostasis. *C. R. Biol.* **2016**, *339*, 462–474. [CrossRef] [PubMed]
95. Nahar, K.; Hasanuzzaman, M.; Alam, M.M.; Fujita, M. Roles of exogenous glutathione in antioxidant defense system and methylglyoxal detoxification during salt stress in mung bean. *Biol. Plant.* **2015**, *59*, 745–756. [CrossRef]
96. Hossain, M.S.; Alam, M.U.; Rahman, A.; Hasanuzzaman, M.; Nahar, K.; Mahmud, J.A.; Fujita, M. Use of iso-osmotic solution to understand salt stress responses in lentil (*Lens culinaris* Medik.). *S. Afr. J. Bot.* **2017**, *113*, 346–354. [CrossRef]
97. Singh, M.; Singh, V.P.; Prasad, S.M. Nitrogen alleviates salinity toxicity in *Solanum lycopersicum* seedlings by regulating ROS homeostasis. *Plant Physiol. Biochem.* **2019**, *141*, 466–476. [CrossRef] [PubMed]
98. Ahmad, P.; Abd_Allah, E.F.; Alyemeni, M.N.; Wijaya, L.; Alam, P.; Bhardwaj, R.; Siddique, K.H. Exogenous application of calcium to 24-epibrassinosteroid pre-treated tomato seedlings mitigates NaCl toxicity by modifying ascorbate–glutathione cycle and secondary metabolites. *Sci. Rep.* **2018**, *8*, 13515. [CrossRef] [PubMed]
99. Ahanger, M.A.; Alyemeni, M.N.; Wijaya, L.; Alamri, S.A.; Alam, P.; Ashraf, M.; Ahmad, P. Potential of exogenously sourced kinetin in protecting *Solanum lycopersicum* from NaCl-induced oxidative stress through up-regulation of the antioxidant system, ascorbate-glutathione cycle and glyoxalase system. *PLoS ONE* **2018**, *13*, e0202175. [CrossRef]
100. Sehar, Z.; Masood, A.; Khan, N.A. Nitric oxide reverses glucose-mediated photosynthetic repression in wheat (*Triticum aestivum* L.) under salt stress. *Environ. Exp. Bot.* **2019**, *161*, 277–289. [CrossRef]
101. Yan, Y.; Pan, C.; Du, Y.; Li, D.; Liu, W. Exogenous salicylic acid regulates reactive oxygen species metabolism and ascorbate–glutathione cycle in *Nitraria tangutorum* Bobr. under salinity stress. *PMBP* **2018**, *24*, 577–589. [CrossRef] [PubMed]
102. Hossain, M.S.; Hasanuzzaman, M.; Sohag, M.M.H.; Bhuyan, M.H.M.B.; Fujita, M. Acetate-induced modulation of ascorbate: Glutathione cycle and restriction of sodium accumulation in shoot confer salt tolerance in *Lens culinaris* Medik. *PMBP Plants* **2019**, *25*, 443–455. [CrossRef] [PubMed]
103. Talaat, N.B. Effective microorganisms enhance the scavenging capacity of the ascorbate-glutathione cycle in common bean (*Phaseolus vulgaris* L.) plants grown in salty soils. *Plant Physiol. Biochem.* **2014**, *80*, 136–143. [CrossRef] [PubMed]
104. Bhuiyan, T.F.; Ahamed, K.U.; Nahar, K.; Al Mahmud, J.; Bhuyan, M.B.; Anee, T.I.; Fujita, M.; Hasanuzzaman, M. Mitigation of PEG-induced drought stress in rapeseed (*Brassica rapa* L.) by exogenous application of osmolytes. *Biocatal. Agric. Biotechnol.* **2019**, *20*, 101197. [CrossRef]
105. Sun, J.; Gu, J.; Zeng, J.; Han, S.; Song, A.P.; Chen, F.; Fang, W.; Jiang, J.; Chen, S. Changes in leaf morphology, antioxidant activity and photosynthesis capacity in two different drought-tolerant cultivars of chrysanthemum during and after water stress. *Sci. Hortic.* **2013**, *161*, 249–258. [CrossRef]
106. Lou, L.; Li, X.; Chen, J.; Li, Y.; Tang, Y.; Lv, J. Photosynthetic and ascorbate-glutathione metabolism in the flag leaves as compared to spikes under drought stress of winter wheat (*Triticum aestivum* L.). *PLoS ONE* **2018**, *13*, e0194625. [CrossRef]
107. Shan, C.; Zhang, S.; Ou, X. The roles of H_2S and H_2O_2 in regulating AsA-GSH cycle in the leaves of wheat seedlings under drought stress. *Protoplasma* **2018**, *255*, 1257–1262. [CrossRef]

108. Nguyen, K.H.; Mostofa, M.G.; Watanabe, Y.; Tran, C.D.; Rahman, M.M.; Tran, L.S.P. Overexpression of *GmNAC085* enhances drought tolerance in *Arabidopsis* by regulating glutathione biosynthesis, redox balance and glutathione-dependent detoxification of reactive oxygen species and methylglyoxal. *Environ. Exp. Bot.* **2019**, *161*, 242–254. [CrossRef]
109. Sreeharsha, R.V.; Mudalkar, S.; Sengupta, D.; Unnikrishnan, D.K.; Reddy, A.R. Mitigation of drought-induced oxidative damage by enhanced carbon assimilation and an efficient antioxidative metabolism under high CO_2 environment in pigeonpea (*Cajanus cajan* L.). *Photosynth. Res.* **2019**, *139*, 425–439. [CrossRef] [PubMed]
110. Sarker, U.; Oba, S. Catalase, superoxide dismutase and ascorbate-glutathione cycle enzymes confer drought tolerance of *Amaranthus tricolor*. *Sci. Rep.* **2018**, *8*, 16496. [CrossRef] [PubMed]
111. Nahar, K.; Hasanuzzaman, M.; Alam, M.M.; Fujita, M. Glutathione-induced drought stress tolerance in mung bean: Coordinated roles of the antioxidant defence and methylglyoxal detoxification systems. *AoB Plants* **2015**, *7*, plv069. [CrossRef] [PubMed]
112. Lima, C.S.; Ferreira-Silva, S.L.; Carvalho, F.E.L.; Neto, M.C.L.; Aragão, R.M.; Silva, E.N.; Sousa, R.M.J.; Silveira, J.A.G. Antioxidant protection and PSII regulation mitigate photo-oxidative stress induced by drought followed by high light in cashew plants. *Environ. Exp. Bot.* **2018**, *149*, 59–69. [CrossRef]
113. Al Mahmud, J.; Biswas, P.K.; Nahar, K.; Fujita, M.; Hasanuzzaman, M. Exogenous application of gibberellic acid mitigates drought-induced damage in spring wheat. *Acta Agrobot.* **2019**, *72*, 1776.
114. Hasanuzzaman, M.; Alam, M.; Rahman, A.; Hasanuzzaman, M.; Nahar, K.; Fujita, M. Exogenous proline and glycine betaine mediated upregulation of antioxidant defense and glyoxalase systems provides better protection against salt-induced oxidative stress in two rice (*Oryza sativa* L.) varieties. *BioMed. Res. Int.* **2014**, *2014*, 757219. [CrossRef]
115. Hasanuzzaman, M.; Hossain, M.A.; Fujita, M. Selenium-induced up-regulation of the antioxidant defense and methylglyoxal detoxification system reduces salinity-induced damage in rapeseed seedlings. *Biol. Trace Elem. Res.* **2011**, *143*, 1704–1721. [CrossRef]
116. Alam, M.M.; Nahar, K.; Hasanuzzaman, M.; Fujita, M. Alleviation of osmotic stress in *Brassica napus*, *B. campestris*, and *B. juncea* by ascorbic acid application. *Biol. Plant.* **2014**, *58*, 697–708. [CrossRef]
117. Alam, M.M.; Hasanuzzaman, M.; Nahar, K.; Fujita, M. Exogenous salicylic acid ameliorates short-term drought stress in mustard (*Brassica juncea* L.) seedlings by up-regulating the antioxidant defense and glyoxalase system. *Aust. J. Crop Sci.* **2013**, *7*, 1053–1063.
118. Jin, X.; Liu, T.; Xu, J.; Gao, Z.; Hu, X. Exogenous GABA enhances muskmelon tolerance to salinity-alkalinity stress by regulating redox balance and chlorophyll biosynthesis. *BMC Plant Biol.* **2019**, *19*, 48. [CrossRef]
119. García-Martí, M.; Piñero, M.C.; García-Sanchez, F.; Mestre, T.C.; López-Delacalle, M.; Martínez, V.; Rivero, R.M. Amelioration of the oxidative stress generated by simple or combined abiotic stress through the K^+ and Ca^{2+} supplementation in tomato plants. *Antioxidants* **2019**, *8*, 81. [CrossRef] [PubMed]
120. Li, J.; Yang, Y.; Sun, K.; Chen, Y.; Chen, X.; Li, X. Exogenous melatonin enhances cold, salt and drought stress tolerance by improving antioxidant defense in tea plant (*Camellia sinensis* (L.) O. Kuntze). *Molecules* **2019**, *24*, 1826. [CrossRef] [PubMed]
121. Ghori, N.-H.; Ghori, T.; Hayat, M.Q.; Imadi, S.R.; Gul, A.; Altay, V.; Ozturk, M. Heavy metal stress and responses in plants. *Int. J. Environ. Sci. Technol.* **2019**, *16*, 1807–1828. [CrossRef]
122. Mahmud, J.A.; Hasanuzzaman, M.; Nahar, K.; Rahman, A.; Hossain, M.S.; Fujita, M. Maleic acid assisted improvement of metal chelation and antioxidant metabolism confers chromium tolerance in *Brassica juncea* L. *Ecotoxicol. Environ. Saf.* **2017**, *144*, 216–226. [CrossRef] [PubMed]
123. Hasanuzzaman, M.; Nahar, K.; Gill, S.S.; Alharby, H.F.; Razafindrabe, B.H.N.; Fujita, M. Hydrogen peroxide pretreatment mitigates cadmium-induced oxidative stress in *Brassica napus* L.: An intrinsic study on antioxidant defense and glyoxalase systems. *Front. Plant Sci.* **2017**, *8*, 115. [CrossRef]
124. Bharwana, S.A.; Ali, S.; Farooq, M.A.; Iqbal, N.; Abbas, F.; Ahmad, M.S.A. Alleviation of lead toxicity by silicon is related to elevated photosynthesis, antioxidant enzymes suppressed lead uptake and oxidative stress in cotton. *J. Bioremed. Biodeg.* **2013**, *4*, 187.
125. Abd_Allah, E.F.; Hashem, A.; Alam, P.; Ahmad, P. Silicon alleviates nickel-induced oxidative stress by regulating antioxidant defense and glyoxalase systems in mustard plants. *J. Plant Growth Regul.* **2019**, 1–14. [CrossRef]

126. Rahman, A.; Mostofa, M.G.; Alam, M.M.; Nahar, K.; Hasanuzzaman, M.; Fujita, M. Calcium mitigates arsenic toxicity in rice seedlings by reducing arsenic uptake and modulating the antioxidant defense and glyoxalase systems and stress markers. *BioMed. Res. Int.* **2015**, *2015*, 340812. [CrossRef]
127. Hasanuzzaman, M.; Fujita, M. Exogenous sodium nitroprusside alleviates arsenic-induced oxidative stress in wheat (*Triticum aestivum* L.) seedlings by enhancing antioxidant defense and glyoxalase system. *Ecotoxicology* **2013**, *22*, 584–596. [CrossRef]
128. Gill, R.A.; Zang, L.; Ali, B.; Farooq, M.A.; Cui, P.; Yang, S.; Ali, S.; Zhou, W. Chromium-induced physio-chemical and ultrastructural changes in four cultivars of *Brassica napus* L. *Chemosphere* **2015**, *120*, 154–164. [CrossRef]
129. Awasthi, J.P.; Saha, B.; Panigrahi, J.; Yanase, E.; Koyama, H.; Panda, S.K. Redox balance, metabolic fingerprint and physiological characterization in contrasting North East Indian rice for aluminum stress tolerance. *Sci. Rep.* **2019**, *9*, 8681. [CrossRef] [PubMed]
130. Nahar, K.; Rahman, M.; Hasanuzzaman, M.; Alam, M.M.; Rahman, A.; Suzuki, T.; Fujita, M. Physiological and biochemical mechanisms of spermine induced cadmium stress tolerance in mung bean (*Vigna radiata* L.) seedlings. *Environ. Sci. Pollut. Res. Int.* **2016**, *23*, 21206–21218. [CrossRef] [PubMed]
131. Nahar, K.; Rahman, M.; Hasanuzzaman, M.; Alam, M.M.; Rahman, A.; Suzuki, T.; Fujita, M. Polyamine and nitric oxide crosstalk: Antagonistic effects on cadmium toxicity in mungbean plants through upregulating the metal detoxification, antioxidant defense and methylglyoxal detoxification systems. *Ecotoxicol. Environ. Saf.* **2016**, *126*, 245–255. [CrossRef] [PubMed]
132. Rahman, A.; Mostofa, M.G.; Nahar, K.; Hasanuzzaman, M.; Fujita, M. Exogenous calcium alleviates cadmium-induced oxidative stress in rice (*Oryza sativa* L.) seedlings by regulating the antioxidant defense and glyoxalase systems. *Braz. J. Bot.* **2016**, *39*, 393–407. [CrossRef]
133. Rahman, A.; Hossain, M.S.; Mahmud, J.A.; Nahar, K.; Hasanuzzaman, M.; Fujita, M. Manganese-induced salt stress tolerance in rice seedlings: Regulation of ion homeostasis, antioxidant defense and glyoxalase systems. *PMBP* **2016**, *22*, 291–306. [CrossRef] [PubMed]
134. Wan, Y.; Wang, K.; Liu, Z.; Yu, Y.; Wang, Q.; Li, H. Effect of selenium on the subcellular distribution of cadmium and oxidative stress induced by cadmium in rice (*Oryza sativa* L.). *Environ. Sci. Pollut. Res.* **2019**, *26*, 16220. [CrossRef] [PubMed]
135. Hussain, I.; Siddique, A.; Ashraf, M.A.; Rasheed, R.; Ibrahim, M.; Iqbal, M.; Akbar, S.; Imran, M. Does exogenous application of ascorbic acid modulate growth, photosynthetic pigments and oxidative defense in okra (*Abelmoschus esculentus* (L.) Moench) under lead stress? *Acta Physiol. Plant.* **2017**, *39*, 144. [CrossRef]
136. Mahmud, J.A.; Hasanuzzaman, M.; Nahar, K.; Rahman, A.; Fujita, M. Relative tolerance of different species of *Brassica* to cadmium toxicity: Coordinated role of antioxidant defense and glyoxalase systems. *Plant Omics J.* **2017**, *10*, 107–117. [CrossRef]
137. Mahmud, J.A.; Hasanuzzaman, M.; Nahar, K.; Rahman, A.; Fujita, M. EDTA reduces cadmium toxicity in mustard (*Brassica juncea* L.) by enhancing metal chelation, antioxidant defense and glyoxalase systems. *Acta Agrobot.* **2019**, *72*, 1722. [CrossRef]
138. Mahmud, J.A.; Hasanuzzaman, M.; Nahar, K.; Bhuyan, M.H.M.B.; Fujita, M. Insights into citric acid-induced cadmium tolerance and phytoremediation in *Brassica juncea* L.: Coordinated functions of metal chelation, antioxidant defense and glyoxalase systems. *Ecotoxicol. Environ. Saf.* **2018**, *147*, 990–1001. [CrossRef]
139. Dias, M.C.; Mariz-Ponte, N.; Santos, C. Lead induces oxidative stress in *Pisum sativum* plants and changes the levels of phytohormones with antioxidant role. *Plant Physiol. Biochem.* **2019**, *137*, 121–129. [CrossRef] [PubMed]
140. Kaya, C.; Akram, N.A.; Sürücü, A.; Ashraf, M. Alleviating effect of nitric oxide on oxidative stress and antioxidant defence system in pepper (*Capsicum annuum* L.) plants exposed to cadmium and lead toxicity applied separately or in combination. *Sci. Hortic.* **2019**, *255*, 52–60. [CrossRef]
141. Zhang, K.; Wang, G.; Bao, M.; Wang, L.; Xie, X. Exogenous application of ascorbic acid mitigates cadmium toxicity and uptake in Maize (*Zea mays* L.). *Environ. Sci. Pollut. Res.* **2019**, *26*, 19261–19271. [CrossRef] [PubMed]
142. Rodriguez, V.M.; Soengas, P.; Alonso-Villaverde, V.; Sotelo, T.; Cartea, M.E.; Velasco, M.P. Effect of temperature stress on the early vegetative development of *Brassica oleracea* L. *BMC Plant Biol.* **2015**, *15*, 145. [CrossRef] [PubMed]

143. Liang, D.; Gao, F.; Ni, Z.; Lin, L.; Deng, Q.; Tang, Y.; Wang, X.; Luo, X.; Xia, H. Melatonin improves heat tolerance in kiwifruit seedlings through promoting antioxidant enzymatic activity and glutathione S-transferase transcription. *Molecules* **2018**, *23*, 584. [CrossRef] [PubMed]
144. Gupta, N.K.; Agarwal, S.; Agarwal, V.P.; Nathawat, N.S.; Gupta, S.; Singh, G. Effect of short-term heat stress on growth, physiology and antioxidative defence system in wheat seedlings. *Acta Physiol. Plant.* **2013**, *35*, 1837–1842. [CrossRef]
145. Vasseur, F.; Pantin, F.; Vile, D. Changes in light intensity reveal a major role for carbon balance in Arabidopsis responses to high temperature. *Plant Cell Environ.* **2011**, *34*, 1563–1576. [CrossRef]
146. Khanna-Chopra, R.; Chauhan, S. Wheat cultivars differing in heat tolerance show a differential response to oxidative stress during monocarpic senescence under high temperature stress. *Protoplasma* **2015**, *25*, 1241–1251. [CrossRef]
147. Khanna, P.; Kaur, K.; Gupta, A.K. Salicylic acid induces differential antioxidant response in spring maize under high temperature stress. *Indian J. Exp. Biol.* **2016**, *54*, 386–393.
148. Jin, S.H.; Li, X.Q.; Wang, G.G.; Zhu, X.T. Brassinosteroids alleviate high-temperature injury in *Ficus concinna* seedlings via maintaining higher antioxidant defence and glyoxalase systems. *AoB Plants* **2015**, *7*, plv009. [CrossRef]
149. Ahammed, G.J.; Wen, X.; Liu, A.; Chen, S. Endogenous melatonin deficiency aggravates high temperature-induced oxidative stress in *Solanum lycopersicum* L. *Environ. Exp. Bot.* **2018**, *161*, 303–311. [CrossRef]
150. Sang, Q.Q.; Shu, S.; Shana, X.; Guo, S.R.; Sun, J. Effects of exogenous spermidine on antioxidant system of tomato seedlings exposed to high temperature stress. *Russ. J. Plant Physiol.* **2016**, *63*, 645–655. [CrossRef]
151. Sgobba, A.; Paradiso, A.; Dipierro, S.; Gara, L.B.; de Pinto, M.C. Changes in antioxidants are critical in determining cell responses to short- and long-term heat stress. *Physiol. Plant.* **2015**, *153*, 68–78. [CrossRef] [PubMed]
152. Damanik, R.I.; Maziah, M.; Ismail, M.R.; Ahmad, S.; Zain, A.M. Responses of the antioxidative enzymes in Malaysian rice (*Oryza sativa* L.) cultivars under submergence condition. *Acta Physiol. Plant* **2010**, *32*, 739–747. [CrossRef]
153. Kim, Y.; Seo, C.W.; Khan, A.L.; Mun, B.G.; Shahzad, R.; Ko, J.W.; Yun, B.W.; Park, S.K.; Lee, I.J. Exo-ethylene application mitigates waterlogging stress in soybean (*Glycine max* L.). *BMC Plant Biol.* **2018**, *18*, 254. [CrossRef] [PubMed]
154. Wang, H.; Chen, Y.; Hu, W.; Snider, J.L.; Zhou, Z. Short-term soil-waterlogging contributes to cotton cross tolerance to chronic elevated temperature by regulating ROS metabolism in the subtending leaf. *Plant Physiol. Biochem.* **2019**, *139*, 333–341. [CrossRef]
155. Sairam, R.K.; Kumutha, D.; Ezhilmathi, K.; Chinnusamy, V.; Meena, R.C. Waterlogging induced oxidative stress and antioxidant enzyme activities in pigeon pea. *Biol. Plant.* **2009**, *53*, 493–504. [CrossRef]
156. Sairam, R.K.; Dharmar, K.; Lekshmy, S.; Chinnusamy, V. Expression of antioxidant defense genes in mung bean (*Vigna radiata* L.) roots under water-logging is associated with hypoxia tolerance. *Acta Physiol. Plant.* **2011**, *3*, 735–744. [CrossRef]
157. Yiu, J.-C.; Liu, C.-W.; Fang, D.Y.; Lai, Y.-S. Waterlogging tolerance of Welsh onion (*Allium fistulosum* L.) enhanced by exogenous spermidine and spermine. *Plant Physiol. Biochem.* **2009**, *47*, 710–716. [CrossRef]
158. Simova-Stoilova, L.; Demirevska, K.; Kingston-Smith, A.; Feller, U. Involvement of the leaf antioxidant system in the response to soil flooding in two *Trifolium* genotypes differing in their tolerance to waterlogging. *Plant Sci.* **2012**, *183*, 43–49. [CrossRef]
159. Jaiswal, A.; Srivastava, J.P. Changes in reactive oxygen scavenging system and protein profiles in maize roots in response to nitric oxide under waterlogging stress. *Indian J. Biochem. Biophys.* **2018**, *55*, 26–33.
160. Woo, S.Y.; Lee, D.K.; Lee, Y.K. Net photosynthetic rate, ascorbate peroxidase and glutathione reductase activities of *Erythrina orientalis* in polluted and non-polluted areas. *Photosynthetica* **2007**, *45*, 293–295. [CrossRef]
161. Seyyednejad, S.M.; Koochak, H.; Vaezi, J. Changes in anti-oxidative enzymes activity, protein content and ascorbic acid level in *Prosopis juliflora* exposed to industrial air pollution. *J. Biol. Today World* **2013**, *2*, 482–492.
162. Lucas, J.A.; Gutierrez-Albanchez, E.; Alfaya, T.; Feo-Brito, F.; Gutiérrez-Mañero, F.J. Oxidative stress in ryegrass growing under different air pollution levels and its likely effects on pollen allergenicity. *Plant Physiol. Biochem.* **2019**, *135*, 331–340. [CrossRef] [PubMed]

163. Wang, J.L.; Zeng, Q.; Zhu, J.G.; Liu, G.; Tang, H.Y. Dissimilarity of ascorbate–glutathione (AsA–GSH) cycle mechanism in two rice (*Oryza sativa* L.) cultivars under experimental free-air ozone exposure. *Agric. Ecosyst. Environ.* **2013**, *165*, 39–49. [CrossRef]
164. Dusart, N.; Gérard, J.; Thiec, D.L.; Collignon, C.; Jolivet, Y.; Vaultier, M.N. Integrated analysis of the detoxification responses of two *Euramerican poplar* genotypes exposed to ozone and water deficit: Focus on the ascorbate-glutathione cycle. *Sci. Total Environ.* **2019**, *651*, 2365–2379. [CrossRef] [PubMed]
165. Muneer, S.; Kim, T.H.; Choi, B.C.; Lee, B.S.; Lee, J.H. Effect of CO, NO_x and SO_2 on ROS production, photosynthesis and ascorbate–glutathione pathway to induce *Fragaria* x *annasa* as a hyperaccumulator. *Redox Biol.* **2014**, *2*, 91–98. [CrossRef] [PubMed]
166. Ma, Y.; Wang, B.; Zhang, R.; Gao, Y.; Zhang, X.; Li, Y.; Zuo, Z. Initial simulated acid rain impacts reactive oxygen species metabolism and photosynthetic abilities in *Cinnamonum camphora* undergoing high temperature. *Ind. Crops Prod.* **2019**, *135*, 352–361. [CrossRef]
167. Alamri, S.A.; Siddiqui, M.H.; Al-Khaishany, M.Y.; Khan, M.N.; Ali, H.M.; Alakeel, K.A. Nitric oxide-mediated cross-talk of proline and heat shock proteins induce thermotolerance in *Vicia faba* L. *Environ. Exp. Bot.* **2019**, *161*, 290–302. [CrossRef]
168. Anee, T.I.; Nahar, K.; Rahman, A.; Mahmud, J.A.; Bhuiyan, T.F.; Alam, M.U.; Fujita, M.; Hasanuzzaman, M. Oxidative damage and antioxidant defense in *Sesamum indicum* after different waterlogging durations. *Plants* **2019**, *8*, 196. [CrossRef]
169. Lin, K.H.R.; Weng, C.C.; Lo, H.F.; Chen, J.T. Study of the root antioxidative system of tomatoes and eggplants under waterlogged conditions. *Plant Sci.* **2004**, *167*, 355–365. [CrossRef]
170. Bhuyan, M.B.; Hasanuzzaman, M.; Al Mahmud, J.; Hossain, M.S.; Alam, M.U.; Fujita, M. Explicating physiological and biochemical responses of wheat cultivars under acidity stress: Insight into the antioxidant defense and glyoxalase systems. *PMBP* **2019**, *25*, 865–879. [CrossRef] [PubMed]
171. Conklin, P.L.; Saracco, S.A.; Norris, S.R.; Last, R.L. Identification of ascorbic acid-deficient *Arabidopsis thaliana* mutants. *Genetics* **2000**, *154*, 847–856. [PubMed]
172. Gao, Q.; Zhang, L. Ultraviolet-B-induced oxidative stress and antioxidant defense system responses in ascorbate-deficient vtc1 mutants of *Arabidopsis thaliana*. *J. Plant Physiol.* **2008**, *165*, 138–148. [CrossRef] [PubMed]
173. Singh, V.P.; Srivastava, P.K.; Prasad, S.M. Differential effect of UV-B radiation on growth, oxidative stress and ascorbate-glutathione cycle in two cyanobacteria under copper toxicity. *Plant Physiol. Biochem.* **2012**, *61*, 61–70. [CrossRef] [PubMed]
174. Shiu, C.-T.; Lee, T.-M. Ultraviolet-B-induced oxidative stress and responses of the ascorbate–glutathione cycle in a marine macroalga *Ulva fasciata*. *J. Exp. Bot.* **2005**, *56*, 2851–2865. [CrossRef] [PubMed]
175. Noshi, M.; Hatanaka, R.; Tanabe, N.; Terai, Y.; Maruta, T.; Shigeoka, S. Redox regulation of ascorbate and glutathione by a chloroplastic dehydroascorbate reductase is required for high-light stress tolerance in *Arabidopsis*. *Biosci. Biotechnol. Biochem.* **2016**, *80*, 870–877. [CrossRef]
176. Zheng, L.-D.; Li, M.; Chow, W.S.; Peng, C.-L. Susceptibility of an ascorbate-deficient mutant of *Arabidopsis* to high-light stress. *Photosynthetica* **2018**, *56*, 427–432. [CrossRef]
177. Choudhury, F.K.; Devireddy, A.R.; Azad, R.K.; Shulaev, V.; Mittler, R. Rapid accumulation of glutathione during light stress in *Arabidopsis*. *Plant Cell Physiol.* **2018**, *59*, 1817–1826. [CrossRef]
178. Akram, N.A.; Shafiq, F.; Ashraf, M. Ascorbic acid-a potential oxidant scavenger and its role in plant development and abiotic stress tolerance. *Front. Plant Sci.* **2017**, *8*. [CrossRef]
179. Ashraf, M.A.; Riaz, M.; Arif, M.S.; Rasheed, R.; Iqbal, M.; Hussain, I.; Salman, M. The role of non-enzymatic antioxidants in improving abiotic stress tolerance in plants. In *Plant Tolerance to Environmental Stress: Role of Phytoprotectants*; Hasanuzzaman, M., Fujita, M., Oku, H., Islam, M.T., Eds.; CRC Press: Boca Raton, FL, USA, 2019; pp. 129–143.
180. Naz, H.; Akram, N.A.; Ashraf, M. Impact of ascorbic acid on growth and some physiological attributes of cucumber (*Cucumis sativus*) plants under water-deficit conditions. *Pak. J. Bot.* **2016**, *48*, 877–883.
181. Xu, Y.; Huang, B. Exogenous ascorbic acid mediated abiotic stress tolerance in plants. In *Ascorbic Acid in Plant Growth, Development and Stress Tolerance*; Hossain, M., Munné-Bosch, S., Burritt, D., Diaz-Vivancos, P., Fujita, M., Lorence, A., Eds.; Springer: Cham, Switzerland, 2017; pp. 233–253.
182. Billah, M.; Rohman, M.; Hossain, N.; Uddin, M.S. Exogenous ascorbic acid improved tolerance in maize (*Zea mays* L.) by increasing antioxidant activity under salinity stress. *Afr. J. Agric. Res.* **2017**, *12*, 1437–1446.

183. Shafiq, S.; Akram, N.A.; Ashraf, M.; Arshad, A. Synergistic effects of drought and ascorbic acid on growth, mineral nutrients and oxidative defense system in canola (*Brassica napus* L.) plants. *Acta Physiol. Plant.* **2014**, *36*, 1539–1553. [CrossRef]
184. Kumar, S.; Kaur, R.; Kaur, N.; Bhandhari, K.; Kaushal, N.; Gupta, K.; Bains, T.; Nayyar, H. Heat-stress induced inhibition in growth and chlorosis in mungbean (*Phaseolus aureus* Roxb.) is partly mitigated by ascorbic acid application and is related to reduction in oxidative stress. *Acta Physiol. Plant.* **2011**, *33*, 2091–2101. [CrossRef]
185. Kobayakawa, H.; Imai, K. Exogenous ascorbic acid scarcely ameliorates inhibition of photosynthesis in rice leaves by O_3. *Plant Prod. Sci.* **2017**, *20*, 83–89. [CrossRef]
186. Wang, J.; Wu, B.; Yin, H.; Fan, Z.; Li, X.; Ni, S.; He, L.; Li, J. Overexpression of *CaAPX* induces orchestrated reactive oxygen scavenging and enhances cold and heat tolerances in tobacco. *BioMed. Res. Int.* **2017**, *2017*, 4049534. [CrossRef] [PubMed]
187. Saeidi-Sar, S.; Abbaspour, H.; Afshari, H.; Yaghoobi, S.R. Effects of ascorbic acid and gibberellin A3 on alleviation of salt stress in common bean (*Phaseolus vulgaris* L.) seedlings. *Acta Physiol. Plant.* **2013**, *35*, 667–677. [CrossRef]
188. Rehman, R.U.; Zia, M.; Abbasi, B.H.; Lu, G.; Chaudhary, M.F. Ascorbic acid and salicylic acid mitigate NaCl stress in *Caralluma tuberculata* Calli. *Appl. Biochem. Biotechnol.* **2014**, *173*, 968–979. [CrossRef]
189. Wang, R.; Liu, S.; Zhou, F.; Ding, C.; Hua, C. Exogenous ascorbic acid and glutathione alleviate oxidative stress induced by salt stress in the chloroplasts of *Oryza sativa* L. *Z. Naturforsch. C* **2014**, *69*, 226–236. [CrossRef]
190. Rady, M.M.; Hemida, K.A. Sequenced application of ascorbate-proline-glutathione improves salt tolerance in maize seedlings. *Ecotoxicol. Environ. Saf.* **2016**, *133*, 252–259. [CrossRef]
191. Terzi, R.; Kalaycıoglu, E.; Demiralay, M.; Saglam, A.; Kadioglu, A. Exogenous ascorbic acid mitigates accumulation of abscisic acid, proline and polyamine under osmotic stress in maize leaves. *Acta Physiol. Plant.* **2015**, *37*, 43. [CrossRef]
192. Xu, Y.; Xu, Q.; Huang, B. Ascorbic acid mitigation of water stress-inhibition of root growth in association with oxidative defense in tall fescue (*Festuca arundinacea* Schreb. *) Front. Plant Sci.* **2015**, *6*, 807. [CrossRef] [PubMed]
193. Zhang, C.; Liu, J.; Zhang, Y.; Cai, X.; Gong, P.; Zhang, J.; Wang, T.; Li, H.; Ye, Z. Overexpression of *SlGMEs* leads to ascorbate accumulation with enhanced oxidative stress, cold, and salt tolerance in tomato. *Plant Cell Rep.* **2011**, *30*, 389–398. [CrossRef] [PubMed]
194. Wang, X.; Wu, L.; Xie, J.; Li, T.; Cai, J.; Zhou, Q.; Dai, T.; Jiang, D. Herbicide isoproturon aggravates the damage of low temperature stress and exogenous ascorbic acid alleviates the combined stress in wheat seedlings. *Plant Growth Regul.* **2018**, *84*, 293–301. [CrossRef]
195. Chao, Y.Y.; Hong, C.Y.; Kao, C.H. The decline in ascorbic acid content is associated with cadmium toxicity of rice seedlings. *Plant Physiol. Bioch.* **2010**, *48*, 374–381. [CrossRef] [PubMed]
196. Alamri, S.A.; Siddiqui, M.H.; Al-Khaishany, M.Y.; Nasir Khan, M.; Ali, H.M.; Alaraidh, I.A.; Alsahli, A.A.; Al-Rabiah, H.; Mateen, M. Ascorbic acid improves the tolerance of wheat plants to lead toxicity. *J. Plant Interact.* **2018**, *13*, 409–419. [CrossRef]
197. Ding, X.; Jiang, Y.; He, L.; Zhou, Q.; Yu, J.; Hui, D.; Huang, D. Exogenous glutathione improves high root-zone temperature tolerance by modulating photosynthesis, antioxidant and osmolytes systems in cucumber seedlings. *Sci. Rep.* **2016**, *6*, 35424. [CrossRef] [PubMed]
198. Cao, F.; Liu, L.; Ibrahim, W.; Cai, Y.; Wu, F. Alleviating effects of exogenous glutathione, glycinebetaine, brassinosteroids and salicylic acid on cadmium toxicity in rice seedlings (*Oryza sativa*). *Agrotechnology* **2013**, *2*, 107. [CrossRef]
199. Yuan, H.; Zhang, Y.; Huang, S.; Yang, Y.; Gu, C. Effects of exogenous glutathione and cysteine on growth, lead accumulation, and tolerance of *Iris lactea* var. chinensis. *Environ. Sci. Pollut. Res.* **2015**, *22*, 2808–2816. [CrossRef]
200. Kim, Y.-O.; Bae, H.-J.; Cho, E.; Kang, H. Exogenous Glutathione Enhances Mercury Tolerance by Inhibiting Mercury Entry into Plant Cells. *Front. Plant Sci.* **2017**, *8*, 683. [CrossRef]
201. Zhou, Y.; Wen, Z.; Zhang, J.; Chen, X.; Cui, J.; Xu, W.; Liu, H. Exogenous glutathione alleviates salt-induced oxidative stress in tomato seedlings by regulating glutathione metabolism, redox status, and the antioxidant system. *Sci. Hortic.* **2017**, *220*, 90–101. [CrossRef]

202. Akram, S.; Siddiqui, M.N.; Hussain, B.N.; Al Bari, M.A.; Mostofa, M.G.; Hossain, M.A.; Tran, L.S.P. Exogenous glutathione modulates salinity tolerance of soybean [*Glycine max* (L.) Merrill] at reproductive stage. *J. Plant Growth Regul.* **2017**, *36*, 877–888. [CrossRef]
203. Zhou, Y.; Diao, M.; Cui, J.; Chen, X.; Wen, Z.; Zhang, J.; Liu, H. Exogenous GSH protects tomatoes against salt stress by modulating photosystem II efficiency, absorbed light allocation and H_2O_2-scavenging system in chloroplasts. *J. Integr. Agric.* **2018**, *17*, 2257–2272. [CrossRef]
204. Romero-Puertas, M.C.; Sandalio, L.M. Nitric oxide level is self-regulating and also regulates its ROS partners. *Front. Plant Sci.* **2016**, *7*, 316. [CrossRef] [PubMed]
205. Yang, H.; Mu, J.; Chen, L.; Feng, J.; Hu, J.; Li, L.; Zhou, J.M.; Zuo, J. *S*-Nitrosylation positively regulates ascorbate peroxidase activity during plant stress responses. *Plant Physiol.* **2015**, *167*, 1604–1615. [CrossRef] [PubMed]
206. Begara-Morales, J.C.; Sánchez-Calvo, B.; Chaki, M.; Mata-Pérez, C.; Valderrama, R.; Padilla, M.N.; López-Jaramillo, J.; Luque, F.; Corpas, F.J.; Barroso, J.B. Differential molecular response of monodehydroascorbate reductase and glutathione reductase by nitration and *S*-nitrosylation. *J. Exp. Bot.* **2015**, *66*, 5983–5996. [CrossRef]
207. Tanou, G.; Filippou, P.; Belghazi, M.; Job, D.; Diamantidis, G.; Fotopoulos, V.; Molassiotis, A. Oxidative and nitrosative-based signaling and associated post-translational modifications orchestrate the acclimation of citrus plants to salinity stress. *Plant J.* **2012**, *72*, 585–599. [CrossRef]
208. Groß, F.; Durner, J.; Gaupels, F. Nitric oxide, antioxidants and prooxidants in plant defence responses. *Front. Plant Sci.* **2013**, *4*, 419. [CrossRef]
209. Demidchik, V.; Shabala, S. Mechanisms of cytosolic calcium elevation in plants: The role of ion channels, calcium extrusion systems and NADPH oxidase-mediated 'ROS-Ca^{2+} Hub'. *Funct. Plant Biol.* **2018**, *45*, 9–27. [CrossRef]
210. Smirnoff, N.; Wheeler, G.L. Ascorbic acid in plants: Biosynthesis and function. *Crit. Rev. Plant Sci.* **2000**, *19*, 267–290. [CrossRef]
211. Gallie, D.R. L-ascorbic acid: A multifunctional molecule supporting plant growth and development. *Scientifica* **2013**, *2013*, 795964. [CrossRef] [PubMed]
212. Makavitskaya, M.; Svistunenko, D.; Navaselsky, I.; Hryvusevich, P.; Mackievic, V.; Rabadanova, C.; Tyutereva, E.; Samokhina, V.; Straltsova, D.; Sokolik, A.; et al. Novel roles of ascorbate in plants: Induction of cytosolic Ca^{2+} signals and efflux from cells via anion channels. *J. Exp. Bot.* **2018**, *69*, 3477–3489. [CrossRef] [PubMed]
213. Kuźniak, E.; Kaźmierczak, A.; Wielanek, M.; Głowacki, R.; Kornas, A. Involvement of salicylic acid, glutathione and protein S-thiolation in plant cell death-mediated defence response of *Mesembryanthemum crystallinum* against *Botrytis cinerea*. *Plant Physiol. Biochem.* **2013**, *63*, 30–38. [CrossRef] [PubMed]
214. Bedhomme, M.; Adamo, M.; Marchand, C.H.; Couturier, J.; Rouhier, N.; Lemaire, S.D.; Zaffagnini, M.; Trost, P. Glutathionylation of cytosolic glyceraldehyde-3-phosphate dehydrogenase from the model plant *Arabidopsis thaliana* is reversed by both glutaredoxins and thioredoxins in vitro. *Biochem. J.* **2012**, *445*, 337–347. [CrossRef] [PubMed]
215. Akter, N.; Okuma, E.; Sobahan, M.A.; Uraji, M.; Munemasa, S.; Nakamura, Y.; Mori, I.C.; Murata, Y. Negative regulation of methyl jasmonate-induced stomatal closure by glutathione in *Arabidopsis*. *J. Plant Growth Regul.* **2013**, *32*, 208–215. [CrossRef]
216. Li, F.; Wang, J.; Ma, C.; Zhao, Y.; Wang, Y.; Hasi, A.; Qi, Z. Glutamate receptor-like channel 3.3 is involved in mediating glutathione-triggered cytosolic calcium transients, transcriptional changes, and innate immunity responses in *Arabidopsis*. *Plant Physiol.* **2013**, *162*, 1497–1509. [CrossRef] [PubMed]
217. Liebthal, M.; Maynard, D.; Dietz, K.J. Peroxiredoxins and redox signaling in plants. *Antioxid. Redox Signal.* **2018**, *28*, 609–624. [CrossRef]
218. Mailloux, R.J.; Treberg, J.R. Protein S-glutathionlyation links energy metabolism to redox signaling in mitochondria. *Redox Biol.* **2016**, *8*, 110–118. [CrossRef]
219. Czerniawski, P.; Bednarek, P. Glutathione S-Transferases in the Biosynthesis of Sulfur-Containing Secondary Metabolites in Brassicaceae Plants. *Front. Plant Sci.* **2018**, *9*, 1639. [CrossRef]
220. Benjeddou, H.; Ahmed, C.B.; Rouina, B.B. Influence of antioxidative enzymes, phytohormones and pigments in alternate bearing of three olive cultivars. *Sci. Hortic.* **2019**, *253*, 17–23. [CrossRef]

221. Brunetti, C.; Fini, A.; Sebastiani, F.; Gori, A.; Tattini, M. Modulation of phytohormone signaling: A primary function of flavonoids in plant–environment interactions. *Front. Plant Sci.* **2018**, *9*, 1042. [CrossRef] [PubMed]
222. Yu, Y.; Wang, J.; Li, S.; Kakan, X.; Zhou, Y.; Miao, Y.; Wang, F.; Qin, H.; Huang, R. Ascorbic acid integrates the antagonistic modulation of ethylene and abscisic acid in the accumulation of reactive oxygen species. *Plant Physiol.* **2019**, *179*, 1861–1875. [CrossRef] [PubMed]
223. Lu, Y.; Chang, X.; Guo, X. Dynamic changes of ascorbic acid, phenolics biosynthesis and antioxidant activities in mung beans (*Vigna radiata*) until maturation. *Plants* **2019**, *8*, 75. [CrossRef] [PubMed]
224. Huang, D.; Huo, J.; Zhang, J.; Wang, C.; Wang, B.; Fang, H.; Liao, W. Protein S-nitrosylation in programmed cell death in plants. *Cell. Mol. Life Sci.* **2019**, *76*, 1877–1887. [CrossRef] [PubMed]
225. Datta, R.; Kumar, D.; Sultana, A.; Hazra, S.; Bhattacharyya, D.; Chattopadhyay, S. Glutathione regulates 1-aminocyclopropane-1-carboxylate synthase transcription via WRKY33 and 1-aminocyclopropane-1-carboxylate oxidase by modulating messenger RNA stability to induce ethylene synthesis during stress. *Plant Physiol.* **2015**, *169*, 2963–2981. [PubMed]
226. Singh, A.P.; Dixit, G.; Mishra, S.; Dwivedi, S.; Tiwari, M.; Mallick, S.; Pandey, V.; Trivedi, P.K.; Chakrabarty, D.; Tripathi, R.D. Salicylic acid modulates arsenic toxicity by reducing its root to shoot translocation in rice (*Oryza sativa* L.). *Front. Plant Sci.* **2015**, *6*, 340. [CrossRef] [PubMed]
227. Cheng, M.C.; Ko, K.; Chang, W.L.; Kuo, W.C.; Chen, G.H.; Lin, T.P. Increased glutathione contributes to stress tolerance and global translational changes in Arabidopsis. *Plant J.* **2015**, *83*, 926–939. [CrossRef] [PubMed]
228. Labrou, N.E.; Papageorgiou, A.C.; Pavli, O.; Flemetakis, E. Plant GSTome: Structure and functional role in xenome network and plant stress response. *Curr. Opin. Biotechnol.* **2015**, *32*, 186–194. [CrossRef]
229. Sevilla, F.; Camejo, D.; Ortiz-Espín, A.; Calderón, A.; Lázaro, J.J.; Jiménez, A. The thioredoxin/peroxiredoxin/sulfiredoxin system: Current overview on its redox function in plants and regulation by reactive oxygen and nitrogen species. *J. Exp. Bot.* **2015**, *66*, 2945–2955. [CrossRef]
230. Cao, S.; Du, X.-H.; Li, L.-H.; Liu, Y.-D.; Zhang, L.; Pan, X.; Li, Y.; Li, H. Overexpression of *Populus tomentosa* cytosolic ascorbate peroxidase enhances abiotic stress tolerance in tobacco plants. *Russ. J. Plant Physiol.* **2017**, *64*, 224–234. [CrossRef]
231. Chin, D.-C.; Kumar, R.S.; Suen, C.-S.; Chien, C.-Y.; Hwang, M.-J.; Hsu, C.-H.; Xuhan, X.; Lai, Z.X.; Yeh, K.W. Plant cytosolic ascorbate peroxidase with dual catalytic activity modulates abiotic stress tolerances. *iScience* **2019**, *16*, 31–49. [CrossRef] [PubMed]
232. Eltelib, H.A.; Fujikawa, Y.; Esaka, M. Overexpression of the acerola (*Malpighia glabra*) monodehydroascorbate reductase gene in transgenic tobacco plants results in increased ascorbate levels and enhanced tolerance to salt stress. *S. Afr. J. Bot.* **2012**, *78*, 295–301. [CrossRef]
233. Shin, S.-Y.; Kim, M.-H.; Kim, Y.-H.; Park, H.M.; Yoon, H.-S. Co-Expression of monodehydroascorbate reductase and dehydroascorbate reductase from *Brassica rapa* effectively confers tolerance to freezing-induced oxidative stress. *Mol. Cells* **2013**, *36*, 304–315. [CrossRef] [PubMed]
234. Yin, L.; Mano, J.; Tanaka, K.; Wang, S.; Zhang, M.; Deng, X.; Zhang, S. High level of reduced glutathione contributes to detoxification of lipid peroxide-derived reactive carbonyl species in transgenic Arabidopsis overexpressing glutathione reductase under aluminum stress. *Physiol. Plant.* **2017**, *161*, 211–223. [CrossRef] [PubMed]
235. Mao, C.; Ding, J.; Zhang, B.; Xi, D.; Ming, F. Os*NAC2* positively affects salt-induced cell death and binds to the *OsAP37* and *OsCOX11* promoters. *Plant J.* **2018**, *94*, 454–468. [CrossRef]
236. Borgohain, P.; Saha, B.; Agrahari, R.; Chowardhara, B.; Sahoo, S.; van der Vyver, C.; Panda, S.K. *SlNAC2* overexpression in *Arabidopsis* results in enhanced abiotic stress tolerance with alteration in glutathione metabolism. *Protoplasma* **2019**, *256*, 1065. [CrossRef]
237. Zhang, Q.; Ma, C.; Xue, X.; Xu, M.; Li, J.; Wu, J.-X. Overexpression of a cytosolic sscorbate peroxidase gene, *OsAPX2*, increases salt tolerance in transgenic Alfalfa. *J. Integr. Agric.* **2014**, *13*, 2500–2507. [CrossRef]
238. Wang, Z.; Li, Q.; Wu, W.; Guo, J.; Yang, Y. Cadmium stress tolerance in wheat seedlings induced by ascorbic acid was mediated by NO signaling pathways. *Ecotoxicol. Environ. Saf.* **2017**, *135*, 75–81. [CrossRef]
239. Liu, F.; Huang, N.; Wang, L.; Ling, H.; Sun, T.; Ahmad, W.; Muhammad, K.; Guo, J.; Xu, L.; Gao, S.; et al. A novel L-ascorbate peroxidase 6 gene, *ScAPX6*, plays an important role in the regulation of response to biotic and abiotic stresses in sugarcane. *Front. Plant Sci.* **2018**, *8*, 2262. [CrossRef]
240. Chin, D.C.; Hsieh, C.C.; Lin, H.Y.; Yeh, K.W. A low glutathione redox state couples with a decreased ascorbate redox ratio to accelerate flowering in *Oncidium* orchid. *Plant Cell Physiol.* **2016**, *57*, 423–436. [CrossRef]

241. Kavitha, K.; George, S.; Venkataraman, G.; Parida, A. A salt-inducible chloroplastic monodehydroascorbate reductase from halophyte *Avicennia marina* confers salt stress tolerance on transgenic plants. *Biochimie* **2010**, *92*, 1321–1329. [CrossRef] [PubMed]
242. Kwon, S.Y.; Choi, S.M.; Ahn, Y.O.; Lee, H.S.; Lee, H.B.; Park, Y.M.; Kwak, S.S. Enhanced stress–tolerance of transgenic plants expressing a human dehydroascorbate reductase gene. *J. Plant Physiol.* **2003**, *160*, 347–353. [CrossRef] [PubMed]
243. Ushimaru, T.; Nakagawa, T.; Fujioka, Y.; Daicho, K.; Naito, M.; Yamauchi, Y.; Nonaka, H.; Amako, K.; Yamawaki, K.; Murata, N. Transgenic *Arabidopsis* plants expressing the rice dehydroascorbate reductase gene are resistant to salt stress. *J. Plant Physiol.* **2006**, *163*, 1179–1184. [CrossRef] [PubMed]
244. Lee, Y.P.; Kim, S.H.; Bang, J.W.; Lee, H.S.; Kwak, S.S.; Kwon, S.Y. Enhanced tolerance to oxidative stress in transgenic tobacco plants expressing three antioxidant enzymes in chloroplasts. *Plant Cell Rep.* **2007**, *26*, 591–598. [CrossRef]
245. Wang, H.S.; Yu, C.; Zhu, Z.J.; Yu, X.C. Overexpression in tobacco of a tomato GMPase gene improves tolerance to both low and high temperature stress by enhancing antioxidation capacity. *Plant Cell Rep.* **2011**, *30*, 1029–1040. [CrossRef] [PubMed]
246. Le Martret, B.; Poage, M.; Shiel, K.; Nugent, G.D.; Dix, P.J. Tobacco chloroplast transformants expressing genes encoding dehydroascorbate reductase, glutathione reductase, and glutathione-S-transferase, exhibit altered anti-oxidant metabolism and improved abiotic stress tolerance. *Plant Biotechnol. J.* **2011**, *9*, 661–673. [CrossRef] [PubMed]

12

Amelioration of the Oxidative Stress Generated by Simple or Combined Abiotic Stress through the K^+ and Ca^{2+} Supplementation in Tomato Plants

María García-Martí [1,2,†], María Carmen Piñero [1,†], Francisco García-Sanchez [1], Teresa C. Mestre [1], María López-Delacalle [1], Vicente Martínez [1] and Rosa M. Rivero [1,*]

1 Department of Plant Nutrition, Campus Universitario de Espinardo, CEBAS-CSIC, Ed 25, Espinardo, 30100 Murcia, Spain; mgmarti@cebas.csic.es (M.G.-M.); mcpinerozapata@gmail.com (M.C.P.); fgs@cebas.csic.es (F.G.-S.); tmestre@cebas.csic.es (T.C.M.); mlopez@cebas.csic.es (M.L.-D.); vicente@cebas.csic.es (V.M.)

2 Programa de Doctorado en Ciencias de la Salud, Campus de los Jerónimos, UCAM (Universidad Católica San Antonio de Murcia), s/n, 30107 Murcia, Spain

* Correspondence: rmrivero@cebas.csic.es

† These authors contributed equally to this manuscript.

Abstract: Abiotic stressors such as drought, heat, or salinity are major causes of yield loss worldwide due to the oxidative burst generated under these conditions. Recent studies have revealed that plant response to a combination of different environmental stressors is unique and cannot be deduced from the response developed to each stress when applied individually. Some studies have demonstrated that a different management of some nutrients in the irrigation solution may provide an advantage to the plants against abiotic stressors. Thus, the aim of this study was to investigate if an increase in potassium (K^+) and calcium (Ca^{2+}) concentration in the nutrient solution may have a positive effect on the amelioration of oxidative stress which occurs under the combination of salinity and heat in tomato plants. Our results indicated that plants irrigated with an increase in K^+ and Ca^{2+} concentrations in the irrigation solution from 7mM (K^+) to 9.8 mM and from 4 mM (Ca^{2+}) to 5.6 mM, respectively, induced a recovery of the biomass production compared to the plants treated with salinity or salinity + heat, and subsequently irrigated with the regular Hoagland solution. This was correlated with a better performance of all the photosynthetic parameters, a reduction in the foliar concentration of H_2O_2 and a lower lipid peroxidation rate, and with a better performance of the antioxidant enzymes ascorbate peroxidase ascorbate peroxidase (APX), dehydroascorbate reductactase (DHAR), glutathione reductase (GR), and NADPH oxidase. Our results showed that these enzymes were differentially regulated at the transcriptional level, showing a higher reactive oxygen species (ROS) detoxification efficiency under salinity and under the combination of salinity and heat, as compared to those plants irrigated with common Hoagland. An increase in K^+ and Ca^{2+} in the irrigation solution also induced a lower Na^+ accumulation in leaves and a higher K^+/Na^+ ratio. Thus, our study highlights the importance of the right management of the plant nutritional status and fertilization in order to counteract the deleterious effects of abiotic stress in plants.

Keywords: abiotic stress combination; heat stress; salinity; photosynthesis; lipid peroxidation; ROS detoxification; antioxidant-related genes; antioxidant enzymes

---

1. Introduction

Tomatoes (*Solanum lycopersicum*) are the second-most important cultivar in the world economically. Southeastern agriculture in Spain is characterized by its production of high-quality agronomical products with a high added value that are commercialized in both national and international markets, resulting in large revenues for the Spanish regions dedicated to tomato cultivation. Unfortunately, adverse environmental conditions, such as salinity (coming from both low-quality irrigation water and salinized soils), scarcity of water, and high temperatures, negatively affect plant production and considerably increase economic losses. These environmental conditions are even more detrimental when they act in concert, resulting in great economic losses worldwide. On the other hand, the climate change predictions expected in the next 50–80 years, as published by the International Panel for Climate Change [1], indicate a worsening of the current environmental conditions that, together with the increase in the world's population, make clear the urgent need for obtaining cultivars and the proper management of crops, in order to increase abiotic stress tolerance and improve yields.

Salinity reduces crop production and negatively impacts the rate of leaf expansion, protein synthesis, energy production, lipid metabolism, opening of stomata, and photosynthesis through osmotic stress caused by soil water deficits, ion toxicity, and nutrient imbalances [2]. Similarly, heat stress affects biomass accumulation in crops of agronomic importance, such as tomato, by impairing the correct functioning of photosystem II (PSII) and reducing the chlorophyll content, resulting in a reduction in plant growth [3,4], which in turn negatively affects final fruit production.

In the last few years, our research group has successfully demonstrated that the combination of two or more abiotic stressors generated a specific plant response that could not be elucidated from the responses obtained when these stressors were applied individually [5–7]. This highlights the importance of studying abiotic stressors in combination rather than separately. Individual as well as combined abiotic stressors induce imbalances in the redox cell homeostasis due to the accumulation of reactive oxygen species (ROS), such as hydrogen peroxide (H_2O_2) [8]. Despite their potential to cause damage, it has been shown that, at non-toxic concentrations, H_2O_2 plays a key role in signal transduction mechanisms, which in turn activate defense responses to various stressors [9].

To remove excess ROS and to maintain redox homeostasis, plants have a well-integrated antioxidant defense system that is comprised of antioxidant compounds (i.e., ascorbate, glutathione, β-carotene, etc.) and enzymes, such as superoxide dismutase (SOD), catalase (CAT), and those involved in the ascorbate-glutathione (AsA–GSH) cycle [10]. Under control conditions, these enzymes are tightly regulated for achieving a high efficiency in ROS detoxification. However, abiotic stressors cause inhibition at transcriptional and post-transcriptional levels, as well as post-translational modifications that translate into a decrease in detoxification efficiency, and thus ROS accumulation and deleterious damage to the cells [8].

Plant growth and metabolism are also dependent on plant nutritional status. In this sense, potassium (K^+) and calcium (Ca^{2+}) nutrition plays a fundamental role. Most biochemical and physiological processes, such as protein synthesis, regulation of plant stomata, water use, control of ionic balance, activation of enzymes, and many other processes, are affected by K^+, positively contributing to the survival of plants exposed to various abiotic stressors [11,12]. In plants subjected to salinity stress, it is common to find a K^+ deficiency [13,14], and a low concentration of K^+ under saline conditions can result in the formation of ROS and cell damage [15]. Also, it has been shown that Ca^{2+} is essential for selective transport by plant cells. An extra Ca^{2+} supply may contribute to cell wall and plasma membrane integrity, and thus may reduce the effect of salinity on plants [16,17]. Moreover, it has also been shown that K^+ nutrition improved with an application of external Ca^{2+} under salinity stress [18]. ROS have been described as playing an important role in the root's response to nutrient deprivation. In this sense, K^+ deprivation induced ROS accumulation in Arabidopsis [12,18,19] and tomato roots [20], with these ROS proposed to participate in the signaling cascade that results in the induction of the genes encoding for the high affinity K^+ transporter HAK5 [12]. The study of salinity and heat stress applied individually has contributed important information to the current knowledge on abiotic stressors, but recent research studies have indicated that a plant's response to a combination of two or more stressors is unique and cannot be elucidated from the study of single stressors [6,7,21–23], so stress combination studies are deemed very necessary. Thus, the aim of the present study was to investigate the beneficial effects of an increase in K^+ and Ca^{2+} concentration in the irrigation solution on the tolerance of tomato plants to the combination of salinity and high temperature, through the amelioration that these elements can exert on the oxidative stress generated under these conditions.

2. Materials and Methods

2.1. Preliminary Experiments on K^+ and Ca^{2+} Doses in the Nutrient Solution

Before the main set of experiments was performed, a preliminary experiment was conducted in order to confirm the hypothesis that an increase in the concentrations of K^+ and Ca^{2+} in the irrigation solution of tomato plants could exert a positive effect on plant biomass production under an abiotic stress combination, and more specifically, under the combination of salinity and heat, two of the most devastating abiotic stressors on plant productivity that act jointly in the southeast of Spain. For this purpose, tomato seeds (*Solanum lycopersicum* L.) cultivar "Boludo" (kindly offered by Monsanto, Torre Pacheco, Murcia, Spain) were sown in vermiculite in a growth chamber under optimal and controlled conditions of light (500 μmol m^{-2} s^{-1}), photoperiod (16/8 h day/night), humidity (60–65%), and temperature (25 °C) (Chamber A, see Table 1 below). When the plants reached a height of 8–10 cm, they were transferred to 20 L containers (a total of 96 plants, two plants per container) and grown in an aerated hydroponic system containing Hoagland solution (Supplementary Table S1, [24]). The plants were grown under these conditions for 10 days in order to acclimate them to this growing system. After this time, half of the plants (48 plants) were transferred to another growth chamber with the same light, photoperiod, and humidity parameters described previously, but with the environmental (air) temperature set at 35 °C (Chamber B, see Table 1 below). At this time, the different treatments (T1, T2 and T3) were applied as follows (n = 12):

Table 1. Nutritional and environmental temperature treatments applied in the preliminary experiment to the tomato plants in order to select the most significant nutritional treatment for further experiments.

Growth Chamber	Treatments	[NaCl]	[K^+]	[Ca^{2+}]
Chamber A (25 °C)	Control	0 mM	7 mM	4 mM
	Salinity	60 mM	7 mM	4 mM
	Salinity + T1	60 mM	8 mM	4.7 mM
	Salinity + T2	60 mM	9.8 mM	5.6 mM
	Salinity + T3	60 mM	11 mM	6.5 mM
Chamber B (35 °C)	Heat	0 mM	7 mM	4 mM
	Salinity + Heat	60 mM	7 mM	4 mM
	Salinity + Heat + T1	60 mM	8 mM	4.7 mM
	Salinity + Heat + T2	60 mM	9.8 mM	5.6 mM
	Salinity + Heat + T3	60 mM	11 mM	6.5 mM

Plants were kept under these conditions for 21 days, after which time the fresh weight and dry weight of whole plants were recorded (Supplementary Figure S1). Based on the results obtained in this preliminary experiment, treatment 2 (T2) (i.e., 9.8 mM K^+ and 5.6 mM Ca^{2+}) was selected for further experiments, as it was the treatment that obtained a better biomass production under the stressors applied.

2.2. Experimental Design, Plant Material and Growth Conditions

Tomato seeds (*Solanum lycopersicum* L.) cv. "Boludo" were sown as described previously. When the plants were 8–10 cm in height. A total of 48 plants were transferred to 120 L tanks (with one plant per tank) and grown in an aerated hydroponic system containing the control Hoagland solution (as described in Supplementary Table S1) for 10 days in order to adapt them to this system. At this time, the plants were divided between two polycarbonate greenhouses (24 plants per greenhouse) located in an experimental station in Santomera (Murcia, Spain) during the spring growing season, and the different treatments were applied for 30 days. Salinity was applied at the same time that plants were separated into the two different greenhouses, as shown in Table 2. Thus, the experiment consisted of six treatments, as follows:

Table 2. Nutritional and environmental temperature treatments applied to tomato plants under greenhouse conditions for the physiological, biochemical and molecular determinations.

Greenhouse	Treatments	[NaCl]	[K^+]	[Ca^{2+}]
Greenhouse A (25 °C)	Control	0 mM	7 mM	4 mM
	Salinity	60 mM	7 mM	4 mM
	Salinity + K/Ca	60 mM	9.8 mM	5.6 mM
Greenhouse B (35 °C)	Heat	0 mM	7 mM	4 mM
	Salinity + Heat	60 mM	7 mM	4 mM
	Salinity + Heat + K/Ca	60 mM	9.8 mM	5.6 mM

Control (25 °C), heat stress (35 °C), salinity stress (25 °C + 60 mM NaCl), salinity combined with increases in K^+ and Ca^{2+} concentrations (25 °C, 60 mM NaCl, 9.8 mM K^+ and 5.6 mM Ca^{2+}), salinity and heat stress combined (60 mM NaCl + 35 °C), and salinity and heat stress combined with the increases in K^+ and Ca^{2+} concentration in the nutrient solution (60 mM NaCl, 35°C, 9.8 mM K^+ and 5.6 mM Ca^{2+}) were set as treatments. Eight tanks per treatment were randomly distributed in both greenhouses (Supplementary photograph 1). The pH of the all the nutrient solutions used in our experiments was maintained between 5.5 and 6.1. Water lost by transpiration was replaced every two days and nutrients were added every week to restore their initial concentrations. To apply the temperature treatments, this parameter was set at a maximum of 25 °C for greenhouse A (control) and 35 °C for greenhouse B. The greenhouse climate control system operated as described in Rodriguez-Ortega et al. [25].

2.3. Plant Sampling

Six plants were used per treatment. Thirty days after starting the treatments, the tomato plants were harvested and divided into leaves, stems, and roots, and their fresh weight (FW) and dry weight (DW) were recorded. Half of the harvested plant material was immediately submerged into liquid N_2 and stored for future procedures. Fully expanded leaves corresponding to the 6th–8th position (middle leaves) were separated and used for the biochemical and molecular measurements. The other half of the plant material was oven-dried for 72 h at 70 °C and weighed to determine the dry weight (DW). Dried plant material was used for ion analysis.

2.4. Ion Analysis and Quantification

Dried plant material was digested with HNO_3:$HClO_4$ (2:1, *v:v*), and Na^+ and K^+ concentrations were determined by atomic absorption spectrometry (Perkin-Elmer 5500, Waltham, MA, USA).

2.5. Leaf Gas Exchange

Net photosynthesis rate (A), stomatal conductance (gs), and transpiration rate (E) were measured at the end of the experiment in the youngest fully-expanded leaf of each plant (six plants per treatment), using a LI-6400XT photosynthesis system (Li-Cor, Inc., Lincoln, NE, USA). Leaf gas exchange was measured as described by Martinez et al. [6]

2.6. H_2O_2 Quantification

H_2O_2 was extracted as described by [26], with some modifications [23,27,28]. The concentration of H2O2 in the extracts was determined as described by Martinez et al. [6]

2.7. Lipid Peroxidation

Malondialdehyde (MDA), as a degradation product of lipid peroxidation, was determined as outlined by Fu and Huang [29] with the modifications listed by Mestre et al. [24], using an MDA extinction coefficient of 155 mM^{-1} cm^{-1}.

2.8. RNA Extraction and qRT-PCR Experiments

Total RNA was isolated from whole tomato leaves, as described by Martinez et al. [6]. The primer sequences used for the different genes tested as well as the primer sequences of the internal controls are available in Supplementary Table S2. The normalization of transcript expression against housekeeping genes, reaction components, and PCR settings were carried out as described by Martinez et al. [6]. Expression rates as well as $\log_2$ values of the different genes can be found in Supplementary Table S3.

2.9. Enzymatic Activities

All the enzymatic activities described were extracted as described by Martinez et al. [6]. Enzymatic activities were expressed as per mg protein [30].

CAT was assayed as described by Chance and Maehly [31]. The activity was expressed as the change in absorbance at 240 nm of a solution of 12.5 mM H_2O_2 in a 50 mM KH_2PO_4 buffer (pH 7.0) at 30 °C. CAT activity was calculated using an extinction coefficient of 39.4 $mM^{-1}{\cdot}cm^{-1}$ [32].

Activities of the Cu/ZnSOD and FeSOD isoenzymes were assayed as described by McCord and Fridovich [33], with some modifications [34]. The reaction mixture and the quantification of Cu/ZnSDO and Fe SOD are fully described in Martinez et al. [6]

Cytosolic APX (cAPX), monodehydroascorbate reductase (MDHAR), and dehydroascorbate reductase (DHAR) activities were assayed as described by Miyake and Asada [35] and the full method can be found in Martinez et al. [6].

GR activity was assayed as described previously by Mestre et al. [24], with the modification of Martinez et al. [6].

NADPH oxidase activity was measured following the method described by Kaundal et al. [36]. The NADPH oxidase-dependent generation of O_2^- in plant tissue has been determined by the reduction of the tetrazolium salt 3′-{1-[(phenylamino)-carbonyl]-3,4-tetrazolium}bis(4-methoxy-6-nitro)benzenesulfonic acid hydrate (XTT) by O_2^-. In the presence of O_2^-, XTT generates a soluble yellow formazan that can be quantified spectrophotometrically at 490 nm. The rate of O_2^- generation was calculated using an extinction coefficient f 2.16×10^4 M^{-1} cm^{-1} [37].

Different absolute values obtained for each treatment and enzyme (Supplementary Table S4) were normalized against its control and $\log_2$ values were calculated (Supplementary Table S5). Enzymatic activities were represented as a heat map of these $\log_2$ values obtained.

2.10. Statistical Analysis

The data were first tested for homogeneity of variance and normality of distribution, and the Tukey HSD (Honestly-significant-difference) multiple range test was used to determine differences between means ($p \leq 0.05$), with $p < 0.05$ *, $p < 0.01$ **, $p < 0.001$ *** and n.s. as not significant. The relative transcript expression assayed by quantitative PCR (qPCR) was calculated using the $2^{-\Delta Ct}$ method. Heat maps for transcript expression and enzymatic activities were created using R.

3. Results and Discussion

In arid and semi-arid regions, the scarcity of good quality water for agriculture necessitates the use of saline water, which has caused scientists to focus on studying the effect of salinity on plants. However, salinity is usually aggravated by the high temperatures that are found in these regions [5]. Our research group has previously shown that the study of the abiotic stressors in combination is essential, as plants showed a specific response to this combination that could not be deduced from the response to the individually-applied stressors alone [5,7]. As shown in Figure 1, the FW and DW of tomato plants were negatively affected by the application of salinity and salinity + heat, with DW reductions of 33% and 38%, respectively, as compared with control plants (Figure 1B). When the irrigation solution was complemented with increases in K^+ and Ca^{2+}, the effects of salinity and the combination of salinity and heat were diminished, obtaining plants with less biomass than control plants, but with reductions of only about 10% (Figure 1B). Thus, plants grown under salinity or under the combination of salinity and heat but with higher concentrations of K^+ and Ca^{2+} in their nutrient solution increased their DW by 33% and 47% respectively, as compared to plants grown under these stressors and irrigated with the normal Hoagland solution. Capula-Rodriguez, et al. [38] highlighted the importance of the use of higher levels of K^+ and Ca^{2+} in the irrigation solution for the mitigation of the effect of the combination of multiple stressors (salinity, alkalinity, and boron) in tomato plants, thus supporting our results. These authors observed that plants grown under the combination of high alkalinity, salinity, and an excess of boron showed improved growth when supplemented with greater concentrations of Ca^{2+} and K^+, which was related to enhanced phosphorous concentration, maintenance of chlorophyll *a* concentration, and/or the partial restoration of the uptake of other nutrients under these stress conditions.

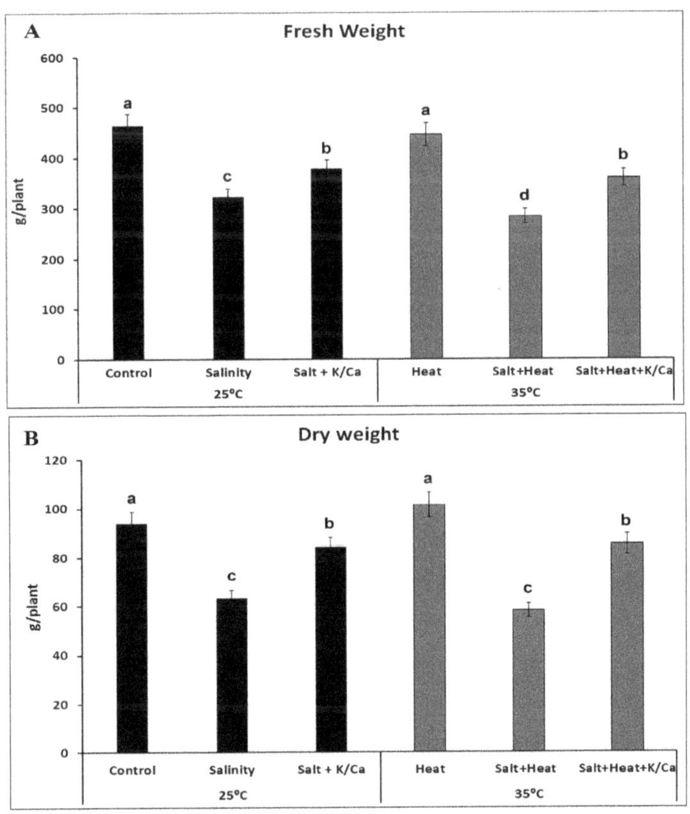

Figure 1. Fresh (**A**) and dry weight (**B**) of tomato plants grown under control, salinity (60 mM NaCl) or salinity and higher concentration of K^+ and Ca^{2+} in the irrigation solution at optimal temperature (25 °C) or heat stress (35 °C). Data represent means ± SE ($n = 6$). Bars with different letters within each panel represent data with significant differences at $p < 0.05$ (Turkey HSD).

In general, salinity problems have been related to an excess of NaCl in the irrigation water, which causes Na^+ toxicity in the plants, modifies the absorption of K^+ and other nutrients by the roots, and exerts severe toxic effects on genes and enzymes, causing detrimental alterations to the plant's metabolism [39–41]. Therefore, it is vital for plants to re-establish cellular ionic homeostasis to maintain correct metabolic functioning and growth. Our ionic analysis in tomato leaves detected that heat, and the combination of salinity and heat, reduced the endogenous concentration of Ca^{2+} and K^+ (Figure 2A,B). Ca^{2+} was reduced by about 18% by heat and the combination of salinity and heat (Figure 2A), whereas K^+ was reduced by these stressors by about 23% (Figure 2B), as compared to plants grown under optimal conditions (control plants). However, when salinity and heat were applied jointly, and the irrigation solution was supplemented with higher concentrations of K^+ and Ca^{2+}, the endogenous Ca^{2+} concentration obtained for the leaves was similar to control plants. Salinity did not induce significant changes in the leaves' endogenous concentration of Ca^{2+} or K^+ under any temperature treatment. However, surprisingly, when plants were grown under salinity and irrigated with higher K^+ and Ca^{2+} concentrations, the endogenous concentration obtained for K^+ was significantly lower than that obtained under control conditions, with a reduction of 12% (Figure 2B).

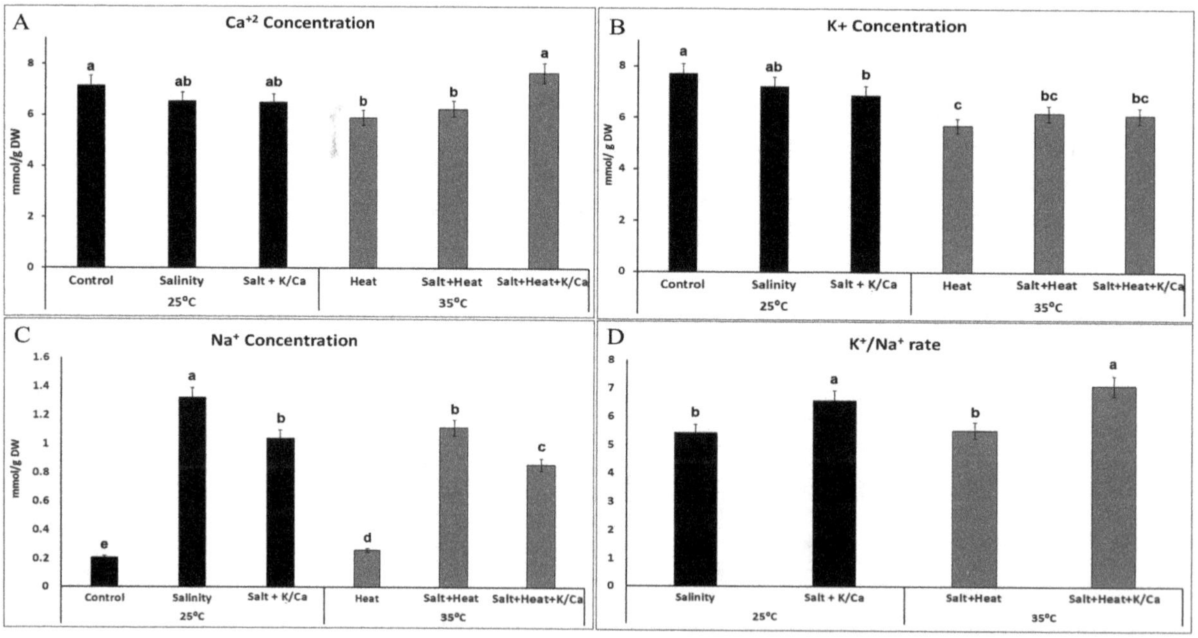

Figure 2. Ca^{2+}, K^+ and Na^+ concentration (**A–C**) and K^+/Na^+ ratio (**D**) in tomato leaves grown under control, salinity (60 mM NaCl) or salinity combined with higher K^+ and Ca^{2+} concentration in the irrigation solution under optimal temperature (25 °C) or under heat stress (35 °C). Values represent means ± SE ($n = 6$). Bars with different letters represent data with significant differences at $p < 0.05$ (Tukey HSD).

The irrigation of plants with a nutrient solution enriched in K^+ and Ca^{2+} and grown under a combination of stressors did not induce significant changes in the K^+ concentration as compared to plants irrigated with the normal Hoagland solution under the same environmental conditions (Figure 2B). These data coincided with those previously observed in a study by Martinez et al., where the combination of salinity and heat caused an inhibition of K^+ uptake. These reductions could be due to the antagonistic effect of K^+ with Na^+ [42]. In order to confirm Na^+ and K^+ homeostasis in our experiments, the endogenous Na^+ concentration of leaves was studied (Figure 2C).

As expected, the Na^+ concentration increased significantly in those treatments where 60 mM NaCl was supplied. The salinity treatment had the highest Na^+ concentration, and the application of salinity combined with heat induced a significantly lower Na^+ accumulation in tomato leaves, with a reduction of 16% compared to salinity alone (Figure 2C). Also, the plants that received a nutrient solution enriched with K^+ and Ca^{2+} and grown under stress conditions showed a lower Na^+ accumulation in the leaves (~22% less) as compared with plants grown under the same conditions but irrigated with the normal Hoagland solution. Thus, our results indicated that endogenous concentrations of Na^+ and K^+ were lower when plants were irrigated with a nutrient solution enriched with K^+ and Ca^{2+} under abiotic stress. The K^+/Na^+ ratio has been extensively related to the salt tolerance in plants [43]. In this sense, our experiments showed that this ratio was higher in those plants that were irrigated with higher concentration of K^+ and Ca^{2+} (Figure 2D) under salinity or under the combination of salinity and heat, which could be related with plants that have an increased tolerance to these stressors. In this light, we found a positive correlation between biomass production and the K^+/Na^+ ratio in plants irrigated with higher K^+ and Ca^{2+} and grown under salinity ($DW_{[Salt+K^+/Ca^{2+}]}$ – K^+/Na^+ ratio$_{[Salt+K^+/Ca^{2+}]}$, $r = 0.971$***) or the salinity and heat combination ($DW_{[Salt+K^+/Ca^{2+}]}$ – K^+/Na^+ ratio$_{[Salt+Heat+K^+/Ca^{2+}]}$, $r = 0.811$***). It has been postulated that some compounds (i.e., osmoprotectants and antioxidant compounds, among others) exert a regulatory function, maintaining a cytosolic K^+ concentration by preventing NaCl-induced K^+ leakage from the cell [12,14]. For example, K^+ has been reported to result in the accumulation of osmolytes and augmentation of antioxidant components in the plants exposed to water and salt stress [44]. Thus, our results may indicate that an increase in K^+ concentration in the nutrient solution may have an influence on the concentration of antioxidants and other beneficial compounds in plants, and therefore on the maintenance of the K^+/Na^+ homeostasis in plants. The increase in K^+ and Ca^{2+} concentrations in the irrigation solution improved tomato salinity tolerance under both optimal temperature and heat stress. The most aggressive stress treatment for tomato plants was salinity and heat; however, an increase in K^+ and Ca^{2+} concentration in the irrigation solution induced a higher recovery of these plants under salinity and heat stress.

At the end of the experimental period and before the plants were harvested, photosynthetic parameters, such as CO_2 assimilation rate, stomatal conductance, transpiration rate, and water use efficiency (calculated from the CO_2 assimilated versus the water transpired) were measured (Figure 3). Changes in these parameters are usually good indicators of stress in plants and they are directly correlated with plant growth. CO_2 assimilation rate (Figure 3A), transpiration rate (Figure 3B), and stomatal conductance (Figure 3C) had a similar behavior in our experiments with plants under stress conditions. As compared to control conditions, the salinity and salinity + heat treatments resulted in a significant reduction of those parameters with respect to control plants, with salinity + heat being more aggressive than salinity itself. These data were consistent with the findings by Pinero, et al. [45] and Rivero et al. [7], who reported that both the salinity and salinity + heat treatments led to a reduction in stomatal conductance, reducing the plant's ability to supply CO_2 to the photosynthetic apparatus, thus decreasing the CO_2 assimilation rate under these stress conditions. On the contrary, it has also been observed that heat stress alone can favor a greater transpiration rate, in order to reduce leaf temperature, which may lead to an increase in the CO_2 assimilation rate. The increase described for those parameters was also observed in these experiments, which were therefore in agreement with the results found by Rivero et al. [7]. Interestingly, when plants were grown under salinity or salinity + heat but supplemented with a nutrient solution enriched with K^+ and Ca^{2+}, CO_2 assimilation, transpiration rate, and stomatal conductance partially recovered to values similar or closer to control levels (Figure 3A–C). These photosynthetic parameters are closely related to growth rate in plants, and our results indicate a positive correlation among them, where a reduction in plant growth (Figure 1) was directly related with an inhibition of these photosynthetic parameters, and the recovery of these by an extra supply of the plants with K^+ and Ca^{2+} implied better growth under abiotic stress conditions. Abass et al. [44] and Bohra and Doerffling [46] found similar results in maize plants and rice, respectively, through the addition of K^+ to a saline soil. Both studies concluded that the addition of K^+ significantly alleviated the harmful effects of salinity, by improving plant growth and gas exchange parameters.

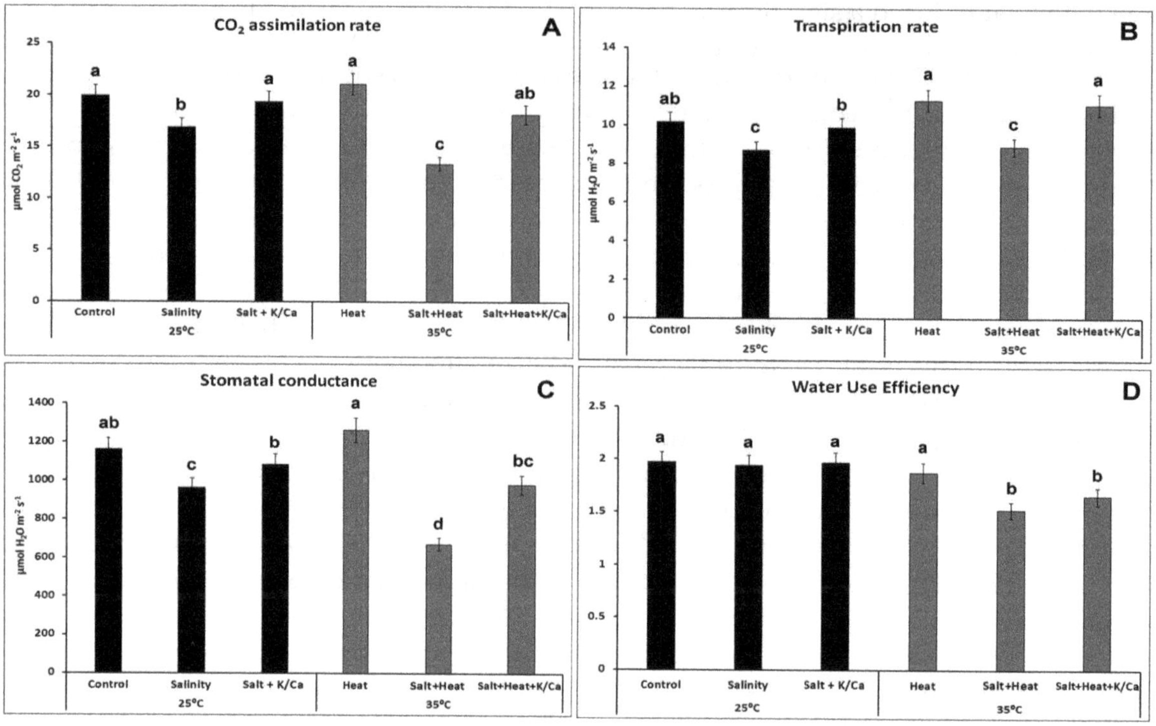

Figure 3. Photosynthetic parameters in tomato leaves grown under control, salinity (60 mM NaCl or salinity combined with higher K^+ and Ca^{2+} concentration in the irrigation solution at optimal temperature (25 °C) or heat stress (35 °C). Values are means ± SE (n = 6). Bars with different letters within each panel represent data with significant differences at $p < 0.05$ (Tukey HSD).

In addition to K^+, Ca^{2+} also plays an important role in the plant cell. Among its numerous cell functions, Ca^{2+} contributes to the cell's signaling processes and helps with the detoxification of ROS [18,47,48]. Bhattacharjee [49] observed that the addition of Ca^{2+} helped to protect against heat-induced oxidative damage. It is widely known that salinity and heat stress can enhance oxygen-induced cellular damage due to a high production of ROS [6,22,50,51], so the importance of enhancing the antioxidant defense plant system to cope with these stressors should be highlighted. In recent studies, it has been observed that plants subjected to different combinations of abiotic stressors accumulated large amount of ROS response transcripts, which defined the importance of the antioxidant machinery on the acclimation pathways during combined stressors [50,52]. Martinez et al. [6] showed that tomato plants subjected to a combination of salinity and high temperature had an overproduction of ROS that caused photosynthesis inhibition, breakdown of photosynthetic pigments, and plant growth reduction. However, although the study of the combined effect of two abiotic stressors has begun and there are many publications that demonstrate the importance of studying stressors in combination, there is little information about the effect of the nutrition management and its possible relationship with abiotic stress combination tolerance in plants. In light of this question, a detailed study of the plant's oxidative metabolism was performed in our plants in order to verify a possible role of the extra supply of K^+ and/or Ca^{2+} on the enhancement of the antioxidant defense pathways in plants.

Under these environmental stress conditions, where water supply to the plant is being compromised (i.e., with regards to salinity, heat or both), the internal cellular CO_2 concentration is restricted due to stomatal closure. This, together with the continuous incident sunlight to the plant, results in a transfer of excess electrons to molecular oxygen, generating a superoxide radical (O_2^{-}•), which is immediately detoxified to H_2O_2 by superoxide dismutases (SODs). In addition, the stress condition generates irreversible damage to the cellular membrane's polyunsaturated fatty acids. Thus, two of the most common oxidative stress markers used to measure oxidative damage to the cells are H_2O_2 concentration and membrane lipid peroxidation, measured as the concentration of MDA derivatives. These two parameters were evaluated in our experiments with tomato plants (Figure 4). Any of the stress conditions used in our experiments resulted in a significant increase in the H_2O_2 concentration with respect to control plants, with the salinity treatment being the one with the highest concentration of this compound, which was found to be almost double the amount found in control plants. On the other hand, when plants were grown either under salinity or the combination of salinity and heat but irrigated with a nutrient solution enriched with K^+ and Ca^{2+}, the concentration of H_2O_2 was significantly lower than that found under the same environmental conditions but irrigated with the normal Hoagland solution. Thus, plants irrigated with higher K^+ and Ca^{2+} and grown under salinity had 20% less H_2O_2 than those grown under salinity but irrigated with the normal nutrient solution. Interestingly, when salinity was combined with heat, the extra supply of K^+ and Ca^{2+} in the nutrient solution resulted in a 30% reduction in H_2O_2 concentration. For the MDA content, the results obtained were similar, with the higher concentration of this compound found in plants under salinity and salinity + heat treatments and irrigated with the normal nutrient solution. Again, the addition of extra K^+ and Ca^{2+} to the nutrient solution resulted in a significant reduction in the level of lipid peroxidation of these plants under these stress conditions. H_2O_2 accumulation and membrane lipid peroxidation leads to oxidative damage to the cells, inhibiting photosynthesis and inducing plant growth inhibition [7,28]. Our results showed that tomato plants grown under these treatments, where H_2O_2 was accumulated to higher concentrations (Figure 4A), showed a higher membrane lipid peroxidation (Figure 4B) and an significant impairment of the photosynthetic parameters (Figure 3), with the concomitant inhibition of plant growth.

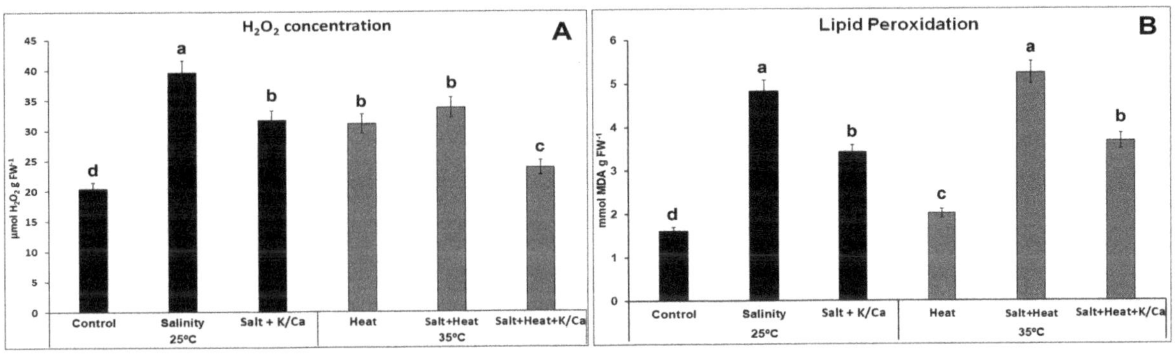

Figure 4. H_2O_2 concentration (**A**) and Lipid peroxidation rate, measured as MDA concentration (**B**) in tomato leaves grown under control, salinity (60 mM NaCl) or salinity combined with a higher concentration of K^+ and Ca^{2+} in the irrigation solution at optimal temperature (25 °C) or under heat stress (35 °C). Values represent means ± SE (n = 6). Bars with different letters within eahc panel represent data with significant differences at $p < 0.05$ (Tuket HSD).

The impairment of the cell antioxidant machinery under abiotic stress conditions can be caused by an overproduction of ROS, an inhibition of the antioxidant enzymes, or both. This inhibition may occur at transcriptional or post-transcriptional levels, or by post-translational modifications of the antioxidant proteins necessary for counteracting the deleterious effects of ROS. In order to assess these impairments, the expression of the main oxidative metabolism-related transcripts and the activity of the main antioxidant enzymes were measured under the different stress treatments used in our study (Figure 4A). After examining both heatmaps (Figure 5A,B) it is clear that tomato plants have a specific antioxidant response under the combination of salinity and heat, and that the response of the genes or enzymes cannot be deduced from the individual responses obtained under salinity or heat applied individually. These results confirm those obtained previously by Rivero et al. [7], Martinez et al. [5], and Martinez et al. [6], and highlight the importance of studying abiotic stressors in combination. The expression levels of the transcripts that code for Fe-SOD and Cu/Zn-SOD were upregulated under all the stress conditions (Figure 5A), and correlated well with the SOD activity found, with the exception of Cu/Zn-SOD activity in plants grown under the combination of salinity + heat and irrigated with the nutrient solution enriched with K^+ and Ca^{2+}, in which case this enzyme was inhibited.

The first cellular line of defense for ROS detoxification is constituted by SODs, which detoxify superoxide radicals into H_2O_2. In this sense, the over-production of H_2O_2 observed under the different stress treatments applied in our experiments was due to an over-expression of Fe-SOD and Cu/Zn-SOD encoding transcripts (Figure 4A), with a concomitant increase in SOD activity. H_2O_2 levels found in the stress treatments of plants supplemented with the K^+/Ca^{2+} enriched nutrient solution were lower than those of plants grown with the normal nutrient solution under the same stress conditions, which may be due to: (1) a lower production of this compound in these treatments, due to the beneficial effects of the extra supply of K^+ and Ca^{2+} under stress conditions, or (2) a more effective H_2O_2 detoxification driven by an overexpression of Fe-SOD encoding transcripts (Figure 5A), with a concomitant increase in SOD activity (Figure 5B, Supplementary Table S4) under these conditions. H_2O_2 accumulation also has a deleterious effect in plant cells, which correlated well with the biomass production obtained in our experiments under all the stress treatments used (Figure 1). Therefore, an effective detoxification system for H_2O_2 must work in coordination in order to avoid the cellular accumulation of this compound. In this sense, CAT and the ascorbate–glutathione cycle act together to detoxify H_2O_2 from the cells to H_2O. In our experiments, CAT encoding transcripts were upregulated under salinity, independently of the nutrient solution used. Also, this transcript was upregulated under heat stress, whereas in the salinity + heat treatment it was downregulated. These results correlate well with the CAT activity obtained under these conditions, concluding that the extra supply of K^+ and Ca^{2+} did not have any effect on the transcription and activity of this enzyme under stress conditions, but the combination of salinity and heat induced the inhibition of its transcription.

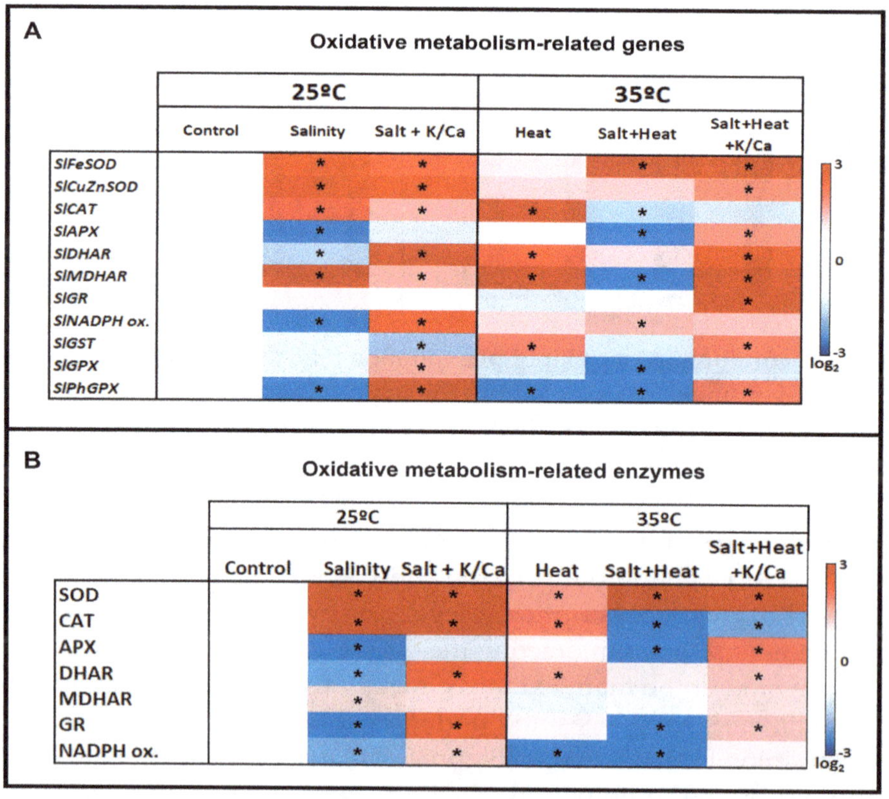

Figure 5. Oxidative metabolism-related gene expression (**A**) and enzymatic activities (**B**) in tomato leaves grown under control, salinity (60 mM NaCl) or salinity and higher concentration of K^+ and Ca^{+2} in the irrigation solution at optimal temperature (25 °C) or heat stress (35 °C). Red color represents a higher relative expression or activity with respect to control plants and blue color represents a lower relative expression or activity. Scale is the $\log_2$ values of the expression after normalization with respect to control plants. Absolute values (gene expression and enzymatic activities) as well as log2 values can be found in Supporting Information Tables S2–S5, respectively. Asterisks are representative of significant differences with respect to control plants ($p < 0.05$).

The ascorbate–glutathione (AsA–GSH) cycle acts in parallel to CAT in the effective detoxification of H_2O_2. In our experiments, the transcription levels of the different transcripts that codify for the antioxidant enzymes from the AsA–GSH cycle were differently affected depending on the stress treatment and the irrigation solution used. Thus, salinity stress downregulated the transcripts that codify for APX, DHAR, and NADPH oxidase, with the concomitant inhibition of their related enzymes. Interestingly, the supplementation of plants with a nutrient solution enriched with K^+ and Ca^{2+} reverted the regulation observed for DHAR and NADPH oxidase under salinity, inducing the upregulation of those genes as compared to control plants. However, the activity of DHAR and NADPH oxidase did not show significant differences with respect to control plants. MDHAR transcripts were upregulated and the activity of this enzyme was also higher under salinity as compared to control plants, independently of the irrigation solution used. Heat stress induced an upregulation of all the transcripts belonging to the AsA–GSH pathway, except for *SlAPX* and *SlGR*, which did not show significant differences with respect to control plants. When salinity and heat stress were applied jointly, a general downregulation of the transcripts codifying for the enzymes belonging to the AsA–GSH pathway was observed, with the concomitant inhibition of their enzymatic activities. However, the feeding of the plants with the nutrient solution enriched with K^+ and Ca^{2+} reverted the expression profile observed under the salinity and heat combination for most of the genes, with an upregulation pattern observed in all of them as compared to the control plants, which correlated with the activities measured for these enzymes. The inhibition of one or more of these enzymes in plants grown under salinity or under salinity + heat may explain the high H_2O_2 concentration found under these treatments, where the detoxification of this compound was not as efficient as compared to control plants. This resulted in the accumulation of this compound, oxidative damage, and plant growth inhibition [7,28]. On the other hand, the supplementation of plants with a nutrient solution containing a higher concentration of K^+ and Ca^{2+} seemed to induce the transcription of all the genes involved in the efficient detoxification of H_2O_2 (i.e., *SlAPX, SlDHAR, SlMDHAR, SlGR, SlNADPH* oxidase), significantly reducing the oxidative damage observed in plants grown under the combination of salinity + heat, and the activation of stress-related genes implied in the signaling mechanisms for ROS detoxification [10]. Thus, K^+ and Ca^{2+} may contribute to the maintenance of the cellular redox homeostasis, protecting photosynthesis and electron transport systems and increasing plant tolerance to abiotic stressors.

GST, PhGPX, and GPX catalyze the GSH-dependent reduction of H_2O_2 and organic peroxides, including lipid peroxides to H_2O or alcohols [44]. In our experiments, the expression of the transcripts *SlGST*, *SlPhGPX*, and *SlGPX* were determined in order to compare them with the level of lipid peroxidation observed under the different treatments and to verify if a supplementation of the plant irrigation solution with K^+ and Ca^{2+} might have any effect on the transcription of these genes. Our results showed that under salinity and salinity + heat, these three genes were downregulated with respect to control plants. However, in plants fed with a nutrient solution enriched with K^+ and Ca^{2+}, an upregulation of the *SlGPX* and *SlPhPGX* (under salinity) and of the *SlGST* and *SlPhGPX* (under salinity + heat) was observed. These results correlated well with the data obtained for lipid peroxidation level in these plants (Figure 4B), where it was observed that K^+ and Ca^{2+} used in higher than recommended concentrations lowered the lipid peroxide level under salinity and salinity + heat as compared with plants irrigated with the normal nutrient solution and grown under the same stress conditions. In this sense, our results may indicate that K^+ and Ca^{2+} could play a role in the transcriptional regulation of the genes that codify for the enzymes involved in the reduction of the H_2O_2 and, consequently, may also play a role in the protection of membrane lipids from peroxidation in plants under abiotic stress conditions.

4. Conclusions

The results obtained in this work highlight the complexity and importance of using the correct supplementation in the plant irrigation solution. We have demonstrated that the plant's tolerance to salinity—and more importantly, to the combination of salinity and heat—can be improved by the proper enrichment of the nutrient solution with K^+ and Ca^{2+}. These nutrients, used in higher concentrations than recommended, helped to control oxidative damage through the transcriptional regulation of the main enzymes implicated in ROS detoxification and the recovery of membrane lipid peroxides. Consequently, a better performance of the photosynthetic apparatus was maintained, leading to optimal growth rates. Our study highlights the importance of plant nutrition in the signaling processes that control plant tolerance responses to abiotic stress combination. However, more studies are needed in order to fully elucidate the role of K^+ and Ca^{2+} in these signaling mechanisms.

Supplementary Materials
Table S1: Nutritional composition of the Hoagland solution used for tomato plants growth, Table S2: Primers used for the quantification of the expression levels of the oxidative metabolism-related transcripts by qPCR, Table S3: Relative expression values of the oxidative metabolism-related transcripts. Values were normalized against control samples and log2 was calculated and shown. Values are means of $n = 3$, Table S4: Absolute activities of the oxidative metabolism-related enzymes. Values obtained were normalized using soluble protein content of each sample and treatment. Values are the means ± SE ($n = 3$), Table S5: Log_2 of the oxidative metabolism-related enzymes. Values obtained in Table S3 were normalized against control and log2 was calculated, Figure S1. Fresh weight (FW) of tomato plants obtained at the end of the preliminary experiment (see Materials and Methods section).

Author Contributions: Conceptualization, M.C.P. and R.M.R.; Data curation, F.G.-S.; Funding acquisition, V.M. and R.M.R.; Investigation, M.G.-M., M.C.P., T.C.M., M.L.-D., V.M. and R.M.R.; Project administration, V.M. and R.M.R.; Resources, M.G.-M., M.C.P., M.L.-D., V.M. and R.M.R.; Supervision, V.M. and R.M.R.; Visualization, M.G.-M., M.C.P., F.G.-S., V.M. and R.M.R.; Writing—original draft, M.G.-M., M.C.P., F.G.-S., M.L.-D. and R.M.R.; Writing—review & editing, M.G.-M., M.C.P. and R.M.R.

Acknowledgments: We sincerely acknowledge Mario G. Fon for his corrections and suggestions for the English grammar in this manuscript.

References

1. IPCC. *Climate Change 2015: Synthesis Report. Contribution of Working Groups I, II and III to the Fifth Assessment Report of the Intergovernmental Panel on Climate Change*; IPCC: Geneva, Switzerland, 2015.
2. Parida, A.K.; Das, A.B. Salt tolerance and salinity effects on plants: A review. *Ecotoxicol. Environ. Saf.* **2005**, *60*, 324–349. [CrossRef] [PubMed]
3. Prasch, C.M.; Sonnewald, U. Signaling events in plants: Stress factors in combination change the picture. *Environ. Exp. Bot.* **2015**, *114*, 4–14. [CrossRef]
4. Rivero, R.; Ruiz, J.; Romero, L. Oxidative metabolism in tomato plants subjected to heat stress. *J. Hortic. Sci. Biotechnol.* **2004**, *79*, 560–564. [CrossRef]
5. Martinez, V.; Rubio, F.; Girones-Vilaplana, A.; Mittler, R.; Mestre, T.C.; Moreno, D.A.; Rivero, R.M. Accumulation of Flavonols over Hydroxycinnamic Acids Favors Oxidative Damage Protection under Abiotic Stress. *Front. Plant Sci.* **2016**, *7*, 838. [CrossRef]
6. Martinez, V.; Nieves-Cordones, M.; Lopez-Delacalle, M.; Rodenas, R.; Mestre, T.C.; García-Sánchez, F.; Rubio, F.; Nortes, P.A.; Mittler, R.; Rivero, R.M. Tolerance to Stress Combination in Tomato Plants: New Insights in the Protective Role of Melatonin. *Molecules* **2018**, *23*, 535. [CrossRef] [PubMed]
7. Rivero, R.M.; Mestre, T.C.; Mittler, R.; Rubio, F.; Garcia-Sanchez, F.; Martinez, V. The combined effect of salinity and heat reveals a specific physiological, biochemical and molecular response in tomato plants. *Plant Cell Environ.* **2014**, *37*, 1059–1073. [CrossRef]
8. Kaur, H.; Sirhindi, G.; Bhardwaj, R.; Alyemeni, M.N.; Siddique, K.H.M.; Ahmad, P. 28-homobrassinolide regulates antioxidant enzyme activities and gene expression in response to salt- and temperature-induced oxidative stress in *Brassica juncea*. *Sci. Rep.* **2018**, *8*, 8735. [CrossRef] [PubMed]
9. Zheng, Y.L.; Yin, X.H.; Ma, H.C. Effects of hydrogen peroxide on seed germination, seedling growth and physiological characteristcs of bombax ceiba after heat shock Pakistan. *J. Bot.* **2018**, *50*, 1327–1333.

10. Ahmad, P.; Alyemeni, M.N.; Ahanger, M.A.; Wijaya, L.; Alam, P.; Kumar, A.; Ashraf, M. Upregulation of antioxidant and glyoxalase systems mitigates NaCl stress in *Brassica juncea* by supplementation of zinc and calcium. *J. Interact.* **2018**, *13*, 151–162. [CrossRef]
11. Wang, M.; Zheng, Q.; Shen, Q.; Guo, S. The Critical Role of Potassium in Plant Stress Response. *Int. J. Mol. Sci.* **2013**, *14*, 7370–7390. [CrossRef]
12. Nieves-Cordones, M.; López-Delacalle, M.; Ródenas, R.; Martínez, V.; Rubio, F.; Rivero, R.M. Critical responses to nutrient deprivation: A comprehensive review on the role of ROS and RNS. *Environ. Exp. Bot.* **2018**. [CrossRef]
13. Qu, C.; Liu, C.; Gong, X.; Li, C.; Hong, M.; Wang, L.; Hong, F. Impairment of maize seedling photosynthesis caused by a combination of potassium deficiency and salt stress. *Environ. Exp. Bot.* **2012**, *75*, 134–141. [CrossRef]
14. Shabala, S.; Cuin, T.A. Potassium transport and plant salt tolerance. *Physiol. Plant.* **2008**, *133*, 651–669. [CrossRef] [PubMed]
15. Gong, X.L.; Chao, L.; Zhou, M.; Hong, M.M.; Luo, L.Y.; Wang, L.; Ying, W.; Cai, J.W.; Gong, S.J.; Hong, F.S. Oxidative damages of maize seedlings caused by exposure to a combination of potassium deficiency and salt stress. *Plant Soil* **2011**, *340*, 443–452. [CrossRef]
16. Martinez, V.; Lauchli, A. Effects of ca2+ on the salt-stress response of barley roots as observed by in-vivo 31p-nuclear magnetic resonance and in-vitro analysis. *Planta* **1993**, *190*, 519–524. [CrossRef]
17. Reid, R.J.; Smith, F.A. The limits of sodium/calcium interactions in plant growth. *Funct. Boil.* **2000**, *27*, 709. [CrossRef]
18. Bacha, H.; Rodenas, R.; Lopez-Gomez, E.; Francisco Garcia-Legaz, M.; Nieves-Cordones, M.; Rivero, R.M.; Martinez, V.; Angeles Botella, M.; Rubio, F. High ca2+ reverts the repression of high-affinity k+ uptake produced by Na^+ in *Solanum lycopersycum* L. (var. Microtom) plants. *J. Plant Physiol.* **2015**, *180*, 72–79. [CrossRef]
19. Schachtman, D.P.; Shin, R.; Berg, R.H. Reactive Oxygen Species and Root Hairs in Arabidopsis Root Response to Nitrogen, Phosphorus and Potassium Deficiency. *Cell Physiol.* **2005**, *46*, 1350–1357.
20. Hernández, M.; Fernandez-García, N.; Garcia-Garma, J.; Rubio-Asensio, J.; Rubio, F.; Olmos, E. Potassium starvation induces oxidative stress in *Solanum lycopersicum* L. roots. *J. Physiol.* **2012**, *169*, 1366–1374. [CrossRef]
21. Zhou, R.; Kong, L.; Wu, Z.; Rosenqvist, E.; Wang, Y.; Zhao, L.; Zhao, T.; Ottosen, C.-O. Physiological response of tomatoes at drought, heat and their combination followed by recovery. *Physiol. Plant.* **2018**, *165*, 144–154. [CrossRef] [PubMed]
22. Zandalinas, S.I.; Mittler, R.; Balfagon, D.; Arbona, V.; Gomez-Cadenas, A. Plant adaptations to the combination of drought and high temperatures. *Physiol. Plant.* **2018**, *162*, 2–12. [CrossRef] [PubMed]
23. Aleman, F.; Caballero, F.; Rodenas, R.; Rivero, R.M.; Martinez, V.; Rubio, F. The f130s point mutation in the arabidopsis high-affinity k transporter athak5 increases K^+ over Na^+ and Cs^+ selectivity and confers na+ and cs+ tolerance to yeast under heterologous expression. *Front. Plant Sci.* **2014**, *5*, 11. [CrossRef]
24. Mestre, T.C.; Garcia-Sanchez, F.; Rubio, F.; Martinez, V.; Rivero, R.M. Glutathione homeostasis as an important and novel factor controlling blossom-end rot development in calcium-deficient tomato fruits. *J. Physiol.* **2012**, *169*, 1719–1727. [CrossRef] [PubMed]
25. Rodriguez-Ortega, W.; Martinez, V.; Rivero, R.; Cámara-Zapata, J.-M.; Mestre, T.; Garcia-Sanchez, F. Use of a smart irrigation system to study the effects of irrigation management on the agronomic and physiological responses of tomato plants grown under different temperatures regimes. *Agric. Manag.* **2017**, *183*, 158–168. [CrossRef]
26. Macnevin, W.M.; Urone, P.F. Separation of hydrogen peroxide from organic hydroperoxides-aplication to polarographic analysis of mixtures. *Anal. Chem.* **1953**, *25*, 1760–1761. [CrossRef]
27. Brennan, T.; Frenkel, C. Involvement of hydrogen-peroxide in regulation of senescence in pear. *Plant Physiol.* **1977**, *59*, 411–416. [CrossRef] [PubMed]
28. Rivero, R.M.; Kojima, M.; Gepstein, A.; Sakakibara, H.; Mittler, R.; Gepstein, S.; Blumwald, E. Delayed leaf senescence induces extreme drought tolerance in a flowering plant. *Proc. Natl. Acad. Sci. USA* **2007**, *104*, 19631–19636. [CrossRef]

29. Fu, J.; Huang, B. Involvement of antioxidants and lipid peroxidation in the adaptation of two cool-season grasses to localized drought stress. *Environ. Exp. Bot.* **2001**, *45*, 105–114. [CrossRef]
30. Bradford, M.M. A rapid and sensitive method for the quantitation of microgram quantities of protein utilizing the principle of protein-dye binding. *Anal. Biochem.* **1976**, *72*, 248–254. [CrossRef]
31. Chance, B.; Maehly, A.C. Assay of catalases and peroxidases. *Methods Enzymol.* **1955**, *2*, 764–775.
32. Aebi, H. Catalase invitro. *Methods Enzymol.* **1984**, *105*, 121–126.
33. Mccord, J.M.; Fridovich, I. The utility of superoxide dismutase in studying free radical reactions. I. Radicals generated by the interaction of sulfite, dimethyl sulfoxide, and oxygen. *J. Boil. Chem.* **1969**, *244*, 6056–6063.
34. Spitz, D.R.; Oberley, L.W. An assay for superoxide dismutase activity in mammalian tissue homogenates. *Anal. Biochem.* **1989**, *179*, 8–18. [CrossRef]
35. Miyake, C.; Asada, K. Thylakoid-Bound Ascorbate Peroxidase in Spinach Chloroplasts and Photoreduction of Its Primary Oxidation Product Monodehydroascorbate Radicals in Thylakoids. *Cell Physiol.* **1992**, *33*, 541–553.
36. Kaundal, A.; Rojas, C.; Mysore, K. Measurement of NADPH Oxidase Activity in Plants. *Bio-Protocol* **2012**, *2*, 278. [CrossRef]
37. Jiang, M.; Zhang, J. Involvement of plasma-membrane NADPH oxidase in abscisic acid- and water stress-induced antioxidant defense in leaves of maize seedlings. *Planta* **2002**, *215*, 1022–1030. [CrossRef] [PubMed]
38. Capula-Rodríguez, R.; Valdez-Aguilar, L.A.; Cartmill, D.L.; Cartmill, A.D.; Alia-Tejacal, I. Supplementary Calcium and Potassium Improve the Response of Tomato (*Solanum lycopersicum* L.) to Simultaneous Alkalinity, Salinity and Boron Stress. *Commun. Soil Sci. Anal.* **2016**, *47*, 505–511.
39. Navarro, J.M.; Garrido, C.; Martinez, V.; Carvajal, M. Water relations and xylem transport of nutrients in pepper plants grown under two different salts stress regimes. *Plant Growth Regul.* **2003**, *41*, 237–245. [CrossRef]
40. Munns, R.; Shabala, S. Salinity stress: Physiological constraints and adaptive mechanisms. In *Plant Stress Physiology*; CABI Publishing: Wallingford, CT, USA, 2017; pp. 24–63.
41. Chakraborty, K.; Basak, N.; Bhaduri, D.; Ray, S.; Vijayan, J.; Chattopadhyay, K.; Sarkar, R.K. Ionic Basis of Salt Tolerance in Plants: Nutrient Homeostasis and Oxidative Stress Tolerance. In *Plant Nutrients and Abiotic Stress Tolerance*; Springer Nature: Singapore, 2018; pp. 325–362.
42. Abbasi, G.H.; Akhtar, J.; Ahmad, R.; Jamil, M.; Anwar-Ul-Haq, M.; Ali, S.; Ijaz, M. Potassium application mitigates salt stress differentially at different growth stages in tolerant and sensitive maize hybrids. *Plant Growth Regul.* **2015**, *76*, 111–125. [CrossRef]
43. Blumwald, E. Sodium transport and salt tolerance in plants. *Curr. Opin. Cell Boil.* **2000**, *12*, 431–434. [CrossRef]
44. Ahanger, M.A.; Tomar, N.S.; Tittal, M.; Argal, S.; Agarwal, R.M. Plant growth under water/salt stress: ROS production; antioxidants and significance of added potassium under such conditions. *Physiol. Mol. Boil. Plants* **2017**, *23*, 731–744. [CrossRef] [PubMed]
45. Pinero, M.C.; Houdusse, F.; Garcia-Mina, J.M.; Garnica, M.; del Amor, F.M. Regulation of hormonal responses of sweet pepper as affected by salinity and elevated co2 concentration. *Physiol. Plant.* **2014**, *151*, 375–389. [CrossRef] [PubMed]
46. Bohra, J.S.; Doerffling, K. Potassium nutrition of rice (*Oryza sativa* L.) varieties under NaCl salinity. *Plant Soil* **1993**, *152*, 299–303. [CrossRef]
47. Malińska, D.; Mirandola, S.R.; Kunz, W.S. Mitochondrial potassium channels and reactive oxygen species. *FEBS Lett.* **2010**, *584*, 2043–2048. [CrossRef] [PubMed]
48. Trono, D.; Laus, M.N.; Soccio, M.; Alfarano, M.; Pastore, D. Modulation of potassium channel activity in the balance of ros and atp production by durum wheat mitocondria—An amazing defense tool against hyperosmotic stress. *Front. Plant Sci.* **2015**, *6*, 1072. [CrossRef] [PubMed]
49. Bhattacharjee, S. Calcium-dependent signaling pathway in the heat-induced oxidative injury in *Amaranthus lividus. Boil. Plant.* **2008**, *52*, 137–140. [CrossRef]
50. Zandalinas, S.I.; Balfagón, D.; Arbona, V.; Gómez-Cadenas, A. Modulation of Antioxidant Defense System Is Associated with Combined Drought and Heat Stress Tolerance in Citrus. *Front. Sci.* **2017**' *8*, 10. [CrossRef]

51. Siddiqui, M.H.; Alamri, S.A.; Al-Khaishany, M.Y.; Al-Qutami, M.A.; Ali, H.M.; Khan, M.N. Sodium nitroprusside and indole acetic acid improve the tolerance of tomato plants to heat stress by protecting against DNA damage. *J. Interact.* **2017**, *12*, 177–186. [CrossRef]
52. Suzuki, N.; Rivero, R.M.; Shulaev, V.; Blumwald, E.; Mittler, R. Abiotic and biotic stress combinations. *New Phytol.* **2014**, *203*, 32–43. [CrossRef]

PERMISSIONS

All chapters in this book were first published in MDPI; hereby published with permission under the Creative Commons Attribution License or equivalent. Every chapter published in this book has been scrutinized by our experts. Their significance has been extensively debated. The topics covered herein carry significant findings which will fuel the growth of the discipline. They may even be implemented as practical applications or may be referred to as a beginning point for another development.

The contributors of this book come from diverse backgrounds, making this book a truly international effort. This book will bring forth new frontiers with its revolutionizing research information and detailed analysis of the nascent developments around the world.

We would like to thank all the contributing authors for lending their expertise to make the book truly unique. They have played a crucial role in the development of this book. Without their invaluable contributions this book wouldn't have been possible. They have made vital efforts to compile up to date information on the varied aspects of this subject to make this book a valuable addition to the collection of many professionals and students.

This book was conceptualized with the vision of imparting up-to-date information and advanced data in this field. To ensure the same, a matchless editorial board was set up. Every individual on the board went through rigorous rounds of assessment to prove their worth. After which they invested a large part of their time researching and compiling the most relevant data for our readers.

The editorial board has been involved in producing this book since its inception. They have spent rigorous hours researching and exploring the diverse topics which have resulted in the successful publishing of this book. They have passed on their knowledge of decades through this book. To expedite this challenging task, the publisher supported the team at every step. A small team of assistant editors was also appointed to further simplify the editing procedure and attain best results for the readers.

Apart from the editorial board, the designing team has also invested a significant amount of their time in understanding the subject and creating the most relevant covers. They scrutinized every image to scout for the most suitable representation of the subject and create an appropriate cover for the book.

The publishing team has been an ardent support to the editorial, designing and production team. Their endless efforts to recruit the best for this project, has resulted in the accomplishment of this book. They are a veteran in the field of academics and their pool of knowledge is as vast as their experience in printing. Their expertise and guidance has proved useful at every step. Their uncompromising quality standards have made this book an exceptional effort. Their encouragement from time to time has been an inspiration for everyone.

The publisher and the editorial board hope that this book will prove to be a valuable piece of knowledge for researchers, students, practitioners and scholars across the globe.

LIST OF CONTRIBUTORS

Ihsan Elahi Zaheer, Zohaib Abbas and Muhammad Rizwan
Department of Environmental Sciences and Engineering, Government College University Allama Iqbal Road, Faisalabad 38000, Pakistan

Shafaqat Ali
Department of Environmental Sciences and Engineering, Government College University Allama Iqbal Road, Faisalabad 38000, Pakistan
Department of Biological Sciences and Technology, China Medical University, Taichung 40402, Taiwan

Muhammad Hamzah Saleem
MOA Key Laboratory of Crop Ecophysiology and Farming System Core in the Middle Reaches of the Yangtze River, College of Plant Science and Technology, Huazhong Agricultural University, Wuhan 430070, China

Muhammad Arslan Ashraf
Department of Botany, Government College University Faisalabad, Faisalabad 38000, Pakistan

Qurban Ali
Key laboratory of Plant Pathology, College of Plant Science & Technology, Huazhong Agricultural University, Wuhan 430070, China

Mohamed A. El-Sheikh, Mohammed Nasser Alyemeni and Leonard Wijaya
Department of Botany and Microbiology, College of Science, King Saud University, 11451 Riyadh, Saudi Arabia

Alicia S. Molina, María V. Pérez Chaca and Fanny Zirulnik
Facultad de Química, Bioquímica y Farmacia (FQByF), Universidad Nacional de San Luis (UNSL), Área Química Biológica, Ejército de los Andes 950, Bloque I, 1er Piso, San Luis, Argentina

Mónica A. Lugo
IMIBIO-CONICET-FQByF-UNSL, Área Ecología, Ejército de los Andes 950, Bloque I, 2do Piso, San Luis, Argentina

Silvina Vargas-Gil
Instituto de Patología Vegetal (IPAVE, CIAP-INTA), CONICET, Camino 60 Cuadras Km. 5.5, C.P. 5119 Córdoba, Argentina

Jorge Leporati
Facultas de Ingeniería y Ciencias Agropecuarias (FICA)-UNSL, Área Matemática, Campus Universitario, Ruta 148 Ext. Norte, 2do Piso Ala Este, Villa Mercedes, San Luis, Argentina

Nuria Ferrol and Concepción Azcón-Aguilar
Estación Experimental del Zaidín, CSIC, Profesor Albareda 1, 18008 Granada, Spain

Carmen Arena, Anna Rita Bianchi, Ermenegilda Vitale, Costantino Parisi, Giulia Guerriero and Anna De Maio
Dipartimento di Biologia, Università degli Studi di Napoli Federico II, Via Cinthia, 80126 Napoli, Italy

Luca Vitale, Carmela Mistretta and Vincenzo Magliulo
Istituto per i Sistemi Agricoli e Forestali del Mediterraneo (CNR-ISAFoM), Via Patacca 85, 80056 Ercolano (NA), Italy

Fernanda Cosme and António Inês
Centro de Química de Vila Real (CQ-VR), Universidade de Trás-os-Montes e Alto Douro (UTAD), Quinta de Prados, 5000-801 Vila Real, Portugal

Rupesh K. Singh and Virgílio Falco
Centro de Química de Vila Real (CQ-VR), Universidade de Trás-os-Montes e Alto Douro (UTAD), Quinta de Prados, 5000-801 Vila Real, Portugal
Departamento de Agronomia, Universidade de Trás-os-Montes e Alto Douro (UTAD), Quinta de Prados, 5000-801 Vila Real, Portugal

Bruno Soares
Departamento de Agronomia, Universidade de Trás-os-Montes e Alto Douro (UTAD), Quinta de Prados, 5000-801 Vila Real, Portugal
CoLAB Vines & Wines, Associação para o Desenvolvimento da Viticultura Duriense (ADVID), Régia Douro Park, 5000-033 Vila Real, Portugal

Piebiep Goufo and Ana A. Oliveira
Departamento de Agronomia, Universidade de Trás-os-Montes e Alto Douro (UTAD), Quinta de Prados, 5000-801 Vila Real, Portugal
Centro de Investigação e Tecnologias Agroambientais e Biológicas (CITAB), Universidade de Trás-os -Montes e Alto Douro (UTAD), Quinta de Prados, 5000-801 Vila Real, Portugal

Isaura Castro and Ana L. Pinto-Sintra
Centro de Investigação e Tecnologias Agroambientais e Biológicas (CITAB), Universidade de Trás-os -Montes e Alto Douro (UTAD), Quinta de Prados, 5000-801 Vila Real, Portugal

Chikahiro Miyake
Department of Applied Biological Science, Faculty of Agriculture, Graduate School for Agricultural Science, Kobe University, 1-1, Rokkodai, Nada, Kobe 657-8501, Japan
Core Research for Environmental Science and Technology (CREST), Japan Science and Technology Agency, 7 Goban, Chiyoda, Tokyo 102-0076, Japan

Md. Sanaullah Biswas
Department of Horticulture, Bangabandhu Sheikh Mujibur Rahman Agricultural University, Gazipur 1706, Bangladesh

Ryota Terada
Faculty of Agriculture, Yamaguchi University, Yoshida 1677-1, Yamaguchi 753-8515, Japan

Juńichi Mano
Science Research Center, Organization of Research Initiatives, Yamaguchi University, Yamaguchi 753-8511, Japan
Graduate School of Science and Technology for Innovation, Yamaguchi University, Yamaguchi 753-8511, Japan

Mohamed A. El-Esawi
Botany Department, Faculty of Science, Tanta University, Tanta 31527, Egypt

Amr Elkelish
Botany Department, Faculty of Science, Suez Canal University, Ismailia 41522, Egypt

Mona Soliman
Botany and Microbiology Department, Faculty of Science, Cairo University, Giza 12613, Egypt

Hosam O. Elansary
Plant Production Department, College of Food and Agriculture Sciences, King Saud University, Riyadh 11451, Saudi Arabia
Floriculture, Ornamental Horticulture, and Garden Design Department, Faculty of Agriculture (El Shatby), Alexandria University, Alexandria 21526, Egypt

Abbu Zaid
Plant Physiology and Biochemistry Laboratory, Department of Botany Aligarh Muslim University, Aligarh 202002, India

Shabir H.Wani
Mountain Research Centre for Field Crops, Sher-e-Kashmir University of Agricultural Sciences and Technology of Kashmir, Khudwani Anantnag 192101, India

Martina Janků, Lenka Luhová and Marek Petřivalský
Department of Biochemistry, Palacký University in Olomouc, Šlechtitelů 27, CZ-78371 Olomouc, Czech Republic

Juan C. Begara-Morales, Beatriz Sánchez-Calvo, María V. Gómez-Rodríguez, Mounira Chaki, Raquel Valderrama, Capilla Mata-Pérez and Juan B. Barroso
Group of Biochemistry and Cell Signaling in Nitric Oxide, Department of Experimental Biology, Center for Advanced Studies in Olive Grove and Olive Oils, Faculty of Experimental Sciences, University of Jaén, Campus "Las Lagunillas", s/n, E-23071 Jaén, Spain

Javier López-Jaramillo
Institute of Biotechnology, Department of Organic Chemistry, Faculty of Sciences, University of Granada, E-18071 Granada, Spain

Francisco J. Corpas
Group of Antioxidants, Free Radicals, and Nitric Oxide in Biotechnology, Food and Agriculture, Department of Biochemistry, Cell and Molecular Biology of Plants, Estación Experimental del Zaidín, CSIC, C/Profesor Albareda 1, E-18080 Granada, Spain

Aleksandra Lewandowska
VIB-VUB Center for Structural Biology, VIB, 1050 Brussels, Belgium
Brussels Center for Redox Biology, 1050 Brussels, Belgium
Structural Biology Brussels, Vrije Universiteit Brussel, 1050 Brussels, Belgium
Department of Plant Biotechnology and Bioinformatics, Ghent University, 9052 Ghent, Belgium
VIB Center for Plant Systems Biology, VIB, 9052 Ghent, Belgium

Trung Nghia Vo, Thuy-Dung Ho Nguyen, Khadija Wahni David Young and Joris Messens
VIB-VUB Center for Structural Biology, VIB, 1050 Brussels, Belgium
Brussels Center for Redox Biology, 1050 Brussels, Belgium
Structural Biology Brussels, Vrije Universiteit Brussel, 1050 Brussels, Belgium

Didier Vertommen
de Duve Institute, Université Catholique de Louvain, 1200 Brussels, Belgium

Frank Van Breusegem
Brussels Center for Redox Biology, 1050 Brussels, Belgium
Department of Plant Biotechnology and Bioinformatics, Ghent University, 9052 Ghent, Belgium
VIB Center for Plant Systems Biology, VIB, 9052 Ghent, Belgium

Mirza Hasanuzzaman and Taufika Islam Anee
Department of Agronomy, Faculty of Agriculture, Sher-e-Bangla Agricultural University, Dhaka 1207, Bangladesh

Masayuki Fujita
Laboratory of Plant Stress Responses, Department of Applied Biological Science, Faculty of Agriculture, Kagawa University, Miki-cho, Kita-gun, Kagawa 761-0795, Japan

M. H. M. Borhannuddin Bhuyan
Laboratory of Plant Stress Responses, Department of Applied Biological Science, Faculty of Agriculture, Kagawa University, Miki-cho, Kita-gun, Kagawa 761-0795, Japan
Citrus Research Station, Bangladesh Agricultural Research Institute, Jaintapur, Sylhet 3156, Bangladesh

Khursheda Parvin
Laboratory of Plant Stress Responses, Department of Applied Biological Science, Faculty of Agriculture, Kagawa University, Miki-cho, Kita-gun, Kagawa 761-0795, Japan
Department of Horticulture, Faculty of Agriculture, Sher-e-Bangla Agricultural University, Dhaka 1207, Bangladesh

Kamrun Nahar
Department of Agricultural Botany, Faculty of Agriculture, Sher-e-Bangla Agricultural University, Dhaka 1207, Bangladesh

Jubayer Al Mahmud
Department of Agroforestry and Environmental Science, Faculty of Agriculture, Sher-e-Bangla Agricultural University, Dhaka 1207, Bangladesh

María Carmen Piñero, Francisco García-Sanchez, Teresa C. Mestre, María López-Delacalle, Vicente Martínez and Rosa M. Rivero
Department of Plant Nutrition, Campus Universitario de Espinardo, CEBAS-CSIC, Ed 25, Espinardo, 30100 Murcia, Spain

María García-Martí
Department of Plant Nutrition, Campus Universitario de Espinardo, CEBAS-CSIC, Ed 25, Espinardo, 30100 Murcia, Spain
Programa de Doctorado en Ciencias de la Salud, Campus de los Jerónimos, UCAM (Universidad Católica San Antonio de Murcia), s/n, 30107 Murcia, Spain

Index

O

P

R

S

T

Printed in the USA
CPSIA information can be obtained
at www.ICGtesting.com
LVHW080051211023
761596LV00010BA/2007